Use of Yeast Biomass in Food Production

Authors

Anna Halász

Assistant Professor and Head
Department of Biochemistry
Central Research Institute of
Food Industry
Budapest, Hungary

and

Radomir Lásztity

Professor and Head
Department of Biochemistry
and Food Technology
Technical University
Budapest, Hungary

CRC Press
Taylor & Francis Group
Boca Raton London New York

CRC Press is an imprint of the
Taylor & Francis Group, an **informa** business

CRC Press
Taylor & Francis Group
6000 Broken Sound Parkway NW, Suite 300
Boca Raton, FL 33487-2742

First issued in paperback 2019

ISBN 13: 978-0-367-45072-4 (pbk)
ISBN 13: 978-0-8493-5866-1 (hbk)

Visit the Taylor & Francis Web site at
http://www.taylorandfrancis.com

and the CRC Press Web site at
http://www.crcpress.com

Library of Congress Cataloging-in-Publication Data

Halász, Anna.
Use of yeast biomass in food production/authors, Anna Halász and Radomír Lásztity.
p. cm.
Includes bibliographical references and index.
ISBN 0-8493-5866-3
1. Single cell proteins--Biotechnology. 2. Yeast fungi--Physiology. 3. Food--Biotechnology. I. Lásztity, Radomír. II. Title.
TP248.65.S56H35 1991
664--dc20 91-42435
CIP

Library of Congress Card Number 91-42435

PREFACE

Although the first records of the use of yeast are concerned with the production of a type of acid beer called "boozah" in 6000 BC Egypt, the technology of microbial cell production on a commercial scale for food and feed has developed only within the past 100 years. Preparation of compressed brewer's yeast began in the late 18th century in Great Britain, The Netherlands and Germany. The conscious growth of microbes for human consumption started in Germany about 1910 with brewer's yeast, which was separated from the beer, washed and dried. The direct food use of yeasts — except the classical use in fermentation and leavening — remained until the fifties of this century limited — and the main usage of microbial biomass was in feeds. The high expectations concerning yeasts grown on hydrocarbons in the 1960s and early 1970s were not realized due to rising costs of petroleum products and supposed safety problems.

In the last decade a growing interest may be observed regarding food usage of yeast biomass. Special yeast species are grown on natural raw materials for food usage. Based on a detailed food analysis showing that yeast is a rich source of protein, essential minerals and vitamins, and according to new trends not only the inactivated dried whole yeast cells are used but the yeast biomass is processed into a lot of valuable products serving as food ingredients.

In the framework of this book a summary will given about the chemistry, molecular biology, production and processing of yeast biomass. The possible applications of whole yeast cells, protein isolates and autolysates (extracts) as food ingredients will be reviewed. The production of yeast based flavors, the use of native yeast for production of commercial enzyme preparations and some other applications in food production will be shortly treated in the Appendix.

The authors hope that this book will be a valuable source of information for food scientists and technologists, microbiologists, nutritionists, economists and decision makers interested in the potential use of yeast biomass.

THE EDITORS

Anna Halász, D.Sc., is head of the Department of Biochemistry of Central Research Institute of Food Industry, Budapest, Hungary, and Associate Professor of Biochemistry, Technical University in Budapest.

Dr. Halász received her M.Sc. degree from Technical University (Faculty of Chemical Engineering) in 1961, and her D.Sc. degree in Chemical Sciences in 1988 from the Hungarian Academy of Sciences.

Dr. Halász is the Chairman of the Section of Food and Agriculture of the Hungarian Biochemical Society, member of the Working Group on Microbiology of Hungarian Scientific Society for Food Industry, a member of the Food Protein Group of Hungarian Academy of Sciences and a member of the Working Group on Yeast of the ICC (International Association for Cereal Science and Technology). She has been the recipient of the Distinguished Researchers Award of the Ministry of Food and Agriculture and she was awarded with the Bronze Medal of the Hungarian Republic. She is also the recipient of the Swiss Federal Foundation Fellowship in Science (1970 to 1971).

Dr. Halász has presented over 100 invited lectures at international and national meetings and at universities and other institutes. She has published more than 100 research papers. Her current major research interests include the biochemistry of yeast production and use of yeast biomass, molecular biology of yeasts and investigation of proteases.

Radomir Lásztity, D.Sc., is Professor and Head of the Department of Biochemistry and Food Technology at the Technical University, Budapest, Hungary. Dr. Lásztity received his M.Sc. degree in Chemical Engineering in 1951 and his D.Sc. degree in Chemical Sciences in 1968.

Dr. Lásztity is chairman of the Food Protein Group of the Hungarian Academy of Sciences, past president, deputy technical director and member of the Executive Committee of the ICC, chairman of the Codex Committee on Methods of Analysis and Sampling of the FAO/WHO Food Standard Program, member of the Working Group on Food Chemistry of The Federation of European Chemical Societies and member of the editorial board of more scientific journals. He was acting vice rector of the Technical University, Budapest from 1970 to 1976.

Among other awards he has received the Bailey Medal and Schweitzer Medal of the ICC, the Sigmond Elek Award for research in food chemistry, the State Prize of the Hungarian Republic and the Golden Medal of the Czechoslavakian Agricultural Academy of Sciences.

Dr. Lásztity's main research activities are chemistry and biochemistry of food proteins, rheology of foods and food analysis. The results of his research work were published in more than 500 scientific papers in foreign and Hungarian scientific journals. He is the author of more than 20 books and textbooks (among them the books: *Chemistry of Cereal Proteins, Amino Acid Composition and Biological Value of Cereal Proteins, Functional Properties of Food Proteins, Gluten Proteins,* and *Principles of Food Biochemistry*).

TABLE OF CONTENTS

PART I

Chapter 5

Extraction of Yeast Proteins, Yeast Protein Concentrates and Isolates

Chapter 8

Use of Yeast Protein Preparations in Food Industry

Chapter 9

Use of Yeast Autolysates and Yeast Hydrolysates in Food Production

Appendix

Other Uses of Yeast Biomass

Part I

Chapter 1

INTRODUCTION

In writing about the potential use of yeast biomass in food production we are thinking about yeasts first as a protein source. The proteins have long been recognized as both quantitatively important and nutritionally vital components of foods. In the last decades substantial interest has been shown in their contribution to food attributes such as texture and behavior in processing and use and thus, ultimately to food acceptability. Keeping in mind the facts mentioned above, it is quite understandable that few subjects in the field of food science and technology have attracted as much active interest in recent years as proteins.

Since the beginning of agricultural production in prehistoric time, the cereals (wheat, rice, corn, rye, barley, sorghum, millet, etc.) were the main agricultural products in most parts of the world playing an important role in the nutrition generally and also in the protein supply. Although amino acid composition of the cereal proteins and the role of essential amino acids were not known, the mixed diet containing different proteins was characteristic of the greatest part of the population allowing a compensation of the effects of the unsatisfactory content of some essential amino acids in cereal proteins. Later, the steady growth of population and the limited increase of agricultural production (especially animal husbandry) changed this situation. In many countries the mixed protein diet changed to a diet based on the overwhelming consumption of cereals resp. cereal products representing 50 to 90% of the total energy and protein supply. The low content of cereal proteins in some essential amino acids combined in many cases with shortage of food leads to a protein resp. essential amino acid deficiency in a lot of countries belonging to the group of developing countries.

As it is known, the major function of proteins is to provide the body with an adequate intake and balance of essential amino acids and nitrogen for the synthesis of other (nonessential) amino acids. The essential amino acids are used for the maintenance of the body, tissue protein synthesis in adults and for net protein gain during growth and development. The daily protein requirements for adult men and women, for infants aged 0 to 6 months and children aged 10 to 12 years are indicated in Table 1. The values included in Table 1 are recommended by the Food and Agriculture Organization (FAO) and World Health Organization (WHO) as "safe protein intakes" for healthy people.[1] Adults require less protein per kilogram than children, since protein turnover decreases with age and growth stops.

Protein (resp. essential amino acid) requirements are influenced by caloric intake. The total amount of energy supplied by a range of foods in the diet, however, determines not only the quantity of the protein in the diet but also its effective utilization.

If the caloric intake is insufficient, a part of the dietary protein is used for the energy production. Results of studies of both children and adults indicate clearly that protein metabolism is very sensitive to energy intake. So it is understandable that various interactions among dietary protein, total energy intake and the proportion of energy derived from non-protein energy sources such as carbohydrate and fats, have important implications for the quantitative estimation of human protein and amino acid requirements.

The recommended quantity of proteins for adults is somewhat controversial. In the framework of studies in developing countries, some adults have been found to remain in nitrogen equilibrium when given a rice protein intake as low as 0.44 g/day/kg.[2] This may be possible because of metabolic adaptation to habitually low protein intakes. Nevertheless, many countries recommend protein intakes higher than those advocated by FAO/WHO. These recommendations are usually 1 g/kg/day and are based on observations that young men fed amounts of egg protein, just sufficient to maintain nitrogen equilibrium, over extended periods of time undergo some biochemical abnormalities. The value finally agreed upon is

TABLE 1
Human Requirements for Essential Amino Acids and Provisional Amino Acid Patterns for Ideal Proteins

Amino acid	Infant requirement (0—6 months; mg/day per kg)	Child requirement (10—12 years; mg/day per kg)	Adult requirement (mg/day per kg)	Provisional ideal pattern (adult; mg/g protein)
Histidine	28	0	0	0
Isoleucine	70	30	10	18
Leucine	161	45	14	25
Lysine	103	60	12	22
Methionine (+cysteine)[a]	58	27	13	24
Phenylalanine (+tyrosine)[b]	125	27	14	25
Threonine	87	35	7	13
Tryptophan	17	4	3.5	6.5
Valine	93	33	10	18
Total essential amino acids	742	261	83.5	151.5
Total protein requirement (egg or milk proteins)	2000	800	550 570:man 520:woman	

[a] Cysteine may supply up to one third the need for total sulfur amino acids.
[b] Tyrosine may supply up to one third the need for total aromatic amino acids.

From FAO/WHO Energy and Protein Requirements, Report of a joint FAO/WHO ad hoc Expert Committee, World Health Organization Techn. Rep. Ser. 522, WHO, Geneva, 1973.

of great importance since it will influence the level of concern directed to the "protein gap" of developing countries, the planning goals for national food supplies, the dietary standards for assistance programs, the information contained in nutrition education programs, how data from food consumption surveys are interpreted, the requirements for nutrition labeling and so on.[3]

To meet the protein requirements of humans not only the quantity of proteins is important, but also the essential amino acid content. The nutritive value of protein (protein quality) depends on the kinds and amounts of amino acids it contains, and represents a measure of the efficiency with which the body can utilize the protein. A balanced or high quality protein contains essential amino acids in ratios commensurate with human needs. Proteins of animal origin generally are of higher quality than those of plant origin. This can be determined by comparing the amino acid contents of various proteins with the FAO reference patterns (Table 2).

The FAO reference pattern was chosen to satisfy the requirements of the young child. The facts mentioned above are the reason why nutritionists prefer a mixture of animal and plant protein sources while only so much is possible to assure a 37% ratio of essential amino acids in proteins of the diet for children. This ratio is much lower for adults (about 15%), because adults do not need amino acids for tissue deposition and because adults are able to efficiently recycle these amino acids. This probably explains why in developing countries diseases of malnutrition attributed first in line to protein deficiency (e.g., kwasiorkor) are more common in children. In the case of adults, a diet of cereals (with protein content of 8 to 10%) can meet the minimal protein requirements if enough is eaten to supply caloric requirements. Looking at the world-wide situation of protein consumption it can be stated that in the developed countries the protein intake is higher than the recommended daily amount and it is recommended to decrease, to some extent at least, the amount of protein from animal sources.[1,4-6] On the other side, there are regions in the world where the people have too little to eat (gap in energy and protein), and also regions where protein shortage per se exists. So in developing countries there is a general need to increase food supplies and protein foods (especially high quality protein foods) in particular. The basis of this need is first in line in the rapid growth rate in world population. According to estimations of experts, about 1 billion humans currently suffer from chronic malnutrition. This situation may become even worse, since the world population is expected to increase from 4.4 to 6.4 billion in a period from 1980 to 2000. Keeping in mind also that an improvement of the average quality (essential amino acid content) of consumed protein (especially in diet of young people) and an increased ratio of animal protein is needed, it is obvious that the world production of protein must be increased. Discussing the possible ways of increasing protein production it is useful first to look at world production of proteins today. Table 3 includes not only the approximate world production and the yield per hectare per year, but also the price per kilogram of the main animal and plant proteins.

Where must we turn for additional quantities of food? On the land we have minimally two possibilities. Firstly, open up new land for the agricultural production and secondly, increase the yield per hectare (probably also increase protein content of the crop). The increase of the land for the agricultural production is really possible in some developed countries (e.g., Canada, U.S.) and also a substantial amount of land can be developed agriculturally in South America and Africa primarily, but care must be taken to ensure that there is not undue loss to erosion from overgrazing and deforestation. Cereals represent the main protein and energy supply for both human and animal nutrition, and attempts to increase the production of cereals have met with considerable success. Genetic selection and improved agricultural practices have led to increased cereal yields and to a higher protein and/or lysine content of wheat, corn and rice grains. For example, in Hungary the yields per hectare of wheat and corn were doubled or trebled in the last 20 years. Very promising results were

TABLE 2
Essential Amino Acid Contents and Nutritional Values of Some Protein Foods

Amino acid (mg/g protein)	Human milk	Cow's milk	Hen's egg	Meat (beef)	Fish (all types)	Wheat grain	Soybeans
Histidine	26	27	22	34	35	25	28
Isoleucine	46	47	54	48	48	35	50
Leucine	93	95	86	81	77	72	85
Lysine	66	78	70	89	91	31[a]	70
Methionine + cysteine	42	33[a]	57	40	40	43	28[a]
Phenylalanine + tyrosine	72	102	93	80	76	80	88
Threonine	43	44	47	46	46	31	42
Tryptophan	17	14	17	11	11	12	14
Valine	55	64	66	50	61	47	53
Total essential amino acids without histidine	434	477	490	445	450	351	430
Protein content (%)	1.2	3.5	12	18	19	12	40
Chemical score (%) (based on FAO ref. pattern)	100	94	100	100	100	56	80
Protein efficiency ratio (PER) (on rats)	4	3.1	3.9	3	3.5	1.5	2.3
Biological value (%) (BV) (on rats)	95	84	94	74	76	65	73
Net protein utilization (%) (NPU) (on rats)	87	82	94	67	79	40	61

[a] Limiting amino acids in diet.

From FAO/WHO Energy and Protein Requirements, Report of a joint FAO/WHO ad hoc Expert Committee, World Health Organization Techn. Rep. Ser. 522, WHO, Geneva, 1973.

TABLE 3
World Production, Yield, and Cost of Important Food Proteins

	Cereals (grains)	Oilseeds	Leguminosae (except oil seeds)	Meats	Milk and dairy products	Eggs
Production (10^6 tons protein/year) (1978—1979)	140	40	8.6	18	15	3
Yield (kg protein/ha per year) (1978—1979)	200—700	500—1200	200—1000	50—200	50—400	—
Cost ($U.S./kg protein)[a] (1978—1979)	1	0.8	1	17	12	10

[a] Reprinted from Fennema, O. R., *Food Chemistry,* Courtesy of Marcel Dekker, New York and Basel, 1985.

achieved in the framework of "green revolution" which has simply shown an approach to the solution of the food problem.[7] High yielding varieties of wheat and rice were introduced in many developing countries and a significant increase of production was achieved, although in terms of production the potential of the high yielding varieties is less than half realized because of the inadequate or improper use of known technology. India for example, where one half the area is irrigated, could produce twice the grain on its present wheat acreage with proper control of weeds, proper time of sowing, proper and adequate use of fertilizers of the right kind and some increase in the use of mechanization for planting and threshing.

In recent years many studies have concerned the improvement of the nutritional value of cereal grains. Very promising results were achieved by breeding new high-lysine varieties of corn, barley and wheat. In many cases simple changes in grain morphology could be the basis of improvement.[8] The embryo of cereal seeds is rich in protein (up to 38%), and the protein may contain about 7% lysine. Selection of larger embryos is particularly important if the whole grain is to be consumed. Also, a selection for a high aleurone cell number could be useful, because these cells are rich in high quality protein. Recent investigations aim at the selection of varieties that can give reasonable yields without irrigation or the addition of fertilizers. One hears a good deal about nitrogen fixation as a means of reducing our reliance on manufactured nitrogen fertilizer. This is a very potent factor with regard to legumes. The possibility of a transfer of this capability to the cereal grasses has aroused a great deal of interest. In the case of maize and other C-4 crops (having an alternative pathway of photosynthesis) this could have marked advantage. In the small grains without this capacity, symbiotic nitrogen fixation could in fact reduce the yields because of the use of plant carbohydrate by the *Rhizobium*. Increase in N-fixation from free living organisms is also unlikely. In all the centuries of adaptation with the billions of microorganisms of this type in the soil, none have been isolated or established over a broad area which have ensured any high degree of fertility.

The increasing use of edible vegetable oils (such as sunflower oil, rape oil, etc.) increased the production of oilseeds. After the removal of oil as a byproduct, a lot of good quality, protein-containing meal is produced being a potential protein source for both food and feed. Nevertheless until now, of all the plant crops giving oil and protein, the soya is the most important. It has been estimated that some 95% of the soybean crop is processed into oil and meal.[9] About one fifth of the world's supply of fats and oils comes from soybean oil, and about 60% of world production of high protein meals is obtained from soybean meal. Most soybean meal is presently utilized for production of swine, poultry and other domestic animals. Although there has been a rapid increase in production of soy flour, which is a basic constituent in edible soy products and also in production of food grade soy protein concentrates and isolates, only 3 to 5% of soybean protein is used directly as human food. The wide use of soy proteins is associated with the high quantity of proteins in soybean (roughly 42%) and high nutritional value of them. The use of other oilseed proteins (in the form of meals, flours, concentrates or isolates) is until now, not solved in a commercial scale.

Potentially sunflower seed proteins and rape seed proteins can be used if some problems associated with the quality of protein preparations could be solved. In the case of sunflower seeds the main problem is the removing of polyphenolic compounds (causing a dark color and off flavor), rape seed protein concentrates have a relatively high level of glucosinolate (an isotiocianate occurring in rape seed), dark color and bitter flavor.[10]

Starchy legumes also have high protein content compared with cereals. Although they are generally poor in sulfur-containing amino acids, the total content of essential amino acids is good. There is also an advantage that due to the microstructure of the seeds an air classification of milled products is possible for preparing protein concentrates. Using such a mechanical procedure for separating protein-rich particles it is possible to avoid some changes occurring during solvent extraction adversely influencing the quality of the protein preparations.

A disadvantage of the legume seeds is the presence of different antinutritive factors. They may have adverse effects on the nutritive value of the product and in some cases, on the health of the consumer. So all the legume seeds contain trypsin inhibitors (reduction of protein digestibility), hemoagglutinins (inhibition of growth) and saponins (reduction of feed or food intake).

The effects of phytic acid on availability of zinc, magnesium, iron and other minerals is also known. Faba bean and related species contain pyrimidine glycosides vicine and convicine. The aglucones are considered to be causative agents of a hemolytic disorder in blood when faba bean is consumed by susceptible individuals.

During the last 25 years considerable interest has been shown in the production of food proteins by microorganisms: bacteria, yeasts, fungi and algae. Microorganisms are technologically attractive in a large part because they offer the promise of increased food production without reliance on traditional agricultural methods. Important features of the production of these unconventional proteins (the generic name adopted for such proteins is single cell protein or SCP) are that they are produced under controlled factory conditions which require very limited space, they are not dependent upon atmospheric conditions and they spare marine fauna and other natural resources. The microorganisms involved reproduce themselves in large fermentation vessels continuously. The microorganisms will grow on many different carbon substrates: natural resources such as carbohydrates, lower alcohols from the petrochemical industry or from fermentation, and hydrocarbons from the oil industry are all potential carbon sources. Moreover, some may serve a secondary role as biological processors of human and industrial wastes.

The cells of bacteria and yeasts are rich in protein and amino acid composition of their proteins is also favorable. The essential amino acid content is high except sulfur-containing amino acids.

Nevertheless, use of SCP directly as human food may be hampered by several serious shortcomings.

A factor that perhaps constitutes a universal and major limitation to the use of SCP as human food is the high nucleic acid content. The prerequisite of further progress is to find a safe process for nucleic acid removal which is economically competitive too.

There are also some other components which may have an adverse effect, especially in bacteria. At the present time bacteria seem to represent the least viable source of SCP. Often lipids of bacteria contain C-17 and C-19 cyclopropane acids or a polymer of hydroxybutiric acid. Bacterial cell walls contain many compounds not present in other organisms. The primary structural component is often a complex polymer of the unusual, carbohydrates muramic acid, along with hexoseamins and both D- and L-amino acids. Gram-negative bacteria have additional layers of lipid, lipoprotein, lipopolysaccharide and proteins as a part of cell wall. Human experiments showed that men fed with a higher amount of bacterial mass suffered from vertigo, nausea, vomiting and diarrhea.[11]

Among the other potential sources of SCP for human nutrition yeast protein seems to be most suitable. The use of yeast in man's diet goes back thousands of years (bread making, beer production, etc.). So consumption of yeast in such quantities which is present, e.g., in bread could be regarded as safe. This is confirmed by experiments of Calloway[11] showing that diets containing yeast as the only protein source resulted in a slightly higher level of fecal nitrogen than a control diet of casein, but the difference was not statistically significant. Similarly, calcium, magnesium and phosphorus excretions exceeded the dietary intake, largely due to increased fecal excretion of these elements. About 8 to 9 g of yeast nitrogen was adequate to maintain balance in healthy male subjects.

Among the other advantages of the use of yeast (yeast protein) that can be mentioned is that production of yeast by fermentation has a long tradition and well-developed commercially and economically applicable technologies are available.

In many cases (e.g., breweries) a lot of yeast is produced as a low priced byproduct which could be used for further processing and application in food production. Yeast biomass can be used not only as a protein source, but also is rich in vitamins. So it can be a raw material for production of vitamin-protein concentrates. Removed nucleic acids may be the basis of production of different nucleotides applicable in various flavor combinations. Production of several enzyme preparations on a commercial scale is also based on yeast as raw material.

In the framework of this book we will discuss first the potential use of yeast biomass as a protein source either in the form of inactive dried yeast and autolysate either as protein concentrate and/or isolate. Other possibilities will be treated shortly in the Appendix.

REFERENCES

1. FAO/WHO, Energy and protein requirements, Report of a joint FAO/WHO ad hoc expert committee, World Health Organization Techn. Rep. Ser. 522, WHO, Geneva, 1973.
2. United Nations University, World Hunger Program Protein — energy requirements under conditions prevailing in developing countries: current knowledge and research needs, Food Nutr. Bull. Suppl., 1. U.N.U., Tokyo, 1979.
3. **Waterlow, J. C. and Payne, P. R.,** The protein gap, *Nature,* 258, 113, 1975.
4. Select Committee on Nutrition and Human Needs, Dietary Goals for the United States, 2nd ed., U.S. Government Printing Office, Washington, D.C., 1977.
5. **Richardson, D. P.,** Consumer acceptability of novel protein products, in *Developments in Food Proteins, Vol. 1.,* Hudson, B. J. F., Ed., Applied Science Publishers, London, 1982, 217.
6. **Schlierf, G.,** Ernährungswissenschaftliche Gesichtspunkte unter Berücksichtigung des Ernährungberichtes, in *Problems in Nutrition and Food Sciences No. 5. Problems with New Foods,* Harmer, W., Auerswald, W., Zöllner, N., Blanc, B., and Brandstetter, B. M., Eds., Wilhelm Mandrich Verlag, Wien-München, 1976, 1.
7. **Anderson, R. G.,** The aftermath of the green revolution, in *Cereals 78: Better Nutrition for the Worlds Millions,* Pomeranz, Y., Ed., AACC Publishers, St. Paul, MN, 1978, 15.
8. **Pomeranz, Y.,** Grain endosperm structure and use properties, in *Developments in Food Science Vol. 5. Progress in Cereal Chemistry and Technology,* Holas, J. and Kratochvil, J., Eds., Elsevier, Amsterdam, 1983, 19.

9. **Pearson, A. M.,** Soy proteins, in *Developments in Food Proteins-2,* Hudson, B. J. F., Ed., Applied Science Publishers, London, 1983, 67.
10. **Sosulski, F. W.,** Rapeseed protein for food use, in *Developments in Food Proteins-2,* Hudson, B. J. F., Ed., Applied Science Publishers, London, 1983, 109.
11. **Calloway, D. H.,** The place of SCP in man's diet, in *Single Cell Proteins,* Davis, P., Ed., Academic Press, London, 1976, 129.

Chapter 2

THE YEAST

I. GENERAL SURVEY

Yeasts are without doubt, both quantitatively and economically the most important groups of microorganisms exploited by man. The total amount of yeast produced annually, including that formed during brewing and distilling practices is in excess of a million tons. No other group of microorganisms has been more intimately associated with the progress and well-being of the human than yeasts.

Although the term yeast is used extensively in scientific literature, it has been difficult to state a precise definition of yeasts on the basis of general morphological and physiological considerations. For example, most yeast cells are colorless, transparent while some produce carotenoid pigments like *Rhodotorula.* Bud formation is widespread however, also not a common characteristic to all yeasts, species of the genus *Saccharomycopsis* multiply exclusively by fission. Cell shape is, in most cases, sound to oval single cell but several species of *Saccharomycopsis* and *Rhodosporidium* are forming true mycelium. The fermentative process, the ability of yeasts to grow in the absence of air, was thought of as a universal property but as the years passed in this respect, there is no identity.

The definition of Lodder:[1] "Yeasts may be defined as microorganisms in which the unicellular form is conspicious and which belong to fungi" seems to be the best one. The taxonomy of yeasts is in a continual state of flux as new microorganisms are discovered and increasing knowledge about the microorganisms results in the yeasts being shifted within subgroups. However, these changes are slow to be put into practice except by taxonomists.

II. CLASSIFICATION OF YEASTS

According to MacMillan and Phaff[2] there are four groups of yeasts:

1. Ascomycetous yeasts or true yeasts — capable of forming ascospores in asci
2. Basidiomycetous yeast — having a lifecycle similar to those of order Ustilaginales or Basidiomycetes
3. Ballistosporogenous yeasts — forcibly discharge spores by the drop excretion mechanism
4. Asporogeneous yeasts, or Deuteromycetes or false yeasts — incapable of producing ascospores ballistospores or sporidia, since sexual lifecycle does not occur or has not been observed so far, these yeasts are members of *Fungi imperfecti*

Industrially important yeasts are classified by Reed and Peppler[3] in two classes of fungi based on their spore-forming capabilities; the Ascomycetes or ascorporogeneous ("true") yeasts and the asexual Deuteromycetes or asporogeneous ("false") yeasts. *Saccharomyces, Candida* and *Kluyveromyces* are the main genera for practical purposes.

Frequently species were named on the basis of fermentation they were associated with like *Saccharomyces vinii* or *Saccharomyces sake.* The morphological and physiological properties are often very similar and separate names have not been justified from a taxonomic point of view.[4] For various industrial processes special characteristics of production strains are of great importance, therefore strain differentiation may be more valuable than species classification. Industrially important yeast species either used or suitable for food industrial purposes are listed in Table 4.

From different yeasts several enzymes and biological active compounds are produced

TABLE 4
Products of Industrially Important Yeast Species

Product	Yeast	Ref.
Baker's yeast	*Saccharomyces cerevisiae*	1, 2, 4, 5
Food yeasts	*Candida utilis*	
	Candida tropicalis	
	Kluyveromyces fragilis	6
	S. carlsbergensis	
	S. cerevisiae	6
Protein	*Saccharomyces cerevisiae*	8
	Torula spec.	7
Alcoholic beverages	*S. cerevisiae*	
	S. bayanus	9
	S. carlsbergensis	
Yeast autolysates and extracts	*S. cerevisiae*	10
	R. rouxii	11
	C. utilis	
	Rhodotorula	
Lipid	*Rhodotorula glutinis*	
	Candida utilis	12
β-Carotene	*Rhodotorula*	13
Nucleotides	*Candida* spec.	14
Mannoproteins	*S. cerevisiae*	15
Glutathione	Brewer's yeast	29
Enzymes		
Amylases	*Schwanniomyces alluvius*	16, 17
Inulinase	*Candida keyfir*	20
	Candida pseudotropicalis v. *lactosa*	18
	K. cicerisporus	
	K. fragilis	19
Lactose	*Candida pseudotropicalis*	19
Protease	*S. cerevisiae, Torulopsis*	21, 22, 23
Pectinases		24, 25
Glucose oxidase	Baker's yeast	29
Lipase		26
Invertase	*S. cerevisiae*	27
α-Galactosidase		28

on a commercial scale like alcoholdehydrogenase, hexokinase, L-lactate dehydrogenase, glucose 6-phosphate dehydrogenase, glycealdehyde-3 phosphate dehydrogenase, inorganic phosphatase, coenzyme A, oxidized and reduced diphosphopyridine nucleotides and mono-, di-, and triphosphates of adenine, cytidine, guanosine and uridine.

III. THE YEAST CELL

Yeast cells have been investigated intensively because of their great importance in practice, but little progress in our understanding in yeast cytology has been made with the ordinary light microscope. Electron micrographs of thin sections have shown the existence of a membranous system and the fine structure of yeast cells. In recent years biochemists could relate many metabolic functions to the ultrastructure and chemical composition of the yeast cell. Conditions of culture influence cytological structure and metabolism of the cell. A schematic cross-section of the yeast cell is shown in Figure 1.

A. CELL WALL AND PLASMA MEMBRANE

The cell is covered by the cell envelope which consists of the plasma membrane and

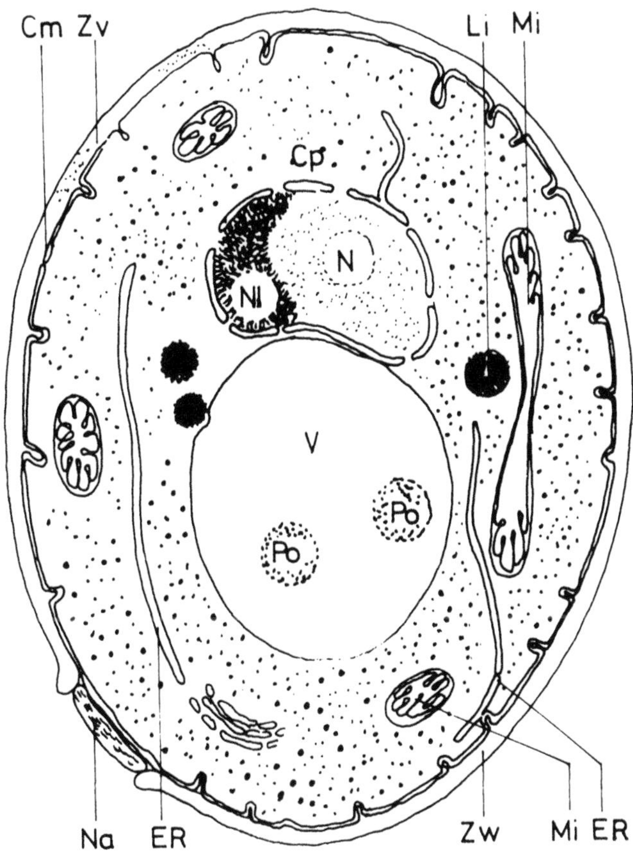

FIGURE 1. Schematic cross-section of the yeast cell: cw — cell wall; cm — cell membrane; cp — cytoplasm; Mi — mitochondrium; V — vacuole; ER — endoplasmatic reticulum; Po — polyphosphate.

the cell wall. According to Suomalainen et al.,[5] the cell wall consists principally of mannanproteins with a certain amount of chitin, the middle layer consists of glucan while the innermost layer has been suggested to contain more protein including enzymic protein.

In yeasts that proliferate by budding bud scars are left on the wall of the cells, the number of them indicates the age of the cell in terms of the reproductive cell cycle. The cell wall forms 15 to 20% of the dry weight of the cell. The proportional weight of the cell wall usually decreases during the growth phase and increases again in the stationary phase. The native cell wall represents a very complex heterogeneous polymer. When it is treated with weak alkali or with certain digestive enzymes, fractions are obtained which are complex macromolecular and structural fragments of the cell wall. In case of stronger acidic hydrolysis glucose, mannose, glucosamine, amino acids, phosphates and lipides are obtained. These compounds originate from the main components of the cell wall, the polysaccharides glucan and mannan, chitin, protein and lipids. Published data for the chemical composition of the cell wall of *Saccharomyces cerevisiae* shows a great variability[5] as carbohydrates vary 60 to 91%, proteins 6 to 13% and lipids 2 to 8.5%. The significant differences may be either a result of different isolation methods and clean up procedures of the different research groups or caused by different growth condition. In *Saccharomyces cerevisiae* the two main polysaccharides are present in about an equal amount. The glucan which is primarily responsible for the shape of the cell wall is a highly branched polysaccharide with β(1-3) and β(1-6) linked glucose residues. The main chain is built up entirely by glucose linked by β(1-6) bonds, the (1-3) linkages are in the side chains. The mannan is alkali soluble and a

TABLE 5
Chemical Composition of Yeast Cell Wall

Components	Content in % of dry matter
Carbohydrates	76—84
Hexoseamine	1—2
Protein	7
Lipids	2

TABLE 6
Overall Composition of Protoplast Membranes in % of Dry Matter

Component	Content (% of dry matter)
Lipids	39.1—5.3
Total N	9.1—1.1
Total carbohydrate	4.0—6.0
Total phosphorus	1.21 ± 0.07
Sterols	6.0 ± 0.5

highly branched polymer of mannose with α(1-6) main chain and α(1-4) and some α(1-3) linked side chains. Chitin is a polymer of β(1-4) linked N-acetyl glucosamine, however not the whole amount of glucosamine content is located in the bud scar region.

The cell wall protein contains all of the common amino acids with high proportions of glutamic acid and aspartic acid. In contrast to the whole-cell protein amino acid composition, cell wall protein has a high content of sulfur amino acids which, especially in case of younger cells, contains SH groups of amino acids. In older cells the protein chains are linked by disulfide bonds. This could be a reason for the resistance to digestive enzymes. The structural protein of the cell wall is mainly bound to polysaccharides to form a complex structure in which glucosamine is suggested as a connecting link between the polysaccharide and protein.[5]

The lipid content of the cell wall varies both with species of yeast and growth conditions. Published data show great variations probably due to incomplete removal of the lipid rich plasma membrane. Some of the lipid content is firmly bound to the cell wall and may play a role in maintaining the ordered structure of the wall.

In fatty acid composition of the cell wall lipids $C_{16:1}$, $C_{18:1}$ and $C_{16:0}$ acids are dominant and there is a lack in short chain fatty acids. A general chemical composition of yeast cell wall is summarized in Table 5.

The plasma membrane has a similar structure in microbes, animals and plants so we can speak about a "unit membrane".[7] The plasma membrane could be obtained by bursting protoplasts in ice cold phosphate buffer. Longley et al.[8] accounted protoplast membranes for 13 to 20% of the dry weight of the yeast cell. Schlegel calculated membrane proportions for only 8 to 15% of dry matter. Typical for this formation is the high lipide content which includes 70 to 90% of the whole cell lipid content. General composition of the plasma membrane lipids is shown in Table 6. Longley et al.,[8] found that neutral lipids of membrane contain a higher amount of $C_{18:1}$ acids than whole cell, the phospholipids of membrane are richer in phosphatidylinositol + phosphatydilserine than the cells.

B. NUCLEUS

The nuclei of yeasts are surrounded by an envelope which is characterized by the presence of many pores.[9] Microtubules are important organelles required for the organization of the cytoskeleton.

C. MITOCHONDRIA

Mitochondria are abundant in all fungi. Their most morphological characteristic is the presence of inner and outer membranes. The size, shape, number and composition of mitochondria vary widely under different conditions of growth. Under anaerobic conditions of growth the unsaturated fatty acids in mitochondrial phospholipids are replaced by fatty acids and the sterol content is significantly reduced.

D. VACUOLES

Vacuoles are important subcellular organelles in yeasts. Both vegetative and reproductive cells contain vacuoles that vary significantly in size. The vacuoles contain a variety of hydrolytic enzymes like proteases and occasionally they contain degrading intracellular substances.

IV. MORPHOLOGY OF YEASTS

Morphology exhibited by a particular yeast is directly associated with its asexual reproduction mechanism. Although the majority of yeasts reproduce by budding, in some it occurs and in others there is a combination of the two processes. Bud formation can occur at different sites on the surface of cells (multilateral), — exclusively of the two opposite sites (bipolar) or at one pole only (monopolar).

Bud fission is the combination of budding and fission. Members of several genera produce true mycelium. The hyphea of some of these filomentous types disaggregates into cells called arthospores. In case pseudomycelium buds do not detach themselves from one another, the chains of cells formed are reminiscent of filamentous fungal growth. Yeast can be distinguished on the basis of their ability to form ascus, mycelium production and budding (Figure 2).

The characteristics of yeast cultures like formation of sediment ring, islets or pellicle in stationary liquid media are easily detectable and valuable in species characterization. Growth mode on media may be a manifestation of hyphal or pseudohyphal growth. Sexual characteristics of ascomycetous yeast are also useful in identification of yeast species. In the life of a typical haploid yeast, budding cells are predominantly haploid and ascus formation occurs immediately after conjugation. Shortly after fusion kariogamy takes place followed by meiosis, and spores are formed in asci. Haploid spores produce upon germination of haploid vegetative cells. The life cycle of such haploid yeasts is shown in Figure 3a. In diploid cells, shortly after spore germination occurs, conjugation is followed by kariogamy. However, spore formation is delayed. The diploid cells or zygotes may bud for many generations, producing additional diploid vegetative cells. The lifecycle of diploid yeast is shown in Figure 3b. However, various yeasts cannot be categorized as strictly haploid or strictly diploid. In some yeast cultures both haploid and diploid vegetative cells may exist together (see References 1 and 2).

Basidomycetous yeasts can be either homothallic or heterothallic and form thick-walled diploid teliospores. Budding cells conjugate and give rise to a dikarigotic mycelial phase. Teliospores are produced either terminally or within the hyphal strains, and kariogamy occurs.

Germination of teliospores results in promycelium formation. Reduction division occurs, and the promycelium becomes septate, producing four cells on which sporidia are born. Sporidia can reproduce as budding yeast cells or conjugate and repeat the cell cycle.

FIGURE 2. Mycelium production and budding of yeast.

Asporegenous yeasts do not produce sexually derived spores such as ascospores or sporidia. All genera reproduce either by budding or fission. Some genera are the perfect form of an asporogenous genera like that of *Kloeckera, Brettanomyces* or *Hanseniospora.* Physiological characteristics are essential in the identification of species. The most important properties are related to fermentation and assimilation of carbon sources, nitrogen utilization, vitamin requirements, tolerance of high salt or sugar concentration and temperature sensitivity. Recently characterization of protein pattern by polyacrylamid gel electrophoresis, immunological properties, analysis of base composition of deoxyribonucleic acid get importance in identification of yeast strains and delimiting relationships among various yeast.

Many yeast are capable of fermenting various carbon sources. There are probably no strictly anaerobic yeasts, and the same carbon sources that can ferment yeast can also be assimilated oxidatively under appropriate conditions. The reverse is not true, in that the ability to assimilate a certain carbon source does not mean that it can necessarily be fermented.

All fermentative yeasts are able to ferment glucose, producing ethanol and CO_2. Some morphological and physiological characteristics of industrially important yeast are summarized in Tables 7 and 8.

V. YEASTS AS HIGHLY EFFICIENT PRODUCERS OF PROTEIN

Yeast are able to build up all the amino acids necessary for protein synthesis. The carbon chain skeleton of the amino acids is derived from metabolites of intermediate metabolic

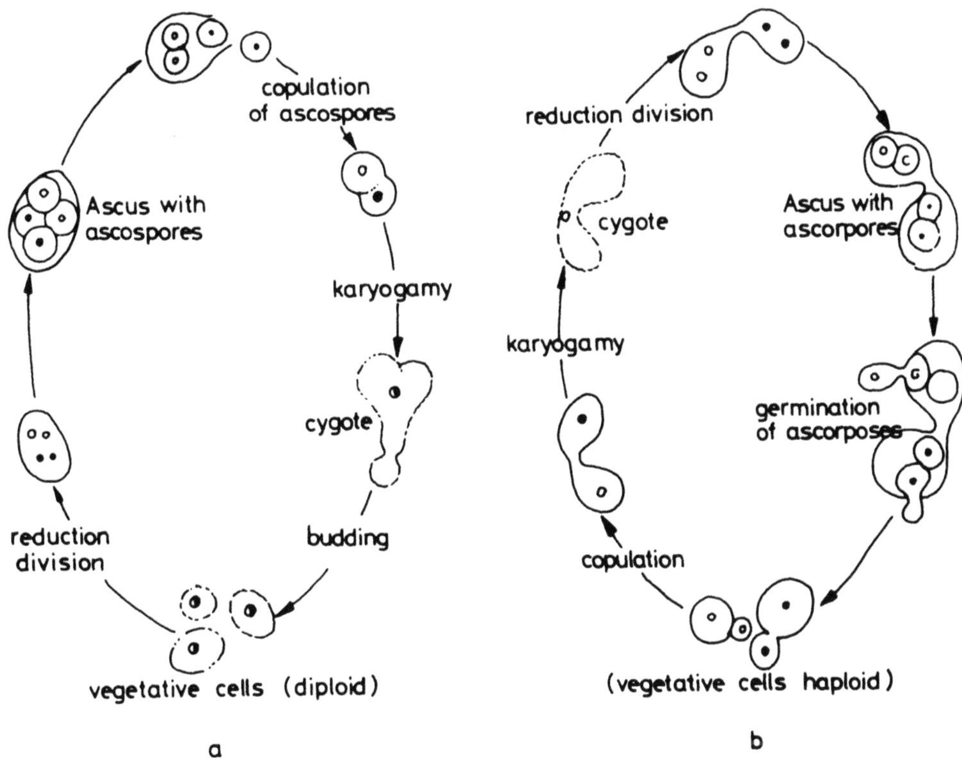

FIGURE 3. Life cycle of yeasts: (a) haploid yeasts; (b) diploid yeasts.

pathways. Amino groups are incorporated either by direct aminization or by transamination. Only a few amino acids could be synthesized by the former reaction (L-glutamic acid, L-alanine, L-asparagine and L-glutamine). The primary NH_3-assimilation is catalyzed by L-glutamate dehydrogenase and L-alanine-dehydrogenase through reductive amination of a ketoacid. All the other amino acids get their amino groups by transamidation of one of the primary amino acids. On the basis of common synthesis-pathways amino acids could be classified into groups as Figure 4 shows.

Proteins, including all enzymes, are synthesized by sequential addition of amino acids to amino acid (peptide) by a complex of enzymes and RNA organized in the ribosome (see Figure 5). The code for ensuring the right sequence of amino acids is carried by a strand of messenger RNA, which in turn is synthesized by copying from a section of DNA in the chromosome of the cell. This process is brought about by DNA-dependent RNA polymerases and is termed transcription.

Ribosomes become attached to the m-RNA, the bases of the m-RNA are "read off", that is translated, three at a time to code for a particular amino acid. The aminoacyl-t RNAs are the active units used by the ribosome to produce the growing peptide (see Figure 5).

According to Saha and Chakraburtty[35] polypeptide chain elongation reaction in yeast requires at least three distinct protein factors. Elongation factor is responsible for bringing in the aminoacyl-t RNA to the ribosomal "A" site. Factor 2 is involved in the peptide chain translocation reaction and factor 3 is uniquely required by the ribosomes. The precise function of elongation factor 3 in translation is still undefined. The protein synthesis takes place largely on membrane-bound ribosomes and a maximum of membrane-bound ribosomes is found between two doublings of cells. The rate of protein synthesis in the cell differs from an exponential mode. The highest rate of protein synthesis of fast growing glucose cells is created within the G_2 phase and thus coincide with the period of a large bud. In slow-

TABLE 7
Morphological Properties of Yeast

Genus	Budding	Pseudomycelium	Mycelium	Pellical	Special characteristics
		Ascomycetous yeasts			
Hansenula	Multilateral (ML)	±		±	
Kluyeromyces	Multilateral (ML)	+	—	(+)	
Saccharomyces	Multilateral (ML)	(±)	—	—	
Schizosaccharomyces	—	±	—	(+)	Reprod. fission
		Asporogenous yeasts			
Brettanomyces	ML	+	—	—	Whey
Candida	ML	+	(+)	—	All produce pseudomycel
Kloeckera	Bipol	±	—	—	Apiculate cells
Rhodotorula	ML	(+)	(+)	—	Carotenoid production
Torulopsis		(+)	—	—	Few pseudomycel
		Basidomycetes yeasts			
Rhodosporoides	Bipol	±	+	—	Perfect form from Rhodotorula

Note: ( ) Property weekly exhibited. ± Approx. half of species.

TABLE 8
Physiological Properties of Yeast

Genus	Fermentation	NO_3 utilization	Without vitamin	Acid production	Spec. properties
Hansenula	M	+	(+)	(+)	Some produce phosphomannan and sphingolipids
Kluyveromyces	+	—	—		
Saccharomyces	+	—	±		All spec. ferment. strongly
Schizosaccharomyces	+	—	—	(±)	No chitin in cell wall
Bretanomyces	+	±	—	+	Produce acetic acid
Candida	+	±	(M)		
Kloeckera	+	—	—	—	All required inositol and pantothenate
Rhodotorula	—	±	(+)	(−)	Carotenoid pigments no fermentation
Torulopsis	M	±	±	±	
Rhodosporidium	—	(+)	±	—	

Note: M = many
F = few

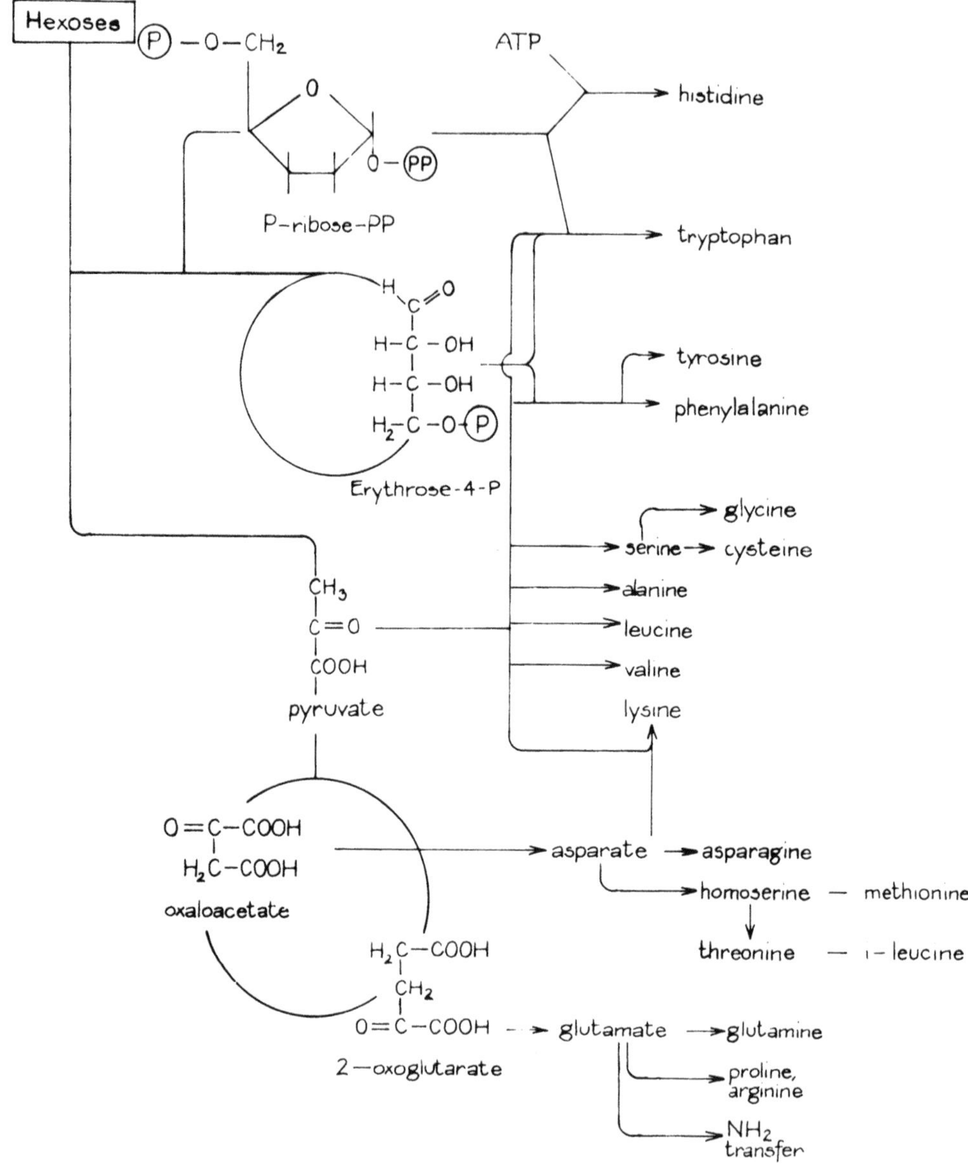

FIGURE 4. Pathways of amino acid biosynthesis.

growing cells the maximum rate of protein synthesis is highest during the final part of the G_1 period. In general, the highest rates of protein synthesis tend to coincide with gross mass volume increments. In fast-growing cells this event is observed prior to cell separation, whereas in slow-growing cells the event is observed as well in daughters following cell separation.

Protein content of yeast biomass varies in average between 40 to 55% on cell dry matter. However, this crude protein (N × 6.25) also includes the nucleic acid content which shows values of 6 to 13%. Data presented in Table 6 illustrate that the protein content of yeast biomass is influenced by strain, substrate composition and fermentation parameters as well.

Under optimal conditions, the doubling time of yeasts is 1.5 to 1.7 h and carbon conversion coefficient on glucose is 1.25. This gives a high rate protein synthesis with a good yield. As yeast strains showing the best growth rate on a given substrate do not always

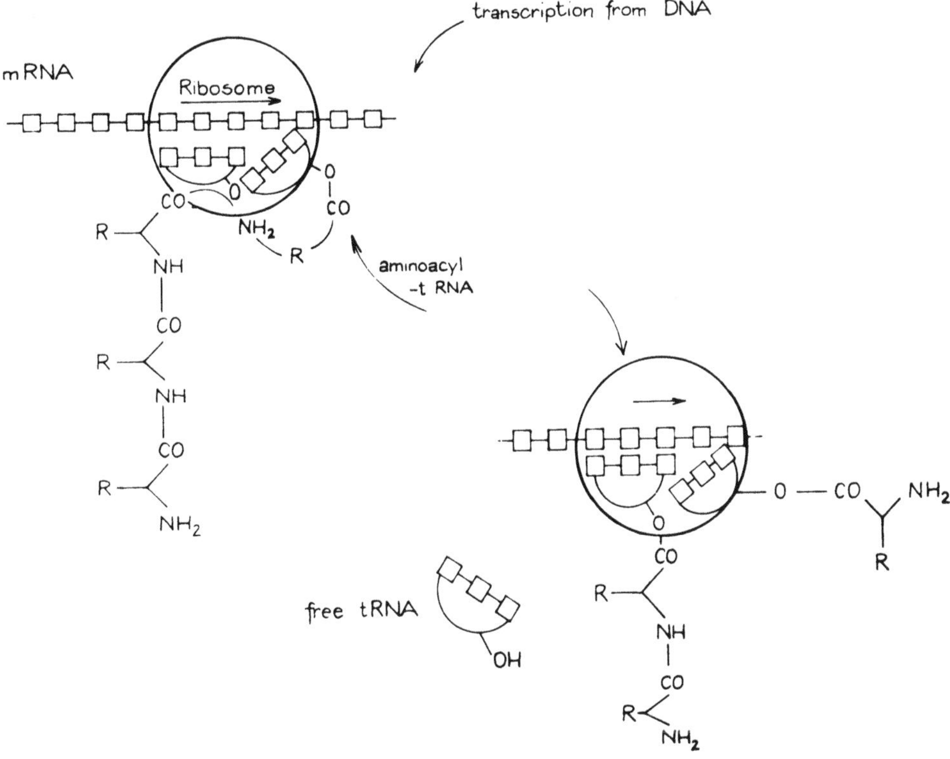

FIGURE 5. Scheme of the protein biosynthesis on the ribosome.

have the highest protein content and best balance in amino acid composition related to human requirements, we have to look for the highest protein production in the fermentation process. The effect of environmental factors on the protein content of yeast will be treated in details in Chapters 3 and 4.

REFERENCES

1. **Lodder, J., Ed.,** *The Yeast — A Taxonomic Study,* North-Holland, Amsterdam, 1970.
2. **MacMillian, J.-P. and Phaff, H. J.,** Yeasts, general survey. in *CRC Handbook of Microbiology,* Vol. II., Laskin, A. C. and Lechevalier, H. A., Eds., CRC Press, Boca Raton, FL, 1978.
3. **Reed, G. and Peppler, H. J.,** Naming and classifying yeasts, in *Yeast Technology,* AVI Publishing, Westport, CT, 1973.
4. **Burrows, S.,** Baker's yeast, in *Economic Microbiology Vol. 4. Microbial Biomass,* Academic Press, London, 1979.
5. **Grigonic, J. I., Tebelshkene, B. V., and Marshtenkene, V. P.,** Increasing of the production of baker's yeast and improving their quality (in Russian). *Czebopek. Kond. Prom.,* No. 2, 28, 1983.
6. **Batt, C. and Sinskey, A. I.,** Use of biotechnology in the production of single cell protein, *Food Technol.,* 38, 108, 1984.
7. **Roskeva, Z., Dukiandjev, S., and Pavlova, K.,** Biochemical characterization of yeast protein isolates, *Die Nahrung,* 30, 402, 1986.
8. **Rooij, I. M. and Hakaart, M. I.,** Yeast extract food flavour, U.K. Patent Application, GB, 2171585 A.
9. **Lacashire, W. E.,** Modern genetics and brewing technology, *Brewer,* 72, 345, 1986.
10. **Rooij, I. M. and Hakaart, M. I.,** Food flavours, European Patent Application, E.B., D., 191513 A 1.

11. **Woodlen, A.,** Flavouring the "natural" way, *Food Eng. Int.*, 11, 33, 1986.

12. **Krumphanzl, V., Gregr, V., Pelechová, I., and Uher, J.,** Biomass and fat production in *Rhodotorula gracilis,* in *Advances of Microbiol Engineering,* Vol. I, Sikyta, B., Prokop, A., and Novák, M., Eds., John Wiley and Sons, New York, 1983.

13. **Costa, I., Martelli, H. L., Silva, I. M., and de Pomeroy, D.,** Production of β-carotene by *Rhodotorula* strain, *Biotechnol. Lett.*, 9, 373, 1987.

14. **Watanabe, S., Kodeira, R., and Kaneko, Y.,** Serial production of nucleoside-5′-triphosphates and uracil from 3′-mono-nucleotides by intact yeast cells, *Eur. J. Appl. Microbiol. Biotechnol.*, 22, 422, 1983.

15. **Cameron, D. R., Cooper, D. G., and Neufeld, R. I.,** The mannoprotein of *S. cerevisiae* is an effective bioemulsifier, *Appl. Environ. Microbiol.*, 54, 1420, 1988.

16. **Calleja, G. B., Levy-Rick, S., Moranelli, J., and Nasim, A.,** Thermosensitive export of amylases in the yeast *Schwannimyces alluvius, Plant Cell Physiol.*, 25, 757, 1984.

17. **Lusena, C. V., Champagne, C. C., and Calleja, G. B.,** Secretion and export of amylolytic activities in *Schwanniomyces alluvius, Can. J. Biochem. Cell. Biol.*, 63, 366, 1985.

18. **deBales, S. A. and Castillo, F. J.,** Production of lactase by *Candida tropicalis* grown in whey, *Appl. Environ. Microbiol.*, 37, 1201, 1979.

19. **Hewitt, G. M. and Grootwassink, J. W. D.,** Simultaneous production of inulase and lactase in batch and continuous cultures of *Kluyveromyces fragilis, Enzyme Microbiol. Technol.*, 6, 263, 1984.

20. **Manzoni, M. and Cavazzoni, V.,** Extracellular inulinase from four yeasts, *Lebesm. Wiss. u. Technol.*, 21, 271, 1988.

21. **Bilinski, C. A., Rossell, I., and Stewart, G. G.,** Applicability of yeast extracellular proteinases in brewing: physiological and biochemical aspects, *Appl. Environ. Microbiol.*, 53, 495, 1987.

22. **Nelson, B. and Young, T. W.,** Yeast extracellular proteolytic enzymes for chill-proofing beer, *J. Inst. Brew.*, 92(6), 599, 1986.

23. **Sturley, S. L. and Young, T. W.,** Extracellular protease activity in a strain of *Saccharomyces cerevisiae, J. Inst. Brew.*, 94, 23, 1987.

24. **Call, H. P., Harding, M., and Emeis, C. C.,** Screening for pectolytic Candida yeasts. Optimization and characterization of the enzymes, *J. Food Biochem.*, 9, 193, 1985.

25. **Waller, W. and Comeau, L. C.,** Characteristics and performance of an acid-resistant fungal lipase, *Rev. Fr. Crops Cras.*, 34, 205, 1987.

26. **Call, H. P., Walter, J., and Emeis, C. C.,** Maceration activity of endopolygalacturonase from *C. macedoniensis, J. Food Biochem.*, 9(2), 193, 1985.

27. **Nilsson, V., Öste, R., and Jagerstad, M.,** Cereal fructans: hydrolysis by yeast invertase *in vitro* and during fermentation, *J. Cereal Sci.*, 6, 53, 1987.

28. **Liljenström, P. L., Tube, P. S., and Korhola, M. P.,** Construction of new alpha-galactosidase producing yeast strains and the industrial application of these strains, European Patent Application EP 0241044 A2, 1987.

Liu, W. H. and Cheng, T. L., Studies in the production of glutathione from brewer's yeast, *J. Chin. Agr. Chem. Soc.*, 25, 318, 1987.

Deák, T., Characterization, classification and identification of yeasts (in Hungarian), I, II, *Söripar*, 31, 63, 70, 1985.

Deák, T., Characterization, classification and identification of yeasts (in Hungarian), III, IV, *Söripar*, 32, 18, 94, 1986.

32. **Suomalainen, H., Nurminen, T., and Oura, E.,** Aspects of cytology and metabolism of yeast, in *Prog. Indus. Microbiol.*, Hockenmell, D., Ed., Churchill Livingstone, 12, 109, 1973.

33. **Schlegel, H. G.,** *Allgemeine Mikrobiologie,* Georg Thieme Verlag, Stuttgart, 1974.

34. **Longley, R. P., Rose, A. H., and Knights, B. A.,** Composition of the protoplast membrane from *Saccharomyces cerevisiae, Biochem. J.*, 108, 401, 1968.

35. **Saha, S. K. and Chakraburtty, K.,** Protein synthesis in yeast. Isolation of variant form of elongation factor 1 from the yeast *Saccharomyces cerevisiae, J. Biol. Chem.*, 261, 12599, 1986.

36. **Kamath, A. and Chakraburtty, K.,** Protein synthesis in yeast. Identification of an altered elongation factor in thermolabile mutants of the yeast *Saccharomyces cerevisiae, J. Biol. Chem.*, 261, 12593, 1986.

37. **Gullov, K., Friis, J., and Bonven, B.,** Rates of protein synthesis through the cell cycle of *Saccharomyces cerevisiae, Exp. Cell. Res.*, 136, 295, 1981.

38. **Ratlege, C.,** Biochemistry of growth and metabolism, in *Basic Biotechnology* Bu'lock, J. and Krismansen, B., Eds., Academic Press, London, 1987, 11.

Chapter 3

CHEMICAL COMPOSITION AND BIOCHEMISTRY OF YEAST BIOMASS

I. GENERAL ASPECTS AND GROSS COMPOSITION

Although the chemical composition of yeast is greatly affected by changes in the medium and culturing conditions, the basic characteristic of the dry matter of yeast biomass composition can be summarized as follows:

- High protein content
- High nucleic acid content
- Low lipid content
- High ash content
- Moderate carbohydrate content
- High vitamin content

Early data about the gross composition of yeasts are summarized in the book edited by Cook.[1] Later in the sixties and early seventies parallel with increased interest in industrial production of SCP, intensive research work had been done in this field and a lot of data were published in scientific journals, review papers[2] and books.[3] Some data about gross composition of different yeasts are summarized in the Table 9.

A. PROTEINS

Protein content of the yeasts is changing in relatively wide ranges depending on the medium and conditions of growth. In a review paper Mauron[2] reported that raw protein content of *Candida utilis* was the lowest (50%) and those of *Candida lipolytica* the highest (65%). In a more recent paper Sarwar,[10] investigating the chemical composition of dried inactive *Saccharomyces cerevisiae* and *Candida utilis* yeast grown on different media, found that the raw-protein content varied from 46.37 to 58.25%. The true protein content calculated on the basis of total amino acid content was found to vary in the range from 36.36 to 43.63% (''as is'' basis). Yeast biomass of four *Candida, Kluyveromyces* and *Saccharomyces* strains grown on cheddar whey permeate was investigated by El-Shamragy et al.[11] *Candida utilis* yeast had the lowest true protein content (30.0%) and *Saccharomyces fragilis* the highest one (40.25%). Soboleva and Popova[12] reported about protein content of 47.8% in *Candida scotii* determined by Lowry method. King-Chin Su et al.[13] studied the protein content of 16 yeast strains prepared from cane molasses. It was found that the raw protein content varied in a range from 57.87 to 66.00%. Yeasts used for food yeast production, (*Candida utilis* and *Torula utilis*) alcohol fermentation, fat production (*Rhodotorula glutinis*), sake making and baking purposes were included in the samples to be investigated. Halász,[21] in a more recent work, studied the changes in protein content in *Saccharomyces fragilis* yeast grown on whey under different conditions. The effect of pH, whey concentration and lactose concentration was controlled. Some of the results are included in the Table 10, 11 and 12. As it is shown and mentioned earlier in the chapter, discussing the problem of nucleic acid content of yeasts, it was once again confirmed that increased protein content and increased growth rate is generally associated with the increased nucleic acid content. Relatively few data are available about the finer composition of protein components. There is a tendency to fractionate the proteins based on solubility similarly to the Osborne-fractionation of plant proteins. The greatest part of the native yeast proteins belongs to the water and/or salt soluble

TABLE 9
Gross Chemical Composition of Different Dried Yeasts

Component	Baker's yeast[4]	Yeast grown on *n*-paraffins[5]	Yeast[6]	Yeast[7]	Saccharomyces cerevisiae[8]	Brewer's yeast[9]
Raw-protein	38.0—60.0	50.5		46.9—53.1	53.0	—
True protein	31.0—48.0	44.0	45.0—49.0	40.9—44.1	46.0	48.0
Nucleic acids	7.0—12.0	6.5	—	6.0—12.0	7.0	—
Lipids	4.0—10.0	10.6	4.0—7.0	2.0—6.0	10.0	1.0
Carbohydrates	25.0—35.0	26.5	26.0—36.0	—	18.0	36.0
Ash	6.0—7.0	9.0	5.0—10.0	5.0—9.5	—	8.0
Water	5.0—8.0	4.4	—	—	—	7.0
Calcium	0.08—0.14	0.1	—	—	—	0.3
Phosphorus	1.9—3.3	2.1	—	—	—	2.4
Magnesium	0.2—0.3	0.15	—	—	—	0.3
Potassium	1.7—2.8	2.0	—	—	—	2.2
Sodium	0.1—0.4	0.14	—	—	—	0.1

TABLE 10
The Effect of pH on the Protein and Nucleic Acid Content of *Saccharomyces fragilis*

pH	Raw protein content (%)	True protein content (%)	Nucleic acid content (%)	Specific growth rate ($k\ h^{-1}$)
3.5	31.90	28.83	5.94	0.228
4.5	42.40	35.17	7.08	0.294
5.5	31.35	25.38	5.85	0.238
6.5	30.80	24.62	5.50	0.211

From Halász, A., Biochemical and Biotechnological Principles of Use of Yeast Biomass in Food Industry, Dr. Sc. thesis, Hungarian Academy of Sciences, Budapest, 1988.

TABLE 11
The Effect of Whey Concentration on the Protein and Nucleic Acid Content of *Saccharomyces fragilis*

Whey concentration expressed in lactose concentration (%)	Raw protein content (%)	True protein content (%)	Nucleic acid content (%)	Specific growth rate ($k\ h^{-1}$)
4.68	43.78	34.50	9.06	0.346
3.08	42.07	34.13	7.76	0.330
2.10	41.58	34.44	6.95	0.320
1.05	39.40	32.81	6.45	0.246

From Halász, A., Biochemical and Biotechnological Principles of Use of Yeast Biomass in Food Industry, Dr. Sc. Thesis, Hungarian Academy of Sciences, Budapest, 1988.

proteins[12] (albumins and globulins), the amount of alcohol- and alkali soluble proteins is lower. Drying of yeast or other treatments causing denaturation of proteins changes the solubility and distribution of the fractions, the amount of alkali soluble fractions increases rapidly and this will be the greatest fraction.[15-17] The solubility of yeast proteins will be discussed in detail in Part II in the chapter dealing with functional properties of yeast proteins.

From the point of view of biological function the yeast proteins can be separated in minimally three groups:

TABLE 12
The Effect of Different Media on the Protein and Nucleic Acid Content of *Saccharomyces fragilis*

Fermentation medium	Raw protein content (%)	True protein content (%)	Nucleic acid content (%)	Specific growth rate (k h^{-1})
Synthetic with 2% lactose	41.60	33.84	7.60	0.160
Whey (2% lactose)	41.70	34.04	7.50	0.303
Whey + 1% lactose[a]	41.53	34.38	7.00	0.277
Whey + 1.5% lactose[a]	41.43	35.00	6.30	0.230
Whey + 1.75% lactose[a]	40.77	34.84	5.80	0.196

[a] Diluted whey + pure lactose; final lactose content 2%.

From Halász, A., Biochemical and Biotechnological Principles of Use of Yeast Biomass in Food Industry, Dr. Sc. Thesis, Hungarian Academy of Sciences, Budapest, 1988.

1. Metabolically active, cytoplasmic proteins (including enzymes and regulatory proteins)
2. Storage proteins
3. Proteins of the cell walls

From a practical point of view among the cytoplasmic, metabolically active proteins the enzyme proteins are the most important. Many of the enzymes can be isolated and different enzyme preparations are produced for practical purposes (see Part II).

Storage proteins of yeasts are poorly investigated. The presence and distribution of these proteins is supported by the experiments published by Wiemken and Nusse[18] associated with the amino acid pool of yeasts. These authors demonstrated the presence of two distinct amino acid pools in the food yeast *Candida utilis*. The first one was released from the cells after treatment of the cells with a basic protein which permeabilized the plasmolemma but left the tonoplast intact. This pool of amino acids had a rapid turnover as demonstrated by pulse-labeling experiments using $^{14}C(U)$-arginine, $^{14}C(U)$-glucose and pool located within cytoplasm. The remaining amino acids showed a slow turnover in pulse-labeling experiments and a high proportion of basic, nitrogen-rich amino acids indicative of storage function.

Application of newer highly effective methods of protein separation such as gel chromatography, reverse phase HPLC, electrophoresis and affinity chromatography is used mainly in isolation and purification of enzymes. The problems and practical solutions associated with isolation and purification of enzymes and peptides from microbial sources, including possibilities of large-scale operations, are summarized in a recent excellent review of Wiseman et al.[19]

Recently gel electrophoresis of yeast proteins is successfully used for the identification of yeast strains, having typical electrophoretic patterns.[14,20]

The proteins of the cell wall were relatively widely understood. The results of investigations before the early seventies are summarized in the review of Phaff.[22] The protein content of yeast cell wall varies in the range from 5 to 15%. However, it must be noted that a lot of data is based on total nitrogen content determination and use of 6.25 factor for calculation of protein content. So such data may be subject to some error due to variable content of nonprotein nitrogen, such as *N*-acetyl-glucosamine. Two types of cell wall proteins are mentioned according to the polysaccharide component of the protein carbohydrate complex: glucan (glucomannan) proteins and mannan protein. The cell wall proteins can be extracted by alkali and precipitated with ammonium sulfate resulting in more glycoprotein fractions. Although glucose and mannose are the main components of the carbohydrate part of the complex, in some fractions glucosamine was also found in a small amount. The high amount of hydroxy-amino acids (serine and threonine) and in some cases of asparagine suggest that both *O*-glycosyl and *N*-glycosyl bonds play roles in the binding of the carbohydrate part to the protein part of the complex.

The amino acid composition of yeast proteins has been extensively studied and a lot of data were published. The general statement is that yeast proteins are rich in lysine and poor in sulfur-containing amino acids (methionine and cysteine). The details of amino acid composition will be discussed in Chapter 5 dealing with the nutritive value of yeast proteins.

B. ENZYMES

Yeast cells are a good source of a lot of enzymes. Some enzymes such as aldehyde dehydrogenase, phosphoglycerate kinase, phosphofructokinase, alcohol dehydrogenase, invertase, etc. are produced on commercial scale. A number of hydrolytic enzymes in yeast are found external to the cytoplasmic membrane, and bound tightly or loosely to certain wall components. Such enzymes are frequently excreted into the medium during growth. Phaff,[22] discussing the facts confirming undoubtedly that some enzymes occur as wall constituents, describes four lines of evidence to prove their location in the wall: (1) their removal in soluble form upon conversion of yeast cells into sphaeroplasts; (2) in the case of oligosaccharidases the products of hydrolysis may be trapped outside the plasma membrane as nontransportable compounds (e.g., oxidation of hexoses by specific oxidases or their phosphorylation by hexokinase + ATP); (3) the *in vivo* pH activity curve is similar to that of the isolated enzyme *in vitro* and (4) sensitivity of hydrolases in a thin film of yeasts to low voltage electron beams of increasing penetrating power. By the last technique it was shown that invertase was located just below the outer layer of the wall.

1. Proteases and RNases of Yeast

The proteolytic enzymes of the yeasts have a two-fold practical importance. The storability of the yeast mass is highly dependent on the proteolytic activity of the yeast cells and also in the production of yeast autolysates the proteolyatic enzymes play an important role. In addition to these practical viewpoints, the deeper investigation of yeast proteases may help the progress in molecular biology of yeast and better knowledge of theoretical principles of yeast fermentation.

In an early publication Lenney[23] reported about two proteases (A- and B-protease) in the yeast. Later, using column chromatography, the researchers[24-28] succeeded in separating two endopeptidases (A and B) and a carboxypeptidase (Y). It was found that endopeptidase B and carboxypeptidase Y belongs to the group of serine proteases. The glycoprotein nature of endopeptidase A and carboxypeptidase Y was revealed.[27] All three enzymes hydrolyzed casein and hemoglobin. Using synthetic peptide substrates the presence of aminopeptidases was also confirmed. The aminopeptidase I is a Zn-dependent enzyme[29] and the aminopeptidase II is able to split peptide bonds of both leucine and lysine. The latter enzyme is inhibited by EDTA.[30] Achstetter[31] and Achstetter et al.[32-34] described an aminopeptidase being active only in the presence of cobalt ions (co-aminopeptidase), a mitochondrium-bound aminopeptidase M. Recently, reports were published about some newer types of proteolytic enzymes occurring in yeasts such as carboxypeptidase-S,[35] proteinase D and E,[36] protease P[33] and protease F.[37]

Investigating the location of proteolytic enzymes within the yeast cells it is accepted by several authors that the endopeptidases A and B and carboxypeptidase Y are located in the vacuoles of yeast cells.[38,39] Carboxypeptidase-S is also located in vacuoles. Cytosol contains the proteinases D and E.[36] Finally, it is evident that protease M is part of mitochondrium.

The yeasts contain natural inhibitors of proteolytic enzymes. The inhibitors are specific ones, and inhibit only the corresponding proteinase. The inhibitors called I_A, I_B and I_C are inhibitors of corresponding proteases A, B and C. The most investigated inhibitors are the proteinase inhibitors occurring in *Saccharomyces cerevisiae* and *Saccharomyces carlsbergensis*.[40,41] The inhibitors are located in the cytosol, so the inactivation of enzyme can occur only after destruction of the cell structure. It is suggested that the mechanism of inhibitory action could be a competitive inhibition.[12]

It was observed by several authors that after disruption of cells and centrifugation of disintegrated mass, the supernatant showed very low proteolytic activity. Incubation at acidic pHs and treatment with urea resulted in activation of proteolytic enzymes.

On the basis of early investigations it was suggested that proteases are originally present in an inactive form which can be activated under influence of some regulatory factors. Later a group of researchers,[43-48] found that the inactive state of proteases is associated with the formation of protein-inhibitor complexes. A decrease of pH causes the destruction of inhibitors and activates the proteases.

Concerning the kinetic action of proteases the recent work of Vorobjev et al.[49,50] can be mentioned. They found that at a high concentration of substrate the reaction type one-by-one is characteristic, where the intermediate product does not accumulate (supposing a two step transformation), while at a low concentration of substrate the proteolysis is slower and the zipper type reaction is dominant. The protease activity of yeast cells is changing depending on growth conditions.[21]

Halász[21] reported about experiments with *Saccharomyces cerevisiae* yeast grown under different conditions. The activity of aminopeptidase, carboxypeptidase Y and S, protease A and B was measured. A more intensive aeration during fermentation resulted in decrease of proteolytic activity. The glucose concentration affected the proteolytic activity. Generally, a maximum activity was observed at 0.5% glucose concentration. Specific protease inhibitors (I_A and I_B) were also investigated. It was found that their activity changed parallel with the corresponding protease activity.

Immunosera against specific proteases (A, B, Y) were used to determine enzyme-protein concentration of the yeast in the different growth phases. Immunoanalytical evaluation of measured data showed that aeration does not influence the synthesis of proteases, only the degree of activation is dependent on intensity of aeration. The increase of the proteolytic activity in the logarithmic phase of growth is due to the *de novo* enzyme synthesis. In the stationary phase the increase of proteolytic activity in comparison to that in inoculum is partly caused by enzyme synthesis and partly by activation of proteases. Proteolytic activities of *S. cerevisiae* and *C. guilliermondii* show similar change with C-source concentration however, enzyme activity of *S. cerevisiae* was significantly higher. Protease activity of *R. glutinis* shows minimum value at 0.5% date syrup concentration and highest value at 0.1%. *R. glutinis* has the lowest proteolytic activity at each C-source concentration, except 0.1%.[65] This difference in protease changes with date syrup concentration could be caused by the different glucose metabolism of *R. glutinis, S. cerevisiae* and *C. guilliermondii* resp.

The ribonucleases of the yeast cell can be divided into several main groups:[51] (1) RNases hydrolyzing RNA (polynucleotide phosphorylases), specific RNA depolymerases and non-specific nucleases; (2) phosphodiesterases and phosphomonoesterases; (3) RNases modifying RNA without degradation and (4) RNA synthetases. From point of view of localization extracellular-, intracellular resp. soluble and membrane-bound enzymes are present in the cell. The typical pathways of degradation of RNA by RNases are shown in Figure 6.

There are known natural RNase inhibitors too. The activation of the RNases using heat treatment (heat shock) is probably associated with the inactivation of heat sensitive inhibitors.[53] The products of hydrolysis can also be classified as natural inhibitors. There are some indications in the literature[53] that some bivalent cations also have an inhibitory effect. Sodium chloride and phosphates activate RNases.[54] The activity of RNases of yeast is changing during growth periods and is dependent on conditions of fermentation.

The effect of glucose concentration and aeration on ribonuclease activity of *S. cerevisiae*,

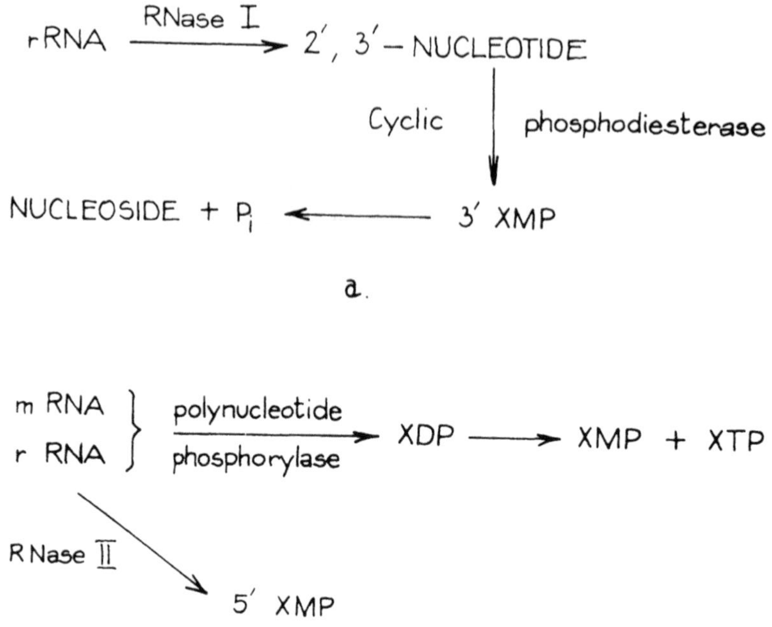

rRNA —RNase I→ 2′, 3′ – NUCLEOTIDE
↓ Cyclic phosphodiesterase
NUCLEOSIDE + P_i ← 3′ XMP

a.

m RNA, r RNA —polynucleotide phosphorylase→ XDP → XMP + XTP
—RNase II→ 5′ XMP

FIGURE 6. Typical pathways of degradation of RNA by RNases.

C. guilliermondii and *R. glutinis* were investigated by Mustafa Kasim.[71] *S. cerevisiae* showed the lowest ribonuclease activity and *R. glutinis* the highest. Maximum values of activities were found at a glucose concentration of 0.8 to 1.0%. The ribonuclease activity of *C. guilliermondii* showed no significant change by increasing aeration intensity. In the case of *R. glutinis,* ribonuclease activity increased gradually with higher aeration intensity. Ribonuclease activity of *S. cerevisiae* showed a similar variation by aeration rate as those of *R. glutinis,* but the absolute values were significantly lower in each case.

C. NUCLEIC ACIDS AND NUCLEOTIDES

High content of nucleic acids is typical for microorganisms and for yeasts too. The amount of nucleic acid is affected by different factors, but it seems that dependence on the rate of growth is the most important. Generally, higher growth rates are associated with higher nucleic acid content. Both ribonucleic acids (RNA) and deoxyribonucleic acid (DNA) occur in the yeast cells but the amount of RNA is much higher. When compared with other cells, yeast appears very rich in ribonucleic acids, and has provided an industrial source of the various molecular species. The amount of RNA may reach one third of the total protein and in some fifty to a hundred times greater than the DNA content.

The DNA of yeasts is located in the nucleus of the cell, however yeast mitochondria have their own DNA too. Similarly to DNA of other cells, yeast DNA is compound exclusively from deoxyribonucleotides containing four bases: adenine (A); guanine (G); thymine (T) and cytosine (C). Methods were elaborated for isolation and separation of nuclear and mitochondrial DNA. The results of the earlier research work associated with yeast DNA were summarized by Mounolou.[55] Due to the fact that DNA content of yeast cell is low in comparison to RNA content, generally the DNA content of yeasts is not determined separately for chemical and nutritional evaluation of yeast biomass or yeast protein preparations. The more intensive investigation of nuclear and mitochondrial DNA is associated with research in molecular biology of yeasts. Some of the questions related to this topic being interesting from the point of view of yeast production will be treated in Chapter 4 of this book.

Similarly to other cells of living organisms, yeast contains messenger RNAs (mRNA), transfer RNAs (tRNA) and ribosomal RNAs (rRNA). Specific RNAs were found in yeast mitochondria. Generally, only the total amount of RNA is determined and taken in account at nutritional evaluation of yeast biomass or yeast protein preparation. The total RNA content is in the first line depending on the growth rate of yeasts, e.g., in a recent study Bueno et al.[56] found that the RNA content of yeast biomass of a strain of *Candida utilis* increased from 8 to 11.93% when growth rate increased from 0.15 to 0.52 h^{-1}. The effect of media on growing is also important,[57,58] e.g., Waldron and Lacroute[25] reported that nucleic acid of *S. cerevisiae* grown in different media ranged from 4.0 to 12.9%. Separation of different types of yeast and investigation of their properties is a subject of intensive research in molecular biology of yeasts. It may be interesting to note that yeast alanyl-tRNA was the first nucleic acid for which the complete sequence was worked out. Some newer data about the base composition of RNA in different yeast strains are shown in Table 13.[10,59]

D. CARBOHYDRATES

The carbohydrate content of yeasts varies in wide range depending on the growing conditions. Normally, one fifth to one third of dry matter consists of different carbohydrates. The greatest part of the carbohydrates is present in the form of different polysaccharides, the amount of mono- and oligosaccharides is low except trehalose. From a morphological point of view the carbohydrates of yeast cells can be divided in two groups: intracellular (cytoplasmic) carbohydrates and cell wall carbohydrates.

In most yeasts energy reserves consist largely of carbohydrates. Almost the whole of the reserve carbohydrate comprises glycogen and trehalose. Baker's yeast usually contains 16 to 20% of glycogen and 6 to 10% trehalose.[60] The amount of glycogen and trehalose is lower in brewer's yeast. Conditions of fermentation (aerobic or anaerobic) and the metabolic state of the cell are the main factors influencing the glycogen content. Abd Allah et al.,[61] investigating the reserve carbohydrates of different *Saccharomyces cerevisiae* strains, found that the glycogen content of yeast cells reached a maximum after 72 hours of fermentation. The trehalose content increased stepwise and reached a maximum at the end of 120 h fermentation.

The maximum values of glycogen resp. trehalose concentration were 21 and 10%, respectively, while the minimums were 0.82 and 1.39%. The same authors investigated the effect of intensivity of aeration on the carbohydrate content. It was concluded that the carbohydrate content of the yeast cells was highly affected by the oxygen concentration in the medium. Higher aeration (until maximum) led to an improvement of the ability of yeast cells to store more carbohydrates, and subsequently increases the yeast stability. The increase of carbohydrate content is generally associated with a decrease of protein content.[21,63] Andreyeva[63] reported about experiments associated with the study of effect of redoxy-potential of fermentation medium on the composition of cells of the *Candida utilis* yeast strain. Small changes in protein (decrease) and carbohydrate (increase) content were observed at higher redoxy-potentials. The existence has been reported of yeast mutants unable to store glycogen in conditions which normal yeasts accumulate large amounts.[60] These mutants, when grown in a medium with 8% glucose instead of the 1% normally used, accumulate glycogen and become indistinguishable from normal cells, therefore, the failure must be at the glucose-transport step. In the mutants, a loss of affinity of the transport system seems to be a plausible explanation of these observations. Glycogen isolated from the *Saccharomyces cerevisiae* is heterogenous as detected by ultracentrifugation studies resulting in two distinct components. The glycogen breakdown in yeast cells is connected with the action of yeast phosphorylase and a yeast isoamylase. As a result of the breakdown, small quantities of glucose and glucose-phosphate can be detected in the yeast cell.

TABLE 13
Base Composition of Yeast RNA (% of Total RNA)[10,59]

Strain	Carbon source	Adenine	Guanine	Cytosine	Uracyl
Saccharomyces cerevisiae	Beet molasses	27.56	27.70	21.68	23.06
Saccharomyces cerevisiae	Cane molasses	33.16	29.28	16.13	21.38
Saccharomyces cerevisiae (autolysate)	Cane molasses	30.73	32.16	15.16	21.90
Candida lypolytica ATCC 8661	C_{15}	26.74	29.08	22.36	21.82
	C_{15}	29.16	28.37	22.47	20.01
	$C_{15} + C_{16}$	24.86	28.78	23.43	22.93
	Glucose	21.39	27.67	27.43	23.51
Candida boidinii CBS 2428	Methanol	33.74	24.28	21.65	20.34
	Ethanol	28.81	23.23	25.37	22.60
	Xylose	29.21	29.44	19.23	21.42
	Glucose	28.23	24.20	25.38	22.19
Candida utilis	Calcium lignosulfate, wood sugars	29.72	30.86	17.14	20.65
Candida utilis	Sulfite waste liquor	29.81	28.80	15.76	22.49
Candida utilis	Ethanol	32.80	27.75	15.88	23.20
Candida utilis (autolysate)	Ethanol	32.68	27.45	16.68	23.59

Trehalose also plays a role in the energy yielding metabolism in yeasts.[60,61] Homogenates of baker's and brewer's yeast contain a trehalase able to split trehalose and give two molecules of glucose. As mentioned above, the trehalose content of the yeast cell is changing depending on the metabolic state of the cell. The level of trehalose depends on two pathways, those of synthesis and degradation. In the absence of any external sugar, when the pool of triphosphates and that of sugar phosphates becomes low, and AMP (being an inhibitor of trehalase enzyme) accumulates, the degradation of trehalose stops but so does trehalose synthesis. In such conditions the concentration of trehalose remains almost steady. In the presence of external assimilable sugar both pathways operate, synthesis holding some initial advantage, so that trehalose accumulates. In the presence of assimilable nitrogen source, when the synthesis of precursors for division starts, again the triphosphate pool is reduced and degradation pathways become predominant so that the stored trehalose gradually disappears.

A great part of the carbohydrates of yeast is located in the cell wall in the form of different polysaccharides. As mentioned in Chapter 2, the cell wall forms 15 to 20% of the dry weight of the cell.

Many species of microorganisms, including yeasts, synthesize exocellular polysaccharides which give rise to capsules of varying thickness. When such microorganisms are grown in liquid media, especially under shaking conditions, the culture media often become extremely viscous due to the release of excess capsular material into the medium. In many species the synthesis of polysaccharides of the cell wall is not subject to the metabolic control, judging by conversions of glucose to extracellular polysaccharides of 42% in the case of phosphomannanes of *Hansenula hostii* yeast species. The composition of the polysaccharides of the cell wall of noncapsulated and capsulated yeast cells is qualitatively the same, both are synthesized by the same enzyme system. Formation of the capsule would then represent surplus production of the external layer of the cell wall.

Although a lot of contradictory data concerning the structure of cell wall polysaccharides of yeast were published in the earlier literature (until 1970) summarized by Phaff[22] and Bartnicki-Garcia,[64] it is now generally accepted that the main polysaccharide of *Saccharomyces cerevisiae* species is the glucan. The glucan isolated from *Saccharomyces* and *Candida* species is a highly insoluble polysaccharide which is regarded as the skeletal support of the cell wall. From a chemical point of view it is a β-1-3 glucan being branched with β-1,6 linkages. The principal role of β-1,3 linkages in this polysaccharide is supported by the observation that the critical enzymes required for yeast autolysis by its own endogenous enzymes are reported to be endo-β-1,3 glucanases which, together with specific and non-specific exo-β-1,3 glucanases, have been detected in *Saccharomyces cerevisiae*. It was also found by several authors that exogenous β-1,3 glucanases prepared from other microorganisms may be effectively used for the rupture of yeast cells.[19,66-70] The exact distribution of β-1,3 and β-1,6 linkages within the glucan molecule is not yet quite clear, probably a different type of glucan molecules from the point of view of branching and lengths of the unbranched regions exists in the cell walls of different yeast strains.[20,64]

Yeast *mannan* discovered at the end of the last century is the second important polysaccharide of the yeast cell wall. It could be extracted from the whole yeast by boiling with 6% sodium hydroxide, and freed from other soluble polysaccharides by precipitation as its insoluble copper complex. The precipitate was then washed with water, dissolved by adding hydrochloric acid and finally precipitated with ethanol. The isolated polysaccharide contained only mannose components. The information about exact structure of the polysaccharide is contradictory. Nevertheless, it seems that there is an agreement that the main chain of the molecules is formed from mannose molecules linked by α-1,6 bonds. It is also generally accepted that a branched structure is characteristic for the mannan molecules. The side chains of the backbone of the macromolecule are also composed from mannose units linked to the backbone with α-1,2 bonds.

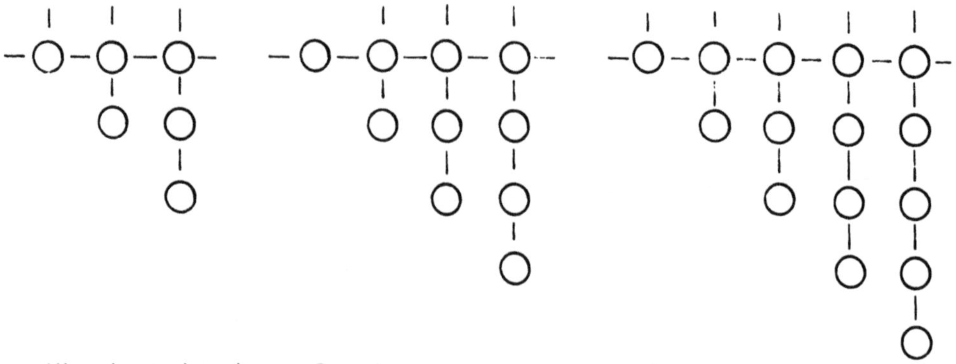

Kloeckera brevis Saccharomyces cerevisiae Fabospora lactis

FIGURE 7. Schematic structure of different types of mannans.

Several authors also found α-1,3 linkages in the side chain. The number and length of side chains is not uniform in the different yeast strains. Most of the data available is connected with the study of the mannans of *Saccharomyces cerevisiae*. The number of α-1,6 linked mannose residues in the backbone of these strains, not substituted in the secondary hydroxyl groups, was very small in the polymer from *Saccharomyces cerevisiae*. Comparing the mannans of different strains of *Saccharomyces cerevisiae* and also one strain grown under different conditions, it was found that differences occur in the number and length of side chains between different strains, however growing conditions had no significant effect.[21] Investigations of the mannans of other yeasts than *Saccharomyces cerevisiae* showed that mannans of different yeasts are similar in structure also having α-1,6 linked backbone and short side-chains linked with α-1,2 bonds. The main differences occur in the side chains as it is demonstrated in Figure 7 showing schematic structure of mannans of three different yeasts. Some of the mannans isolated from yeasts contain phosphorus.

The results of the investigations connected with phosphorus content are summarized in the review paper of Phaff.[22] The amount of phosphorus found in different mannans vary in the range from 0.04 to 4.4%. It seems most likely that in the species of *Saccharomyces,* the phosphate is linked to C-6 of mannose, whereas in *Kloeckera* strain C-3 or C-4 of the mannose in the backbone appears to be involved.

Not all of the mannans which have been extracted with alkali from yeast cells or cell walls, and which were subsequently purified via their insoluble copper complexes, contain exclusively mannose as the component sugar. It was found that some yeast strains such as four *Schizosaccharomyces* species, four *Torulopsis* species and *Candida lypolytica* contain galactomannans. The galactose molecules are linked to the side chains of the mannan.

The third polysaccharide which was detected in yeast cell wall is the chitin. This polymer, at least in the exoskeleton of crustaceans where it exists in a highly insoluble form, consists of long unbranched molecules made up of *N*-acetyl-D-glucosamine residues linked together by β-1,4 bonds. The chitin content of the yeast cell wall is much lower than that of glucan and mannan. Although only limited data are available, the cell walls of ascomycetous yeast (e.g., *Saccharomyces*) and their anascosporogenous forms (e.g., *Candida*) contain mainly glucan and mannan and low quantities of chitin. In contrast, basidiomycetous yeasts (*Sporobolomyces*) and related forms (*Rhodotorula*) contain more chitin and less glucan. The chitin content is generally determined measuring the nitrogen content of hexoseamins. In Table 14 the hexoseamine nitrogen content of different yeast strains as determined by Smith and Palmer[72] is shown.

As mentioned earlier, cell wall contains also about 10% protein. The protein is present

TABLE 14
Hexoseamine-Nitrogen Content of Yeasts

Yeast strain	Hexoseamine-nitrogen (% of total dry matter)	Hexoseamine-nitrogen (% of total nitrogen)
Candida (produced by BP Proteins Ltd)	0.11	1.3
Candida (produced by BP Proteins Ltd)	0.15	1.4
Candida (grown on hydrocarbon substrate)	0.06	0.6
Candida (Grown on hydrocarbon substrate)	0.24	2.8
Tropina (BP)	0.21	2.3

From Smith, R. H. and Palmer, R., *J. Sci. Food. Agric.*, 27, 763, 1976. With permission.

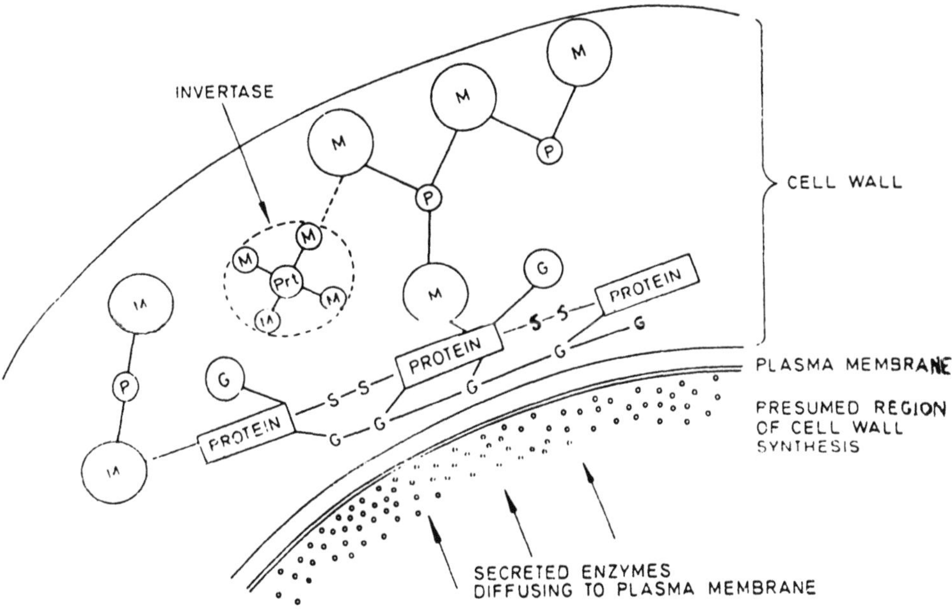

FIGURE 8. Diagrammatic scheme of yeast cell wall. (From Hough, J. S. and Maddox, I. S., *Proc. Biochem.*, 5(5), 50, 1970. With permission.)

in the form of complexes with polysaccharides. In Figure 8 a diagrammatic scheme of yeast cell wall is shown as proposed by Hough and Maddox.[66]

From a nutritional point of view, an interesting question is the dietary fiber content of yeast. As it is known, dietary fiber is a mixture of different compounds (mainly polysaccharides) remaining intact after specific enzymatic treatment. The insoluble part of the dietary fiber consists of cellulose and lignin and the soluble one contains pectic substances, gums and mixed (1→3 and 1→4) glucans.[73] Due to the fact that the hypocholesterolemic benefits of soluble fiber have been substantially documented[74] and the health benefits of β-glucans were recently reviewed by Klopfenstein,[75] a growing interest exists concerning the dietary fiber content of different foods. Recent data about dietary fiber content of yeast were reported by Sarwar et al.[10] Dietary fiber content of six different *Saccharomyces cerevisiae* and *Candida utilis* strains varied in the range from 15.42 to 17.79%. The amount of cellulosic (insoluble) components was low (2.47 to 5.49%) and this of noncellulosic components was much higher (11.19 to 12.97%). Similar results were published by Saló.[76]

E. LIPIDS

The total lipid content of yeasts varies widely with the species. The majority of the yeasts contain about 7 to 15% lipids (calculated on dry weight), but there is a smaller class, sometimes referred to as the "fat yeasts" which contain much more lipid, ranging from around 30 to over 60% of the dry weight. The lipids of microbial cells occur in different forms having different biological roles:[77] as lipid droplets (granula) in the cytoplasm, as depot (storage) lipids, mainly triglycerides; functional constituents of cytoplasma membrane (up to 25% lipid) mainly phospholipids and structural elements of cell wall (lipoproteins, liposaccharides, phospholipids). The lipid content and composition is dependent on the growth conditions. Results of earlier work in this field are reviewed by Hunter and Rose.[78] The total lipid content of yeast cells increases as the culture ages until the stationary phase of growth is reached. Thereafter, following the exhaustion of glucose in the medium, there is a rapid decrease in the lipid content of the cells. Similar results were reported by Biacs et al.[79] investigating different industrial yeast strains (baker's yeast, brewer's yeast and yeast used in spirit production). Alteration in the composition of *Saccharomyces* yeast (brewer's yeast) during repeated beer fermentation was studied by Biacs et al.[80] A stepwise decrease of total lipid content (from 24 to 1.0%) was observed.

Growth temperature is known to affect the lipid content of yeasts: decrease of the temperature of growth is generally accompanied with an increase of total lipid content. The lipid content also varies with sodium chloride concentration in the medium. As the concentration of sodium chloride in the medium was increased from 0 to 10% (W/v), the lipid content of *Candida albicans* increased from 0.32 to 6.29%.[78]

The biosynthesis of lipids is supported by increased aeration, low temperature and an endogenous supply of easily bioconvertible substrates, like liquid hydrocarbons, where a preference is given to assimilate them directly to the corresponding fatty acid. In most cases, low temperatures favor the synthesis of unsaturated lipids like phospholipids and triglycerides containing monounsaturated fatty acids. Polyunsaturated fats are relatively easily incorporated by yeast cells.[81]

The lipids extracted from yeast cells contain a wide range of components having different biological functions:

1. Depot lipids:
   - Triglycerides
   - Mono- and diglycerides
   - Fatty acids
2. Structural lipids:
   - Phospholipids
   - Glycolipids
   - Sterols
3. Surface lipids:
   - Wax esters
   - Hydrocarbons
   - Complex lipids

The fatty acid composition of total lipids of yeasts is characterized by a high content of unsaturated fatty acids. Changes in growth conditions affect the fatty acid distribution. Sultanovics et al.[82] studied the changes in the fatty acid composition of yeast lipids depending on their accumulation in the biomass and on the conditions of yeast growth. The work was concentrated on the correlation between the content of lipids in yeast cells and the accumulation of diunsaturated and polyunsaturated fatty acids. A correlation was established between the fatty acid composition of lipids and the level of their accumulation in the

TABLE 15
The Effect of N-source on the Composition of Lipids of Yeast Strain *R. gracilis* K-76

N-source	C:N	Fatty acid content of the biomass (mg/g)	Degree of unsaturation
$NH_4H_2PO_4$	15:1	76.1	131.6
	60:1	90.9	122.0
$(NH_4)_2SO_4$	15:1	114.3	120.0
	60:1	86.4	118.8
H_2N CON H_2	15:1	122.3	100.6
	60:1	101.0	105.5
Corn extract	15:1	244.2	78.8
	60:1	340.8	84.2

From Yu, A. S., Rozhdestvenskaya, M. V., and Nechayer, A. P., *Mikrobiologiya,* 56(2), 232, 1987. With permission.

TABLE 16
Distribution of the Main Fatty Acids in Yeast Lipids During Repeated Fermentation

Number of fermentation	$C_8 + C_{10} + C_{12} + C_{14}$	$C_{16} + C_{18}$	$C_{16:1} + C_{18:1}$	$C_{18:2}$
1. Brewing process	10.5%	44.2%	33.4%	6.1%
2. Brewing process	22.6%	33.4%	28.7%	8.5%
3. Brewing process	26.2%	31.3%	27.1%	8.6%
4. Brewing process	33.1%	33.1%	20.7%	8.6%
5. Brewing process	35.2%	30.4%	19.6%	7.1%
6. Brewing process	39.0%	22.2%	24.8%	5.1%

From Biacs, P., Grúz, K., and Klupács, S., *Microbial Associations and Interactions in Food,* Kiss, I., Deák, T., and Incze, K., Eds., Akadémiai Kiadó, Budapest, 1984, 301. With permission.

biomass. The correlation was shown to depend on certain factors of the growth medium and on the condition of cultivation. Changes in the cultivation parameters were found to influence the rate at which storage lipids were formed whereas differences in the chemical composition of synthesized lipids were mainly caused by changes in structural lipids and in the ratio between their amount and the content of storage lipids. Some results of investigations are shown in Table 15. Biacs et al.[80] investigated the changes of fatty acid composition and the ratio of different lipid groups in brewer's yeast during repeated fermentation. Some of the results are shown in Tables 16 and 17.

The most effective means of altering the unsaturation of fatty acids appears to be with oxygen. Comparing the fatty acid composition of *Saccharomyces cerevisiae* (baker's yeast) and *Rhodotorula rubra* at different levels of aeration during fermentation Biacs and Holló[83] confirmed the predominance of unsaturated fatty acids in both strains. The level of unsaturated fatty acids increased as the oxygen tension was raised. *Saccharomyces cerevisiae* had a much lower level of polyunsaturated fatty acids. Due to the fact that some wild yeast strains have a very high amount of polyunsaturated fatty acids, a rapid method for the determination of wild yeast contamination in baker's yeast by lipid analysis was elaborated.[84] Table 18 shows some data about main fatty acids of baker's yeast and contaminating other yeast strains.

The triglyceride content of yeast is generally 20 to 50% of the total lipid content. Mono-

TABLE 17
Main Lipid Classes and their Fatty Acid Composition in Yeast Lipids During Repeated Fermentation

Lipid class	$C_8 + C_{10} + C_{12} + C_{14}$	$C_{16} + C_{18}$	$C_{16:1} + C_{18:1}$	$C_{18:2}$
Phospholipids				
1. Brewing process	18.5%	37.0%	32.9%	6.2%
6. Brewing process	35.2%	30.6%	25.5%	5.2%
Sterol-esters				
1. Brewing process	16.3%	10.6%	66.2%	2.2%
6. Brewing process	47.8%	18.3%	26.4%	0.6%
Triglycerides				
1. Brewing process	30.9%	25.4%	32.7%	7.1%
6. Brewing process	54.3%	22.6%	16.9%	5.2%
Free fatty acids				
1. Brewing process	16.7%	46.7%	27.9%	4.4%
6. Brewing process	60.8%	21.4%	10.2%	2.9%

From Biacs, P., Grúz, K., and Klupács, S., *Microbial Associations and Interactions in Food,* Kiss, I., Deák, T., and Incze, K., Eds., Akadémiai Kiadó, Budapest, 1984, 301. With permission.

TABLE 18
Main Fatty Acids of *Saccharomyces cerevisiae* and Wild Yeasts Occurring in Baker's Yeast

Yeast species	Main fatty acids in the fatty acid composition %					Polyene fatty acids, %
	$C_{16:0}$	$C_{16:1}$	$C_{18:1}$	$C_{18:2}$	$C_{18:3}$	
Saccharomyces cerevisiae[a,b,c]	15.6	43.4	26.9	—	—	—
	8.9	57.0	30.3	1.5	—	1.5
	15.6	25.8	37.5	2.0	0.8	2.8
Candida krusei[d]	15.0	6.2	48.2	14.9	13.7	28.6
C. tropicalis[d]	21.8	5.4	28.6	26.2	4.4	30.8
C. pulcherrima[d]	21.0	7.4	41.4	25.6	1.4	27.0
C. utilis[b]	23.8	16.5	19.6	23.9	5.2	29.1
Candida 107[e]	21.0	3.0	36.0	28.0	—	28.0
Pichia membr.[d]	12.3	14.8	40.9	23.6	7.7	31.3
C. mycoderma[d]	14.1	17.1	41.1	18.6	4.7	23.3
Torulopsis cand.[d]	27.9	3.7	42.5	11.9	2.7	14.6
Rhodotorula rubra[c]	22.2	—	61.0	10.2	2.7	12.9

Note: [a] Hunter and Rose,[84] continuous cultivation.
[b] Biacs,[85] pure baker's yeast from factory.
[c] Biacs and Holló,[82] laboratory-scale cultivation.
[d] Kaneko et al.,[86] plating on YM-agar medium.
[e] Ratledge,[87] cultivation on glucose.

and diglycerides were also detected in yeasts in a lower quantity (1 to 15%). A significant amount of free fatty acids occurs in yeast cells (1 to 20%).

Phospholipids are biologically important and characteristic components of yeast lipids. Their quantity ranges from 15 to 60% of the total lipids. The main phospholipid components are the following ones:

Phosphatidylcholine	40—50%
Phosphatidyletanolamine	15—30%
Phosphatidylinositol and phosphatidylserine	8—15%
Sphingolipids	1%

Minor phosphatides (phosphatidic acid, diphosphatidyl glycerol and lysophospatides are also present.[89] The glycolipid content of yeasts is low. These compounds occur mainly in the extracellular lipids. Small amounts of cerebrosides were detected.

Sterols are characteristic components of yeast lipids. Their amount changes from 1 to 10% (dry matter basis). Ergosterol (in free form or as fatty acid ester) is the typical sterol of yeasts. The intensity of aeration influences the ergosterol content. Addition of sterols to the fermentation medium increases the sterol content of yeast cells. Basarova and Löblova,[96] investigating different brewer's yeasts, found an ergosterol content ranging from 0.12 to 0.90% (dry matter basis) depending on the conditions of growing. Some yeast strains such as *Rhodotorula, Cryptococcus, Sporobolomyces,* synthesize carotenoid pigments. The pigment production is dependent on time and growth rate.[90,91]

F. VITAMINS

As it is generally known, yeast are rich sources of several vitamins, especially vitamins of B-group. From a biochemical point of view it is interesting that different yeast strains synthesize vitamins in different amounts and in many cases yeast strains are unable to fully supply their own vitamin needs. So it is not surprising that several vitamins belongs to the group of growth factors of yeasts which must be added to the media of fermentation. The most common growth factors, taken up alone or together with others, are biotin, panthoteinic acid, inositol, thiamin, nicotinic acid and pyridoxine.

All top and bottom yeasts, and a large number of other yeasts need biotin. Most *Saccharomyces* species need panthotenic acid for growth. To some extent baker's and brewer's yeast are able to synthesize inositol, however, in the cultivation of baker's yeast inositol must be added if optimum yields are to be produced. Only a few strains of *Saccharomyces* require thiamin addition for growth. The ability of yeasts to synthesize nicotinic acid is limited under anaerobic conditions. Most strains of *Saccharomyces cerevisiae* and *Saccharomyces carlsbergensis* do not require pyridoxine, other strains need this growth factor. All yeast are capable of synthesizing riboflavin. When baker's yeast cells are transferred from anaerobic to aerobic culture conditions during industrial propagation, the riboflavin content rises and reaches its maximum in the semi-aerobic growth phase. Folic acid is synthesized by all yeasts. Early data about the vitamin content of yeasts are shown in Table 19.

Some newer data reported by Sarwar et al.[10] are shown in Table 20. Total folacin content in the *Saccharomyces cerevisiae* yeast products was lower than in *Candida utilis* yeast products. For each yeast species, the autolysed products showed significantly lower values, indicating that autolysis reduced initial amounts of folacin present in the respective yeasts. *Candida utilis* grown on calcium lignosulfate/wood sugars had the highest folacin content. Similar results were reported by Perloff and Buttrum[94] for baker's dry active yeast.

Biotin of the products ranged from 0.84 to 1.85 μg/g for both species of yeast. Autolysis had no effect on biotin content. Higher biotin content was found in *Saccharomyces cerevisiae* grown on cane and/or sugarbeet molasses.

Saccharomyces cerevisiae species had lower pantothenic acid content than the *Candida utilis* species. Autolysis lowered the pantothenic acid content. Schay and Wegner[95] reported about vitamin content of *Torula yeast.* The following amounts were found: thiamine 9.5, riboflavin 44.1, vitamin-B_6 79.1, vitamin-B_{12} 0.01, biotin 0.36, folic acid 21.5, niacin 450, pantothenic acid 189 μg/g. Small differences were found between the vitamin content of yeasts grown on sucrose and grown on ethanol.

TABLE 19
Vitamin Content of Yeasts (μg/g)

Yeast	Vitamin B_1	Vitamin B_2	Vitamin B_6	Panthothenic acid	Niacin	Folic-acid	Biotin
Candida arborea	32	46—69	0.3	—	157—580	15—19	—
Endomyces vernalis	15—34	—	—	—	—	—	—
Hansenula suaveolena	8.5	54	—	180	590	1.7	—
Mycotorula lypolytica	5.3	59	—	—	600	3.1	118
Oidium lactis	20—29	40—55	—	—	195—248	6—15	—
Saccharomyces carlsbergensis	31	—	—	—	—	—	—
Saccharomyces ellipsoideus	38	—	—	—	—	—	—
Saccharomyces cerevisiae	29—90	—	—	118—198	190—585	19—35	0.5—1.8
Saccharomyces logos	7—30	—	—	—	—	—	—
Torula utilis	6—53	26—62	35	86—180	210—535	4—31	1.1—1.9
Willia anomala	10—30	—	—	—	—	57	—
Zygosaccharomyces sp.	39	—	—	—	—	—	—
Baker's	9—40	—	16—65	180—330	29—30	15—80	—
Enriched baker's	650—750	—	—	—	200—700	—	—
Brewer's	50—360	36—42	25—100	100	310—1000	3	—
Dried yeast[92]	2—20	30—60	40—50	30—200	—	—	—
BP Protein-Vitamin concentrate	3—16	75	23	150—192	—	—	—

From Eddy, A. A., *The Chemistry and Technology of Yeasts*, Cook, A. H., Ed., Academic Press, New York, 1958, 157. With permission.

TABLE 20
Folacin, Pantothenic Acid and Biotin in Inactive Dried Yeast Products (as is basis)

Product	Vitamin content (μg/g)		
	Total folacin	Panthothenic acid	Biotin
Saccharomyces cerevisiae (grown on cane molasses)	36.2	127	1.85
Saccharomyces cerevisiae (autolysate)	22.4	106	1.34
Candida utilis (grown on calcium lignosulfate/wood sugars)	75.2	151	1.30
Candida utilis (grown on sulfite waste liquor)	53.4	126	0.84
Candida utilis (grown on ethanol)	52.8	291	0.85
Candida utilis (autolysate)	26.5	173	1.18

From Sarwar, G., Shah, B. G., Mongeau, R., and Hoppner, K., *J. Food Sci.*, 50, 353, 1985. With permission.

TABLE 21
Macro-Element Content of Different Yeasts (mg/g)

Yeast	P	Ca	Mg	K	Na	Al	Fe
Saccharomyces fragilis[a]	13.0	53.2	1.1	0.4	20.4	0.5	0.3
Kluyveromyces[a] *marxianus*	25.8	48.1	1.3	2.9	19.8	0.9	0.1
Candida tropicalis[a]	29.1	42.1	1.6	0.4	16.7	2.2	0.3
Candida utilis[a] ATCC 9226	25.2	52.2	1.4	0.4	19.5	0.8	0.1
Candida utilis[a] ATCC 9950	23.7	49.1	1.1	0.4	17.5	0.9	0.3
Candida albicans[a]	15.4	51.1	1.6	0.5	19.5	1.2	0.2
Saccharomyces cerevisiae (grown on cane molasses)	36.0	0.60	1.56	28.0	0.54	—	0.07
Saccharomyces (autolysate)	48.0	0.18	0.18	18.0	170.0	—	0.01
Candida utilis (grown on calcium lignosulphate/wood sugars)	58.0	4.64	2.36	22.0	0.23	—	0.12
Candida utilis (grown on sulfite waste liquor)	55.0	0.89	1.83	19.0	0.50	—	0.10
Candida utilis (grown on ethanol)	68.0	0.05	2.62	33.0	1.00	—	0.11
Candida utilis (autolysate)	69.0	0.09	2.47	33.0	1.04	—	0.05

[a] Grown on cheddar cheese whey permeate.

Berndorfer and Telegdy Kováts[97] reported about tocopherol, tocopherylquinone and ubiquinone content of yeasts. The contents of these compounds varied as follows: tocopherol 2.15 to 8.68 mg/100 g, tocopherylquinone 0.63 to 1.22 mg/100 g and ubiquinone 15.12 to 59.41 mg/100 g. Hughes and Tore investigated the α-tocopherylquinol and α-tocopherylquinone of yeasts. Both compounds were found in nine yeast strains. The levels of quinones and hydroquinones averaged 3 nmol of each compound per g of packed cells.

G. MINERALS

Although mineral content of yeasts is changing in wide range, the average ash content of yeast cells is relatively high. Typical values lie about 10% of the dry matter. Phosphorus, potassium, magnesium, calcium and sulfate are the main components of the ash. Some data about the mineral content of different yeasts are shown in Tables 21 and 22. The results shown in the tables vary in a very wide range especially in the case of microelements. All the data concerning phosphorus and iron content are comparable to the yeasts investigated. The big variations in calcium content can be explained by the fact that yeast samples, having

TABLE 22
Micro-Elements of Yeasts

Yeast	Content of elements (mg/kg)										
	Mn	**Cu**	**B**	**Zn**	**Mo**	**Co**	**Cd**	**Cr**	**Ni**	**Pb**	**Si**
Saccharomyces fragilis	9.81	70.11	0.08	70.50	1.23	8.43	0.84	67.43	40.99	3.83	2.30
Kluyveromices maxianus	3.41	76.25	0.74	108.70	0.94	5.42	1.77	10.37	4.21	10.54	4.68
Candida tripicalis	9.47	50.38	1.79	76.72	1.76	10.65	1.41	46.18	26.60	10.73	1.57
Candida albicans (ATCC 20402)	5.12	44.78	2.09	96.64	1.12	6.98	1.83	17.72	8.88	10.67	5.97
Candida utilis (ATCC 9226)	3.97	42.46	0.79	70.24	0.79	4.76	1.23	10.48	6.15	7.46	5.95
Candida utilis (ATCC 9950)	5.32	33.29	3.74	45.53	1.42	8.90	0.85	63.01	31.09	5.45	2.85
Saccharomyces cerevisiae (grown on cane molasses)	4.0	4.0	—	70.0	—	—	—	—	—	—	—
Saccharomyces cerevisiae (autolysate)	0.2	0.5	—	80.0	—	—	—	—	—	—	—
Candida utilis (grown on calcium lignosulfate)	27.0	4.0	—	100.0	—	—	—	—	—	—	—
Candida utilis (grown on sulfite waste liquor)	153.0	12.0	—	80.0	—	—	—	—	—	—	—
Candida utilis (grown on ethanol)	5.0	1.0	—	70.0	—	—	—	—	—	—	—
Candida utilis (autolysate)	5.0	1.0	—	70.0	—	—	—	—	—	—	—

high calcium content, were grown on media with a considerable amount of calcium (whey, calciumlignosulfate, sulfite waste liquor). A similar situation is with the samples containing more magnesium. There is no acceptable explanation for high potassium content of yeasts grown on ethanol. High sodium content of *Saccharomyces cerevisiae* autolysate is caused by addition of sodium chloride. The data included in both tables confirm the significant effect of growing conditions on the quantity and composition of ash.

REFERENCES

1. **Cook, A. H., Ed.,** *The Chemistry and Biology of Yeasts,* Academic Press, New York, 1958.
2. **Mauron, J.,** Haben Einzellerproteine noch eine Zukunft?, in *Probleme der Ernährungs — und Lebensmittelwisselschaft-5, Probleme um neue Lebensmittel,* Harner, R., Auerswald, W., Zöllner, M., Blanc, B., and Brandstetter, B., Eds., Wilhelm Mautrid Verlag, Wien, 1978, 125.
3. **Rose, A. H. and Harrison, J. S., Eds.,** *The Yeasts,* Vol. 2, Academic Press, London, 1971.
4. Research Institute of Hungarian Spirit Industry, The yeast in foods, Report No. 3, Budapest, 1987.
5. **Mauron, J.,** Technology of protein synthesis and protein-rich foods, *Bibl. Nutr. Dieta,* 18, 24, 1973.
6. **Guzman-Juarez, M.,** Yeast protein, in *Developments in Food Protein-2,* Hudson, B. J. F., Ed., Applied Science Publishers, London, 1983, 263.
7. **Kihlberg, R.,** The microbe as a source of food, *Ann. Rev. Microbiol.,* 427, 1972.
8. **Sergeev, V. A., Solosenko, V. M., Bezrukov, M. G., and Saporovskaja, M. B.,** Vergleichscharakteristik von Isolaten der Gesamteiweisse der Hefe in Abhängigkeit von den Bedingungen ihrer Abscheidung, *Acta Biotechnol.,* 4(2), 105, 1984.
9. **Diezak, J. D.,** Yeast and yeast derivatives: applications, *Food Technol.,* 41(2), 122, 1989.
10. **Sarwar, G., Shah, B. G., Mongeau, R., and Hoppner, K.,** Nucleic acid, fiber and nutrient composition of inactive dried food yeast products, *J. Food Sci.,* 50, 353, 1985.
11. **El-Samragy, Y. A., Chen, J. H., and Zall, R. R.,** Amino acid and mineral profile of yeast biomass produced from fermentation of cheddar whey permeate, *Proc. Biochem.,* 23(1), 28, 1988.
12. **Soboleva, G. A. and Popova, E. A.,** Fractional composition of the proteins of yeasts grown on cellulosic hydrolysates (in Russian), *Prikl. Biochim. Mikrobiol.,* 22(3), 368, 1986.
13. **Kung-Chin Su, Ming-Chang Hsie, and Hung Chao Lee,** Amino acid composition of various yeasts prepared from cane molasses *Proc. Int. Soc. Sugar Cane Technol.,* 13, 1934, 1968.
14. **Levaux, J. Y., Bonix, M., and Cuinier, G.,** Differenciation fine des levures por electrophorese et immunoelectrophorese des proteines cytoplasmiques, in *Current Developments in Yeast Research,* Stewart, G. G. and Russel, I., Eds., Pergamon Press, Toronto, Canada, 1981, 387.
15. **Medvedeva, E. I. and Selitch, E. F.,** Study of the composition of alkali-soluble protein in fodder yeasts (in Russian), *Prikl. Biochim. Mikrobiol.,* 5(2), 164, 1969.
16. **Tkachenko, V. V., Rilkin, Sz. Sz., Skidchenko, A. N., and Sterkin, V. A.,** The influence of growth conditions on the distribution of protein fractions in microorganisms (in Russian), *Mikrobiologija,* 40(4), 651, 1971.
17. **Sadkova, N. P., Valjkovskij, D. G., and Kozlova, N. I.,** Isolation and some characteristics of glutelin fractions of *Candida* yeasts strains (in Russian), *Biochimija,* 39(6), 1252, 1974.
18. **Wiemken, A. and Nurse, P.,** Isolation and characterization of the amino acid pools located within the cytoplasm of *Candida utilis, Plants,* 109, 293, 1973.
19. **Wiseman, A., King, D. J., and Winkler, M. A.,** The isolation and purification of protein and peptide products, in *Yeast Biotechnology,* Russel, I. R. and Stewart, L. F., Eds., Allen and Unwin, London, 1987, 343.
20. **Halász, A. and Mátrai, B.,** Characterization of yeast strains important to food industry by PAGE (in Hungarian), *Élelmez. Ipar,* 41, 300, 1987.
21. **Halász, A.,** Biochemical and Biotechnological Principles of Use of Yeast Biomass in Food Industry, Dr. Sc. thesis, Hungarian Academy of Sciences, Budapest, 1988.
22. **Phaff, H. J.,** Structure and biosynthesis of the yeast cell envelope, in *The Yeast,* Vol. 2., Rose, A. H. and Harrison, J. S., Eds., Academic Press, New York, 1971, 135.
23. **Lemy, I.,** A study of two yeast proteinases, *J. Biol. Chem.,* 221, 919, 1956.
24. **Hayashi, R., Oka, Y., Boi, E., and Hata, T.,** Occurrences and activation of latent yeast proteinases, *Agric. Biol. Chem.,* 31, 1102, 1967.

25. **Doi, E., Hayashi, R., and Hata, T.,** Purification of yeast proteinases. Part II. Purification and some properties of yeast proteinases C, *Agric. Biol. Chem.*, 31, 160, 1967.
26. **Hata, T., Hayashi, R., and Doi, E.,** Purification of yeast proteinases. Part I. Fractionation and some properties of the proteinases, *Agric. Biol. Chem.*, 31, 150, 1967.
27. **Hata, T., Hayashi, R., and Doi, E.,** Purification of yeast proteinases. Part III. Isolation and physico-chemical properties of yeast proteinase A and C, *Agric. Biol. Chem.*, 31, 357, 1967.
28. **Hayashi, R., Oka, Y., Doi, E., and Hata, T.,** Activation of intracellular proteinases of yeast. I. Occurrence of inactive precursors of proteinases B and C and their activation, *Agric. Biol. Chem.*, 32, 359, 1968.
29. **Matiele, P. and Wiemken, A.,** The vacuole as the liposome of the yeast cell, *Arch. Microbiol.*, 56, 148, 1967.
30. **Achstetter, T.,** Proteolysis in eucaryotic cells, identification of new proteolytic enzymes in yeast, *J. Biol. Chem.*, 259, 13334, 1984.
31. **Achstetter, T., Ehmann, C., and Wolf, D.,** New proteolytic enzymes in the yeast, *Arch. Biochem. Biophys.*, 207, 445, 1981.
32. **Achstetter, T., Ehmann, C., and Wolf, D.,** Aminopeptidase-Co, a new yeast peptidase, *Biochem. Biophys. Res. Commun.*, 109, 341, 1982.
33. **Achstetter, T., Ehmann, C., and Wolf, D.,** Proteolysis in eucaryotic cells: aminopeptidases and dipeptidyl-aminopeptidases of yeast revisited, *Arch. Biochem. Biophys.*, 226, 292, 1983.
34. **Achstetter, T., Ehmann, C., Osaki, A., and Wolf, D.,** Proteolysis in eucaryotic cells: proteinase Ysc, a new peptidase, *J. Biol. Chem.*, 259, 13344, 1984.
35. **Wolf, D. and Ehmann, C.,** Carboxypeptidase S from yeast, regulation of its activity during vegetative growth and differentiation, *FEBS Lett.*, 91, 59, 1978.
36. **Empter, O. and Wolf, D.,** Vacuoles are not the sole compartments of proteolytic enzymes in yeast, *FEBS Lett.*, 166, 321, 1984.
37. **Wolf, D.,** Proteinases, proteolysis and regulation in yeast, *Biochem. Soc. Trans.*, 13, 279, 1984.
38. **Wiemken, A.,** Eigenschaften der Hefervacuole, Thesis No. 4340, ETH, Zürich, 1969.
39. **North, M.,** Comparative biochemistry of proteinases of eucaryotic mikroorganism, *Microbiol. Rev.*, 45, 308, 1982.
40. **Wolf, D.,** Control of metabolism in yeast and other lower eukaryotes through action of proteinases, *Adv. Microbiol. Physiol.*, 21, 267, 1980.
41. **Wolf, D. and Holzer, H.,** Proteolysis in yeast, in *Microorganisms and Nitrogen Sources,* Payne, W., Ed., John Wiley & Sons, Chichester, England, 1980, 431.
42. **Tschesche, H.,** Structure and function of natural inhibitors as antagonists of proteinase activities, *Eur. F. Biochem.*, 120, 1, 1982.
43. **Saheki, T. and Holzer, H.,** Proteolytic activities in yeast, *Biochim. Biophys. Acta,* 384, 203, 1975.
44. **Holzer, H.,** Possible mechanism for a selective control of proteinase action, *Adv. Enzymol. Regul.*, 12, 1, 1974.
45. **Holzer, H.,** Possible mechanisms for a selective control of proteinase action, *Proc. 9th FEBS Meeting,* Vol. 32, Akadémiai Kiadó, Budapest, 1975, 181.
46. **Lenney, J.,** Three yeast proteins that specifically inhibit yeast proteases A, B and C, *J. Bacteriol.*, 122, 1265, 1975.
47. **Magni, G., Drewniak, M., Santarelli, J., and Huang, L.,** Reexamination of the activation of yeast proteinase inhibitors., *Biochem. Int.*, 12, 557, 1986.
48. **Babajan, T., Bezrukov, M., and Kalumyane, K.,** Metabolites — inductors of the autolysis of yeast *Saccharomyces cerevisiae* (in Russian), *Acta Biotechnol.*, 5, 279, 1985.
49. **Vorobjev, M., Paskonova, E., Vitt, S., and Belikov, S.,** Kinetic description of proteolysis. II., *Nahrung,* 30, 995, 1986.
50. **Vorobjev, M., Vitt, S., and Belikov, V.,** Kinetic description of proteolysis. III., *Nahrung,* 31, 331, 1987.
51. **Egami, F. and Nakamura, H.,** *Microbial ribonucleases. Molecular Biology — Biochemistry and Biophysics,* Vol. 6, Springer Verlag, Berlin, 1965, 1.
52. **Ohta, S., Maul, S., Sinskey, J., and Tannenbaum, S.,** Characterization of heat-shock process for reduction of the nucleic acid content of *Candida utilis, Appl. Microbiol.*, 22, 415, 1981.
53. **Shetty, J., Weaver, R., and Kinsella, J.,** A rapid method for the isolation of ribonuclease from yeast *(Saccharomyces carlsbergensis), Biochem. J.*, 198, 363, 1980.
54. **Shetty, J., Weaver, R., and Kinsella, J.,** Ribonuclease isolated from yeast *(Saccharomyces carlsbergensis), Biotechnol. Bioeng.*, 23, 953, 1981.
55. **Mounolou, J. C.,** The properties and composition of yeast nucleic acids, in *The Yeasts* Vol. 2, Rose, A. H. and Harrison, J. S., Eds., Academic Press, New York, 1971, 309.
56. **Bueno, G. E., Otero, M. A., Klibansky, M. M., and Gonzales, A. C.,** Nucleic acid reduction from yeast. Activation of intracellular RNase, *Acta Biotechnol.*, 5, 91, 1985.
57. **Litchfield, J. H.,** Single cell proteins, *Science,* 219, 740, 1983.

58. **Waldron, C. and Lacroute, F.,** Effect of growth rate on the amounts of ribosomal and transfer ribonucleic acids in yeasts, *J. Bacteriol.*, 122, 855, 1975.
59. **Beretta, M. G., Potenza, D., Scolastico, C., Manachini, P. L., and Zennaro, E.,** RNA-base or whole cell nucleotide composition of yeast, *Lebensm. Wiss. Technol.*, 14, 82, 1980.
60. **Sols, A., Gancedo, C., and Dela Fuente, G.,** Energy-yielding metabolism in yeasts, in *The Yeast*, Vol. 2, Rose, A. H. and Harrison, J. S., Eds., Academic Press, London, 1971, 271.
61. **Abd-Allah, M. B., Rizk, I. R. S., Ramadan, E. M., and Abu Salem, F. M.,** Reserved carbohydrates of different *Saccharomyces cerevisiae* strains, *Die Nahrung*, 30(5), 507, 1986.
62. **Harrison, J. S.,** Baker's yeast, in *Biochemistry of Industrial Microorganisms*, Rainbow, C. and Rose, A. H., Eds., Academic Press, New York, 1971, 9.
63. **Andereyeva, E. A.,** Physiological-biochemical changes of *Candida utilis* yeast strain depending on the redoxy-potential (in Russian), *Mikrobiologija*, 43(5), 780, 1974.
64. **Bartnicki-Garcia, S. and McMurrough, I.,** Biochemistry of morphogenesis in yeast, in *The Yeasts*, Vol. 2, Rose, A. H. and Harrison, J. S., Eds., Academic Press, London, 1971, 441.
65. **Kassim Mustafa, M. and Halász, A.,** The effect of date syrup concentration on growth rate, protein, RNA content and protease activity of *S. cerevisiae*, *C. guilliermondii* and *R. glutinis*, *Acta Aliment.*, 18, 177, 1989.
66. **Hough, J. S. and Maddox, I. S.,** Yeast autolysis, *Process Biochem.*, 5(5), 50, 1970.
67. **Phaff, J. J.,** Enzymatic yeast cell wall degradation in food proteins, Feeney, R. R. and Whitaker, J. R., Eds., *American Chemical Society*, Washington, D.C., 1977, 244.
68. **Asenjo, J. A. and Dunnil, P.,** The isolation of lytic enzymes from *Cytophaga* and their application to the rupture of yeast cells, *Biotechnol. Bioeng.*, 23, 1045, 1989.
69. **Chisti, Y. and Murray, M.-Y.,** Disruption of microbial cells for intracellular products, *Enzyme Microbiol. Technol.*, 9, 194, 1986.
70. **Ryan, E. and Ward, O. P.,** The application of lytic enzymes from *Basidiomycetes aphyllophoroles* in production of yeast extract, *Process Biochem.*, 23(1), 12, 1988.
71. **Mustafa Kasim, M.,** SCP-Production: Effect of Carbon Source Concentration and Aeration on Proteinase and RN-ase Activity of Different Yeast Species, Ph.D. thesis, TU Budapest, 1989.
72. **Smith, R. H. and Palmer, R.,** A chemical and nutritional evaluation of yeasts and bacteria as dietary protein sources for rats and pigs, *J. Sci. Food Agric.*, 27, 763, 1976.
73. **Newman, R. K., Newman, C. W., and Graham, H.,** The hypocholesterolemic function of barley β-glucan, *Cereal Foods World*, 34(10), 883, 1989.
74. **Chen, W. J. L. and Anderson, J. W.,** Hypocholesterolemic effects of soluble fibers, in *Dietary Fiber: Basic and Clinical Aspects*, Vahouny, G. V. and Krichevsky, D., Eds., Plenum Press, New York, 1986.
75. **Klopfenstein, C.,** The role of cereal β-glucans in nutrition and health, *Cereal Foods World*, 33, 865, 1988.
76. **Salo, M. L.,** The carbohydrate composition and solubility of Pakilo Protein and two yeasts, *Acta Agric. Scand.*, 27, 77, 1977.
77. **Biacs, P. and Gruiz, K.,** Lipid analysis of microbial biomass, *Oleagineaux*, 39(10), 491, 1984.
78. **Hunter, K. and Rose, A. H.,** Yeast lipids and membranes, in *The Yeasts*, Vol. 2, Rose, A. H. and Harrison, J. S., Eds., Academic Press, London, 1971, 211.
79. **Biacs, P., Ujváry, H., Orcsi, É., Barkóczay, A., and Szladecskó, K.,** Data on the lipid composition of industrial yeast strains, Proc. XVIth Ann. Meeting Hungarian Biochem. Soc., Rosdy, B., Ed., Debrecen, Hungary, 1986, 79.
80. **Biacs, P., Grúz, K., and Klupács, S.,** Alteration in the composition of *Saccharomyces* yeast during repeated beer fermentation, in *Microbial Associations and Interactions in Food*, Kiss, I., Deák, T., and Incze, K., Eds., Akadémiai Kiadó, Budapest, 1984, 301.
81. **Biacs, P.,** Chemistry of Plant and Microbial Lipids, D.Sc. thesis, Hungarian Academy of Sciences, Budapest, 1985.
82. **Yu, A. S., Rozhdestvenskaya, M. V., and Nechayev, A. P.,** Changes in the fatty acid composition of yeast lipids depending on their accumulation in the biomass and on the conditions of yeast growth, *Mikrobiologiya*, 56(2), 232, 1987.
83. **Biacs, P. and Holló, J.,** Changes in the growth rate and fatty acid composition of yeasts grown in aerated systems, in EUCHEM Conf. Metabol. Reactions Yeast Cell Anserobic and Anaerobic Conditions, Oksaanen, J. and Suomalainen, H., Eds., Helsinki, 1977, 67.
84. **Biacs, P. A.,** Rapid method for the determination of wild yeast contamination in baker's yeast by lipid analysis, *Acta Aliment.*, 8(1), 57, 1979.
85. **Hunter, K. and Rose, H.,** Lipid composition of *Saccharomyces cerevisiae* as influenced by growth temperature, *Biochim. Biophys. Acta*, 260, 639, 1972.
86. **Biacs, P. A.,** Wild yeast contamination and the ratio of linoleic and linolenic acid content of commercial baker's yeast, in Proc. 5th Int. Specialized Symp. Yeasts, Novák, E. K., Deák, T., Török, T., and Zsolt, J., Eds., Keszthely, Hungary, 1977, 119.

87. **Kaneko, H., Hoschara, M., Tanaka, M., and Itoh, T.,** Lipid composition of 30 species yeasts, *Lipids*, 11, 837, 1976.
88. **Ratledge, C.,** Microbial conversions of n-alcanes to fatty acids: a new attempt to obtain economical microbial fats and fatty acids, *Chem. Ind.*, No. 25, 843, 1970.
89. **Franzke, V., Göbel, R., and Füst, M.,** Zur Kentniss der Lipide aus Hefebiomassen, *Lebensmittelindustrie*, 28(11), 495, 1981.
90. **Biacs, P. and Kövágó, Á.,** Investigation on the carotene production of *Rhodotorula* yeast (in Hungarian), *Szeszipar*, 20, 151, 1972.
91. **Simpson, K. L., Chichester, C. O., and Phaff, H. J.,** Carotenoid pigments of yeast, in *The Yeasts*, Vol. 2, Rose, A. H. and Harrison, J. S., Eds., Academic Press, London, 1971, 493.
92. **Eddy, A. A.,** Aspects of the chemical composition of yeast, in *The Chemistry and Technology of Yeasts*, Cook, A. H., Ed., Academic Press, New York, 1958, 157.
93. **Roth, W.,** Eiweiss für die Zukunft, *Chimia*, 26(11), 589, 1972.
94. **Perloff, B. P. and Buthrum, R. R.,** Folacin in selected foods, *J. Am. Diet. Assoc.*, 70, 161, 1977.
95. **Schay, L. K. and Wegner, G. H.,** Improved fermentation process for producing *Torula* yeast, *Food Technol.*, 39, 61, 1985.
96. **Basarova, G. and Löblova, L.,** Ergosterol in spent brewer's yeast (in Czech), *Kvasny Prum.*, 33(3), 65, 1987.
97. **Berndorfer, E. and Telegdy Kováts, L.,** Bioquinone content of some yeast strains, *Szeszipar*, 19, 20, 1971.
98. **Hughes, P. E. and Tore, S. B.,** Occurrence of α-tocopheryl-quinone and tocopherylquinol in microorganisms, *J. Bacteriol.*, 151, 1397, 1982.

Chapter 4

YEAST PRODUCTION

For thousands of years man has made profit on the products synthesized by yeasts. In the beginning it was bread leavening, beer and wine fermentation, alcohol production later on yeasts provided man with other products such as glycerol, enzymes, coenzymes and vitamin. There is a great interest in utilizing yeasts for upgrading waste materials into single-cell protein, autolysate and yeast extracts. This heterogeneous collection of organisms includes 39 genera but only three genera are commercially cultivated — *Saccharomyces, Candida* and *Klyuveromyces.*[1] Today, yeasts are acquiring increasingly more attention for their nutritional and flavors benefits, the high protein and vitamin content. Dried yeast is an attractive natural supplement for improving the nutritional profile of human food. By modifying processing parameters during their growth in the fermentors uniquely flavored yeasts can be produced for enhancing snack foods, processed meats and cheese containing products.[1]

I. CELL REPRODUCTION

Yeasts of industrial interest reproduce asexually by budding. Ascosprogeneous yeasts under certain environmental conditions can reproduce sexually, leading to the formation of ascospores.

A. BUDDING

In the budding process a small localized area of the cell wall protrudes outwardly; cytoplasm from the parent cell streams into the area, nuclear material replicates by mitosis and one of the daughter nuclei moves into the distended protuberance. The developing bud continues to increase in size until it attains about a third of the size of the mother cell. The cell wall then closes off the newly formed yeast cell, leading a bud scar on the mother cell and a birth scar on the new yeast cell. Cells may produce as many as 12 to 15 bud scars.[2] When bud formation of *Saccharomyces cerevisiae* is not limited by nutrients or overcrowding according to data published by Miller[2] the number of bud scars varied between 9 and 43 with an average of 24 buds. Bud formation requires about 1.5 to 2 h for completion under ideal conditions. Aged cells, however require longer times for duplication. The process of formation of new cell, called cell cycle, may be divided in three phases: G_1, G_2 and S. The cell cycle of yeast begins with an unbudded cell whose shape approximates a prolate spheroid. The cell grows on volume during this G_1 phase of the cycle. But despite recent work, we are left with the question why is protein synthesis needed for the initiation of a DNA synthesis.

Studying the event, synchronous cell cultures have been investigated and results show a continuous increase of total protein during the whole cell cycle. However, this continuous increase could encompass periodic patterns of synthesis of individual proteins. Indeed it is now widely accepted that histones are synthesized periodically during the S period.[3] But it might be that many of the cell proteins are synthesized in steps at different parts of the cycle and the sum of these steps gives the continuous increase in total protein. The initiation of the S phase coincides with the emergence of buds and start of DNA synthesis. According to Carter[4] S phase lasts approximately 30 min and followed by the G_2 phase. At the end of the G_2 phase the nucleus migrates to the junction of parent cell and bud. Nuclear division occurs without the nuclear membrane breaking down, the nucleus elongates such that part of it lies within the parent cell and part within the bud and then it divides. Time for nuclear division has been calculated as 15 min. Nuclear division is not immediately followed by cell division. Slater et al.[5] have calculated that in cells growing at 30°C with a 2.2 h mass doubling time 0.52 h separates nuclear division and cell division.

It is attractive to think there is a decondensation of the chromosomes during G_1 until they reach a fully extruded state at the start of S. When S is completed, the chromosome then condenses again during G_2 until they are apparent.[3] The effect of growth rate on the cell cycle of yeast has been examined only comparatively recently. The results of Meyenburg[6] were the first to show that the lengthening of the cell cycle at slow growth rates is largely due to an expansion of the unbudded phase (G_1) of the cell cycle although the time cells spend in the budded phase also increases slightly with increasing mass doubling time. Hartwell and Unger[7] found that at slow growth rates every 10 min increase in the mass doubling time increased the budded interval by 1.7 min. Studies on cells growing at different growth rates have revealed that there are marked differences in the cycle times of parents and daughters at slow growth rates.[7-9] At fast growth rates the bud almost reaches the size of the parent cell at cell division and the parent and daughter from division have similar cycle times. At slow growth rates the bud at division is much smaller than the parent cell and takes a much longer time to produce a bud than the parent in the subsequent cell cycle.

B. SEXUAL REPRODUCTION

Ascomycetous yeasts can reproduce sexually through a cycle that involves altering the haploid and diploid states of their life-cycles and leads to the production of sexual spores in an ascus (the old cell wall). Yeasts may be classified sexually as heterothallic or homothallic. Heterothallic yeasts require compatible mating types for conjugation. Homothallic yeasts in contrast, have the potential of converting from one mating type to another,[10] and are capable of self-fertilization. However, in each case mating is restricted to unbudded cells. Budded cells of the opposite mating type mixture complete cell division before mating.[11] When haploid cells of the opposite mating type are mixed together the proportion of budded cell decreases rapidly and that of unbudded cells increases before mating. This appears to be brought about by the action of hormone-like molecules secreted by the two haploids α-cells produce α-factor, a small polypeptide of 13 amino acids which arrest *a* cells in the G_1 phase of the cell cycle. Similarly *a*-cells produce *a*-factor which transiently arrests cells of mating type α in G_1.[4] The mutual arrest of cells of *a* and α mating types provides a mechanism whereby cells can be synchronized prior to mating.

II. ENERGY-YIELDING METABOLISM

Metabolism is a matrix of two closely interlinked but divergent activities. Anabolic processes are concerned with the building up of cell materials. Anabolic processes do not occur spontaneously, they must be driven by an energy flow provided by energy-yielding catabolic processes. Most common of these catabolic processes is the degradation of carbohydrates to CO_2 + H_2O. Yeasts are recognized as economically important because of two energy-yielding reactions: fermentation and oxidative respiration.

A. GLUCOSE AND OTHER CARBOHYDRATES

Two important pathways of sugar metabolism used by yeast are the hexose diphosphate pathway (also referred as Embden-Meyerhof, Parnas (EMP) pathway or glycolysis pathway) and the hexose-monophosphate (also known as pentose-phosphate pathway). The EMP pathway converts glucose into pyruvate without loss of carbon, reducing two molecules NAD^+ coenzyme and generating two molecules of ATP. The overall reaction is

$$\text{glucose} + 2NAD^+ + 2H_2PO_4^- \rightarrow 2\ \text{pyruvate} + 2NADH_2 + 2ATP \qquad (1)$$

The pyruvate is further substrate to aerobic reactions and also a key compound for anabolic metabolism. Under anaerobic conditions pyruvate or its transformation products also act as reoxidizing agent for the $NADH_2$.

The final steps of the pathway involve the decarboxylation of pyruvate to acetaldehyde and the $NADH_2$ mediated reduction of acetaldehyde to ethanol. In most organisms between 68 to 80% of glucose is metabolized in the EMP pathway[12] in case of *S. cerevisiae* more than 95% in the absence of oxygen.[1] The pathway results in the synthesis of ATP and the liberation of free energy, in addition to the production of ethanol (Equation 2).

$$1 \text{ glucose} \rightarrow 2\text{EtOH} + 2\text{CO}_2 + 2\text{ATP} + 5\text{-}6\text{Kcal} \qquad (2)$$

The fermentative process operates even under aerobic conditions. The hexose-monophosphate pathway as an oxidative process converts glucose into pentose and CO_2, reducing two molecules of $NADP^+$ (a coenzyme related to NAD^+) to NADPH. This pathway yields only one half net energy than EMP pathway.

The EMP pathway is important for ATP synthesis, the main role of the hexose-monophosphate is generation of reduced coenzyme $NADH_2$ (Equation 3) which

$$1 \text{ glucose-6P} \rightarrow 6\text{CO}_2 + 12\text{NADPH}_2 \qquad (3)$$

is used during aerobic respiration in the synthesis of ATP. This mechanism also operates under aerobic conditions, accounting for 6 to 30% of glycolysis in baker's yeast and 30 to 50% in *C. utilis.*[13]

1. Tricarboxylic Acid Cycle

Pasteur demonstrated that air stimulates yeast growth and suppresses ethanol production from fermentation by causing the yeasts to switch to an oxidative pathway, a respiration process. This phenomenon is called Pasteur effect and is characteristic of baker's yeast. Respiration proceeds by an oxidative pathway known as tricarboxylic acid pathway (TCA), Krebs cycle, or citric acid pathway. In this pathway acetyl CoA derived from pyruvate (from fermentation) is oxidized by enzymes to form CO_2 and reduced coenzymes. In eukaryotic cells the reactions of TCA and energy production are carried out in the mitochondria. As the mitochondrial process begins with the transport of pyruvate into the mitochondrion it is convenient to include the reactions linking pyruvate to the TCA cycle. Pyruvate is converted to acetyl-CoA by a multi-enzyme complex. The subsequent metabolism of acetyl-CoA is through the reactions of TCA cycle shown in Figure 9. The overall reaction is

$$\text{acetyl-CoA} + 3\text{NAD}^+ + \text{FAD} + \text{GDP} \rightarrow 2\text{CO}_2 + \text{Coenzyme-A}$$
$$+ 3\text{NADH} + \text{FADH}_2 + \text{GTP} \qquad (4)$$

The coenzymes NAD^+ and FAD which become, respectively, NADH and $FADH_2$ are then reoxidized to the original coenzymes by oxidative phosphorylation which produces 3 mol of ATP from each mole of NADH and 2 mol of ATP from the reoxidation of $FADH_2$.

2. Glyoxylate Bypass

If an organism grows on a C_2 compound or on fatty acid or hydrocarbon that is degraded primarily into C_2 units, the TCA cycle is insufficient to account for its metabolism, as C_2 compounds cannot be converted to pyruvate. Acetyl-CoA can be generated directly from acetate, if this is used as carbon source, or from C_2 compound more reduced than acetate, i.e., acetaldehyde or ethanol. The manner in which acetate units are converted to C_4 compounds is known as glyoxylate bypass (see Figure 10).

B. FATTY ACIDS AND HYDROCARBONS

The ability to utilize fatty acids or oils and fats is more common among yeast than hydrocarbon utilization. But both subrates are well known as sole sources of carbon in the

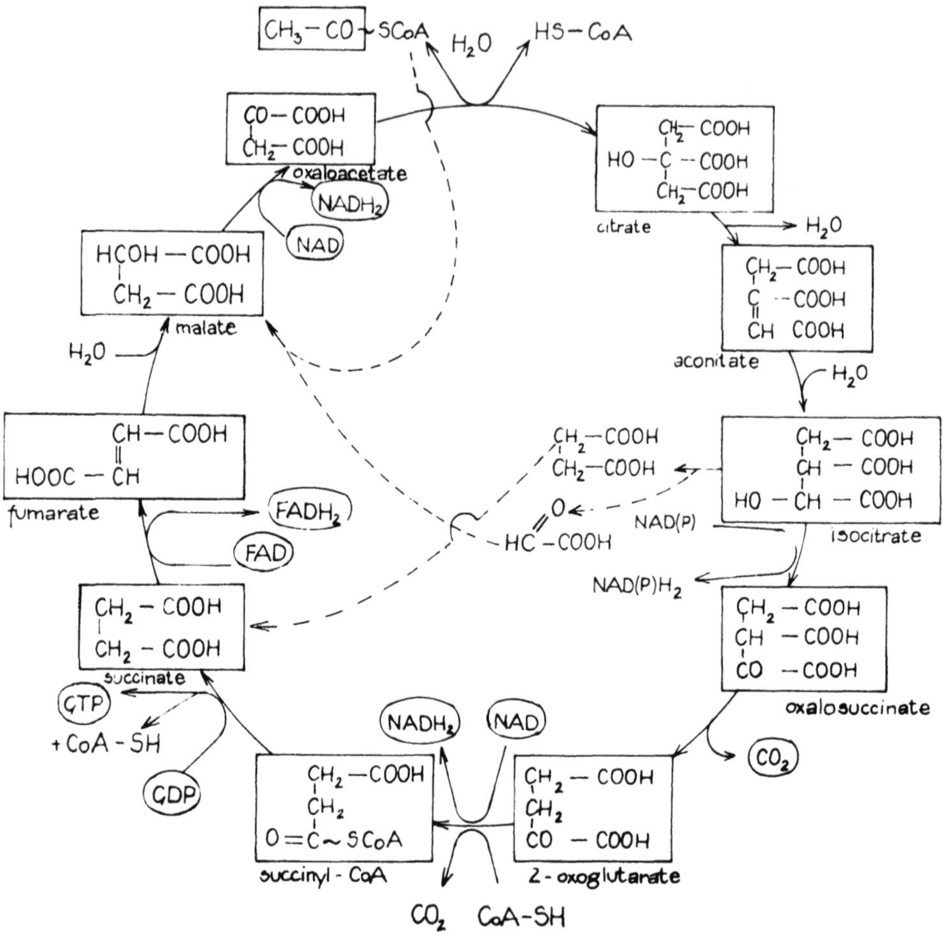

$CH_3-CO\sim SCoA$
H_2O
$HS-CoA$
$CO-COOH$ / CH_2-COOH oxaloacetate
$NADH_2$
NAD
CH_2-COOH / $HO-C-COOH$ / CH_2-COOH citrate
H_2O
CH_2-COOH / $C-COOH$ / CH $COOH$ aconitate
H_2O
$HCOH-COOH$ / CH_2-COOH malate
H_2O
CH_2-COOH / CH_2-COOH
O / $HC-COOH$
$CH-COOH$ / $HOOC-CH$ fumarate
$FADH_2$
FAD
CH_2-COOH / $CH-COOH$ / $HO-CH-COOH$ isocitrate
NAD(P)
NAD(P)H_2
CH_2-COOH / CH_2-COOH succinate
CH_2-COOH / $CH-COOH$ / $CO-COOH$ oxalosuccinate
GTP
+ CoA - SH
GDP
$NADH_2$
NAD
CO_2
CH_2-COOH / CH_2 / $O=C\sim SCoA$ succinyl - CoA
CH_2-COOH / CH_2 / $CO-COOH$ 2 - oxoglutarate
CO_2 CoA-SH

FIGURE 9. The tricarboxylic acid cycle.

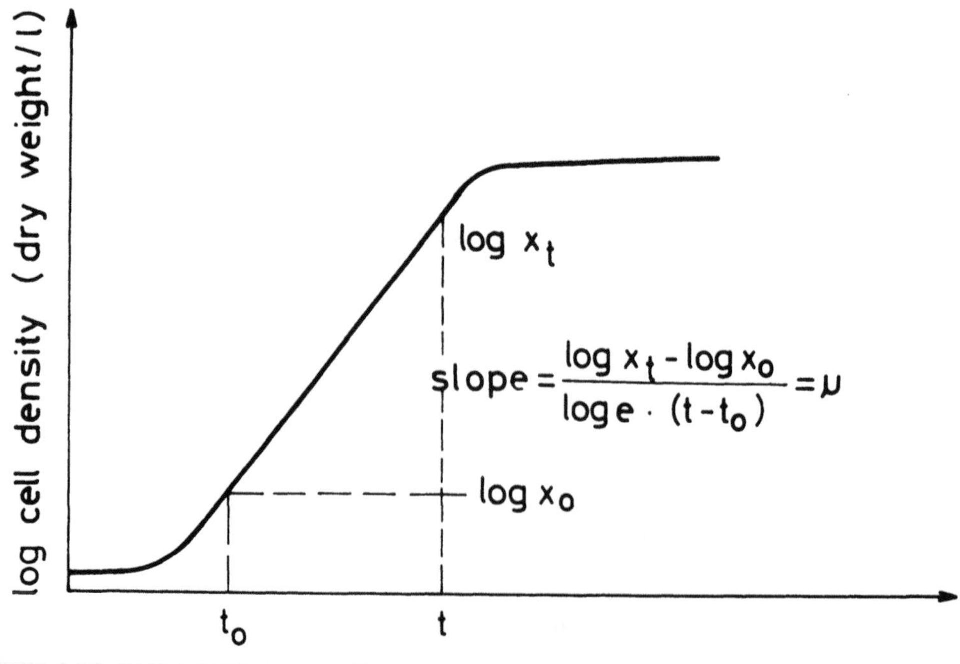

FIGURE 10. The glyoxylate bypass.

production of single cell protein (Simek et al.[14]). For the utilization of oils and fats, the organism must hydrolyse the ester linkage using its own lipase. This yields fatty acids plus glycerol. Glycerol is utilized by the EMP pathway. Acids are converted into Coenzyme-A-thiolester which is degraded by cyclic sequences. At each turn of the cycle 1 mol of acetyl-CoA is released. The β-oxidation cycle is continuous until the substrate for the final reaction is the C_4 compound, aceto-acetyl-CoA, which then gives 2 mol of the acetyl-CoA. In case of an odd numbered carbon chain, the degradation proceeds until propionyl-CoA is reached, which is converted to pyruvate. In case of hydrocarbon substrate *n*-alkanes are oxidized to fatty acids via fatty alcohols, fatty acids are further degraded by β-oxidation cycle.

C. REOXIDATION OF REDUCED COENZYMES

In the metabolism of glucose and in the TCA cycle oxidation of various intermediates is linked to the reduction of a limited number of cofactors (NAD^+, $NADP^+$, FAD) to the corresponding reduced form (NADH, NADPH and $FADH_2$). The reducing power of these products is released by a complex reaction sequence which in aerobic systems is linked eventually to reduction of atmospheric O_2. During this sequence, ATP is generated from ADP and inorganic phosphate (P_i) at two or more specific points in the electron transport chain, depending on the nature of the original reductant.[12] The overall reactions may be written:

$$\text{NADPH} + 3\text{ADP} + 3\text{P}_i + 1/2\text{O}_2 \rightarrow \text{NADP}^+ + 3\text{ATP} + \text{H}_2\text{O} \quad (5)$$

$$\text{NADH} + 3\text{ADP} + 3\text{P}_i + 1/2\text{O}_2 \rightarrow \text{NAD}^+ + 3\text{ATP} + \text{H}_2\text{O} \quad (6)$$

$$\text{FADH} + 2\text{ADP} + 2\text{P}_i + 1/2\text{O}_2 \rightarrow \text{FAD} + 2\text{ATP} + \text{H}_2\text{O} \quad (7)$$

The production of biologically utilizable energy in the form of ATP occurs in membranes of mitochondria. The main components of the electron transport chain are flavoproteins, quinones and cytochromes which are able to become oxidized and reduced, releasing electrons to the next carrier. Each carrier has a different redox potential increasing stepwise from −320 mV for NADH/NAD^+ to about +800 mV for the final reaction 1/2 O_2/H_2O.

D. ANAEROBIC ENERGY PRODUCTION

In anaerobic metabolism ATP production is directly coupled to an energy yielding reaction: this is termed substrate level phosphorylation. In yeasts such as *Saccharomyces cerevisiae* the substrate which reoxidizes reductants such as NADH or NADHP is acetaldehyde. Most of the pyruvate generated from glucose is converted into ethanol:

$$\text{pyruvate} \rightarrow \begin{matrix}\text{2-hydroethyltiamine}\\ \text{phosphate}\end{matrix} \rightarrow \text{acetaldehyde} \xrightarrow[\text{NADH+H}^+ \;\downarrow\; \text{NAD}^+]{} \text{ethanol} \quad (8)$$

As 2 mol of pyruvate are produced from 1 mol of glucose, the production of ethanol can reoxidize both moles of NADH produced in the triose phosphate dehydrogenase reaction and the overall stoichiometry is therefore:

$$\text{glucose} + 2\text{ADP} + 2\text{P}_i \rightarrow 2\text{ ethanol} + 2\text{ATP} \quad (9)$$

The net production of ATP provides energy for the yeast cells to grow but the yield per mole glucose transformed is less than 5% of that which occurs under aerobic conditions.

Any glucose that is metabolized via pentose phosphate pathway can produce only 1 mol of pyruvate from each mole of glucose, and only with the simultaneous generation of 2 mol of NADPH and 1 mol of NADH. These additional reducing equivalents must be reoxidized by being coupled to other reactions as the amount of produced pyruvate is not sufficient. Prime among these other reactions is the formation of fatty acids. By adding bisulfite to yeast growing anaerobically glycerol can be formed. The bisulfite forms a complex with acetaldehyde, so that ethanol can no longer be produced, and the microorganism seeking an alternative means to reoxidize NADH, reduces triose phosphate to glycerol.[12] The maintenance of redox balance is of critical importance in the anerobic growth of yeast.

III. PHYSIOLOGY OF GROWTH

The microbial cell is able to reproduce itself from the very simplest of nutrients. The number of pathways which a cell must use is enormous and the number of enzymes which catalyze these reactions is more than 2000. The cell carries out all its metabolic activities in a balanced manner so that end products are neither over- nor underproduced. Control of cell metabolism begins by the regulation of nutrient uptake. Most nutrients are taken up by specific transport mechanisms so that they may be concentrated within the cell from dilute solutions outside. Such active transport systems require an input of energy.

Disaccharides such as maltose and sucrose are assimilated by active transport. While monosaccharides such as glucose, fructose, mannose — by facilitated diffusion, that means no direct energy input is involved. According to Berry and Brown[18] maltose can be concentrated in the cell to 15 times of extracellular concentration however, free glucose is not normally detected in the cell even under conditions of rapid glucose assimilation, rather GLU-6P than free glucose is accumulated in the cell. Glucose is phosphorylated by hexokinase I, II or glucokinase. The difference in the function of these three enzymes is not clear since loss of any one or two does not directly affect the uptake of glucose. This supports the model of glucose assimilation which involves the glucose transport into the cell by facilitated diffusion and the maintenance of the diffusion gradient by the rapid removal of the free sugar in the yeast cell by phosphorylation. The high levels of hexokinase activity which occur in *Saccharomyces cerevisiae* also support this model. The low level of free glucose indicates that it is the transport process which is rate limiting during the metabolism of glucose.

A. MODES OF GROWTH

The need for aeration to obtain high yields of baker's yeast was recognized early. The importance of oxygen and glucose in the regulation of the growth characteristics in yeast has been recognized for many years. The reduction in the rate of glycolysis when yeast is grown in the presence of oxygen, the Pasteur effect, the repression of mitochondrial function when yeast is grown in high levels of glucose, referred to as the Crabtree[15] effect, and the glucose effect or catabolite repression are important mechanisms in the regulation of glucose metabolism in growing yeast cell.[16]

On the control of mitochondrial development four different states can be recognized with respect to the oxygen and carbohydrate status of the growth conditions: namely aerobic catabolite — repressed and derepressed status and anaerobic catabolite — repressed and derepressed states. These different states exhibit different levels of mitochondrial enzymes but also have different nutritional requirements and different growth characteristics.

At high glucose concentrations yeast shows a diphasic growth. After an initial lag phase during the exponential growth glucose is fermented to ethanol. During this period aerobic respiration is repressed. The second lag phase is followed by the aerobic ethanol metabolism with CO_2 + H_2O as an end product. In this growth phase RQ drops to 1.0 or less and mitochondrial functions develop.[17]

Not all yeast are sensitive to high glucose concentration. For example, species of *Candida, Rhodotorula, Torulopsis,* do not produce ethanol when grown on high glucose media in aerobic conditions. They respire aerobically and give a high yield of biomass. However, in aerobic conditions or reduced aeration they are still capable of fermenting glucose to ethanol. Glucose sensitive and insensitive yeasts also exhibit different growth characteristics in continuous culture. Glucose-sensitive yeasts show aerobic respiration and corresponding higher yield at lower dilution rates. But at a higher D value RQ rises, ethanol begins to accumulate and biomass yield drops to a lower level. Steady states can also be established at higher dilution rates during which yeast exhibits ethanol fermentation as well as aerobic respiration.

Glucose repression can be induced by high levels of glucose or by high influx of glucose even when the actual concentration of the sugar in the medium is as low as 10 to 100 mgl^{-1}.[18] However, other medium components like low levels of Cu + Fe can also cause glucose repression at low dilution rates. Whereas when glucose-insensitive yeast exhibit classical growth kinetics in continuous culture that means biomass concentration remains at a constant level until dilution rate exceeds μ_{max} when washout occurs. Rieger[19] assumed that the occurrence of aerobic ethanol formation resulted from the limited respiratory capacity of glucose-sensitive yeast and saturated respiratory capacity, causing ethanol formation before any repression of mitochondrial functions could be noticed. By glucose pulse experiments it was shown that primarily an overflow reaction on the level of pyruvate is a consequence of saturated oxidative capacity.

Long-term adaptation of mitochondrial functions to the conditions of aerobic ethanol formation was observed by Käppeli et al.[20] by dilution rate shift experiments; however both oxidative and fermentative glucose catabolism occurred simultaneously. The corresponding state was therefore, referred to as oxidoreductive or respirofermentative glucose metabolism.

B. COORDINATION OF METABOLISM AND GROWTH

1. Batch Cultivation

The cell always attempts to optimize its internal biochemistry so that it can make the most efficient use of preformed carbon and nitrogen compounds, maximizing the energy yield, minimizing energy expenditure and growing as rapidly as it is able. Under conditions when given a supply of all essential nutrients the yeast grows. In batch culture, the cells multiply in a closed system until either some nutrient becomes exhausted, until some product accumulates to inhibit further growth, or the number of cells reaches such a level that there is no further space available for new ones.

During growth the cell changes in size and the various main constituents alter in relative amounts Figure 11. In the initial exponential growth period, cells are growing at a maximum rate and RNA content also increases to the maximum value. The rate at which an organism can grow is expressed either as the doubling time (t_d) or generation time (t_g), it means the time taken for one cell to become two. Growth rate is also characterized by the specific growth rate (μ) which is the rate of synthesis of new cell material expressed per unit weight of existing cell material. These two values are related by the following equation:

$$\mu = \frac{1}{x} \cdot \frac{dx}{dt} = \frac{d(\ln x)}{dt} = \frac{\ln 2}{t_d} \tag{10}$$

x = cell concentration, t = time, $\mu = 0.69/t_d$.

In batch culture the μ value varies throughout, due to the continually decreasing concentration of nutrients, in many practical situations with aerobic organisms the rate of supply

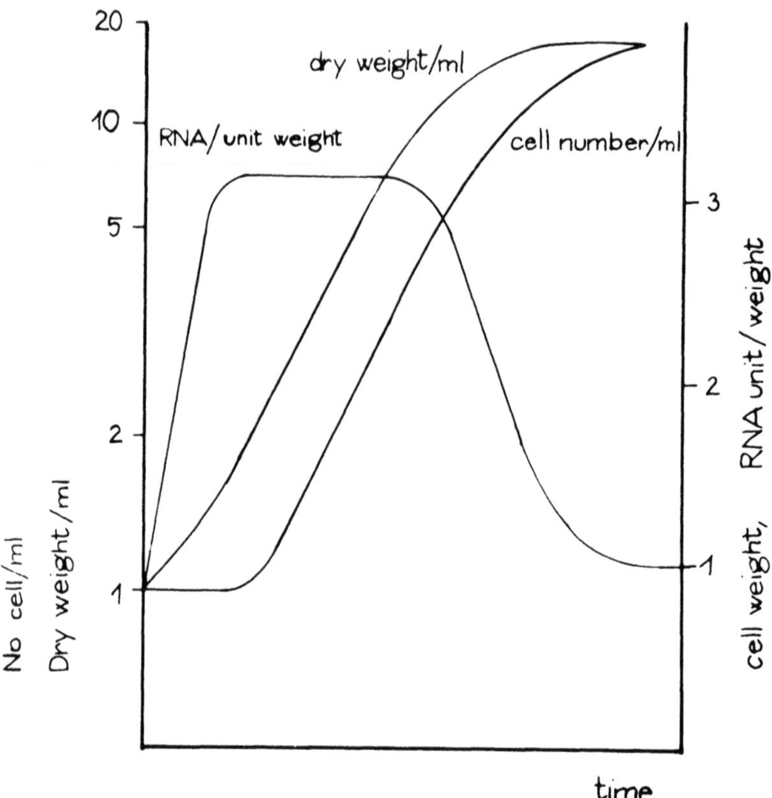

FIGURE 11. Idealized representation of the changes in cell size (i.e., cell weight), cell numbers, total dry weight and chemical composition during growth of microorganisms in batch culture. (The scale on left is logarithmic.)

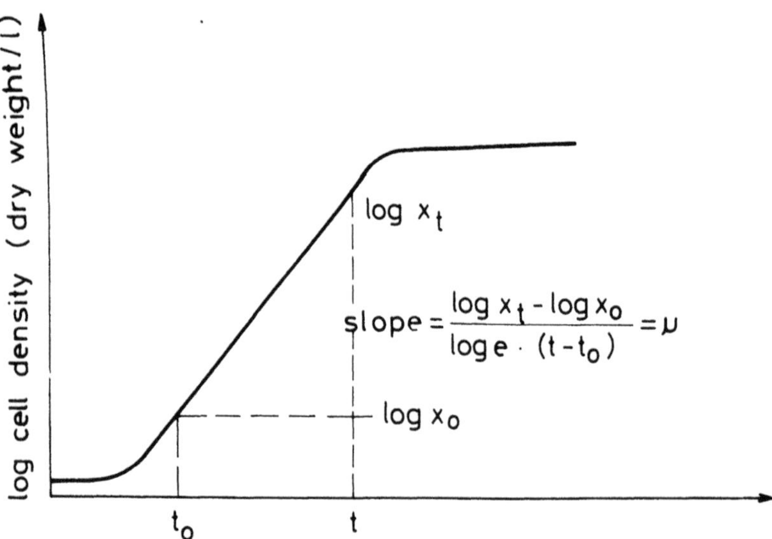

FIGURE 12. Growth curve of cell in batch culture.

of oxygen eventually governs the rate of their growth. In batch culture, cell growth follows a growth curve as shown in Figure 12. The first phase is the lag period, derivation of which is mainly influenced by the inoculum and suitability of the growth medium. In case of an old inoculum (cells are in the stationary growth phase) cells have to adapt RNA-, ribosome-

and enzyme synthesis to the new growth conditions. If the carbon and energy source differs from the medium from those of the precultivation, adaptation will involve synthesis of new enzymes too. The exponential (or logarithmic) growth phase is characterized by a maximum constant division rate. This division or growth rate is species specific and depends on environmental conditions (substrate, temperature, etc.). The stationary growth phase cell concentration reaches a maximum value. As growth rate is influenced by substrate concentration, cell division rate decreases before complete substrate exhaustion in the late exponential period. Not only lowering of substrate concentration but also increase of cell density or toxic metabolites and low oxygen partial pressure initiate a decrease in growth rate and parallel to this introduce the stationary phase. Only very sensitive cells show a rapid cell death but commonly reserve carbohydrates or proteins that serve as energy sources for cell maintenance. About death phase there is relatively little information. Living cell numbers can decrease even exponentially and intracellular lytic enzymes may cause autolysis. For practical purposes most important characteristics of cell growth in a batch culture are the duration of lag phase, growth rate or doubling time and the yield. Yield is the difference between initial and maximum cell mass. For economical aspects the relation of yield to substrate consumption is especially important and is characterized by yield — coefficient (Y). If we refer Y to the mole of substrate we will get the mole yield coefficient (Ym) which enables us to relate yield to ATP (Y_{ATP}). Y_{ATP} was calculated for *Saccharomyces cerevisiae* and for anaerobic fermentation a value of $Y_{ATP} = 10$ was found.[21]

Yield coefficient of *S. cerevisiae* varies significantly with the mode of growth.[17] Most probable values for the oxidative pathway are 0.49 g/g while for reductive biomass synthesis is only 0.05 g/g on glucose substrate.

The exponential growth rate is a measure for cell growth rate in the exponential growth phase and can be calculated from cell concentrations at two distinct points

$$\mu = \frac{\ln x - \ln x_0}{t - t_0}$$

2. Continuous Cultivation

In batch cultures cell concentration, substrate content and other state variables change continually. The existent methods of continuous cultivation can be subdivided into two groups regarding mechanism providing the stability of growth according to Minkewich et al.[22] In the first group, represented mostly by chemostat, the stability is maintained by the culture itself. The methods of the second group (turbidostat, pH auxostat) are based on the use of special automatic devices to control different growth-linked parameters. Chemostat operates successfully within the region of growth limitation by any substrate. Methods of the second groups are operable on the plateau of specific growth rate μ or at the excess substrate inhibition. When μ approaches μ_{max} but residual substrate concentration is low, chemostat is close to washout and turbidostat-type systems are close half due to nutrient limitation. The relation of cell density, substrate concentration, doubling time in steady state cultures to dilution rate is shown in Figure 13.

Dependence of growth rate on substrate concentration follows a saturation curve (Figure 14). At substrate concentration equal to K_s is growth rate half of the maximum growth rate ($\mu = \mu_{max}/2$). At low dilution rates the substrate concentration in the vessel and flow-out is almost zero. Only when D is near to D_{max} part of the substrate is washed out and at least S in flow-out reaches the value of input concentration. Cell concentration is over a wide region of D (between zero and D_{crit}) almost constant as in this region an increase of D causes a reduction of cell doubling time. The stability of steady state in a chemostat depends on the relation of growth rate on substrate concentration. Growth rate is held at a low value. The chemostat as a self-regulating system is easy to operate.

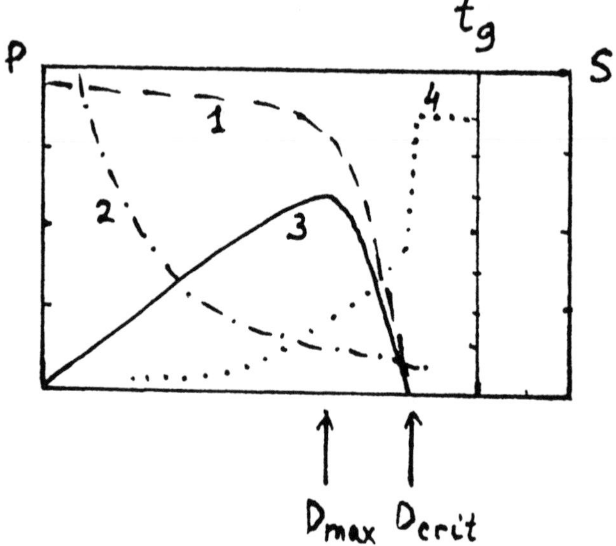

FIGURE 13. Relation of cell density, substrate concentration, doubling time in continuous steady state cultures to dilution rate. (1) cell density; (2) doubling time (tg); (3) biomass productivity (P) and (4) substrate concentration (S). D — dilution rate, D_{max} — dilution rate at maximal biomass productivity and D_{crit} — critical dilution rate.

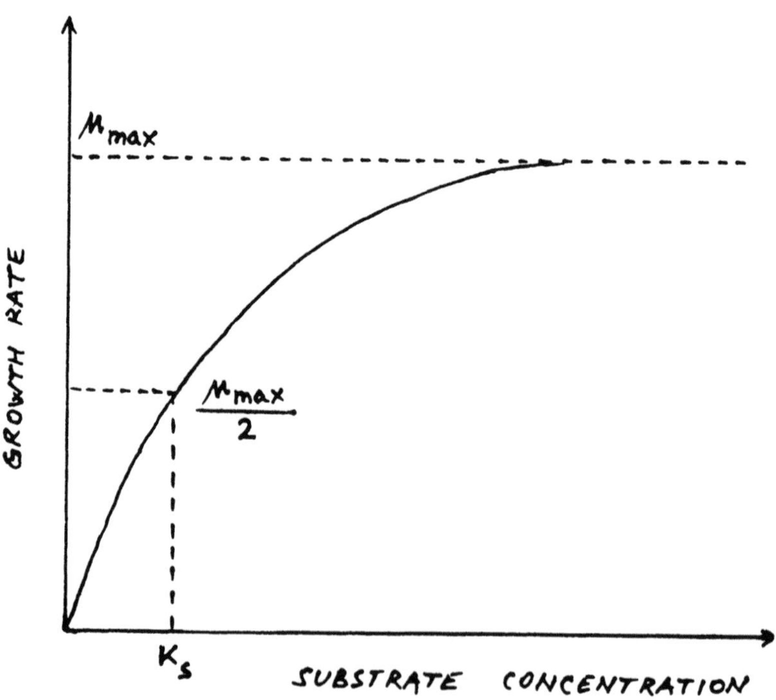

FIGURE 14. Dependence of growth rate on substrate concentration.

Turbidostat systems are working at constant cell concentration or turbidity values. Substrate input is regulated by a turbidimeter. In the culture vessel substrate is at excess and cells double near to their maximal growth rate. However, operation of a turbidostat needs more expense in technique than a chemostat. Bistat is the method of continuous cultivation developed on the basis of pH-auxostat. It provides the stability of cultivation in each point of μ (S) dependence. Theoretically, pH-auxostat is operable at low residual concentrations S of any substrate. However, to maintain a low S value the interrelation between alkalinity and substrate concentration S_o of the medium should be adjusted with a very high accuracy. When S_o becomes insufficient for the culture to acidify the medium from its high pH to the set point pH, the controller stops feeding. Therefore, the substrate limitation of the pH-auxostat means to halt. When pH-auxostat is halted because of substrate limitation, the feeding is usually restored by addition of the substrate or an acid into the fed medium reservoir.

Another way is to feed the limiting substrate into the fermentor using another medium flow independent of the pH controller, i.e., a "Chemostat" controller.[22] The substrate level in the broth then depends on the additional flow value but never approaches zero, and the process proceeds without halting. A combination of chemostat and pH-auxostat flows forms a new cultivation method further referred to as a bistat.[22]

In fed batch cultures the cell concentration, liquid volume in the vessel and other state variables change significantly during the course of operation. A fed batch culture can be mathematically described according to Takamatsu et al.[23] as follows:

$$x = \left(\mu - \frac{F}{V}\right) \times \underline{\underline{\Delta}}f_1 \tag{11}$$

$$S = -\mu X + \frac{F}{V}(S_o - S)\underline{\underline{\Delta}}f_2 \tag{12}$$

$$V = F\underline{\underline{\Delta}}f_3 \tag{13}$$

$$y = g(X,S,V,F,y)\underline{\underline{\Delta}}f_4 \tag{14}$$

where X, S, V, F, S_o are the biomass concentration, the substrate concentration, the liquid volume, the input flow rate and the substrate concentration in the feed, respectively. Parameters μ and V are the specific growth rate and substrate uptake rate which are functions of S or occasionally y and X; y represents all other state variables except X, S and V. Such variables could be for example, an inhibitory substance, an intermediate product in the medium or an inorganic substance needed for growth. The objective of the control system for this fed-batch process is to make the actual X (t) follow the derived curve $X^x(t)$. However, in industrial processes direct measurement of biomass concentration is not acceptable so indirect measurement like ethanol production have been reported.[23] The problem of controlled fed-batch fermentation is to control the cell concentration or the specific growth rate so as to follow the desired pattern by manipulating the feed rate of the substrate in the system where biomass concentration or growth rate is indirectly observed.

3. Substrate Inhibition

The dependence of cell growth rate from substrate concentration in the Monod kinetics implies that growth rate continuously increases with substrate concentration up to maximum value. In practice the growth rate usually begins to decline above some particular value of the substrate concentration. This relationship is expressed by the following equations:[24]

$$\mu = \frac{\mu_m S}{K_s + S + (K_i S)^{2-}} \tag{15}$$

$$\mu = \mu_m \frac{S}{K_s + S}(1 - K_i I) \tag{16}$$

$$\mu = \frac{\mu_m S}{K_s + S} \cdot \frac{K_i}{K_i + I} \tag{17}$$

$$\mu = \frac{\mu_m S}{K_s + S} \exp(-K_i I) \tag{18}$$

where I is the inhibitor concentration. All of the expressions given above are first order with respect to cell concentration but K_i coefficients do not have the same meaning in each equation. The choice between them is partly a matter of convenience and partly requires reference to actual observations made over a wide range of values of the inhibitor concentration. In general, one should use the expression which fits close to the observed facts.

IV. KINETICS OF MICROBIAL CELL GROWTH

The growth cycle of microorganisms is divided into phases on the basis of changes in the growth rate. These phases as described by Monod[25] are as follows:

1. Lag phase
2. Acceleration phase
3. Logarithmic phase
4. Retardation phase
5. Stationary phase
6. Accelerating death phase
7. Logarithmic death phase

In individual cases any or even several of these phases may be absent, and occasionally a more complex pattern of growth is observed.

Each of these phases was discussed by Clifton,[26] Lamanna and Maletta.[27] During the logarithmic growth phase the mean generation time remains constant. The specific growth rate may be defined as the increase in microbial mass per unit time per unit microbial mass. Thus for the logarithmic phase the specific growth rate k (h^{-1}) could be expressed as

$$k = \frac{1}{x} \cdot \frac{dc}{dt} 2 \cdot \frac{d\ln x}{dt} \tag{19}$$

where x = concentration of microorganism.

The time necessary to double the microbial mass (t_g) may be derived from the previous equation as follows:

$$t_g = \frac{\ln 2}{k} = \frac{0.692}{k} \tag{20}$$

It is generally assumed that the specific growth rate is related to the concentration of an essential growth substance by a relatively sample equation according to Monod[25]

TABLE 23
Growth Rate Equation and Integral Form in Each Phase of Microbial Cell Growth in the Case of the Four-Phase Curve

Phase	Growth rate equation	Integral form
I. Induction	$\frac{dx}{dt} = o$	$x = o$
II. Transient	$\frac{dx}{dt} = KØx$	$x = x_o + (x_L - x_o)\left(\frac{t - t_o}{t_L - t_o}\right)^2$
III. Exponential	$\frac{dx}{dt} = KX$	$x = x_c \cdot e^{k(t - t_c)}$
IV. Declining	$\frac{dx}{dt} = Kx_c\left(\frac{x_m - x}{x_m - x_c}\right)$	$x = x_m - (x_m - x_c)\exp\left[-k\frac{x_c}{x_m - x_c}\right](t - t_c)$

Note: x = cell concentration.
x_o = cell concentration at the boundary point between the induction phase and the transient phase.
x_L = cell concentration at the boundary point between the transient phase and the exponential growth phase.
x_c = critical cell concentration at the boundary point between the exponential growth phase and declining growth phase.
x_m = theoretical maximum growth concentration.
K = coefficient of growth rate.
$Ø$ = coefficient of consumption activity.
t = time.

$$k = k_{max}\frac{S}{K_s + S} \tag{21}$$

where k_{max} = the maximum specific growth rate, S = the concentration of the limiting nutrient and K_s = the concentration of a nutrient required for half of the maximum specific growth rate, a Michaelis-Menten constant.

Originally the previous equation was applied to cultures where the sole limiting nutrient was glucose or some similar source of carbon and energy. Where substrates are present in such low concentrations that the cell growth rate is limited by both, a small increase in either concentration will increase the growth rate according to Sinclair. We can write:

$$\mu = \mu_m \frac{S_1}{K_{s1} + S_1} + \frac{S_2}{K_{s2} + S_2} \tag{22}$$

This equation, as shown by Lamanna and Maletta[27], has a limitation. The data at low concentrations are not adequate to establish whether the culture actually does pass through the origin or whether it originates from some positive value on the abscissa. Perret[28] has pointed out another important limitation of the Monod equation: it applies to the limiting and not to the transient specific growth rate. The nutrient concentration varies during batch growth, and often it falls so rapidly that the concentrations of the components within the cell system do not have time to reach their exponential state levels. Thus, the specific growth rate that is observed for a given nutrient concentration will be higher than the true specific growth rate.

Kono,[29] Kono and Asai[30] derived the growth rate equations of microorganisms from the aspect of chemical reaction kinetics, including the new concepts of critical concentration and coefficient of consumption activity. They found that the calculated values of cell concentration, as based on this equation, (Table 23) show quite good agreement with observed

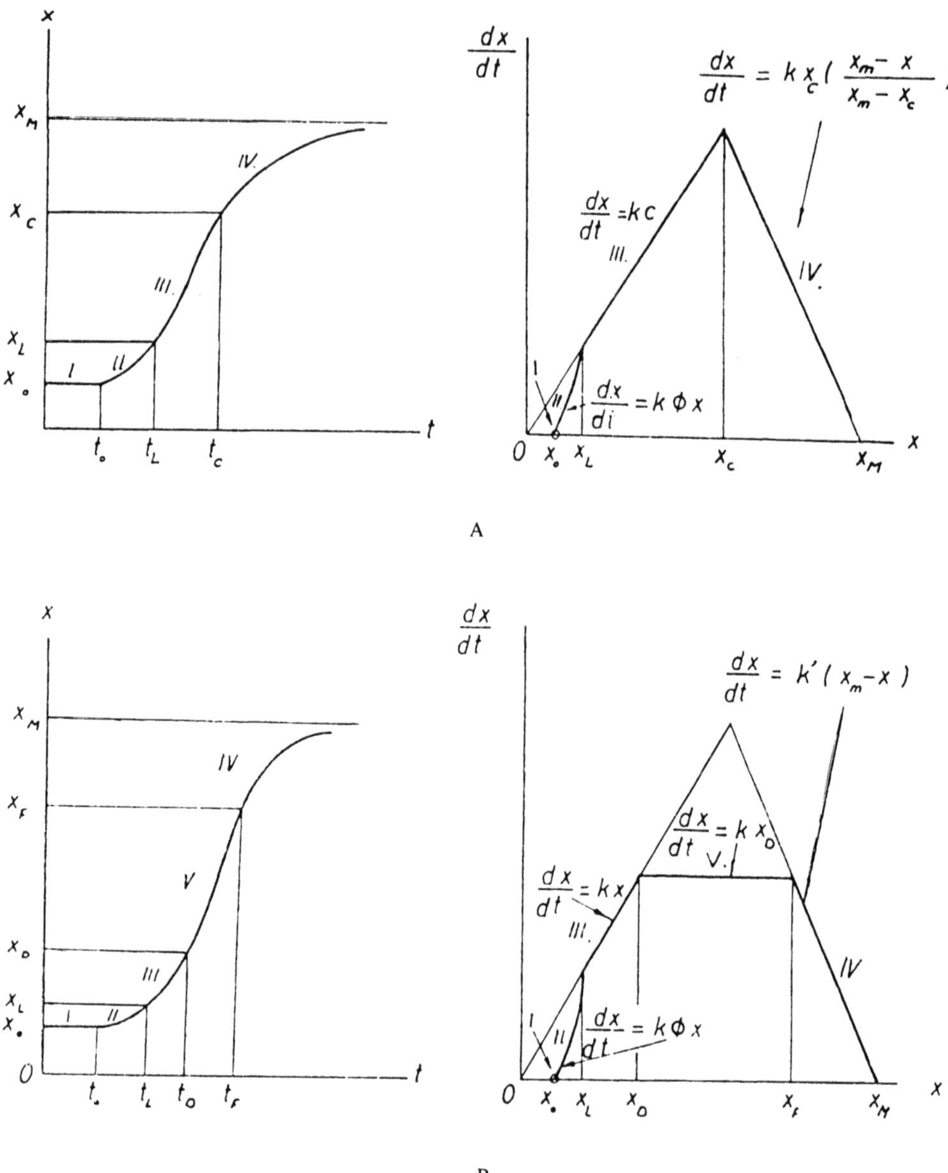

FIGURE 15 (A). Typical time course of microbial cell growth (left) and graphical expression of growth rate equation (right): (I) — induction phase, (II) — transient phase, (III) — exponential growth phase, and (IV) — declining growth phase. (B) Typical time course of microbial cell growth including constant growth phase (left) and graphical expression of growth rate equation (right): (I) — induction phase, (II) — transient phase, (III) — exponential growth phase, (V) — constant growth phase and (IV) — declining growth phase.

values in the full course of microbial cell growth including the lag period in batch cultivation. Applying this equation to continuous cultivation, they derived a general equation to calculate cell concentration in continuous cultivation.

Growth curves of four phases are not found to be suitable for the description of the growth of every microorganism. In the case of several microorganisms the existence of a phase was found in which the rate of growth remained constant and this growth curve can be derived into five phases[30] as given in Table 23 and Figure 15.

A. THE CONTINUOUS CULTURE

The continuous culture of microorganisms is a technique of increasing importance. The essential feature of this technique is that the medium is continually fed into the fermentor and the broth and suspended cells are continually removed in such a way the volume during the operation is kept constant. The contents of the vessel are well mixed and microbial growth takes place under steady-state conditions which means constant growth rate and constant environment.

Factors such as pH, temperature, nutrients concentration, metabolic products and oxygen are kept constant and can be controlled independently, resulting in a uniform product.[31,32] The most important feature of continuous culture is the dilution rate, which is the flow rate to the volume of the liquid in the vessel. Dilution rate effects the cell concentration, growth rate, yield, nucleic acid and other constituents of this cell.[32,33,34,35]

The merits of continuous culture have been well established in improving the productivity of biomass production.[36] Among the advantages of continuous culture, relative to batch, is the reduction in the size and cost of equipment necessary to produce a given output, in addition to the lower cost of running.[37] It also insures better uniform production in quality and quantity. In the design of continuous culture, kinetic constants (μ_{max}, K_s) determined on data of batch cultures are used for calculation or predicted cell concentrations at different dilution rates. Predicted values, determined by graphic and mathematical methods using the equations of Kono show good coincidence with experimental data measured at different dilution rates.[38]

V. ENVIRONMENTAL EFFECTS ON BIOMASS PRODUCTION

A. EFFECTS OF TEMPERATURE

The individual reactions which occur within the cell are influenced by temperature in the usual way, i.e., the rate constant for the reaction is given by the Arrhenius equation in the form:

$$K = A\exp(-E/RT) \tag{23}$$

where E is the activation energy and T the absolute temperature. Since a very large number of cooperating reactions influence the growth and product formation within the cell, temperature will have a complex effect overall. An additional complicating factor arises from the effect of temperature on the formation of the enzyme proteins, in which very rapid changes in activity take place over a small range of temperature.[24]

It is normal to express the parameters in the rate equations as Arrhenius functions of temperature

$$r_x = \frac{\mu_m(T)S}{K_s(T) + S} \times \mu_m(T) = A_1\exp(-E_1/RT) - A_2\exp(E_2/RT) \tag{24}$$

$\mu m\ (T) = A_1 \exp(-E_1/RT) - A_2 \exp(E_2/RT)$.

Any microorganism will grow more rapidly as the temperature is increased but only until the optimal temperature and the limits of temperature tolerance are varying. Most often used incubation temperatures for yeasts are in the range of 30 to 35°C.[22,39-42] Knight[43] et al., and Nour-El Dien[44] observed an optimal growth of *K. fragilis* in whey at temperatures ranging from 35 to 40°C. The higher specific growth rate at higher temperature in the range between 30 to 45°C does not influence yields and cell biomass produced according to results of Castillo and Sanchez.[45] Growth of yeast cells is greatly reduced at 20°C and there are no yeasts able to grow at temperatures of 60 to 70°C.[46] Thermotolerant strains are getting more and more attractive for industrial fermentations as they enable us to reduce cooling costs.

B. EFFECT OF pH VALUES

According to Sinclair,[24] pH has less effect on biological activities than does temperature because the cell is well able to regulate its internal hydrogen ion concentration in the face of adverse external concentrations, however the maintenance energy required for this process is obviously affected. The pH of the external medium may also have an effect on structure and permeability of cell membrane.

Hydrogen ion concentration in some fermentations has the role of minimizing contamination by other microorganisms. On average, pH of the medium varies between 4.0 and 5.0 for yeast propagation.[44,47,49] Yamada et al.[50] found that pH level affected the yield and crude protein content of *C. tropicalis* grown on hydrocarbon. The yield of yeast cell increased by increasing the pH from 4 to 7, after which it dropped sharply.

C. EFFECTS OF DISSOLVED OXYGEN

Probably no microorganism is indifferent to the presence of oxygen. For many it is essential for life, for some it is toxic even in small quantities, while in others it brings about fundamental changes in metabolism. Because of the differences in metabolism that leads to great increases in available energy when cells are grown aerobically the supply of oxygen is most important for efficient use of the carbon source. Under anaerobic conditions a part of carbohydrate is not oxidized completely and accumulates as ethanol, acetic acid, lactic acid, etc. and is lost as a potential source of energy.[46]

Oxygen acts primarily as a terminal acceptor of electrons from the respiratory chain. It also acts as a yeast growth factor as oxygen appears to be involved in the synthesis of oleic acid and ergosterol which stimulates yeast growth under anaerobic conditions.[51] Lie et al.[52] underlined the importance of small amounts of oxygen in beer fermentation. The main effect of the small amount of oxygen made available to the yeast is an increase in the level of unsaturated lipids, which are essential to the functioning of the cell membranes. Harding and Kirshop[53] found in their experiments that the use of oxygen-saturated rather than air-saturated worts resulted in a stimulation of fermentation in worts. Investigations of Babij et al.[54] stated, that in case of the highly aerophylic yeast *Candida utilis* high oxygen concentrations increased the level of polyunsaturated fatty acids. The rapid increase in linoleate was accompanied by a proportional decrease in oleate until a common level of about 25% of total fatty acids present in the cell.

The essential role of oxygen in the synthesis of unsaturated fatty acids and sterols in yeast is due to the oxygenases involved in the synthesis of them.[55] In aerobic fermentation aeration and agitation should be maintained above a critical value to give dissolved oxygen of 40% saturation.[56] Similar findings were published by Miskiewicz[47] that dissolved oxygen values higher than D_{ocrit} do not cause an essential rise in specific growth rate.

Oura[57] investigated the effect of dissolved oxygen in the region of 10 ppm to almost zero. At high oxygen levels the yield of cell mass calculated per carbon source in the culture medium was found to be 52%, at low dissolved oxygen concentrations yield decreased to 10 to 15%. The amount of cytochromes is a good indicator for the degree of aeration of the medium.

The anaerobic metabolism is clearly reflected in the activity of alcohol dehydrogenases, whereas malate dehydrogenase activity is a good indicator of a shift toward aerobic metabolism. Oxygen requirement of the yeast cell depends upon its physiological state. Kono and Terui[58] and Brighton[59] pointed out that with growing cells, as compared to resting cells, the rate of respiration reaches its highest level at the start of the logarithmic phase.

Effective aeration depends on a number of factors such as the amount of air, fineness of air dispersion, rate of agitation, size of fermentor and volume of medium. Paca and Gregor[60] investigated the effect of O_2 partial pressure (PO_2) on biomass production and physiological states of *C. utilis*. According to their data PO_2 up to $2.48 \cdot 10^{-3}$ MPa led to an increase in biomass concentration and growth yield but excess oxygen ($PO_2 \geq 2.94 \cdot 10^{-3}$ MPa) caused a decrease for both values. It is obvious when exceeding a definite upper limit of PO_2 in aeration gas the cell physiology is changed due to an overoxidation.[61] A very interesting finding is a fact that at a PO_2 this high dissolved oxygen never exceeded 4 mgl^{-1}.

The volumetric coefficient of oxygen absorption into culture media in tower fermentors and sparged stirred fermentors are affected considerably by the presence of surface active or antifoam agent added to the media.[62]

Aeration could also effect protein yield and amino acid composition of yeast. Vananuvat and Kinsella[63] found that the amino acid content of yeast grown in batch culture at higher agitation speed was higher than those grown at lower speed (700 rpm, 600 rpm resp.). Akashi et al.[64] found that high oxygen tensions had an inhibitory effect on the growth and ability to produce arginine. Methionine synthesis of methionine rich yeast strains is decreased at high oxygen absorption rates (higher than 200 mmol $O_2l^{-1}h^{-1}$).[45,65]

D. THE EFFECT OF CO_2 CONTENT

CO_2 at low concentrations may be required by *Saccharomyces cerevisiae,* for example for the generation of 4-carbon compounds. This observation is generally not exploited since most of the studies have been concerned with the inhibitory effects of CO_2 on yeast growth.[66] Parker et al.[67] investigated the effect of CO_2 concentrations in comparison to 100% air in batch cultures of *S. cerevisiae.* Data presented show that even 2.5% CO_2 causes a significant increase of yeast concentration. Optimum CO_2 content proved to be 5%. The positive effect of CO_2 was also stated by Nour-El Dien[44] who controlled aeration and agitation speed by CO_2 content of outlet gas in continuous culture of *S. fragilis* grown in whey medium.

E. THE EFFECT OF MEDIUM COMPOSITION

1. Effect of Carbon Source Concentration

The degradation of sugars under aerobic growth conditions via fermentation, the so called Crabtree effect, was studied for several yeast species. Studies were carried out under well-defined aerobic growth conditions in batch and continuous culture. Wöhrer et al.[68] proposed to quantify the Crabtree effect as the decrease of the respiratory capacity due to anaerobic fermentation. *Saccharomyces carlsbergensis, S. cerevisiae* and *Schizosaccharomyces pombe* show a high percentage of fermented glucose to the total amount of glucose utilized (75%). *Brettanomyces bruxellensis, Brettanomyces lambicus* only ferment 20% of utilized glucose while *Kluyveromyces fragilis* shows the lowest Crabtree effect. *S. cerevisiae* has a diauxic growth at glucose concentrations causing Crabtree effect.[6,69] *Brettanomyces lambicus* growth characteristics in batch culture show a triphasic cultivation. In the first phase, ethanol and acetic acid are formed during the degradation of glucose. After exhaustion of the monosaccharide, ethanol is oxidized to acetic acid, but ethanol cannot be utilized for growth. In the third phase, accumulated acetic acid is further oxidized to CO_2 and H_2O. *Schizosaccharomyces pombe* has a monophasic growth type in batch culture and in contrast to *S. cerevisiae* ethanol cannot be utilized for growth. Ethanol production and endogeneous respiration of the accumulated alcohol are in separate growth phases which are well characterized by the different RQ values. In continuous culture at low dilution rates *S. pombe* exhibits a pure respiratory metabolism at higher values of dilution rate and the fermentative metabolism increases.

Kluyveromyces fragilis shows a low Crabtree effect, and ethanol formed during aerobic fermentation is reutilized before glucose is exhausted. This weak Crabtree effect is also observed in continuous culture. Only small amounts of ethanol are released at high glucose concentrations. *C. utilis* cytochromes are also markedly repressed by high glucose concentration however, respiration is only marginally affected. Low glucose grown cells (glucose conc. 0.09 gl^{-1}) show weak ethanolic fermentation capability.[70] Petrikovich et al.[71] stated that with increasing concentrations of glucose, the structure and metabolism of the cell undergoes a unidirectional change regardless of how it is fed to the culture (continuous or pulse feeding). In response to an increased concentration of glucose, periplasmic space in the cell, amount of glycogen, the length and diameter of mitochondria and the size of the mitochondria increase.

In studying mitochondria of organisms inhibited by glucose it was found that the total and specific activity of a number of oxidizing enzymes were at a very low value and a decrease in specific mitochondrial membrane proteins and phospholipids has been detected. Experiments of Luzikov et al.[72] with *S. cerevisiae* demonstrated that in the course of glucose repression the physical state of inner mitochondrial membrane changes drastically. The changes are correlated with the decrease in the relative content of unsaturated fatty acid residues in the phospholipid molecules. In the repressed yeasts glycolysis is the principal mechanism for producing energy. A decrease for cell yield at increasing glucose concentrations (0.1 to 1.0%) was observed not only for *S. cerevisiae* well known for Crabtree effect but also in case of *C. guilliermondii* and *Rhodotorula glutinis*.[73] However, for *S. cerevisiae* glucose repression occurs even at very low concentrations (up to 30 to 100 mgl^{-1}). Wöhrer and Röhr[74] concluded for their experiments that critical glucose concentration is not a constant value, but varies considerably to the nutritional state of the yeast cells.

2. Effect of Nitrogen Source

Yeast cells are able to synthesize all precursors necessary for their macromolecular components such as DNA, RNA and protein from basic nutrients such as glucose, ammonia, inorganic phosphate and sulfate. Protein and lipid contents of microorganisms reflect the composition of the medium and growth conditions. Ash content of SCP products also varies depending on the mineral content of the growth medium.[75]

Saita and Slaughter[76] reported that in a defined medium of both D-glucose and maltose the fermentation rate of *S. cerevisiae*, a brewing strain, was dependent on ammonium concentration. The ability of several amino acids to activate glycolysis followed the same order as their effectiveness as sole sources of nitrogen. It seems that ammonium concentration does not stimulate fermentation through direct activation of glycolytic enzymes, but through its function as substrate for protein synthesis. Hill and Thommel[77] observed that with low ammonium concentrations, RNA content declines and glycogen reserves accumulate. These effects are important in the commercial production of yeast, and it is advantageous to develop an ammonium concentration control system applicable to yeast fermentation. Kole et al.[78] used the feed-forward control technique to control the concentration of ammonium during the fed batch cultivation of *S. cerevisiae*. They found at increasing ammonium concentration (from 38 to 300 m*M*), both the ammonium consumption and the protein content of the cells increased, while the biomass yield decreased. At 230 and 300 m*M*, the ammonium and glucose consumption appears to support production of some, until now, unidentified extracellular products.

3. Effect of Bioactive Compounds

It is well known that some yeast strains require substances, such as biotin and yeast extract for cell division. The biotin concentration seemed to be correlated with the overall micronutrient content, and in the absence of an essential nutrilite yeasts would grow slowly after a long incubation period. For more rapid growth several nutrilites should be present. Under conditions of adequate aeration the biotin requirement is about $0.25\mu g/g^{-1}$ yeast dry matter.[44] Harju et al.[79] reported that in case biotin was the only vitamin supplemented (30 μgl^{-1}), the yield of *K. fragilis* increased from 22 to 31% but the specific growth rate remained very low. Thus, not yeasts in general but *Kluyveromyces fragilis* and some other wild strains in particular can do without any growth factor, including biotin. Requirements of yeast extract to improve the growth varies (0.1 to 3.0 ng/m^3).[75]

Wasserman et al.[175] established that the addition of 0.1% yeast extract produced a two-fold increase in cell count. Castillo and Sanchez[45] found that yeast extract of a concentration of 0.1% and above in 2% whey gave maximum utilization of lactose and cell yield of *Kluyveromyces fragilis*. Nour-El Dien[44] could not detect any growth stimulation when adding biotin to whey medium or mineral medium (2% lactose content). The stimulating effect of yeast extract biotin in mineral and whey medium was confirmed by others.[173] Not only growth rate and yield but also lactose utilization was improved. Growth factors could be several amino acids purines and vitamins[21] and also unsaturated fatty acids for yeasts under anaerobic conditions.[17]

VI. GENOTYPIC MODIFICATIONS OF YEAST

A. MUTATIONS AND MUTAGENIC AGENTS

It is certain that cell information is stored in its DNA and that cell protein at any one time represents a partial transcription and translation of this memory store.[80] The change in the structure of this memory produce changes which are inherited. These changes are called mutations. Grick et al.[81] classified mutations into two main types:

1. Microlesions which include substitutions of base pairs, which is the replacement of one pair by another either by transition or transversion or frame shift mutation, either by addition of one base, deletion of one base or addition of a frame base.
2. Macrolesions which include deletions, duplications or rearrangements.

Mutations are caused by many factors among which chemicals and radiation are the most important.

Chemical mutagenic agents include:

1. Analogs, which are substances closely related to A, G, C or T bases which can be incorporated in the DNA without hindering its replication, like 5-bromo-uracil inducing transitions (i.e., GC $\rightarrow$ AT)
2. Nitrous oxide induces transitions of

$$A \rightarrow G \text{ and } C \rightarrow U \text{ or } AT \rightarrow GC$$

   Another effect of nitrous oxide is the covalent binding of the two strands of the DNA, one cross-link for four deaminations
3. Alkylating agents
4. Acridines

Nadson and Fillippov[82] first discovered the mutagenic action of radiation on yeast cells. The biological mechanism of the radiation mechanism is that one or more of the reaction(s) are damaged which leads to a change in the cell properties. Drake[83] studied UV induced mutations in bacteriophage T4. He reported that half of the induced mutants were reverted by base analogs and therefore corresponded to base pair substitution. The other half of the induced mutants were reverted by acridines, which showed that they were of the frame shift type. In DNA pyrimidines appear to be more sensitive than purines. There are two main classes of photoproducts: hydrates and dimers. Pyrimidines can fix water molecules on 5, 6 double bond to form hydrates. Thymine also gives rise to 5,6 dihydrothymine. These compounds could be important in inducing mispairing and subsequent mutation. UV photoproducts in DNA also include pyrimidine dimers (mostly TT, but TC and CC are also possible) of the cyclobutane type. They usually arise from a pair of adjacent pyrimidine in the same DNA strand. Other dimers could also result from crosslinking between the strands of the DNA double helix.[81] Ionizing radiation (X-ray) alters nucleic acids both directly and indirectly. In direct effect, the chemical bonds of the bases of deoxyribose and of the sugar phosphate linkages are ruptured. In indirect effect, ionizing radiations produce free radicals either from water (H^+, OH^-) or from organic molecules. The free radicals attack the constituents of the DNA. The destructive effect is greater in the presence of oxygen. UV, in contrast to X-ray and other high energy radiations, produces only atomic excitations but no ionizations.[84]

The relationship between the dose and the mutation frequency has been studied.[85] It was found that the frequency of mutation increases linearily with decreasing survivors. Therefore, the cells derived from samples which show 0.01 to 0.1% survivors were usually examined for mutations.[86]

B. INDUCED GENOTYPIC CHANGES IN YEAST

As it is known, methionine is the limiting amino acid in yeast protein, and it is evident that the value of SCP as a feed, food or food supplement would be greatly enhanced if mutants with higher methionine levels could be produced and isolated. Okanishi and Gregory[85] isolated mutants of *Candida tropicalis* with up to 41% higher methionine content than the parent. Ken-Ichi et al.[87] found pool methionine rich mutants among mutants resistant to ethionine (an analog of methionine) in an *n*-paraffin-assimilating yeast *Candida petrophilum*. One of the mutants had 40% more methionine than the parents. Using *N*-methyl-*N*-nitrosoguanidine Young and Smith[88] produced mutants with 268% higher methionine content compared to parent strain. Okanishi and Gregory[86] isolated the methionine rich mutants, making use of the fact that mutants with increased levels of sulfur-containing amino acids produced smaller colonies on sulfur-deficient media than did the parental strain. This approach was based on the assumption that methionine-rich mutants would have higher sulfur requirements for growth than would their parent. Delgado et al.[89] induced auxotrophic mutants of *Candida utilis* utilizing *N*-methyl-*N*-nitrosoguanidine as a mutagenic agent.

Momose and Gregory[90] showed that the amino acid pattern in the cellular protein of a yeast can be changed greatly by specific temperature sensitive mutations. They reasoned that it should be possible to isolate mutants with temperature sensitive mutations in either amino-acyl-transfer ribonucleic acid synthetases, transfer ribonucleic acids, or the ribosome itself, such that at a nonpermissive temperature a culture would begin to synthesize all of its proteins with methionine substituting in part for a more common amino acid(s).

De Zeeuw[91] discussed that point mutations arising from the base transitions and transversions in structural gene triplets will dictate amino acid substitutions in specific proteins. Yanofsky[92] studied tryptophan synthetase substitutions and showed that there is leeway in the specifications for biological activity. Serine and alanine can both substitute for glycine at certain sites and impart normal activity. De Zeeuw[91] gave details of the point mutations that can lead to a substitution by methionine. The pathways are clear even though it is not known whether the particular codons capable of transition or transversions to AUG are major or minor. The possible point mutations may be summarized as follows:

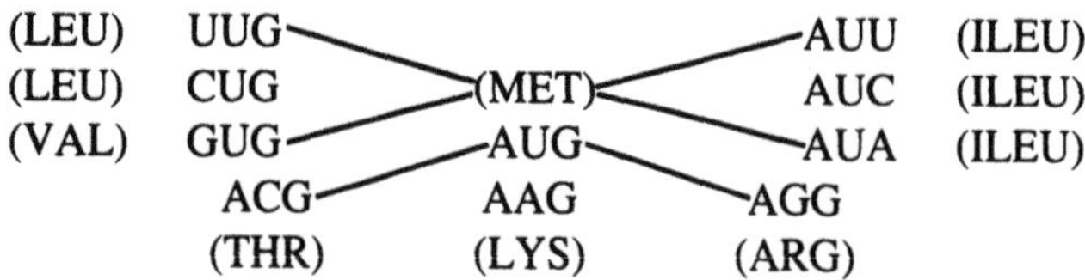

Most of the eukaryotic organisms show diploid phasis during their lifecycle. The diploid form has a greater stability and higher independence of changes in the medium. During the diploid phase the gene organization reaches optimum yield in mutation. For this reason, diploid cells have a greater selectivity value than haploid cells. Investigations of Kondratyeva et al.[93] confirmed that the *Candida utilis* diploid strain has greater variations induced by UV irradiation in the activity of biomass synthesis as compared with the parent haploid culture. Clones with an activity of synthesis greater than the mean population appear frequently in the diploid strains. Mathematical analysis has confirmed the significance of the results and the hypothesis according to which the frequency of variants are more active in biomass synthesis sizes after the action of UV irradiation.

C. THE GENETIC MANIPULATION OF INDUSTRIAL YEAST STRAINS

Hybridization, protoplast fusion and transformation have been studied as means to genetically manipulate brewing and related yeast strains. Hybridization is a secondary technique for verifying the gene composition of recombinants produced by fusion and transformation.[94] Further, it can be used to study gene dosage and gene suppression effects.[95] Fusion is more empirical than manipulative since little control can be exerted over the genotypic make-up of the fusion product. Transformation with native DNA is a more subtle technique then fusion since the introduction of a single trait into strains can be controlled. 2 μm DNA plasmids are present in all brewing yeast examined but the growth temperature of some strains is critical before significant numbers of plasmid can be detected. A good brewing yeast must be effective in wort utilization, must give the required flavor and must be easily removed after fermentation. In strain breeding mutagenesis has been ignored because of difficulties connected within mutagenic treatment of polyploid brewing yeasts. However, some laboratories have achieved considerable success with mutagenic treatment of brewing strains.

1. Hybridization

Hybridization cannot be used directly as a means to manipulate brewer's yeast strains because of their polyploid nature. Nevertheless this technique made an invaluable contribution to yeast improvement. Hybridization has been used in combination with more novel techniques to verify the composition of recombinants. Stewart et al.[56] and Russel et al.[97] used hybridization to study the genetic control of flocculation and sugar uptake of brewing yeast.

An acceptable yeast strain for brewing purposes must be able to utilize the following sugars: glucose, fructose, sucrose, maltose and maltobiose. The sugar content of worts contains 50 to 55% maltose and 10 to 15% maltobiose. The ability of yeasts to utilize these compounds is vital and is dependent on the correct genetic complement. Both sugars have independent uptake mechanisms but inside the cell, they are hydrolyzed to glucose units by α-glucosidase systems. The maltose/maltobiose uptake system is controlled by the polymeric MAL genes. Strains containing multiple MAL 2 show a faster rate of fermentation compared to production strains. The gene dosage effect is more pronounced in more concentrated worts.

2. Fusion

Techniques that show greatest potential and promise as aids in the genetic manipulation of brewer's yeast strain are protoplast fusion and transformation. Both of these methods have a total disregard for ploidy and mating type and have great applicability to brewer's yeast strains because of their polyploid nature and absence of mating type characteristics.

a. Protoplast Fusion

Spontaneous protoplast fusion of yeast (*Saccharomyces* and *Candida*) was first reported by Müller.[98,99] Ferenczy et al.[100] were the first to report a controlled protoplast fusion and complementation with the yeast-like filamentous fungus *Geotrichum candidum*. Mutants, usually with auxotrophic, antibiotic resistance, temperature-sensitive, respiration-deficient, morphological and/or color markers are used for fusion. Fusion frequency calculation is based upon the frequency of complementation, although the two events are never the same, as fusion also occurs between noncomplementary identical partners. The frequency is usually determined by comparing the number of complemented cells (i.e., cell growing on incomplete medium when auxotrophic partners are used) to that of the noncomplemented ones (colonies growing only on complete medium). The presumed fusion products have to be characterized by using cytological, biochemical and genetics methods.[101]

The first step of this genetic manipulation is the removal of the cell wall with lytic enzyme, which could be microbial in origin like the commercial available zymolase or extracts of snail gut. After enzymatic digestion of the cell wall only the membrane around the cytoplasm exists. Such spheroplasts are osmotically sensitive and to keep them intact the medium must have a high osmotic pressure which is usually provided by sorbitol, or KCl. Traces of cell wall lytic enzymes have to be removed by washing, only then follows the protoplast fusion in the presence of a fusion agent which consists of polyethylene glycol (PEG) and calcium ions in buffer by mixing the protoplast of two yeast strains with different genetic characteristics. Then aggregation of protoplasts is induced. Ferenczy et al.[100,102] used centrifugal force and later Ferenczy et al.[103] applied intensive aggregation of protoplasts in cold KCl osmotic stabilizer to induce their fusion. Kao and Michaluck[104] used PEG as a fusogenic agent of plant protoplasts and Ferenczy et al.[105] revealed that PEG preparations with molecular weights of 4000 to 6000 were optimal in inducing fusion in the concentration range of 25 to 40% in the presence of 10 to 100 m*M* $CaCl_2$.

Weber and Spata[106] found that the stability of interspecific and intraspecific fusion hybrids were different. Interspecific and intergeneric fusion hybrids show a high instability. The effect of agglutinant agents like PEG with Ca^{2+} ions as stimulants of the fusion process could be enhanced by an electric field pulse in the μs-range.[107,108] Weber et al.[109] reported successful application of this technique both for intraspecific and intergeneric fusion.

After interchange of the DNA material the fused protoplast (hybrid) must regenerate its cell wall and begin cell division. Stewart et al.[110] succeeded in producing several brewer's yeast hybrids which contain genetic characteristics of both parents. Intergeneric fusion has also been attempted; an auxotrophic strain of *K. lactis* was fused with a haploid strain of *S. cerevisiae*. After cell wall regeneration a number of hybrids were found which were able to utilize both lactose and maltose, however these recombinants proved to be unstable and reverted to their constituent fusion partners. But also the stable *Saccharomyces* hybrids give unsatisfactory results in brewing experiments.

The protoplast fusion technique provides a useful method for improving industrial yeasts lacking normal sexual exchange. Protoplast fusion is widely accepted in genetic, physiological and biochemical studies. Protoplast fusion of *Candida tropicalis, S. cerevisiae* and *Schizosaccharomyces pombe* has been carried out by Fournier et al.,[111] Gunge and Tatinra[112] and Sipiczky and Ferenczy[113] in the seventies. De von Broock et al.[114] reported on the successful fusion of *S. cerevisiae* and *S. lypolytica*, and found that polyploidy is a rather common phenomenon at least in the interspecific fusion of *S. cerevisiae* protoplast carrying the same mating type. Protoplast fusion is used for cross-breeding of yeast strains each having valuable industrial properties worth combining in a single strain.[79] For example, the fusion of an amylolytic strain and a strain using sugars in preparing proteins by culturing yeasts on an amylaceous medium. A hybrid or recombinant obtained from these two strains may allow the production of proteins on an amylaceous medium to be improved, particularly if the strain using sugars has a high protein content.[115] Fusion technique is also suggested for the preparation of polyploid strains which are much larger than the parent strains, which are useful when separating the strains from the culture medium, particularly such strains

(*Candida tropicalis*) growing on hydrocarbon products which cannot be assimilated in food. As in the fusion patterns the genome of both parents becomes integrated; it would be difficult to introduce a single trait into the strain.

b. Transformation

This technique offers a means of overcoming the nonspecificity of fusion. This method which is also a nonsexual way to achieve genetic recombination — involves the implantation of DNA material originated from the donor cell into the recipient cell. Donor DNA can be taken from another yeast strain, genes or even artificially synthesized DNA. The donor DNA is taken up into the gene pool by the recipient cell and the recipient takes on some of the genetic characteristics of the donor strain. The initial transformation studies used native, purified DNA from donor cells and the transformation procedure has many common steps with protoplast fusion. The recipient must be converted into the protoplast state and incubated with purified DNA in the presence of PEG and Ca^{2+} ions. After a short incubation of 15 min the protoplasts are transferred in agar medium to regenerate cell wall and cell division. The regenerated colonies are screened for successful transformations.

The objective of several laboratories at present is to genetically engineer a yeast to efficiently convert lignocellulosic substrates into useful byproducts. The most advanced method used in genetic engineering is recombinant DNA. This technique allows for the construction of hybrid molecules of genetic information by joining pieces of DNA isolated from different biological systems. Main tasks required for the construction of a cellulolytic yeast include:

1. Source of gene — donor organism which is well characterized in cellulose synthesis, their expression and biomass synthesis
2. Construction of gene branch — high molecular weight DNA with a cloning vector to allow rapid scanning of gene sequences
3. Selection of recipient — more than one host is required; preferred most is *Saccharomyces cerevisiae* and *E. coli* and it should be able to gene transfer, gene expression and export pathways should be known
4. Construction of gene probes — assay for enzymatic activity, anti-enzyme antibody production, use AA sequence of enzyme to construct oligonucleotide probes
5. Gene selection — colony assay using probes from (4)
6. Gene characterization — determination of regulatory elements to provide dues for gene stabilization and expression at the extra cellular level
7. Gene transfer and expression — efficient utilization of the substrate requires: enzyme stability, coordination of expression with fermentation functions, relaxed repression by metabolites

An important reason to study the genetics of brewery yeast is that knowledge is important for breeding. Carlsberg lager production yeast seems to consist of at least one set of chromosomes that is closely related to *Saccharomyces cerevisiae* laboratory strain plus at least one set which is homologous.

VII. YEAST PRODUCTION ON CARBOHYDRATES

As estimated, 2 million t of single cell protein, mostly yeast are produced annually in the world. Production is mainly from cane and beet molasses with about 500,000 t from hydrolysed wood wastes and corn crash and from sulfite-waste liquor.[116] From an economical point of view high price for protein and use of low-cost carbohydrates are of critical importance so the technology to produce SCP on carbohydrates has been extended to the purpose of pollution control.

Except for brewer's or distiller's yeast in which the biomass is a byproduct, all commercial SCP processes are aerobic as anaerobic yields of cells from carbohydrates being very much lower than aerobic yields. Since the carbon source and aeration together represent the major cost of yeast biomass production, the efficiency of their utilization in cell synthesis is of utmost importance. Biomass production is generally optimized for protein content however, the yield coefficient and productivity are also of great importance. With carbohydrates as substrates, changes in metabolism toward a less efficient utilization of the substrate at high growth rates are frequently observed, even when the carbohydrate is the growth limiting factor. In case of *Saccharomyces cerevisiae* glucose concentrations higher than 6.10^{-3} mol · 1^{-1} or growth rates higher than 0.25 h^{-1} shifts to the anaerobic dissimilatory pathway with ethanol and carbon dioxide production and low cell yields.[6,117,118]

Productivity and the quantity of biomass produced per unit size of plant determines the capital investment required for a given output of product. Since fermentors and ancillary equipment comprise the major cost of biomass plants, fermentor productivity is a critical component in process design. It is for this reason that continuous rather than batch cultures are invariably used in modern processes. Productivity in continuous culture is the product of dilution rate (D h^{-1}) and yeast concentration in the effluent (x), D · x. The physiological limit is thus the maximum growth rate of the yeast strain used and the maximum concentration of organism that can be obtained. However, in practice productivities are an order of magnitude less than the theoretical maximum of about 100 glh.$^{-1}$ The reasons for this lie in the difficulties of substrate (especially oxygen supply) to the cells.

A. PRODUCTION OF BAKER'S YEAST

Baker's yeast (*Saccharomyces cerevisiae*) is one of the largest products of the fermentation industry as it is produced at a rate of over 1.8 million ton (30% solids) annually.[119] The manufacture of yeast specifically for supply to bakers was started in the middle of 19th century and the use of pure yeast culture was recommended. Baker's yeast can, in fact, be regarded as a specialized form of single cell protein.

Proposals have been made to increase the protein content of bread by adding inactive yeast — but in such a case modifications have to be made to the dough recipe to correct adverse effects on loaf characteristics.[120] This procedure has the effect of improving the nutritional value of the lysine deficient cereal protein. However, problems of altered texture and flavor might be a serious disadvantage. The main reasons for the disadvantageous effect on loaf volume is the high proteolytic activity of baker's yeast[121] and the glutathione content.[122]

Nutritional studies have uncovered an unsuspected property of yeast in breadmaking. Phytic acid in flour has an adverse nutritional effect since it strongly binds essential trace metals. Yeast fermentation results in a substantial decrease in phytic acid and consequent improvement in nutritional value of the bread.[120]

In the single cell production the main consideration is the yield of biomass although its protein and nucleic acid contents and digestibility must also be taken into account. Although, from an economic point of view, the yield of yeast per unit of substrate is of great importance. For these reasons, every effort must be made to control the process so that it conforms as closely as possible to the predetermined pattern. Baker's yeast is sensitive to the Crabtree effect and fed batch fermentation is used to keep sugar concentration at a low value, and prevent glucose repression.[123] To optimize the feed rate several control systems were developed based on RQ and oxygen uptake, online measurement of yeast content and dissolved oxygen tension.[124,125] The multivariable adaptive control technique provides both good cell production and high yield.

Baker's yeast production begins with pure culture slants used as inoculum for yeast propagation in Pasteur flask on sterilized medium containing beet or cane molasses as carbon source and is supplemented with N + P salt and several growth factors. Growth of baker's

yeast proceeds at 30°C for up to two days. Then the entire content of the flask is used to inoculate the first of a series of progressively larger fermentors. During these phases, key parameters such as incremental feeding, pH, temperature and aeration are strictly controlled in accordance with requirements of the yeast strain. The trade stage of the process requires 10 to 20 h for completion and has 3.5 to 5% cell solids in the broth.[126]

Yeast cells are harvested in a centrifugal separator, repeatedly washed with water and reseparated. The resulting "yeast cream" is cooled and filtered on a rotary vacuum filter. The filtered wet yeast is stable at 0°C until it is further processed either as compressed or as active dry yeast. Instant active dry yeast has a similar production as active dry yeast except for the drying process, in place of a continuous belt dryer, dehydration is accomplished with an airlift fluid-bed dryer.[128] Disadvantages are expensive drying costs, sensitivity to rehydration with added water and low activity on a dry weight basis compared to compressed yeast. The fluid-bed method of drying has recently found considerable favor since it is potentially much more rapid than band or drum drying. Drying times as short as 20 min can be achieved with minimal loss of yeast activity. Fluid-bed dryings can employ yeast pretreated with an emulsifier (e.g., sorbitan monostearate) to allow drying to lower (50%) most stable moisture levels, and also to facilitate "wetting" under instant-type usage conditions.[125] Fatty acid esters and polyols give protection at the reconstitution stage. The effect is associated with the lower leakage of nucleotides during reconstitution. An addition of $CaCl_2$, glucose and polyethyleneglycol to the rehydration medium causes a decrease in cell permeability, assessed as the loss of potassium ions and nucleotides as well as the total loss of intracellular compounds.[130]

1. Strain Breeding Development of New Yeast Strains

Although such techniques as hybridization have been known by yeast geneticists for several decades, a new technique called protoplast fusion enables microbiologists to create new and improved yeast strains with high osmotic tolerance and high activity in lean dough. Frozen bread dough methods have been extensively adapted in the baking industry and in instore bakeries. The most important problem encountered in frozen dough processing is how to maintain the viability and gasing power of frozen yeast. Oda et al.[31] selected 11 yeast strains suitable for frozen dough from over 300 *Saccharomyces* strains. All of these were identified as *Saccharomyces cerevisiae* from morphological, cultural and physiological characteristics. The selected yeast cells accumulated a higher amount of trehalose than did the commercial baker's yeast cells. This property seems to be related to the higher viability of these yeast cells in frozen dough. Hino et al.[132] isolated freeze-tolerant yeast from banana peel and identified as *Kluyveromyces thermotolerans* and *Saccharomyces cerevisiae.*

As it was shown by studying *Saccharomyces cerevisiae* synchronous and asynchronous cultures, their cells in the G_1 and G_2 phases are most resistant to the damaging of dehydration, dehydration and freezing-thawing, but the cells in the S and M phases are better capable of repairing the damages. In terms of the concept that the periods of extrophy and endotrophy alternate in the cell cycle of microorganisms, cell resistance in the G_1 and G_2 phases may be hypothetically attributed to the highly osmotic cytoplasm of the cells in these phases whereas cell active repair in the S and M phases stems from the presence of intracellular energy reservoir accumulated by the end of G_1 and G_2 phases.[133] The cell cycle not only influences the damaging effect of freezing but the freeze rate also. Slow freezing was shown to be optimal for *Saccharomyces cerevisiae* strains.[134]

B. BREWER'S YEAST

Brewer's yeasts include selected strains of *Saccharomyces cerevisiae* for producing lager beer. Lager beer is produced by the bottom fermenting yeast *Saccharomyces uvarum,* formerly named *Saccharomyces carlsbergensis. Saccharomyces uvarum* settles to the bottom of the fermentor at the end of the alcohol-generating process, leaving lager beer as a clear liquid. The bottom fermenting brewer's yeast is capable of fermenting entire raffinose molecules.[127]

Brewer's yeasts are obtained either from pure cultures or, as practiced by most large breweries today, recovered and prepared for pitching fresh nutrient. The propagation process involves serial transfer of yeast on wort until obtaining an adequate inoculum for use in the plant. Recovered yeast can be recycled up to 20 times before it needs to be replaced by pure culture. There is however, considerable variation in this practice as some brewers replace the yeast after only a few recycles. Between recyclings, the yeast is washed with acidic solution to eradicate bacteria, then cooled hand held overnight.

Spent brewer's yeast is processed into yeast extracts for use as flavorants or into dried yeast for use as nutritional supplements. Spent brewer's yeast is generated in high amounts, the U.S. alone produces about 75 million pounds of spent brewer's yeast solids annually, of which only a minor amount is recycled through the brewing process. According to the National Formulary (NF XIII) of the American Pharmaceutical Association, dried yeast is defined as a byproduct from the brewing of beer washed free of beer prior to drying.[126]

1. The Effect of Medium Composition on Metabolism and Composition of Brewer's Yeast

The maltose concentration effects the metabolism of brewer's yeast. The Crabtree effect is characteristic for yeasts of *Saccharomyces* genus when the glucose concentration is more than 6.1^{-3M}. After exhaustion of glucose the yeast oxidizes the ethanol. *Saccharomyces carlsbergensis* shows a diauxic growth in 0.15% glucose containing medium. Generation time of the first exponential phase is a 2 h diauxic lag period 8 h and the second growth phase has t_g 10 h.[135] In the case of maltose substrate at 0.15% sugar concentration a single inflexion point of 1 h shows the change from glucose fermentation to ethanol respiration. Both phases are characterized by a generation time of 5 h. At 2% maltose concentration only there is a significant diauxic growth. At the lower maltose concentration the sugar content is only partly utilized at the point of diauxic, this means that ethanol is assimilated before this time.

Flocculation of brewer's yeast is of great importance, the ability of which is influenced by medium composition. Bonaly et al.[136] stated that nitrogen has an action mainly in the time of flocculation. With regard to the sugar source, it was observed that there is a strain specificity as one has a more marked sensitivity to monosaccharides and the other to oligosaccharides. The ratio of K^+/Ca^{2+} in the medium appears to be a determining factor.

Rose et al.[137] studied the amino acid uptake from the wort. Specific alterations to the lipid composition of the yeast plasma membrane showed that when the membrane is enriched in linoleyl acyl residues uptake of arginine and lysine was affected. Since uptake of unsaturated fatty acids from wort by yeasts aids retention of viability, this effect of fatty-acyl unsaturation on amino acid uptake has relevance in brewery fermentations. Aeration of pitching yeast results in high sterols and unsaturated fatty acid content. Aerated yeast has a shorter fermentation time by about 25% compared with the anaerobic pitching yeast. A further advantage is that vicinal diketone concentrations are strongly reduced after the primary fermentation. The yeast yield after the primary fermentation is increased by about 10% and no off-flavors are detected in beers produced by aerated pitching yeast.[138]

a. Strain Breeding

Starch is the principal raw material for beer and fermentation ethanol production, therefore, a significant interest in the research on amylolytic enzymes exists. The conventional production of ethanol from starch requires an initial pretreatment step whereby the action of amylolytic enzymes on the starch molecules produces low molecular weight sugars which are readily fermentable to ethanol by *Saccharomyces* spp. As *Saccharomyces* spp. are the microorganisms responsible for over 96% of the total industrial ethanol fermentation and starch is the principal raw material commercially utilized for this purpose. It is suggested that a *Saccharomyces* sp., possessing the ability to hydrolyze starch entirely, i.e., able to synthesize and secrete α-amylase and glycoamylase with debranching activity in addition to its fermentation efficiency and tolerance to ethanol, would be most advantageous for the production of potable (beverage) and industrial ethanol.

In an attempt to genetically construct a strain of *Saccharomyces* sp. with the characteristics mentioned above, several producers are currently being employed, ranging from classical hybridization techniques, spheroplast fusion, mutation and recombinant DNA technology. Traditional procedures for hybridizing strains of *Saccharomyces* have met with little success in producing novel or improved strains for brewing. Successes in the transformation of yeast with a recombinant plasmid[139] offer one such prospect: by first cloning a desirable gene into a suitable yeast plasmid vector it should be possible to construct hybrids which differ from the parental brewing strain only in a single specific characteristic.

Tubb et al.[140] used the "rare mating" technique to hybridize dextrinfermenting haploids of *Saccharomyces* with a nonsporulating asexual lager yeast. Aim of their research work was to produce a brewing yeast able to ferment wort dextrin and thereby produce low carbohydrate beer. Hybrids ferment brewer's wort at rates similar to that of the brewing parent. In addition, they produced extracellular amyloglucosidase and were able therefore to hydrolyze maltodextrins and to produce low-carbohydrate beers. However, such beers possessed an objectionable herbal phenolic aroma.

The lifecycle of *Saccharomyces* normally alternates between haploid and diploid states. An ascus containing four or fewer haploid spores is produced and mating occurs between cells carrying different mating-type alleles designated *a* and α. This mating reaction is initiated by cell agglutinaton involving complementary peptide factors located on the cell surfaces. Thus hybridization employs the lifecycle of the yeast and a novel diploid can be produced by using haploids with different genetic makeups and opposite mating types.

There is a species of the genus *Saccharomyces, S. diastaticus,* which is a species closely related to *Saccharomyces cerevisiae* that is able to utilize dextrin and starch due to its ability to produce the extracellular enzyme glycoamylase. Using classical hybridization techniques a diploid strain containing the DEX and STA genes in the homozygous condition has been constructed and its fermentation rate studied in the brewer's wort. The initial fermentation rate of this strain was slower than a production ale brewing strain, however it fermented the wort to a greater extent due to the partial hydrolysis of the dextrins by the action of glycoamylase. Thus, *Saccharomyces diastaticus* strains possess the capacity to produce beer with a high degree of fermentability, which is desirable in the production of low-carbohydrate beer. However, Sills et al.[141] noted that beer produced by this strain has a characteristic phenolic off-flavor (POF). Constructing strain from a haploid which is DEX positive and POF positive by mating with a DEX negative and POF negative (brewing *Saccharomyces*) haploid the tetrad dissection of the resultant diploid gives the possibility of selection on DEZ positive and POF negative haploid. However, beer produced from this haploid has a rather winey taste and a slightly sulfury character. Although hybridization is a very important technique in a yeast strains improvement program, it has limitations with the most brewer's yeast strains. The latter are often polyploid and even aneuploid and as a consequence, do not possess a mating type.

Spheroplast fusion, on the other hand, is a technique that displays a total disregard for ploidy and mating type and consequently has an applicability for many industrial strains because of their polyploid nature and absence of mating type characteristics. The treatment of whole cells with lytic enzymes results in an osmotically fragile spheroplast which can be fused in the presence of PEG and Ca^{2+} anions. Selecting against the parental types, the resulting hybrid strain can be regenerated in the medium containing an osmotic stabilizer which is essential to provide osmotic support for the spheroplasts. Fusion hybrids of *Saccharomyces diastaticus* strain and *Saccharomyces uvarum (carlsbergensis)* show an increased fermentation rate when compared to either fusion partner. Thus almost 93% of glucose is utilized in 40 h and 12% ethanol is produced. The constructed strain is less sensitive to osmotic pressure but has lower ethanol tolerance.[141]

Rezessy-Szabó et al.[142] constructed a yeast strain which is more suitable for a given beer fermentation technology whereby the quality of beer is improved and economy of production is increased. The aim of strain breeding was to decrease the production of undesirable aromas (vicinal diketons), to increase flocculation ability as well as to construct yeast strains with killer activity for the purpose to repress the contaminating wild strains. By protoplast fusion technique new strains of increased flocculation and of killer factor were produced, from which brewer's strains with good fermentation ability were selected. The technique of *DNA transformation* — more particular, plasmid mediated transformation — prevents an opportunity to transfer the brewing yeast-specific selected pieces of DNA molecules (genes) of interest.[143] Since the first successful transformation there have been numerous reports of native DNA and plasmid-mediated transformations, most of which involve DNA transfer into yeast spheroplasts in the presence of PEG and calcium ions followed by regression of the spheroplasts.

The use of plasmid vector systems is the natural progression from transformation with native DNA. Genes of interest like α-amylase and glycoamylase could be spliced out from donor DNA by partial digestion with a specific restriction endonuclease and ligated onto a suitable plasmid vector. Most species of *Saccharomyces* carry an extrachromosomal DNA plasmid, the 2 μm plasmid, which is present in approximately 50 to 100 copies per haploid cell and do not seem to have any known functions[144] as strains lacking it do not show any obvious phenotypic deficiencies.[93] Brewing yeast strains have also been found to contain 2 μm plasmids, although the copy number in these polyploids is believed to be much higher. It was found that growth temperature had a dramatic effect on the concentration of the plasmid, the higher the temperature (up to 37°C) the higher the concentration of the plasmid.[141] Although the 2 μm plasmid does not carry any selectable markers, it has been very useful as part of a chimeric yeast-bacterial plasmid where the origin of replication of the 2 μm plasmid allows autonomous replication of the chimeric plasmid in the yeast host.

One of the problems in plasmid-mediated transformation, besides stability, has been the low frequency of transformation. Of the DNA presented to recipient spheroplasts, only a small fraction is incorporated into the spheroplast. This can be enhanced significantly with the use of lipid bilayer vesicles called "liposomes".[145] Liposomes have been reported to transport a wide variety of substances including drugs, numerous enzymes, DNA and RNA, into various cells. Russel et al.[146] found that when ^{14}C labeled DNA was incubated with yeast spheroplasts, less than 1% of the label appeared in the spheroplasts treated with DNA alone, but 9 to 25% of the label was recovered from spheroplasts treated with DNA encapsulated in liposomes made with a 1:4:5 ratio of stearylamine:phosphatidylcholine:cholesterol. Moreover, the age of the spheroplasts was found to be an important factor in the uptake/binding DNA. Thus the 24 h spheroplasts had the highest DNA uptake capability while older ones had a drastically decreased DNA-uptake capability.

In summary, it can be concluded that in strain improvement programs such as increasing the spectrum of carbohydrate utilization by brewing yeasts different genetic methods are available such as DNA transformation, protoplast fusion and hybridization. In the past few years, as the potential commercial exploitations of recombinant DNA technology have been investigated yeast has become the focus of attention. Haploid and diploid yeasts are being studied for their role as host systems along with the industrially proven polyploid yeast.

C. WINE YEAST

Yeast for wine, distilled beverages and sake include numerous strains of *Saccharomyces* selected for their alcohol production (18 to 20%) and flavor.[147,148] Since several decades, pure cultures grown on slants have been used in wine production. Through inoculation of sterilized or pasteurized must, the pure culture is built up into the amassment of cells required for charging a plant-size fermentor. When wine production is completed, the wine yeast settles with the sediment to the bottom of the fermentor. Unlike brewer's yeast and distiller's yeast which are recovered, wine yeasts are not recycled since their recovery is too expensive.

Wine yeasts are available in compressed or active forms. These products are produced from yeasts grown on molasses wort using the same propagative scheme as that employed for baker's yeast.[1] The production of high concentration ethanol is limited by the inhibitory effect of ethanol on the growth and productivity of the fermenting microorganism by the substrate concentration which affects osmotic pressure[149] and by the synergistic effect of ethanol and substrate.[150,151] Substrate inhibition can be avoided by the stepwise addition of fermentable sugar; the inhibitory effect of ethanol on yeasts is a complex of different effects and therefore scarcely avoidable.[152]

Jimenez and Benitez[153] investigated selected wine yeasts for their ethanol and sugar tolerance and for their fermentative capacity. Growth and fermentation rates were increasingly inhibited by increasing ethanol and glucose concentrations, "flor" yeasts being the least inhibited. By adding glucose, stepwise ethanol production rate was accelerated except in flor yeasts. Breeding of wine yeasts belongs to the most rapidly developing field of biotechnology. The aim of the strain improvement program is to construct wine yeast strains which are able to produce good quality wine and are tolerant for high level sulfurous acid, sugar and ethanol and have a good sedimentation ability. Saddekni, Maráz and Deák[154] made wine yeast breeding by protoplast fusion. The aim of their work was to construct new yeast strains which have good fermentation ability at low as well as high temperatures, producing good quality wine in open-air containers, are tolerant for high concentration of alcohol and sugar and have good flocculation ability.

VIII. SCP PRODUCTION ON WHEY

Whey, the byproduct of cheese making, is high in biochemical demand and represents a serious problem for municipal waste treatment. Gross composition of dried whey is about 70% lactose, 9 to 14% crude protein and a comparatively high ash content of about 9%. Because of the low protein content, whey powder is not a high grade food, despite its good amino acid composition and vitamin content. The amount of whey derives from curds of various types of cheese and varies as follows (kg of whey per kg of cheese): hard cheeses, 11.3, soft cheeses, 7.5 to 9.5; cottage-type cheeses, 4.0 to 9.0; casein, 3.0.[155]

The volume of world-wide whey production is about 100 billion pounds annually. One third to one half of the whey produced in the U.S. is actually utilized. According to Vananuvat and Kinsella[156] 7 million pounds of native whey are utilized world-wide as human food. In some countries like the U.S. about one quarter of the whey is processed into lactose[157] and protein. Poor solubility, low sweetness, the laxative effects and the problem of human lactose intolerance limits its applicability. The largest amount of lactose is still probably used for penicillin production. Small quantities of whey are used for soft drinks and protein fractions. Considering the early development of feed yeast in Germany it is understandable that in this area whey too has been considered as a raw material. In Austria, yeast was produced from whey in 1940 by Messrs, Harmer KG in Spillern. On a larger scale it was produced in Linz.[105] Yeast yields were 35 and 30%, respectively, based on available lactose.

In the Waldhof process *Candida utilis* was used in batch culture and yields of 50% dry matter, based on available lactose, were obtained but in continuous cultivation the yield dropped significantly and an average of 35% was quoted.[155] In Czechoslovakia Tomisek and Gregr[158] adopted a semi-continuous process by which 150 m^3 of whey were fermented in three 20 m^3 fermentors per day. A mixed culture of *Candida utilis* and *Candida pseudotropicalis* was used and yields of 50 to 51% dry matter, based on lactose, were quoted. In Hungary Simek et al.[159] developed a technology for SCP production on whey. To improve cell growth and yield, molasses was added in small concentrations to supplement growth with growth factors. Up to the early 1960s yields of yeast and rate of production were not to the degree that might reasonably be expected. The high sugar content of whey, about 4.5% and a dilution rate of 0.25 h^{-1} would result in a productivity of 5.6 $gl^{-1}h^{-1}$ with an approximately equal oxygen demand.[155] However, up to the late 1960s commercial-scale fermentors for biomass production were not more efficient than an equivalent of 2 to 2.5 $gl^{-1}h^{-1}$. Fermentors developed in the meantime are able to overcome this shortage of oxygen supply and cell density of 15 gl^{-1} is achieved in 90 m^3 equipment in continuous cultivation.[160] More knowledge about respirative and fermentative metabolism of lactose utilizing yeast strains, strain selection and breeding resulted in efficient technologies.

A. STRAIN SELECTION

Several authors[161-167] confirmed that yeast strains belonging to *Kluyveromyces fragilis* and *Kluyveromyces lactis* resp. are most suitable for SCP production on whey. A screening program for industrial biomass production cannot be based merely on the ability to utilize lactose. To find out the best SCP producer strain-specific growth rate, protein and nucleic acid content and yield coefficient are the most important characteristics which have to be taken into consideration. High growth rate is not only important to keep the fermentor volume as small as possible, but also to have a high degree of certainty that they will not be overrun by other types of organisms.

Al Omar[33] investigated *Kluyveromyces fragilis, Torulopsis spherica* and *Candida utilis* for lactose utilization. *Torulopsis spherica* had best results at 30°C while *Kluyveromyces fragilis* had better yields at 35°C. The highest lactose utilization and yield of biomass was obtained by the combination of *Torulopsis spherica* and *Kluyveromyces fragilis.*

Yeast developed by Kroger[160] through hybridization and natural selection technique is able to also metabolize galactose. The Nutrisearch Company began producing the new baker's yeast at its Winchester, KY plant in 1983. The yeast is osmotolerant to both salt and sugar.

Mahmoud and Kosikowski[169] compared five *Kluyveromyces* strains grown in concentrated whey to produce single cell protein and ethanol. *Kluyveromyces fragilis NRRL Y 2415* produced the highest yield of alcohol (9.1% v/v) and *Kluyveromyces bulgaricus* gave highest yield of biomass 13.5 gl^{-1}.

Food grade yeast may be produced by harvesting the yeast grown on whey substrate. The protein quality of the finished product is good, although slightly low in sulfur-containing amino acids, as is true for most single cell proteins. By varying the conditions of the fermentation, increased amounts of ethanol can be produced and recovered but at the expense of cell yield. However, energy considerations could also effect the kind of SCP process. Aeration costs could be important hence an anaerobic fermentation process, in which both alcohol and SCP are produced, could be the process of choice. The cost of production is dependent primarily on two factors: the cost of whey and the capital investment. Efficiency demands a plant that is highly automated, instrument controlled and capable of having an annual capacity of at least 4000 to 10,000 ton of finished product. For such a production, raw material is required in large amounts and such a facility must be near a large cheese-producing area. Whey is a clean, wholesome, food-grade substance in excess supply and a potential environmental pollutant. By the processes developed it can be converted to a useful and needed high protein material.

TABLE 24
Effect of Whey Content Expressed as Lactose Concentration on Specific Growth Rate, Yield Coefficient, Protein and Nucleic Acid Content of *S. fragilis*

Lactose concentration (%)	Specific growth rate (K h^{-1})	Generation time (h)	Yield coefficient (%)	Protein content: Crude protein (%)	Protein content: True protein (%)	Nucleic acid content (%)
4.68	0.346	2.00	23.30	43.78	34.50	9.06
3.08	0.330	2.10	31.56	42.07	34.13	7.76
2.10	0.320	2.16	32.81	41.58	34.44	6.95
1.05	0.246	2.82	30.22	39.40	32.81	6.45
F test	h.s.	—	h.s.[a]	h.d.	h.s.[a]	h.s.[a]
L.S.D. ($p \geq 95\%$)	0.044	—	1.89	0.12	0.23	0.13

[a] h.s. = highly significant ($p \geq 99\%$).

From Nour-El Dien, H., Hálasz, A., and Lengyl, Z., *Acta Aliment.*, 10, 11, 1981. With permission.

B. THE EFFECT OF MEDIUM COMPOSITION

Lactose content of the permeate from the deproteinizing ultrafiltration unit is approximately 5%. Whey also has a relatively high crude protein content (0.8 to 1.0%) from which about 45 to 50% is protein nitrogen, 30 to 40% peptide nitrogen, 2 to 5% on free amino acids, 1 to 2% purine nitrogen and about 10% urea nitrogen.

1. The Effect of Whey (Lactose) Concentration

The effect of whey concentration has been investigated in the range of 1.0 to 4.7% for *Kluyveromyces fragilis* in batch culture.[166] The whey medium was prepared as follows:[116] spray-dried whey (Répcelaki Sajtüzem factory) was reconstituted, (the composition of this powdered whey was 65% lactose, 96% total solids and 10 to 15% protein) with distilled water to give a lactose concentration of approximately 4.7% (w/v). Reconstituted whey was autoclaved 15 min at 1.25 kg cm^{-2} pressure, and the precipitated proteins removed by centrifugation (5000 rpm). The clear supernatant was adjusted to pH 7.0 with concentrated KOH (56.1 gl^{-1}) and autoclaved again. The precipitate was removed by centrifugation. The medium was adjusted to pH 4.5 and autoclaved in sterilized flasks.

Inocula yeast were precultured in mineral medium with 4.5% lactose as sole carbon and energy source, in petri dishes for 24 h at 30°C then transferred to 150 cm^3 mineral medium in 500 cm^3 shaking flasks and cultured on a reciprocal shaker at 30°C for 20 h. After centrifugation, the sedimented cells were resuspended into mineral medium to obtain a cell density of 640 mgl^{-1} (optical density cc. 1.0 at 430 nm).

Batch cultivations carried out in a column fermentor in different lactose concentrations showed that the specific growth rate for *Saccharomyces fragilis* decreased with decreasing lactose concentration (Table 24). At the higher whey (lactose) concentrations growth did not enter the stationary phase after 10 h cultivation so calculated yield values for this concentration are not correct but show the trend that yield coefficient decreased from about 22 to circa 10%. Experiments concerning lactose content of whey medium carried out by reconstituting different amounts of dried whey in water and supplemented with 0.5% NH_2PO_4 and 0.5% $(NH_4)_2SO_4$ and 150 μgl^{-1} biotin in batch culture showed that the specific growth rate of *S. fragilis* decreased with decreasing lactose content.[166] These results show no significant difference in specific growth rate at lactose concentrations of 4.68, 3.08 and 2.10%. On the other hand, there is a highly significant difference between K values obtained at lactose concentration 1.05% and those at 2.10, 3.08 and 4.68%.

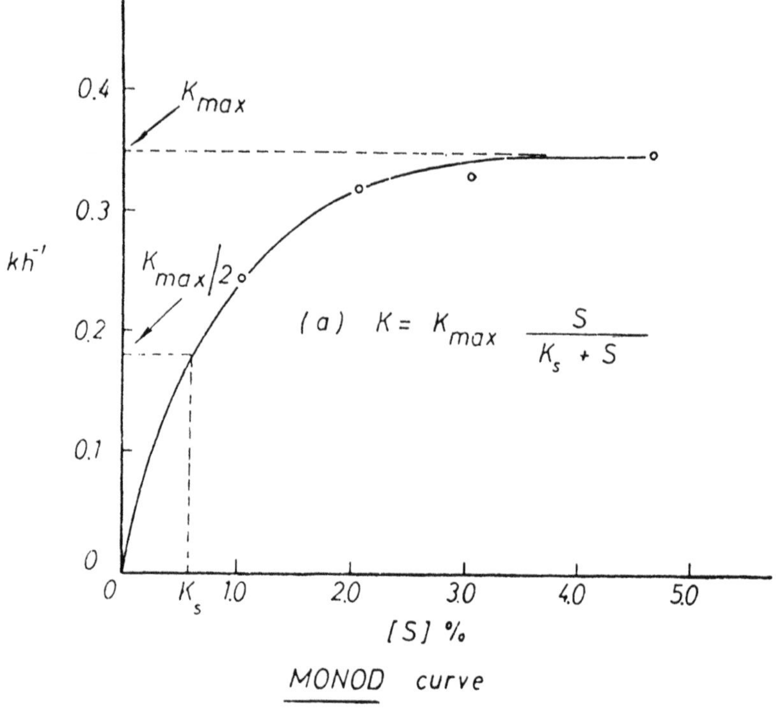

FIGURE 16. Relation between lactose concentration and specific growth rate according to Monod equation.

According to Monod's equation, the relation between lactose concentration (S) and specific growth rate (K):

$$K = K_{max} \frac{S}{K_s + S} \tag{25}$$

Equally by using Lineweaver and Burk transformation and equation

$$\frac{1}{K} = \frac{K_s}{K_{max}} \cdot \frac{1}{S} + \frac{1}{K_{max}} \tag{26}$$

These relations are shown in Figure 16 and 17.

It was observed, that the best lactose concentration was 2%. A lactose concentration higher than 2% is not necessary and has no significant effect on the specific growth rate, but values lower than this limit the growth rate. Comparison of yield coefficients however, show highest efficiency at 1.05% lactose and lowest at 4.68% lactose content but still, the best value is much lower in case of laboratory column fermentors. In stirred tank laboratory fermentors where oxygen absorption rate is much faster, yield coefficient for *Saccharomyces fragilis* on 2% whey substrate was 50%, which is in agreement with results of Bernstein et al.[168] Generation times calculated from specific growth rates and the number of doublings, within an 8 h fermentation period in the agitated fermentor are also similar to Bernstein et al.[168] These results are a further statement for the importance of oxygen supply to keep low alcohol production by the Pasteur effect.

Experiments of Nour-El Dien et al.[166] confirmed that supplementation of deproteinized whey medium with 0.5% each of $(NH_4)_2SO_4$ and K_2HPO_4 caused a slight increase in specific growth rate (0.32 or 0.35 h^{-1}) and there was some increase in net dry weight after 9 h

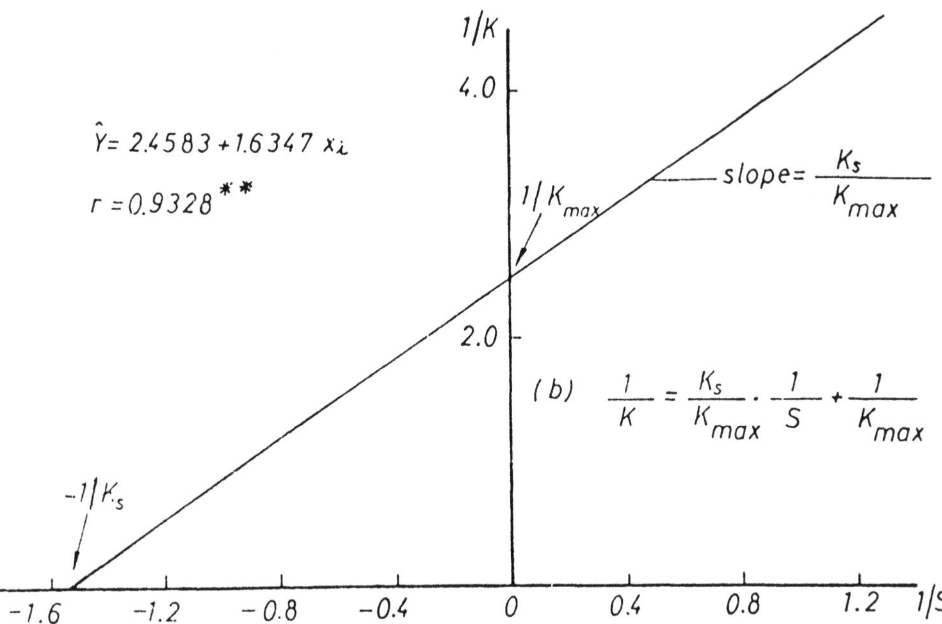

FIGURE 17. Relation between lactose concentration and specific growth rate (Lineweaver — Burk transformation).

cultivation (5.05 and 6.52 gl^{-1} resp.). In the technology developed by researchers of Kroger Company (U.S.) supplementation with mineral salts is not mentioned and published results on final cell density (15 gl^{-1}) on 5% whey medium could be caused either by low efficiency of aeration or low C:N ratio.

Mahmoud and Kosikowski[169] compared different nitrogen sources supplemented at 1% to a highly concentrated (30 to 32%) whey medium and found that $(NH_4)_2SO_4$ had no effect, urea caused a decrease in cell yield but peptone caused a supplementation improved cell concentration by circa 40%. Bernstein et al.[168] added anhydrous ammonia, phosphoric acid and also yeast extract to the 8% lactose-containing whey medium in batch conditions. The extremely high protein content of yeast biomass could be explained by the better N supply and bioactive value of yeast extract.

Several publications underline the importance of growth factors for SCP production on whey medium. Biotin has been reported as a vital growth factor[159,169-173] for lactose utilizing yeast. Experiments of Nour El-Dien and Halász[173] indicate no significant effect in specific growth rate, ribonucleic acid, protein content and yield at different concentrations of biotin or without the addition of biotin (Table 25). Supplementing biotin (150 μgl^{-1}) with 2% lactose mineral medium had no significant effect in comparison to the control value. But both specific growth rate and yield coefficient were significantly higher when *Saccharomyces fragilis* was grown on whey medium containing 2% lactose (Table 26). These findings could be explained by the high content of vital growth factor in whey.[174] This hypothesis was confirmed[173] in a fermentation experiment where the following growth media have been compared:

1. Mineral medium + 2% lactose fortified with 150 μgl^{-1} biotin
2. Medium 1 + 0.5% yeast extract (powder)
3. Whey medium containing 2% lactose
4. Whey medium containing 1% lactose + 1% pure lactose
5. Whey medium containing 0.5% lactose + 1.5% pure lactose
6. Whey medium containing 0.25% lactose + 1.75% pure lactose

TABLE 25
The Effect of Biotin Supplementation of Different Concentrations to Whey Medium on Specific Growth, Yield and Protein and Nucleic Acid Content of *S. fragilis*[165,173]

Medium	Specific growth rate (K h^{-1})	Generation time (h)	Yield coefficient (%)	Protein content		Nucleid acid content (%)
				Crude protein (%)	True protein (%)	
1	0.298	2.33	32.03	42.40	34.34	7.90
2	0.305	2.27	33.43	42.40	34.03	8.20
3	0.305	2.27	33.78	42.50	34.25	8.10
4	0.307	2.26	32.97	42.90	34.65	8.10
F test	n.s.[a]	—	n.s.[a]	n.s.[a]	n.s.[a]	n.s.[a]

Medium — (1) Whey medium 2% lactose + 0.5%$(NH_4)_2SO_4$ + 0.5% K_2HPO_2, (2) whey medium 2% lactose + 0.5%$(NH_4)_2SO_4$ + 0.5% K_2HPO_2 + 150 μg l^{-1}, (3) whey medium 2% lactose + 0.5%$(NH_4)_2SO_4$ + 0.5% K_2HPO_2 + 450 μg l^{-1} and (4) whey medium 2% lactose + 0.5%$(NH_4)_2SO_4$ + 0.5% K_2HPO_2 + 750 μg l^{-1}.

[a] n.s. — not significant.

TABLE 26
The Effect of Medium Composition on Specific Growth Rate, Yield, Protein and Nucleic Acid Content of *S. fragilis*[165,173]

Medium	Specific growth rate (K h^{-1})	Generation time (h)	Yield coefficient (%)	Protein content		Nucleic acid content (%)	PNC (%)
				Crude protein (%)	True protein (%)		
1	0.156	4.44	17.44	41.60	33.20	8.23	20.19
2.	0.160	4.33	20.63	41.65	33.23	8.25	20.23
3	0.303	2.29	31.77	42.20	34.13	7.90	19.12
4	0.308	2.25	32.30	42.40	34.38	7.85	18.91
F test	h.s.[a]	—	h.s.[a]	n.s.[b]	h.s.[a]	n.s.[b]	h.s.[a]
L.S.D. ($p \geq 95\%$	0.038	—	1.89	—	0.38	—	0.32

Medium — (1) Mineral medium 2% lactose, (2) mineral medium 2% lactose + 150 μg l^{-1} biotin, (3) whey medium 2% lactose + 0.5% $(NH_4)_2SO_4$ + 0.5% K_2HPO_4 and (4) whey medium 2% lactose + 0.5% $(NH_4)_2SO_4$ + 0.5% K_2HPO_4 + 150 μg l^{-1} biotin.

[a] h.s. = highly significant ($p \geq 99\%$).
[b] n.s. = not significant.

Media (3) to (6) were also supplemented with N + P at the same concentration as the mineral medium. Results summarized in Table 27 show that the specific growth rate of *Saccharomyces fragilis* decreases when decreasing the whey content (original lactose in whey medium 2, 1, 0.5 and 0.25% resp.) in spite of adding pure lactose to obtain 2% lactose in all media. It can also be seen, from results in Table 28, that the addition of yeast extract to mineral medium (0.5%) produced about a three-fold increase in specific growth rate (from 0.131 to 0.401). Wasserman et al.[175] reported that the addition of 0.1% yeast extract to whey produced a two-fold increase in specific growth rate. As it can be seen in Figure 18 and Table 29 best lactose utilization is also in the mineral medium containing 0.5% yeast

TABLE 27
Effect of Whey Content at a Constant Lactose Concentration (2%) in Whey Medium, on the Growth Rate, Yield Coefficient Protein and Nucleic Acid Content of *S. fragilis*[165,173]

Media	Specific growth rate (k h^{-1})	Generation time (h)	Yield coefficient (%)	Protein content: Crude protein (%)	Protein content: True protein (%)	Nucleid acid content (%)	PNC
1	0.160	4.33	21.12	41.60	33.84	7.60	18.66
2	0.303	2.28	33.53	41.70	34.04	7.50	18.35
3	0.277	2.50	30.88	41.53	34.38	7.00	17.22
4	0.230	3.01	30.57	41.43	35.00	6.30	15.52
5	0.196	3.55	27.00	40.77	34.84	5.80	14.53
F test	h.s.[a]	—	h.s.[a]	n.s.[b]	n.s.[b]	h.s.[a]	h.s.[a]
L.S.D. ($p \geq 95\%$	0.036	—	0.42	—	—	0.18	1.84

Note: Media — (1) Mineral medium 2% lactose, (2) whey medium 2% lactose, (3) whey medium of 1.0% lactose content + 1.0% added lactose, (4) whey medium of 0.5% lactose content + 1.5% added lactose and (5) whey medium of 0.25% lactose content + 1.75% added lactose.

[a] h.s. = highly significant ($p \geq 99\%$).
[b] n.s. = not significant.

extract which is in agreement with the findings of Castello and Sanchez.[45] By lowering the whey content in the growth medium, the lactose utilization was decreased, but even the 2% whey-lactose medium was not as effective as mineral medium supplemented with 0.5% yeast extract.

Yield coefficients calculated on biomass produced per lactose utilized however, were about the same for the 2% whey medium and for yeast extract supplemented medium despite the significantly higher yeast content (8.5 gl^{-1}) in the latter one. Explanation for this contradiction is the relatively low oxygen transfer of the column fermentor which results in a lower repression of alcohol production at higher cell concentration and thus about an equal yield coefficient despite faster lactose utilization and cell production. The biological value of synthesized biomass expressed in true protein value is significantly higher in the case of mineral medium supplemented with yeast extract, all the other crude protein and RNA values are comparable, especially the calculated true protein concentrations. The very specific effect of yeast extract was the aim of a study, experiments were carried out to determine the growth promoting effect of amino acids, vitamin B, bases of nucleic acids and components of yeast extracts.

The supplementation with different growth stimulating compounds resulted in different efficiency as shown in Table 28. Highest net dry weight, lactose utilization and yield coefficient gives whey medium + yeast extract. According to Table 28 and 29 yeast extract containing mineral medium gives higher lactose utilization and cell production after 8 h propagation than whey medium in shaken culture and better aerated cultures in a column fermentor. The low oxygen transfer in shaken cultures results in higher ethanol production so each medium yield coefficient is lower than in better aerated column fermentor cultivated cultures. The positive effect of vitamin B supplementation reported by several authors[45,174] is significant in comparison to mineral medium in growth rate, yield coefficient, lactose utilization and biomass production as well. However, vitamin B addition is not as effective as yeast extract or original growth factors in whey (whey contains substantial amounts of vitamins).[176] Although yeast extract powder is a rich source for several amino acids,[173] the

TABLE 28
Effect of Vital Growth Factors, Amino Acids, Vitamin B and Yeast Extract on the Growth Parameters of *S. fragilis* in Column Fermentor[165,166,173]

Medium	Specific growth rate K (h^{-1})	Generation time[b] (h)	Net dry weight after 8 h (mg ml^{-1})	Lactose utilized after 8 h (%)	Yield coefficient[a] (%)	Nucleic acid (%)	Protein content: Crude protein N × 6.25 (%)	Protein content: True protein $TN-N_{RNA}$ 6.25 (%)	Protein nitrogen coefficient (%)
1	0.131	5.29	2.17	4.53	17.44	8.70	45.62	36.74	19.45
2	0.134	5.17	2.15	51.06	17.92	8.60	44.80	36.02	19.60
3	0.146	4.75	2.32	55.32	17.85	8.60	46.77	37.99	18.85
4	0.401	1.73	8.59	97.87	36.87	9.70	48.85	38.94	20.20
5	0.303	2.29	5.97	99.03	29.12	7.00	44.50	37.35	16.01
6	0.450	1.54	9.23	99.03	45.02	8.80	49.67	40.68	18.11
F test	h.s.[c]	h.s.[c]	h.s.[c]	h.s.[c]	h.s.[c]	h.s.[c]	h.s.[c]	h.s.[c]	h.s.[c]
L.S.D. ($p \geq 95\%$)	0.021	0.180	0.631	1.78	1.34	0.14	0.99	.071	0.58

Note: Media — (1) Mineral medium 2% lactose, (2) mineral medium 2% lactose + amino acids, (3) mineral medium 2% lactose + vitamin B, (4) mineral medium 2% lactose + 0.5% yeast extract, (5) whey medium 2% lactose and (6) whey 2% lactose + 0.5% yeast extract.

[a] Yield coefficient = milligram cells produced per milligram lactose utilized, multiplied by 100.
[b] Generation time = ln 2/K (where K = specific growth rate).
[c] h.s. — highly significant.

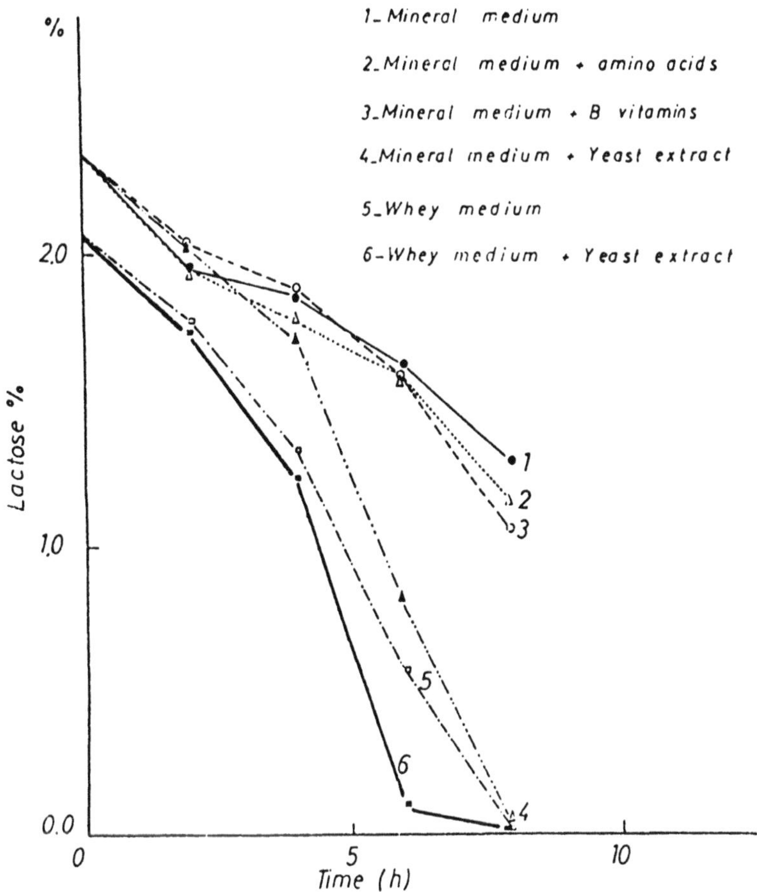

FIGURE 18. Lactose utilization by *Saccharomyces fragilis*.

unique effect of yeast extract supplementation might also be a result of the amino acids. However, investigations on the growth effect of amino acids utilized in amounts equivalent in yeast extract powder indicate that the growth in whey and whey supplemented with yeast extract was better than in amino acid fortified mineral medium.

Bases of nucleic acids such as adenine, guanine, cytosine and uracyl at equivalent amounts supplemented with yeast powder added separately and together did not show any effect on the growth of *Saccharomyces fragilis*. With disc diffusion technique a slight positive effect of alanine, cysteine and histidine was obtained.[173] Glattli and Blanc[177] reported that the addition of alanine is necessary for good yeast growth. The effect of the amino acid, vitamin B, yeast extract supplementation on biomass production in lactose-containing mineral medium, N + P complementation and yeast extract addition to whey medium in a column fermentor is shown in Table 32. From these results it is obvious that the addition of yeast extract to either mineral or whey media increases the specific growth rate (from 0.131 to 0.401 h^{-1} and 0.303 to 0.450 h^{-1} resp.). Supplementation of amino acids or vitamin B instead of yeast extract, resulted in the specific growth rate remaining very low. It is clear from the results that the growth of *Saccharomyces fragilis* showed a high yield when the mineral or the whey media were supplemented with yeast extract (36.87 and 45.02% resp.). However, addition of amino acids or vitamin B to the mineral medium has no effect on the yield coefficient as compared to the control. Shaken culture supplementation of vitamin B not only increased growth rate but also yielded coefficient and lactose utilization. In the better aerated column fermentor this difference could not be detected,[173] but a slight im-

TABLE 29
Effect of Vitamin B and Yeast Extract on the Growth of *S. fragilis* as Dry Weight and Residual Lactose in Shaken Cultures[165,166,173]

	Fermentation time (O h)		(20 h)						Protein content		
Medium	Net dry weight (mg ml^{-1})	Lactose content (mg ml^{-1})	Net dry weight (mg ml^{-1})	Lactose content (mg ml^{-1})	Net dry weight after 20 h (mg ml^{-1})	Lactose utilized after 20 h (%)	Yield[a] coefficent (%)	Nucleic acid (%)	Crude protein N × 6.25 (%)	True protein TN-N_{RNA} 6.25 (%)	PNC[b] (%)
1	0.48	19	1.050	11.50	0.57	39.47	7.6	7.7	45.8	37.94	16.81
2	0.48	19	2.250	6.00	1.77	68.42	13.62	7.9	45.7	37.63	17.29
3	0.48	19	3.690	0.00	3.21	100.00	16.90	9.8	49.2	39.69	19.92
4	0.48	19	3.180	0.00	3.33	100.00	17.52	7.0	44.5	17.38	15.73

Note: Media — (1) Mineral medium 2% lactose, (2) mineral medium 2% lactose + vitamin B 1 ml/l^{-1}, (3) mineral medium 2% lactose + 0.5% yeast extract powder (Difco) and (4) whey medium 2% lactose + N and P in the same percent in mineral medium.

[a] Yield coefficient milligram cell produced per milligram lactose utilized, multiplied by 100.

[b] $$\text{PNC (protein nitrogen coefficient)} = \frac{\text{NAN (nucleic acid nitrogen)}}{\text{total N}} \times 100.$$

provement of true protein content was seen. The addition of yeast extract also slightly increased the protein content of yeasts.

Mahmoud and Kosikowski[169] stated that supplementation of peptone resulted in the highest biomass yield with minimum alcohol production. This result is in agreement with the positive effect of yeast extract. As neither amino acids nor supplementing them together with RNA bases or vitamins has a comparable vital growth effect on lactose-utilizing yeast, it seems to be that some small peptides are responsible for the improvement of growth rate and yield coefficient.

IX. BIOMASS PRODUCTION ON DATE SYRUP

Dates are shown to be an interesting substrate for utilization in SCP production. Some of the required characteristics of yeasts to be grown on date syrup in an SCP production process include:[178]

1. Good nutritive value
2. High protein content
3. Ease of harvesting
4. Rapid growth
5. Efficient use of date sugar
6. Ability to grow at high temperature and low pH

Manufacturing costs of SCP are very dependent upon the cost of the carbon and energy source. Various estimates for the cost of hydrocarbon substrates, as a percentage of total manufacturing cost, range from 13 to 57% depending upon the process and the year of the estimate.[179] SCP production costs from methanol show a much greater dependency upon increases in substrate costs than in the case of *n*-paraffins.[180] For waste-type materials, raw material costs are a lower percentage of total costs than for hydrocarbons, methanol or ethanol. Moo-Young[181] cites a range of 17 to 26% of total costs for sulfite waste liquor and bagasse. Carbohydrates and wastes that contain carbohydrates are the lowest cost substrates for yeast SCP production. The future of yeast protein production for food and feed purposes either as yeast powder or as protein concentrates and isolates depends on the cost of raw material.

Carbon source and nitrogen concentrations for yeast growth should be adjusted to provide a C:N ratio in the range of 7:1 to 10:1 to favor high protein contents. At higher C:N ratios many yeasts, particularly those of the genus *Rhodotorula,* accumulate a substantial portion of the cell weight in the form of lipids. Anhydrous ammonia or ammonium salts are suitable nitrogen sources for yeasts of interest in SCP production. The prices of carbon sources occupy 25 to 44% of the total cost of SCP production. The cost and efficient use of carbon sources are crucial for economical production of SCP.[180,181] As mentioned, carbohydrates and wastes that contain carbohydrates are the cheapest substrates for yeast SCP production.

The reducing sugars of the date syrup comprised about 95.5% of total sugar content. The major sugars present are glucose 48.7%, fructose 45.2% and sucrose 6.1%. Cason et al.[182] found that fructose is utilized slower than glucose when the two sugars are fermented separately. Date syrup is shown to be an interesting substrate for *S. cerevisiae* and *C. utilis* for utilization in SCP production.[178,128,184]

A. EFFECT OF DATE SYRUP CONCENTRATION ON YEAST GROWTH

Kassim and Halász[183] determined growth curves for *S. cerevisiae, Rhodotorula glutinis* and *Candida guilliermondii* at different date syrup concentrations (0.1, 0.3, 0.5, 0.8 and 1%) medium was supplemented with 0.5% yeast extract. Figure 19 shows the Lineweaver-Burk plot of substrate and growth relationship for the investigated yeast strains. *S. cerevisiae*

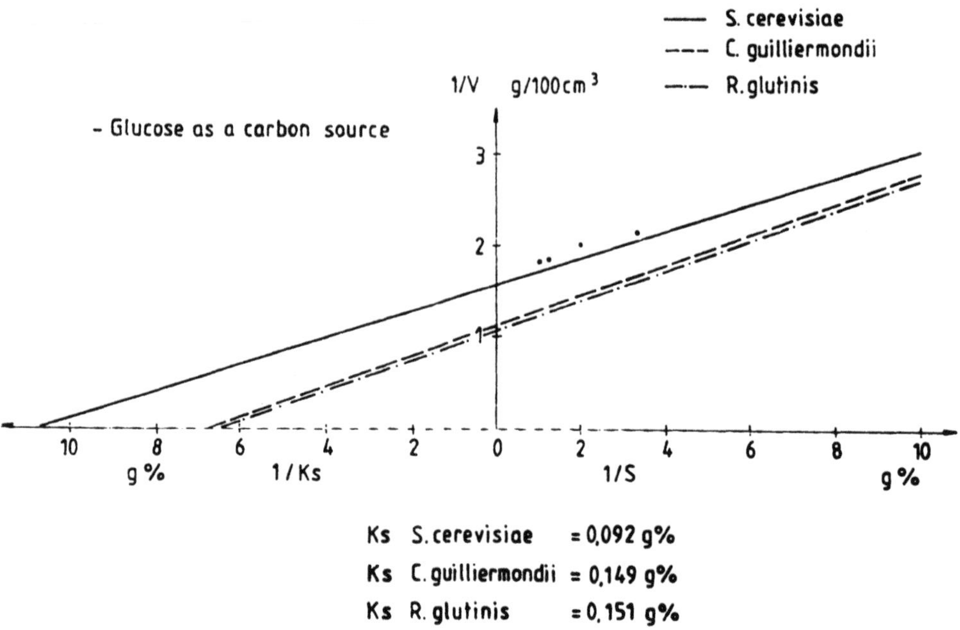

FIGURE 19. Growth rate and substrate concentration relationship of *S. cerevisiae*, *Rhodotorula glutinis* and *Candida guilliermondii* grown on date syrup.

has the lowest μ_{max} value and *R. glutinis* the highest. K_s values are in the range of 0.092 and 0.151. This means that even the lowest initial glucose content is over the saturation substrate concentration. However, the residual concentrations in each case are significantly below it, the discrepancy is greatest for *R. glutinis* and lowest for *S. cerevisiae*. The slope of the growth curves increased for each strain by increasing date syrup concentration Figure 20. For none of the investigated strains was a significant difference found in the growth curve between 0.5, 0.8 and 1% date syrup concentration at $p \leq 95\%$. The growth curve expressed in Δ OD time of *R. glutinis* and *C. guilliermondii* at each date syrup concentration was found to be better than those of *S. cerevisiae*.

B. EFFECT OF DATE SYRUP CONCENTRATION ON BIOMASS YIELD AND PROTEIN CONTENT

Kassim[73] investigated the effect of date syrup concentration in the range of 0.1 to 1.0% date syrup content medium was supplemented with 0.5% yeast extract. Fermentation was stopped after 4.5 h for *S. cerevisiae* and after 4 h in *C. guilliermondii* and *R. glutinis*. Figure 21 shows *S. cerevisiae* biomass produced at a different date syrup concentration. The biomass yield decreases with increasing substrate content. 0.1% date syrup level resulted in maximal yield 51.58 g yeast biomass/100 g substrate (dry material), while 1.0% date syrup level had a lower amount 27.92 g yeast biomass/100 g substrate (dry material). Yield values change significantly with date syrup concentration. Also in the case of *C. guilliermondii* the biomass yield gradually decreases by increasing date syrup concentration. *R. glutinis* had the highest biomass yield at 0.1% date syrup concentration compared to other date syrup concentrations. The differences in biomass yield between 0.1% date syrup concentration and the other investigated are significant. It is clear from the results that the biomass yield decreases with increasing date syrup concentration for the three investigated yeast strains. This may be explained by the Crabtree effect or catabolic repression due to excess of glucose (>0.05 g/l) and ethanol production.[184] Figure 22 shows the protein content of the investigated yeast strains propagated at different date syrup concentrations. The protein content of *S. cerevisiae*

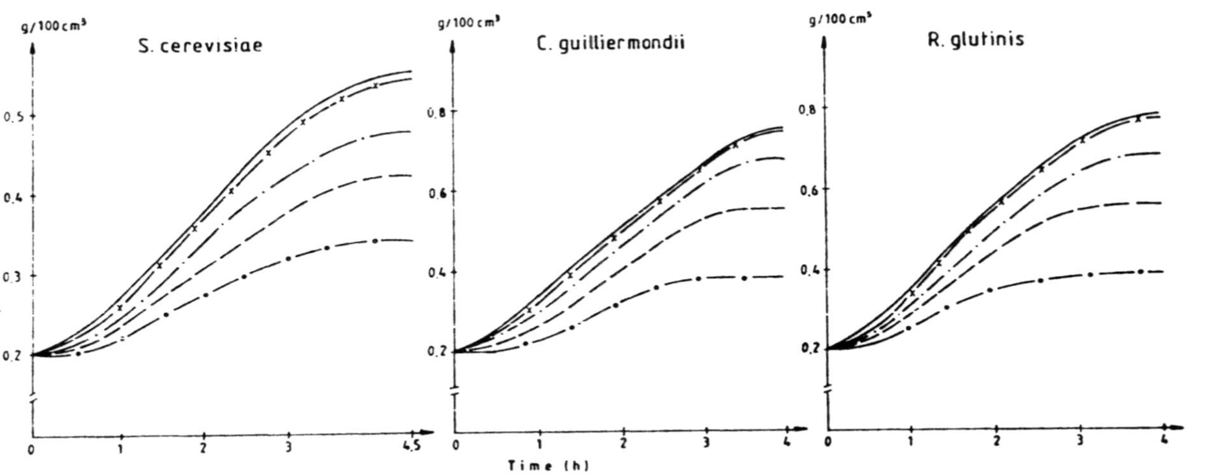

FIGURE 20. Growth curves of *S. cerevisiae, R. glutinis* and *Candida guilliermondii* grown on date syrup.

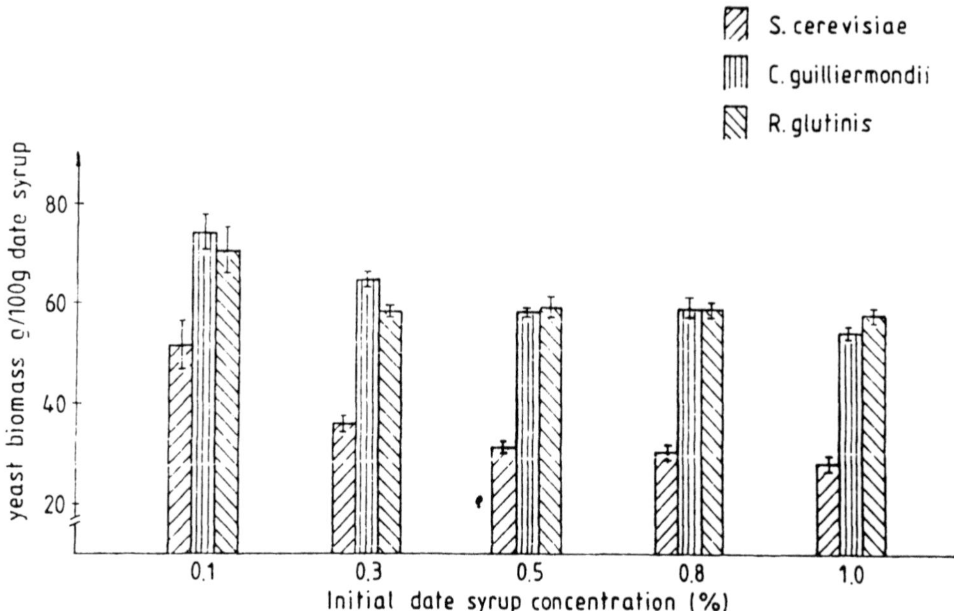

FIGURE 21. Effect of date syrup concentration on biomass yield of *S. cerevisiae*, *R. glutinis* and *C. guilliermondii*.

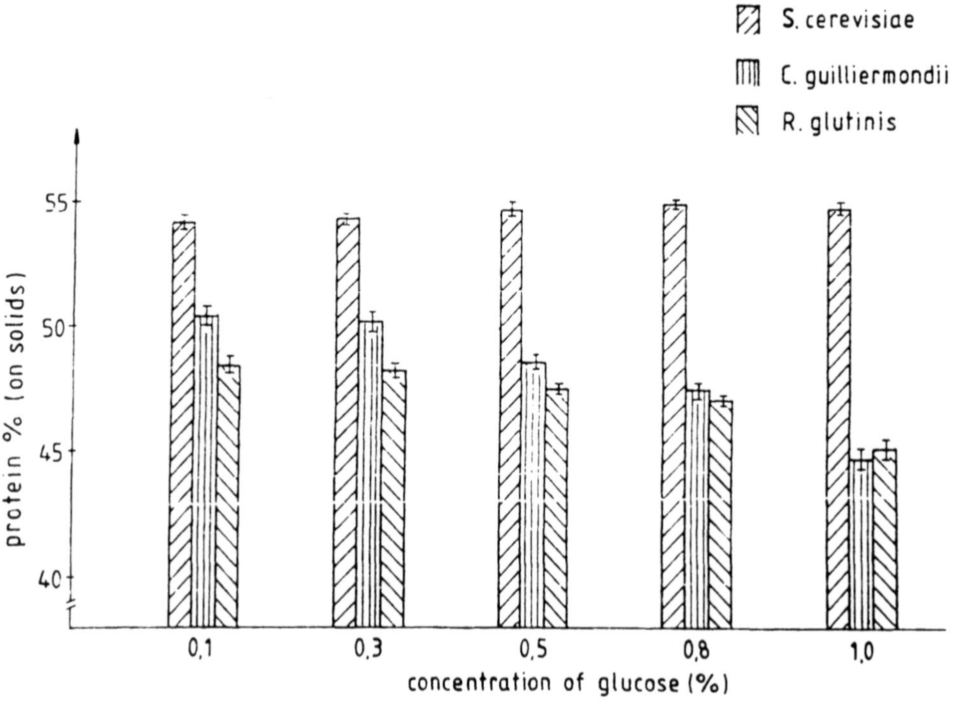

FIGURE 22. *Protein content* of *S. cerevisiae*, *R. glutinis* and *C. guilliermondii* propagated at different date syrup concentrations.

TABLE 30
Effect of Date Syrup Concentration on Total Synthesized Protein (Protein % × Yeast Concentration g/100 cm³) of Yeast Strains

	Protein % × yeast concentration g/100 cm³		
Concentration	***S. cerevisiae***	***C. guilliermondii***	***R. glutinis***
0.1	11.06 ± 1.16	19.2 ± 1.4	19.51 ± 1.21
0.3	22.80 ± 1.98	26.7 ± 1.58	27.59 ± 1.38
0.5	25.98 ± 0.98	32.39 ± 2.03	32.12 ± 1.26
0.8	29.58 ± 1.34	34.32 ± 1.38	34.3 ± 1.1
1.0	30.14 ± 0.89	34.42 ± 1.09	33.87 ± 1.49

From Mohammed, Kassim Mustafa and Hálasz, A., *Acta Aliment.*, 18(2), 177, 1989. With permission.

increases from 50 ± 0.28% to 54 ± 0.98% when the date syrup content was increased from 0.1 to 0.3%. Further date syrup addition had no significant effect on protein concentration. *C. guilliermondii* protein content shows no significant change with increasing carbon source content (Figure 22).

In *R. glutinis* a gradual decrease of protein content could be observed (Figure 22) by increasing date syrup content.

If we calculate the whole assimilated protein from cell concentration and protein content we can see that the amount of protein synthesized by the cells increases at higher date syrup concentrations for each strain. The total synthesized protein (protein % × yeast concentration) at each date syrup concentration is found to be higher for *R. glutinis* and *C. guilliermondii* than for *S. cerevisiae* (Table 30). Date syrup could be an economic carbon source in several developing countries. Its carbohydrate content is a good carbon source for yeasts. For economical reasons *Candida guilliermondii* and *Rhodotorula glutinis* are better for biomass production growth rate and protein production is higher than for *S. cerevisiae*.

X. BIOMASS PRODUCTION ON ETHANOL

Yeast for food uses are traditionally produced from carbohydrates including molasses, malt, sulfite liquor, whey and starch hydrolysates. Food grade ethanol, a noncarbohydrate, approved for the production of food yeasts as this carbon source has many merits. Most important advantages of ethanol are

1. It is accepted as a food ingredient
2. Soluble in water
3. Easily transported and stored
4. Can be obtained as a uniform, pure food grade ingredient both from renewable and nonrenewable resources
5. Has no residue or impurity problems
6. Supports the growth of food yeast *Candida utilis*

Ethanol is a very pure material and acceptable as a raw material for food manufacture. It can be easily stored and transported and is completely soluble in water. Ethanol is not markedly inhibitory to microorganisms as many species of yeast can use it readily as a source of carbon and energy. In another sense however, this can be looked upon as a disadvantage in that it poses a potential threat of contamination in the fermentation. Typical yields for ethanol-based biomass production are about 0.79 dry cell matter/g substrate. A

TABLE 31
Cell Growth on Ethanol (Wt %)

	Dilution rate h^{-1}	
O_2 limited	0.25	0.4
Cell yield (wt%)	83.8	82.8
Protein (wt%)	45.0	47.3
EtOH limited cell yield (wt%)	71.6	70.3
Protein (wt%)	46.8	45.7

further advantage is that ethanol is in a partially oxidized state so the fermentation requires much less oxygen than for highly reduced substrates. Its primary disadvantage is its relatively high cost. Despite this concern, ethanol has been receiving attention as a potential substrate for SCP production and is even used in practice for food yeast production. The material, tradenamed Torutein, is being sold in the U.S. as a nutritional supplement and flavor enhancer for such processed foods as meat patties, pasta, baked goods, frozen pizza and sauces. The Milani Foods division of Alberto Culver Corporation, for example, uses very small amounts in a low calorie salad dressing to "enliven the spices" and to reduce costs, calories and cholesterol by partly substituting this for expensive dried egg yolk.[210] From the technical and patent literature it is obvious that numerous organizations in various countries have been actively looking at ethanol as a fermentation substrate.

During the last three decades, many new aerobic fermentation processes for the production of food and feed ingredients were developed to the pilot and semi-works stages. Amoco's pure culture yeast process developed by Standard Oil of Indiana is one of the few to be commercialized, with the 1975 completion of the Hutchinson, MN plant of Amoco Foods Company and its successor, Pure Culture Products, Inc. both affiliates of Standard Oil of Indiana. Amoco's process is a continuous aerobic fermentation involving the growth of a selected *Candida utilis* strain on ethanol substrate with a cell yield of 80 to 83% on ethanol. The pathway of the growth of *Candida utilis* on ethanol was studied by Wattew et al.[211] The developed mathematical model can be used for controlling and optimizing the process. Yeast cultivation is commercially carried out by either batch or continuous processes. There are many advantages for operating the fermentation in the continuous mode, including ease of control, greater uniformity in product quality, reduced fermentor volume, higher productivity and generally reduced costs. However, contamination and culture instability are the drawbacks.

A. AMOCO'S PURE CULTURE PROCESS

In the continuous aerobic fermentation process yeast is produced in a vertical stirred tank. Agitation is provided by twin flat bladed mixers which provide mixing and high shear for gas contacting. During fermentation ethanol is maintained at a 220 ppm level in the broth. Dilution rate is in the range of 0.25 to 0.50 h^{-1} typically about 0.40 h^{-1}. The concentration of dissolved oxygen is maintained within the range from 0.1 to 0.3 ppm under oxygen-limiting conditions and may range as high as 1 to 4 ppm when operating under ethanol-limiting conditions. The yield of yeast cells based on ethanol substrate consumed, is generally within the range from 65 to 84%, the higher yields being achieved under oxygen limiting rather than ethanol-limiting conditions[212] (see Table 31). The initial slow growth of the yeast is superceded after a few hours by rapid exponential growth which is thereafter maintained in the fermentor by withdrawal of fermentation broth containing about 3 wt% cell concentration suspended in the liquor. Fermentation cost is influenced by pressure and oxygen utilization. At higher pressure the cost is relatively insensitive to pressure. In this range the compression costs increase due to the higher head pressure, while agitator costs decrease due to the lower mixing requirements to achieve the same oxygen transfer rate. The agitation costs to achieve the desired capacity become very large at low pressures due to the K_{La}. As the oxygen utilization is increased, compression costs decrease due to lower total compressor flow rates, while agitation costs increase. The food yeast obtained from the pure culture process is being used as a food ingredient for flavor, functional and/or nutritional purposes. Its high lysine and high threonine content make it an ideal supplement for cereal flours. It is a functional additive to improve baking quality in baked foods. Food application properties of the yeast can be modified by fermentation[213] and post-fermentation operations.[214]

B. LINDE AG PROCESS

The Linde process shows some technical novelities which result in good economy, plant availability, operatability and product purity. The main features are

1. Utilization of a very productive yeast strain; due to the utilization of yeast sterile operation of the fermentor is not required
2. Utilization of highly oxygen enriched or, preferably 95% pure oxygen instead of air for gasing the fermentors
3. Modular assembly
4. Thickening step in stage 1, which is simple and safe to operate

Due to the fact, that fermentation is operated with addition of sterile oxygen instead of air there is no danger of contamination from this side. Pure oxygen is toxic for most microorganisms and therefore sterile. Due to the high oxygen demand of the highly concentrated yeast broth (75 gl^{-1}) the dissolved oxygen is rapidly consumed and thereby reaches low concentrations which are not critical for the yeast cells, but cause very good growth conditions. Also, the protein content of the product increases substantially if oxygen is provided in optimal amounts. The oxygen transfer rate 26 g $O_2l^{-1}h^{-1}$ is high due to the high oxygen partial pressure in the gas phase. Total gas hold up is low thus reducing coalescence of gas bubbles. Less gas hold up means less foam generation. The data result in a very good productivity (greater than 20 up to 50 kg/m^3)[5] culture fluid per hour — is considerably higher than published figures for comparable technical processes. The dry yeast produced on ethanol has 60 to 65% crude protein, 8.5 to 9.5% RNA, 27 to 31% carbohydrate and 8 to 10% ash per dry mass.[215]

C. STRAIN BREEDING

Mitsubishi Petrochemical Company has been developing *Candida* strains which can grow at higher temperatures and at lower values of pH. These conditions would reduce contamination potential and the higher temperature would also reduce cooling requirements. Chistyakova et al.[216] investigated 57 yeast strains belonging to different genera *Candida, Saccharomyces, Kluyveromyces, Hansenula, Pichia, Saccharomycopsis, Lodderomyces* and *Debaromyces*. For the screening experiments synthetic medium containing 1.0% ethanol and 0.4% yeast autolyzate was used. Biomass yield at 37°C protein content amino acid composition and lipid composition was measured for the yeast strain able to grow at temperatures higher than 37°C. Six of the investigated *Candida* strains and two *Kluyveromyces* strains were able to grow even at 45°C and four further strains up to 40°C. Eight out of the twelve thermotolerant yeasts had highest biomass production at 37°C (in the investigated temperature range of 28 to 46°C). *Hansenula fabiani 728* gives twice the amount of cell concentration at 37°C than at 28°C.

Protein content varies between 30.8 and 43.9% (according to Lowry). RNA values are in the range of 3.3 up to 7.1%. The essential amino acid composition of the investigated strains also shows great differences even for lysine (from 6.8 to 10.4% on protein content).

The best methionine source proved to be *Candida 78* with a methionine concentration of 1.6% on protein. Lipid content shows values from 5.2 to 10.5% lipid composition is quite good as odd number carboxylic acids content is below 5% of the sum of carboxylic acids. Most importantly, essential carboxylic acid linoleic acid shows highest concentration in *Candida scotti 780* and *Candida utilis 405* (30 and 36% resp.).

XI. YEAST BIOMASS PRODUCTION ON OTHER SUBSTRATES

A. METHANOL

Because of its low cost and solubility in water as well as other advantages, methanol is an attractive raw material for single cell protein production. Although only a few types of microorganisms are able to utilize methanol for growth one can select both yeast and bacteria with this capability. In the seventies considerable attention was given to methanol as a carbon source for the production of single cell protein. Extensive investigations have been performed on the isolation of strains and metabolisms of methanol. Sahm and Wagner[217] isolated the methylotrophic yeast strain *Candida boidinii*. Fructose accelerates the adaptation of the yeast to methanol and a shortening of the lag phase was observed in fermentation. As methanol is toxic at higher concentrations, in batch cultivation methanol concentration should not exceed 0.5% (w/v). Reuss et al.[218] investigated the effect of methanol concentration on yeast growth in the range of 0.1 to 0.8% (w/v) methanol concentration and found that growth rate increased up to 0.5% methanol followed by a decline due to substrate inhibition. In continuous culture an effective control strategy is inevitable to prevent substrate inhibition.[218]

One of the disadvantages of methanol over more traditional fermentation substrates such as carbohydrates is an increased heat load from its utilization during fermentation. One approach to minimize this problem is to utilize thermotolerant or thermophilic microorganisms to permit the fermentation to be conducted at elevated temperature. Levine and Cooney[219] isolated *Hansenula polymorpha DL-1* a thermotolerant yeast with an optimal growth temperature range of 37 to 41°C when grown in a methanol mineral salts medium. Protein content of the yeast is about 47 to 50% depending on the growth conditions. The maximum cellular yield on methanol was 0.37 g/g methanol.

B. CARBOHYDRATE CONTAINING-WASTES AND WASTE WATERS

Shannon and Stevenson[220] investigated the growth of *Candida steatolytica, Candida utilis, Saccharomyces cerevisiae* and *Saccharomyces uvarum* on selected brewery wastes. Trub press liquor proved to be the best substrate for all the organisms. Dry cell weights ranged from 8.94 to 10.56 g/l^{-1}. *Candida steatolytica* produced the largest dry cell mass among the investigated yeasts. The protein content of biomass was highest when trub press liquor was the substrate. *Candida steatolytica* had the lowest protein content in grain press liquor and trub press liquor although it produced essentially the same quantity of total protein due to its greater yield. *Candida utilis* had best protein concentration on grain press liquor (27.11%) and *Saccharomyces cerevisiae* on trub press liquor (32.91%). Overall there appeared to be little difference in total protein production between the yeasts. Biomass produced on brewery wastes had consistently lower protein content than those grown under optimal conditions.

Continuous fermentation of the food and fodder yeast *Candida pseudotropicalis* on deproteinated potato waste water supplemented with glucose hydrol was investigated by Kramarz et al.[221] Nearly 90% of carbon, 80% of nitrogen and 67% phosphorus sources present in the substances were assimilated. The obtained biomass contains at least 62% of raw protein and not more than 7.35% of RNA. Yield of biomass in relation to reducing substances was 90.7% on potato waste water and 85.7% on potato waste water enriched with hydrol.

Sanchez-Marroquin[222] reported that *Saccharomyces carbajali, Candida parapilosis* and *Candida utilis* are most adequate to convert agave-juice into biomass. Agave-juice has a usual solid content of 2.0 to 5.0°Brix. Saponins and other impurities should be eliminated. Protein content of the biomass was 47.3 to 50.5% in agave-juice for the investigated strains. The mixed yeast culture of *Saccharomyces carbajali* and *Candida utilis* has a high digestibility (81 to 86%) and a high methionine content (1.9% on protein) and is rich in vitamin B. Nucleic acid content is 7% and can be reduced by heat treatment to 2%.

C. BIOMASS PRODUCTION ON POTATO STARCH

Starch is a renewable resource that is cheap (often a waste product), reliable and available year-round in surplus quantities. It is nontoxic, of predictable and constant quality and universally accepted for human and animal consumption. SCP production from starch may be indirectly achieved by most microbes after hydrolytic pretreatment of starch into its component saccharides by mineral acid or enzymes. Alternatively, associative cultures of a slow growing, amylase-producing organism and a fast biomass converter have been used.[223] A simpler and more direct strategy calls for an amylolytic organism that is itself an efficient biomass converter. A number of comparative surveys of amylolytic yeasts have been made.[224,225] Calleja et al.[223] reported the optimal conditions of growth of the amylolytic and fermentative yeast *Schwanniomyces alluvius* in a medium containing soluble starch as the sole carbon source. Fully aerated cultures of this strain grew on 4% soluble potato starch at a doubling time of 1.5 h at 30°C. A yield of 51% on starch was obtained similar to those on glucose. About 53% of the dry weight of the cell was made up of protein. Also *Saccharomyces castelii, Saccharomyces occidentalis* and *Saccharomyces persornii* proved to be efficient in starch conversion.

D. BIOMASS PRODUCTION ON OIL AND FAT

Processing of animals for human food produces large quantities of fats as waste or low grade products but there is also a fat overproduction caused by the changes in nutritional habits. Such fats are potentially useful as feedstock for growth of commercially valuable microorganisms. Successful laboratory experiments and scaling up at the pilot plant confirmed that animal fat as the sole carbon source gives a good growth rate, biomass yield and high cell concentration. Solid fats as a carbon source raise similar problems in fermentation technology like *n*-paraffins and could be overcome in a similar way. Best results were obtained with *Saccharomyces fragilis*.

Tan and Gill[227] reported biomass production on olive oil as a carbon source. *Saccharomyces lypolytica's* lipase acts preferentially on oleyl residues at the 1,3 positions of glycerides. It would seem that the metabolic consequences of triglyceride and alkane utilization are likely to be similar, but the substantially different physical systems are presented to microorganisms by liquid cultures with the two types of substrate as a carbon source. Unlike alkane substrates, dispersion of substrate and effective substrate-cell contact can be readily achieved with fatty oils, but use of low pH (<5.0) must be avoided with fatty substrates if growth rates are not to be adversely affected. Oxygen limitation results in protein content decrease. Cell composition during exponential growth was 42% protein and 2% fat. Yield coefficients for triglyceride were near unity.[227]

XII. SPECIAL ASPECTS OF PRODUCING METHIONINE-RICH YEAST STRAINS

As it is discussed in detail in Chapter 2 and Chapter 5 of this book yeast biomass is rich in protein and lysine. However, although the amino acid composition of different yeast strains varies with species and growing conditions it is valid that the methionine content is low for all yeast. As it is known, methionine is one of the most important essential amino acids not only as an important building stone of body proteins but also as an important

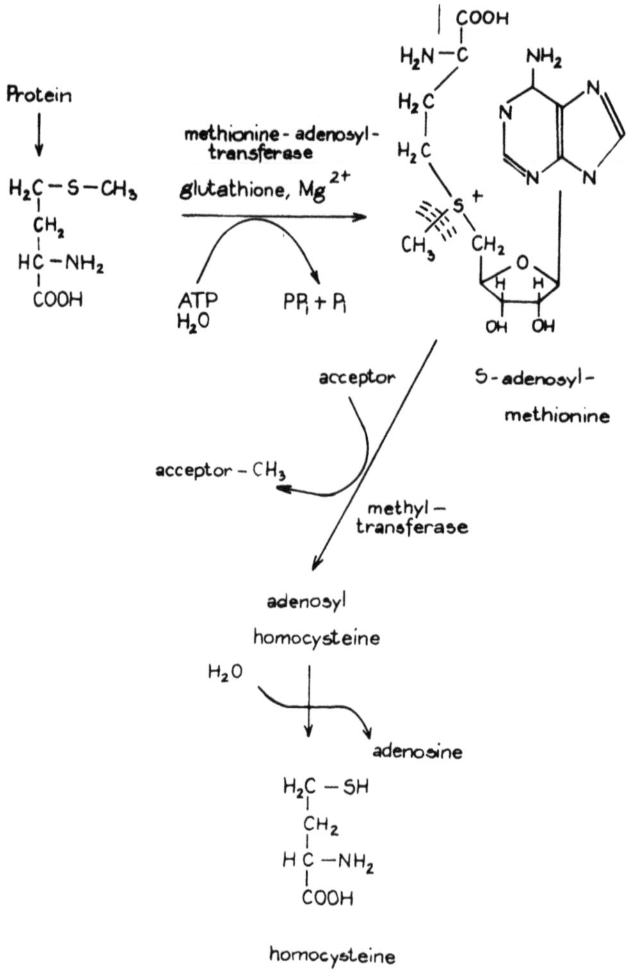

FIGURE 23. Methionine as methyl group donor.

methyl group donor[185] in the human and animal organism (see Figure 23). Methionine metabolism is coupled with the metabolism of other sulfur-containing amino acids (cysteine, cystine) and some important metabolites such as tauro-cholate playing an important role in digestion of lipids. The toxic effect of ingested ethanol can be significantly lowered by methionine.[186]

The daily methionine requirement of an adult is about 2.8 g, infants and children need much higher quantities, which are very often not fully covered in some countries where the protein supply is unsatisfactory. The deficient methionine and tryptophan supply contributes to the frequent incidence of various diseases (hepatic disease, anemia, etc.). The simplest way to eliminate the deficiency could be an enrichment of the food with synthetic methionine. Such supplementing of yeast proteins results in a protein source having nutritional value comparable to high quality animal (meat, milk, egg) proteins. However, according to the feeding trials, the free amino acid shows a more rapid resorption and of this, only 30 to 40% of the added methionine is utilized.[187,188] Thus, the production of a biomass with increased protein-bound methionine content constitutes a more effective means for supplementing methionine deficient food and feed. From a theoretical point of view three potential ways can be used for the production of yeast biomass with elevated methionine content:

1. Selection of mutants producing higher levels of methionine

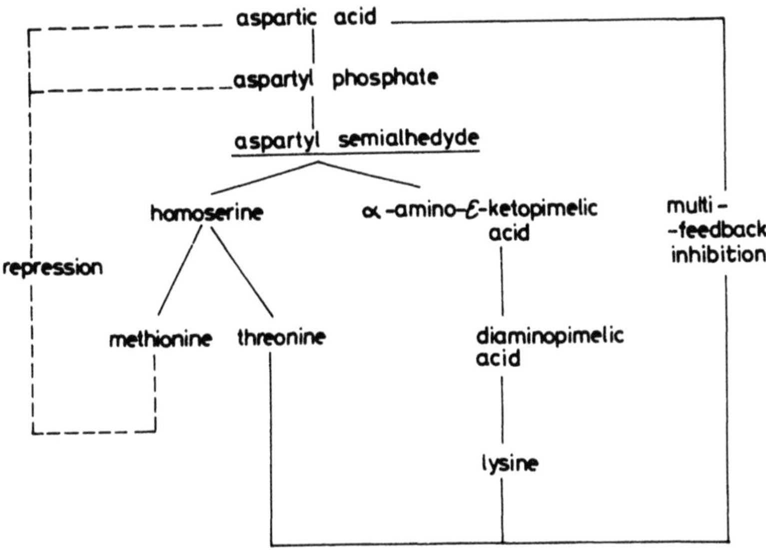

FIGURE 24. Ramified biosynthetic pathway of methionine, threonine and lysine formation.

2. Optimization of the conditions of fermentation
3. Genetic engineering of yeasts

In addition, a chemical modification of isolated yeast proteins using plastein reaction for incorporating new methionine molecules into the protein is also possible.

The intensive research work associated with the possibility of finding yeast strains producing higher amounts of methionine started in the early seventies. Komatsu et al.[189] successfully produced a mutant rich in methionine from *Candida petrophilum* ATCC 20226. The methionine content of the mutant was about 40% higher as compared to the strain. Okanashi and Gregory[86] produced, from *Candida tropicalis,* a mutant with a methionine concentration higher by 41%. The increase in methionine concentration was practically entirely the result of the rise of free methionine concentration. Komatsu et al.[189] reported that during active cell growth, in a shaken culture, the pool methionine concentration was nearly unchanged but decreased in the declining phase of proliferation. It is worth giving consideration to the proper selection of the parent strain for mutant production as the methionine content of yeasts varies between wide limits, 0.17 to 1.0%. It was reported the same for bacteria. Among yeasts, mainly the genus *Rhodotorula* was found to be rich in limiting amino acids.

The key enzyme of methionine synthesis is homoserine-*O*-trans-acylase. This is repeatedly de-repressed in feedback-resistant mutants. The enzyme is resistant against methionine, *S*-methyl-methionine, ethionine and norleucine and it is less sensitive to *S*-adenosyl-methionine than the wild strain.[190] In feedback-resistant yeast mutants, pool methionine concentration increases significantly and in numerous cases, the amount of methionine excreted into the surroundings increases as well. In strains defective to *S*-adenosyl-methionine synthesis, the enrichment of pool methionine can be observed too, the formed compound being the metabolite of methionine. The precursor of methionine formation is aspartic acid semi-aldehyde, similar to the biological synthesis of threonine, lysine and isoleucine (Figure 24). The first specific precursor of methionine is *O*-succinyl-homoserine formed through homoserine acylation catalyzed by homoserine-*O*-trans-succinase. Feedback inhibition is exerted on the enzyme by *S*-adenosyl-methionine and *S*-methyl-methionine. The mutants resistant to ethionine, *S*-methyl-methionine and norleucine are able to produce methionine in excess and are therefore, well utilizable for mutant selection, Gree et al.[191] found accu-

TABLE 32
Amino acid Content of *Candida tropicalis* ATCC 1369 and the Mutant Obtained by Treatment with Nitrous Acid

Amino acid	w/w% Parent	w/w% Mutant	Increase or decrease
Aspartic acid	3.24	3.44	+6.2
Threonine	1.90	2.07	+8.9
Serine	1.78	1.77	−0.6
Glutamic acid	5.12	4.43	−13.5
Proline	1.20	1.14	−5.0
Glycine	1.52	1.50	−1.3
Alanine	2.00	2.32	+16.0
Cystine	0.52	0.49	−5.8
Valine	1.64	1.89	+15.2
Methionine	0.42	0.49	+16.7
Isoleucine	1.44	1.66	+15.3
Leucine	2.19	2.38	+8.2
Tyrosine	1.19	1.24	+4.2
Phenylalanine	1.35	1.39	+3.0
Histidine	0.62	0.90	+45.2
Lysine	2.31	2.76	+19.5
Arginine	1.69	2.40	+42.0
Tryptophan	0.28	0.24	−14.3
Total	30.41	32.51	

From Hálasz, A., Baráth, Á. and Szalma-Pfeiffer, I., 5th Int. Symp. on Amino Acids, February 21, 1977, Budapest, Abstr. G13. With permission.

mulation of free methionine in the case of low *S*-adenosine-methionine synthetase level. The selection of mutants producing the desired product (methionine) in greater quantities is a well-proved method in strain improvement work. Another requirement toward the mutants to be considered in selection is that the cells should not excrete the excess amino acid into the culture medium. The work associated with the increase of methionine content of yeast biomass starts with the selection of existing strains on the basis of their methionine content. It is recommended to use only such strains in further research work which have methionine content higher than 0.6%.[192] Such strains were successfully used in the different experiments.[193,194]

The most common methods of using mutagens for selection of new mutants of yeast were shortly reviewed in the previous part of this chapter. In the newer experimental work with yeast strains generally the treatment with ultraviolet light (UV), ethionine, or nitrous acid are used. Recent experiments of Tarka and Fields[195] produced mutants of three yeast strains (*Candida tropicalis* ATCC 1369, *Candida tropicalis* ATTCC 9968 and *Candida utilis* ATCC 9950). All yeast have been used previously by others[196] in the production of food yeast and yeast proteins. UV light and nitrous acid mutants (1369/F and 1369/O) had the highest increase in methionine of the mutants tested. In terms of an increase in methionine, the mutant produced by nitrous acid had the highest (71% increase). Since growth occurred in media without methionine added, the cells were able to synthesize methionine as well as to accumulate it from the growth medium. There was no difference in carbon assimilation tests between the parents and mutants.

All of the essential amino acids (for an adult and for growth except tryptophan) increased as shown in Table 32. The greatest increases came in histidine followed by arginine, lysine, methionine, isoleucine and valine. Lesser increases occurred in threonine, leucine and phe-

TABLE 33
Methionine, Protein and Nucleic Acid Content of Four UV-Treated *Candida guilliermondii* 812 strains Compared to the Parent Strain

Duration of UV treatment (s)	Methionine content (w/w% of dry mass)	Total protein content (w/w% of dry mass)	Nucleic acid content (w/w% of dry mass)
Parent	0.70 ± 0.07	59.30 ± 3.0	10.8 ± 1.0
120	0.83 ± 0.08	41.20 ± 2.1	6.2 ± 0.6
180	0.80 ± 0.08	52.00 ± 2.5	5.6 ± 0.6
240	0.85 ± 0.09	43.00 ± 2.1	6.4 ± 0.6
300	0.88 ± 0.09	39.00 ± 2.0	7.3 ± 0.7

From Halász, A., Biochemical Principles of the Use of Yeast Biomass in Food Production, D.Sc. thesis, Hungarian Academy of Sciences, Budapest, 1988.

nylalanine. There was no increase in the amount of methionine excreted. All methionine was retained in the yeast cell as determined by the microbiological method used. Ethionine had no significant effect on the methionine production. Muayad et al.[197] studied the efficiency of different mutagenic treatments in producing methionine-rich mutants from yeasts. Four industrially important yeast strains (*Candida guilliermondii, Saccharomyces lactis, Kluyveromyces fragilis* and *Kluyveromyces marxianus*) were treated with UV light, gamma radiation and nitrous oxide, aiming at producing methionine-rich mutants.

From the 2080 UV-treated strains tested, 14 auxotrophic mutants were isolated. From these only one Arg^- and one Lys^- mutant were successfully maintained, the rest of the mutant strains back-mutated after 1 to 3 weeks. From the total number of colonies tested on sulfur-deficient media, 25 colonies showed bigger sizes on the sulfur-rich media. They were checked for their methionine content and only four of them were found to have a methionine content higher than that of the parent strain (Table 33).

Kluyveromyces fragilis 732 and 1068 and *Kluyveromyces marxianus* 857 were also treated with UV light. A slight increase in the methionine content of *Kluyveromyces marxianus 857* was impaired by a 30% increase in its nucleic acid content. *Kluyveromyces fragilis 1068* showed an 80% increase in its methionine content compared to the parent strains, the nucleic acid content was not changed. No increase in the methionine content of *Kluyveromyces fragilis 732* was observed. No methionine-rich mutants were obtained by gamma irradiation. Treatment with nitrous acid was also ineffective under conditions of the experiments.

Halász et al.,[198] in a recent study, reported about experiments with different yeast strains (Table 34) using mild mutagenic agents such as UV radiation and nitrite treatment to improve their methionine content. The production of yeast mutants was tried with two kinds of mutagenic agents; with nitrite and UV irradiation, respectively. Genetic manipulation was performed according to Bowien and Schlegel[199] applying nitrite concentration and treating time established by us. Following mutagenic treatment, a 72 h incubation in a shaken culture was applied for the development of mutation and the enrichment of mutants. For the separation of methionine-rich mutants, the colonies spread on malt culture medium were transferred to sulfate-deficient and sulfate-rich synthetic glucose culture media by replica plating technique. The colonies, being of greater diameter on sulfate-rich culture medium than in sulfate-deficient conditions, were inoculated onto sulfate-rich agar plants. Comparisons were drawn between the methionine content of these mutants in relation to the parent strain. Mutants were selected using an amino acid antagonist. Norleucine was added to culture medium and the beginning and extent of growth were studied in a shaken culture, in comparison to the control without norleucine. For the selection of the mutant, the inhibiting concentration determined for the parent strain was used as an indicator. Keeping in mind

TABLE 34
RNA Protein- and Methionine Content of Yeasts Treated With Mutagenic Agents

Name of the yeast	RNA content	Protein content	Methionine content
		(% of dry matter)	
Rhodotorula glutinis CBS 315	—	47.7	0.8
Candida guilliermondii CBS 5256	8.0	46.5	0.5
Candida utilis CBS 5609	6.0	41.1	0.65
Saccharomyces carlsbergensis	7.0	47.1	0.5

From Halász, A., D.Sc. thesis, Hungarian Academy of Sciences, Budapest, 1988.

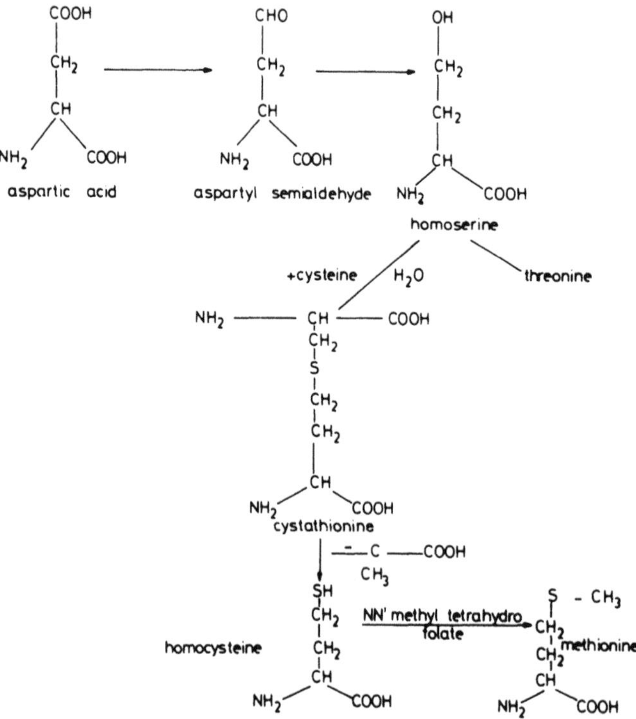

FIGURE 25. Role of methyl donor in methionine biosynthesis.

that another special requirement of methionine synthesis is — beside sulfate — the methyl donor (Figure 25), for selection *S*-methyl-methionine sulfonium chloride (vitamin U) was also applied.

The methionine-, protein- and nucleic acid content of mutants of *Rhodotorula glutinis CBS 315* selected after treatment with nitrous acid are shown in Table 35. It seems only one mutant has higher methionine content than the parent strain, but the increase is moderate. UV treatment also resulted in only one mutant being rich in methionine (1.0%). Protein-, RNA- and methionine content of NO_2-treated *Saccharomyces carlsbergensis* mutants is

TABLE 35
Methionine and Protein Concentrations of the Mutants Obtained by NO_2 Treatment from *Rhodotorula glutinis* CBS 315

Sign of sample	Methionine	Protein
	% related to solids	
	$\bar{x}_1$[a]	$\bar{x}_2$[b]
CBS 315	0.80	44.7
315/1	0.65	—
315/2	0.50	—
315/3	0.75	42.7
315/4	0.50	40.3
315/5	0.90	38.8
315/6	0.80	41.1

[a] $\bar{x}_1$ — Mean values of three replicates; standard deviation of the method: ±10%.
[b] $\bar{x}_2$ — Mean values of three replicates; standard deviation of the method: ±5%.

From Halász, A., Biochemical Principles of the Use of Yeast Biomass in Food Production, D.Sc. thesis, Hungarian Academy of Sciences, 1988.

TABLE 36
Methionine, Protein and RNA contents of *S. carlsbergensis* Mutants Obtained by NO_2 Treatment

Sign of sample	Methionine	Protein	RNA
	% related to solids		
	$\bar{x}_1$[a]	$\bar{x}_2$[b]	$\bar{x}_3$[c]
S. carlsbergensis	0.5	47.1	7.0
S_{34}	0.3	32.7	4.6
S_{34}	0.4	31.8	4.0
SC_{1021}	0.25	32.2	5.8
SC_{1022}	0.25	31.7	4.0

[a] $\bar{x}_1$ — Mean values of three replicates; standard deviation of the method ±10%.
[b] $\bar{x}_2$ — Mean values of three replicates; standard deviation of the method ±5%.
[c] $\bar{x}_3$ — Mean values of three replicates; standard deviation of the method ±5%.

From Halász, A., Biochemical Principles of the Use of Yeast Biomass in Food Production, D.Sc. thesis, Hungarian Academy of Sciences, Budapest, 1988.

shown in Table 36. All the mutants showed lower methionine content than the parent strains. The treatment with UV light was more effective. As it can be seen from Table 37, four of the selected mutants had a higher methionine content than the parent strain. The best effects of the treatment with $NaNO_2$ and UV light were observed in experiments with *Candida guilliermondii* (5256 and 812) and *Candida utilis* strains (Tables 38, 39 and 40).

TABLE 37
Methionine, Protein and RNA Contents of UV of *S. carlsbergensis*

Sign of sample	Methionine	Protein	RNA
	% related to solids		
	$\bar{x}_1$[a]	$\bar{x}_2$[b]	$\bar{x}_3$[c]
S. carlsbergensis	0.5	47.1	7.0
SC_1	0.65	38.6	7.0
S_2	0.32	42.4	6.2
S_3	0.50	52.8	9.8
S_4	0.80	54.8	11.6
S_5	0.50	49.2	9.7
S_6	0.68	33.6	4.2
S_7	0.25	36.0	5.8
S_8	0.48	54.0	8.0
S_9	0.43	46.0	8.6
S_{10}	0.75	62.2	7.6

[a] $\bar{x}_1$ — Mean values of three replicates; standard deviation of the method: ±10%.
[b] $\bar{x}_2$ — Mean values of three replicates; standard deviation of the method: ±5%.
[c] $\bar{x}_3$ — Mean values of three replicates; standard deviation of the method: ±5%.

From Halász, A., Biochemical Principles of the Use of Yeast Biomass in Food Production, D.Sc. thesis, Hungarian Academy of Sciences, Budapest, 1988.

TABLE 38
Methionine, Protein and RNA Contents of NO_2 Mutants of *Candida utilis* CBS 5609

Sign of sample	Methionine	Protein	RNA
	% related to solids		
	$\bar{x}_1$[a]	$\bar{x}_2$[b];	$\bar{x}_3$[c]
C. utilis CBS 5609	0.65	41.1	6.0
FT_3	0.77	42.6	7.0
FT_4	0.75	42.4	7.0
FT_5	0.38	31.6	4.0
FT_6	0.40	43.8	6.0
FT_7	0.50	43.8	6.0
FT_8	0.50	39.1	6.1
FT_9	0.55	33.8	5.0

[a] $\bar{x}_1$ — Mean values of three replicates; standard deviation of the method: ±10%.
[b] $\bar{x}_2$ — Mean values of three replicates; standard deviation of the method: ±5%.
[c] $\bar{x}_3$ — Mean values of three replicates; standard deviation of the method: ±5%.

From Halász, A., Biochemical Principles of the Use of Yeast Biomass in Food Production, D.Sc. thesis, Hungarian Academy of Sciences, Budapest, 1988.

TABLE 39
Methionine, Protein and RNA Contents of the Mutants Obtained by NO_2-Treatment from *Candida guilliermondii* CBS 5256

Sign of sample	Methionine	Protein	RNA
	% related to solids		
	$\bar{x}_1$[a]	$\bar{x}_2$[b]	$\bar{x}_3$[c]
C. guilliermondii CBS 5256	0.50	47.2	8.0
1 × mutated 1	0.65	47.2	13.4
1 × mutated 2	0.70	44.5	7.6
2 × mutatated from B_1	0.80	45.7	10.0
2 × mutated from B_2	0.90	47.5	9.9
2 × mutated from C_4	0.75	50.4	9.6
2 × mutated from D_2	0.65	42.1	9.6

[a] $\bar{x}_1$ — Mean values of three replicates; standard deviation of the method: ±10%.

[b] $\bar{x}_2$ — Mean values of three replicates; standard deviation of the method: ±5%.

[c] $\bar{x}_3$ — Mean values of three replicates; standard deviation of the method: ±5%.

From Halász, A., Biochemical Principles of the Use of Yeast Biomass in Food Production, D.Sc. thesis, Hungarian Academy of Sciences, Budapest, 1988.

TABLE 40
Methionine, Protein and RNA Contents of UV-Mutants of *Candida guilliermondii* CBS 812

Sign of sample	Methionine	Protein	RNA
	% related to solids		
	$\bar{x}_1$[a]	$\bar{x}_2$[b]	$\bar{x}_3$[c]
Candida guilliermondi CBS 812	0.50	44.3	10.8
120 s	0.83	41.2	6.8
180 s	0.80	52.0	5.6
240 s	0.85	43.0	6.3
300 s	0.88	39.0	7.3

[a] $\bar{x}_1$ — Mean values of three replicates; standard deviation of the method: ±10%.

[b] $\bar{x}_2$ — Mean values of three replicates; standard deviation of the method: ±5%.

[c] $\bar{x}_3$ — Mean values of three replicates; standard deviation of the method: ±5%.

From Halász, A., Biochemical Principles of the Use of Yeast Biomass in Food Production, D.Sc. thesis, Hungarian Academy of Sciences, Budapest, 1988.

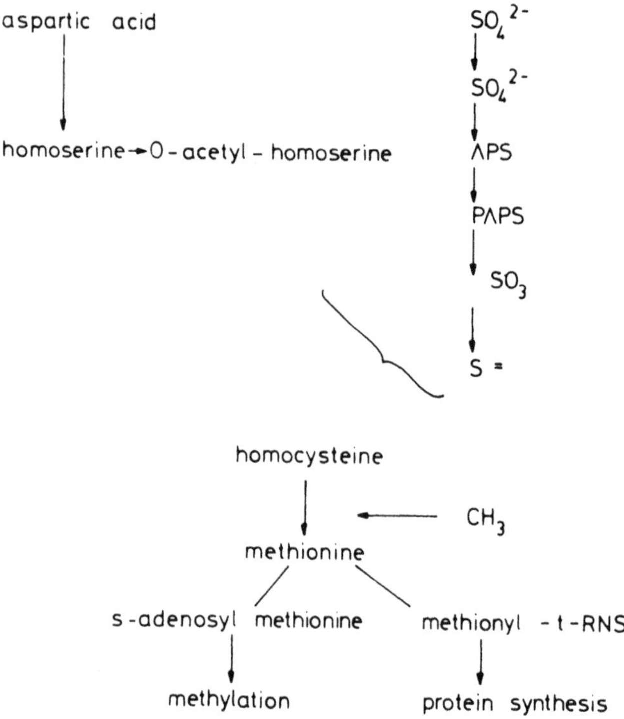

FIGURE 26. Sulfur source side of methionine biosynthesis.

As a first step, increased sulfate requirement was utilized for the separation of methionine-rich mutants from the cells of the parent strain and from the mutants changed in other directions, respectively. The sulfur required for methionine synthesis is taken up by the yeast cells from the sulfate content of the culture medium and utilized, reduced in several steps to sulfite, then to sulfide for homocysteine synthesis (Figure 26). Methionine is then formed by methylization of homocysteins. Taking the increased sulfur utilization as a base, the growth of colonies streaked after mutation was investigated in sulfate-deficient and sulfate-rich culture media. This selection method proved to be successful. Nevertheless, the authors[198] were not satisfied by the efficiency of the selection method, as after methionine determination only about 25% of the assumedly favorable cells proved to be really methionine-rich mutants, and some showing negative changes for methionine were detected as well.

To improve the mutant separation work, two solutions were studied: utilization of a methionine antagonist and adding a methionine homolog. According to the literature it is a general experience that the methionine-rich mutant shows a better tolerance towards the antimetabolite and this phenomenon is often used in selection work.[200] From the methionine analogs, ethionine was used the most frequently as this proved to be suitable in the selection of methionine-rich mutants and in the stable maintenance.[201,202] Lawrence[190] observed a resistance against norleucine in methionine-rich mutants. Shaken colonies were prepared for the assays and growth was measured after 24 and 48 h. Norleucine was added in concentrations of 0.001, 0.002, . . . 0.01%, in a total of 10 concentrations. Based on the absorbance and pH changes of the colonies, no inhibiting effect was observed in the starting of growth, nor in its degree. The inhibiting norleucine concentration was investigated in the range of 0.01 to 1% for the strains CBS 5256 and CBS 315. Significant inhibition was observed only with 1% norleucine. The starting of growth was delayed for *Rhodotorula glutinis* CBS 315, while for *Candida guilliermondii* CBS 5256 the cell density was only one tenth of those formed at 0.01 to 0.05%. Norleucine sensitivity was compared for *Candida guilliermondii* strain CBS 5256 and mutant B_1. Based on values related to control growth, of the untreated cells a definitely slower growth could be observed for the parent strain and also for the mutant in the presence of 1% norleucine. The mutant showed a higher sensitivity at this concentration than the parent strain and its inhibition was observed already at 0.5% norleucine.

Another special requirement of methionine synthesis besides sulfate is the methyl donor, therefore it can be assumed that the requirement of the methionine-rich mutant for this would also be higher than that of the initial strain. *S*-methyl-methionine sulfonium chloride (vitamin U) is capable of playing the role of methyl donor and to replace the methionine in this respect.[202] In experiments aimed at the improvement of selection, Halász et al.[198] added vitamin U to the sulfate-rich culture medium in concentrations of 10 to 60 $\mu g/cm^3$ and the sizes of the colonies developed were compared to those produced on plain culture medium. With the increase in added vitamin U concentration, the colonies gradually increased. This could be observed for the original strains (CBS 2556 and CBS 315) as well as for the corresponding mutants (B_1 and CBS 315 UV_1). However, the stimulating on mutants was more expressed.

Sulfate utilization of *S. carlsbergensis* was investigated by Halász et al.[198] $Na^{35}SO_4$ was used in experiments and the quantitative distribution of sulfur-containing amino acids formed from the sulfate was determined from the radio chromatogram of the hydrolysates obtained after hydrolysis of 2, 5, 10, 15 and 20 h. As it can be seen from Figure 27, each hydrolysate showed three radioactive peaks, one of them for methionine and another for cysteine. The size of the nonidentified third peak decreased with increasing duration of hydrolysis. At the same time, the methionine content increased. The sum of activity percentages represented by the two peaks was constant.

The nonidentified peak was assumed to be methionine-sulfoxide. This was in agreement with the changes with hydrolysis time in the amount of the substance to be identified and with its transformation to methionine, further with the R_f value of methionine-sulfoxide.

Methionine-utilization of *S. carlsbergensis* was also investigated. The culture medium used for the trials contained, as a sole sulfur source, S-labeled methionine in a quantity similar to the sulfate content of the sulfate culture medium. Based on the combined activity of nutrient liquid and washing solution, and on the labeled sulfur detectable in yeast, only 45% was incorporated from the methionine present in the culture medium. This incorporation ratio was lower than that found with the sulfate medium even when taking into consideration that the cell growth was less in the methionine-containing medium. According to the radiochromatogram of the yeast hydrolysate, 45% of the uptake activity was present in the form of methionine in yeast. This proves that the proportion of sulfur containing-amino acids of *S. carlsbergensis* does not depend on whether the sulfur source was sulfate or methionine.

The effect of oxygen absorption rate on the methionine content of some methionine-rich mutants was also studied.[198,203] The results clearly showed that the aeration rate had a great effect on the specific growth rate of yeast strains examined. It is also obvious that not only the amount of air passed through the media is important, but the oxygen absorption rate, vigorousity of mixing (which determines the air bubble surface area available) and the volume and depth of the fermentor; therefore different fermentors showed different oxygen absorption rates at similar aeration levels.

The importance of the oxygen absorption rate was clear on the specific growth rate when a high rate of aeration (1250 dm^3h^{-1} dm^{-3}) was used, but still lower specific growth rates were obtained compared to lower aeration levels of 60 $dm^3h^{-1}dm^{-3}$. This is easily explained by knowing the oxygen absorption rate. The protein content was found to be stable at different levels of aeration except in one case, in which the parent strain *Candida guilliermondii* 812 was cultivated in the "Biofer" fermentor. Only in this case, a higher aeration rate produced a biomass of constantly lower protein content compared to that at lower levels. Different fermentors showed different protein content even when the oxygen absorption rates

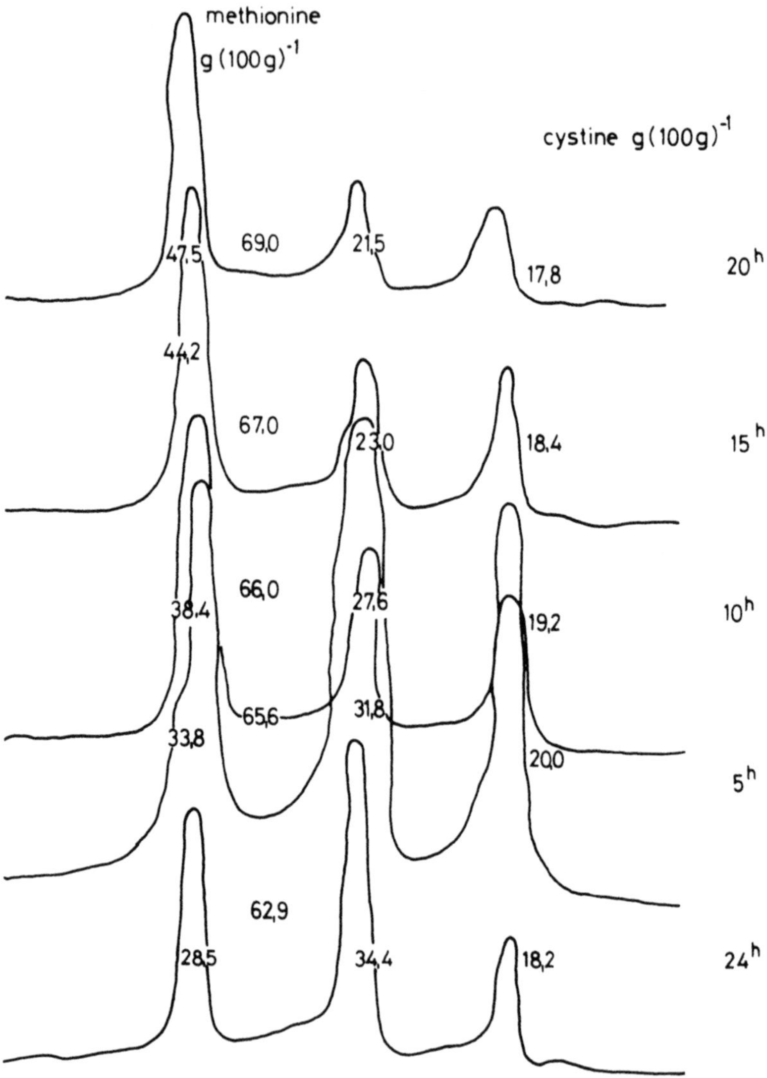

FIGURE 27. Radiogram of the hydrolysis products of *S. carlsbergensis* grown on media containing $Na_2S^{35}O_4$.

were similar in both cases. Methionine content was found to be affected by the aeration rate and the oxygen absorption rate. Different fermentors showed different results. It could be suggested that up to certain limits, aeration does not affect the methionine content, when this limit is exceeded, methionine contents start dropping. This could add to the conclusion that low aeration levels with high oxygen transfer rates would result in a yeast biomass with its highest methionine content (Figure 28).

In conclusion, it can be stated that the methionine-rich mutants produced with gentle mutagenic agents as nitrite and UV irradiation have an increased sulfate and methyl donor requirement. They show a higher sensitivity toward norleucine than the parent strain, thus the increased methionine concentration cannot be attributed to feedback resistance to the de-repression of homoserine-*O*-transsuccinase. The increased lipoic acid content of the mutants and the sensitivity toward the oxygen supply of the media result in lower methionine and lipoic acid concentrations at higher O_2 transfer rates. This indicates that for the mutants produced the sulfate reduction step became more effective and this results in an increased

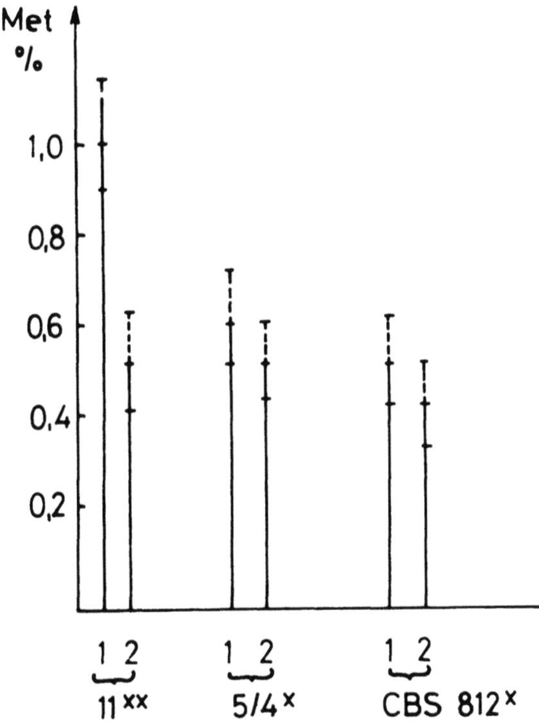

1, 2 = Oxygen transfer rate = 90, 200 mmol $O_2 l^{-1} h^{-1}$

xx significant difference $P \leq 1\%$
x significant difference $P \leq 5\%$

FIGURE 28. Influence of oxygen transfer rate on the methionine content of *Candida guilliermondii 812* and mutants 11 and 5/4 (Propagation in laboratory fermenter CHEMAP).

methionine content. At the same time it is noteworthy that, in contrast to the mutants produced with powerful mutagens, in our assays the increase of methionine concentration occurred not in the higher concentration of free amino acids but in methionine in peptide-linkage. Both from a biochemical and practical point of view the work connected with the sulfate transport in yeasts may be interesting.[204-208] Sulfate transport is the first step in the biosynthesis of methionine and cysteine. It was stated[208] that sulfate, sulfite and thiosulfate incorporation in the yeast *Candida utilis* is inhibited by extracellular sulfate, sulfite and thiosulfate and by sulfate analogs selenate, chromate and molybdate. The three processes are blocked if sulfate, sulfite, thiosulfate, cysteine and homocysteine are allowed to accumulate endogenously. Incorporation of the three inorganic sulfur oxyanions is inactivated by heat at the same rate. Mutants previously shown to be defective in sulfate incorporation are also affected in sulfite and thiosulfite uptake. Revertants of these mutants selected by plating in ethionine — supplemented minimal medium recovered the capacity to incorporate sulfate, sulfite and thiosulfate. These results demonstrate the existence of a common sulfate, sulfite and thiosulfate incorporating system in this yeast.

Protoplast fusion technique was used in production of methionine-rich mutants from yeast by Halász[198] and Halász et al.[196] The freeze-dried enzyme of *Helix pomatia* was used

TABLE 41
Main Characteristics of the Mutants and Fusion Hybrids of *Kluyveromyces fragilis*

Strain	Methionine content % (dry matter basis)	Protein content % (dry matter basis)	RNA content % (dry matter basis)	Methionine content % (protein basis)
Mutanta of *Kluyveromyces fragilis*				
1223 VM 37 *a* His	0.52	41.0	5.0	1.25
1224 Cap *a* His	0.55	45.1	6.8	1.2
1225 K9 alpha Met	0.46	45.4	6.2	1.06
1226 K9-01 alpha Met	0.60	48.0	7.1	1.23
		Fusion hybrids		
1226 + 1223 4	0.61	49.6	6.5	1.23
11	0.97	53.8	6.8	1.56
12	0.95	54.7	6.7	1.76
15	0.67	55.6	12.2	1.25
1226 + 1224 2	0.61	53.2	6.2	1.14

From Halász, A., Biochemical Principles of the Use of Yeast Biomass in Food Production, D.Sc. thesis, Hungarian Academy of Sciences, Budapest, 1988.

to digest the cell walls.[209] The digestion of cell walls, the formation of protoplast and the cell aggregation were followed by microscopic examination. Protoplast fusion between auxotrophic markers of *Saccharomyces lactis* 571 and 290 resulted in two fusion hybrids with a methionine content 70 to 75% higher than that of the parent strains, but their protein content dropped by 24 to 30%. Biotin content was nearly doubled in one of the fusions (Table 44). In the protoplast fusion experiments with *Kluyveromyces lactis* auxotrophic markers three mutants were used: 1223 His^-, 1224 His^- and 1226 Met^-. The fusion of 1223 His^- + 1226 Met^- resulted in the selection of four colonies numbered 4, 11, 12 and 15 which showed a methionine content higher than that of the parents. The protein content of all selected fusion hybrids was higher than that of the parents but they kept a nucleic acid content close to that of their parents (Table 41). The amino acid pattern determination of four fusion hybrids showed that the lysine content increased in three of them. Histidine increased very significantly in all fusions with a noticeable decrease in leucine and isoleucine. In the case of hybrid No. 11, the value of leucine aval isoleucine dropped below the level recommended by FAO/WHO, but still the methionine content was well above the FAO/WHO reference value (Figure 29).

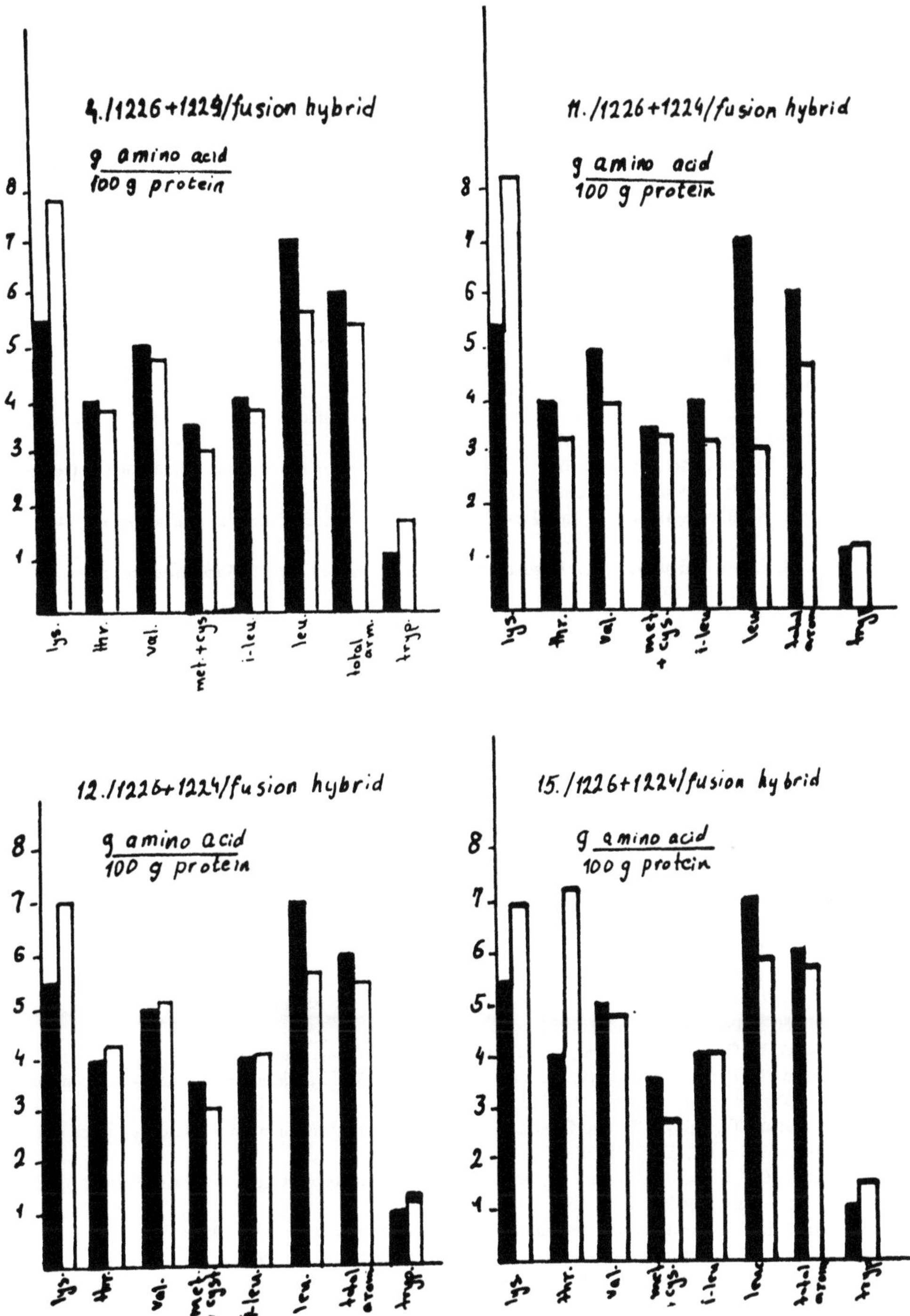

FIGURE 29. Comparison of the amino acid content of fusion hybrids (□) with the FAO/WHO reference values (■).

REFERENCES

1. **Dziedak, J. D.,** Yeast and yeast derivatives: definitions, characteristics and processing, *Food Technol.*, 41, 104, 1987.
2. **Miller, M. W.,** Yeasts in *Prescott & Dunn's Industrial Microbiology,* 4th ed., Reed, R., Ed., AVI Publishing, Westport, CT, 1982, 15.
3. **Mitchison, J. M.,** Changing perspectives in the cell cycle, in *The Cell Cycle,* John, P. C. L., Ed., Cambridge University Press, New York, 1981, 1.
4. **Carter, B. L. A.,** The control of cell division in *Saccharamoyces cerevisiae,* in *The Cell Cycle,* John, P. L. C., Ed., Cambridge University Press, New York, 1981, 565.
5. **Slater, M. L., Sharrow, S. O., and Cart, J. J.,** Cell cycle of *Saccharomyces cerevisiae* in populations growing at different rates, *Proc. Natl. Acad. Sci. U.S.A.*, 74, 3850, 1977.
6. **Meyenburg, H. K.,** The budding cycle of *Saccharomyces cerevisiae, Pathol. Microbiol.*, 31, 117, 1968.
7. **Hartwell, L. H. and Unger, M. W.,** Unequal division in *Saccharomyces cerevisiae* and its implications for the control of cell division, *J. Cell. Biol.*, 75, 422, 1977.
8. **Adams, J.,** The interrelationship of cell growth and division in haploid and diploid cells of *Saccharomyces cerevisiae, Exp. Cell Res.*, 106, 267, 1977.
9. **Carter, B. L. A. and Jagadish, M. N.,** The relationship between cell size and cell division in the yeast, *Saccharomyces cerevisiae, Exp. Cell Res.*, 112, 15, 1978.
10. **Trivedi, N., Jacobson, G. K., and Tesch, W.,** Baker's yeast, *CRC Crit. Rev. Biotechnol.*, 24(1) 75, 1986.
11. **Reid, B. J. and Hartwell, L. A.,** Regulation of mating in the cell cycle of *Saccharomyces cerevisiae, J. Cell Biol.*, 75, 355, 1977.
12. **Ratledge, C.,** Biochemistry of growth and metabolism, in *Basic Biotechnology,* Bulock, J. and Kristiansen, B., Eds., Academic Press, London, 1987, 11.
13. **Reed, G. and Peppler, H. J.,** Biochemical aspects of yeast, in *Yeast Technology,* AVI Publishing, Westport, CT, 1973, 33.
14. **Széchenyiné Márton, É., Simek, F., and Kárpáti, Gy.,** Production de levures sur matieres grasses d'origine animale et vegetale pour l'alimentation humaine, *Int. Aliment. Agric.*, 93, 415, 1976.
15. **Daeken, R. H.,** The Crabtree effect: a regulatory system in yeast, *J. Gen. Microbiol.*, 44, 149, 1966.
16. **Fiechter, A., Fuhrmann, G. F., and Käppeli, O.,** Regulation of glucose metabolism in growing yeast cells, *Adv. Microbiol. Physiol*, 22, 123, 1981.
17. **Sonnleitner, B. and Käpbeli, O.,** Growth of *Saccharomyces cerevisiae* controlled by its limited respiratory capacity: formulation and verification of a hypothesis, *Biotechnol. Bioeng.*, 28, 927, 1986.
18. **Berry, D. R. and Brown, C.,** Physiology of yeast growth, in *Yeast Biotechnology,* Berry, D. R., Russell, I. P., and Stewart, C. G., Eds., Allen, & Unwin, London, 1981.
19. **Rieger, M.,** The respiratory capacity of *Saccharomyces cerevisiae,* in *Current Developments in Yeast Research,* Stewart, C. G. and Russell, I., Eds., Pergamon Press, New York, 1981, 369.
20. **Käppeli, O. and Sonnleitner, B.,** Regulation of sugar metabolism in *Saccharomyces* type yeast, *CRC Crit. Rev. Microbiol.*, 4(3), 299, 1986.
21. **Schlegel, H. G.,** *Allg. Mikrobiol.*, 3.Aufl., Georg Thieme Verlag, Stuttgart, 1974.
22. **Minkievich, I. G., Krunistkaya, A. Yo., and Eroshin, V. K.,** Bistat-a novel method of continuous cultivation, *Biotechnol. Bioeng.*, 33, 1157, 1989.
23. **Takamatsu, T., Shioya, S., and Okada, Y.,** Profile control scheme in a baker's yeast fed-batch culture, *Biotechnol. Bioeng.*, 27, 1675, 1985.
24. **Sinclair, C. G.,** Microbial process kinetics, in *Basic Biotechnology,* Bulock, J. and Kristiansen, B., Eds., Academic Press, London, 1987, 75.
25. **Monod, J.,** The growth of bacterial cultures, *Ann. Rev. Microbiol.*, 3, 371, 1959.
26. **Clifton, C. E.,** *Introduction to Bacterial Physiology,* McGrowstin, H., New York, 1957, 414.
27. **Lamanna, L. and Maletta, M. F.,** *Basic Bacteriology,* Williams & Wilkins, Baltimore, MD, 1953, 853.
28. **Perret, C. J.,** A new kinetic model of a growing bacterial population, *J. Gen. Microbiol.*, 22, 589, 1960.
29. **Kono, T.,** Kinetics of microbial cell growth, *Biotechnol. Bioeng.*, 10, 105, 1968.
30. **Kono, T. and Asai, T.,** Kinetics of continuous cultivation, *Biotechnol. Bioeng.*, 11, 19, 1969.
31. **Rhodes, A. and Fletcher, D. L.,** *Principles of Industrial Microbiology,* Pergamon Press, London, 1966, 282.
32. **Mateles, R. I.,** Application of continuous culture, in *Single Cell Protein I.*, Mateles, R. I. and Tannenbaum, R. S., Eds., MIT Press, Cambridge, 1968, 208.
33. **Alroy, Y. and Tannenbaum, S. R.,** The influence of environmental conditions on the macromolecular composition of *C. utilis., Biotechnol. Bioeng.*, 15, 239, 1973.
34. **Dostalek, M. and Molin, N.,** Studies of biomass production of methanol oxidizing bacteria, in *Single Cell Protein II.*, Tannenbaum, R. and Wang, I. C., Eds., MIT Press, Cambridge, 1975.

35. **Vananuvat, P. and Kinsella, J. E.,** Production of yeast protein from crude lactose by *Saccharomyces fragilis*. Batch culture studies, *J. Food Sci.*, 40, 336, 1975.
36. **Imrei, F. and Vlitose, A.,** Production of fungal protein from carob, in *Single Cell Protein II.*, Tannenbaum R. and Wang, I. C., Eds., MIT Press, Cambridge, 1975, 223.
37. **Elsworth, R., Telling, R. C., and East, D. N.,** The investment value of continuous culture, *J. Appl. Bacteriol.*, 22, 138, 1959.
38. **Zetelaki, K.,** Kinetic constants of growth for use in the design of continuous culture, *4th FEMS Symp.*, Vienna, March 28, Abstr. B7, 1977.
39. **Bechtle, R. M. and Clayton, T. J.,** Accelerated fermentation system of cheese whey, developing the system, *J. Dairy Sci.*, 54, 1965, 1971.
40. **Jeffreys, G. A.,** Two-stage whey fermentation ideal for small .plant wastes, *Food Ing. INTIL*, October, 34, 1978.
41. **Al-Omar, M.,** Utilization of Whey Nutrients for Yeast Production and Pollution Control, Ph.D. thesis, Louisiana State University, 1977.
42. **Michalski, H. J. and Kopec, L. T.,** Multistream continuous ethanol fermentation on molasses and sugar media in system, *4th FEMS Symp.*, Vienna, 03.28, B9, 1977.
43. **Knight, S., Smith, W., and Mickie, J. B.,** Cheese whey disposal using *Saccharomyces fragilis yeast* cultures, *Dairy Prod. J.*, 1, 17, 1972.
44. **Nour-El Dien, H.,** Cheese Whey as Substrate for Single-Cell Protein Production (Yeast Biomass), Ph.D. thesis, Hungarian Academy of Sciences, Budapest, 1980.
45. **Castillo, F. J. and Sanchez, S. B.,** Studies on the growth of *Kluyveromyces fragilis* in whey for production of yeast protein, *Acta Cient. Venezolana*, 29, 113, 1978.
46. **Snyder, H. E.,** Microbial sources of protein, *Adv. Food Res.*, 18, 85, 1970.
47. **Miskiewicz, T.,** Control of substrate concentration in baker's yeast process with dissolved oxygen as nutrient feed indicator, *J. Ferment. Technol.*, 59, 411, 1981.
48. **Miller, S. R.,** Nutritional factors in single cell protein, in *Single Cell Protein, II*, Mateles, R. I. and Tannenbaum, R. S., Eds., MIT Press, Cambridge, 1968.
49. **Muayad Abdulrahman, K.,** Production of Methionine Rich Biomass for Human Consumption, Ph.D. diss., Hungarian Academy of Sciences, Budapest, 1983.
50. **Myzuchenko, L. A., Kantere, V. M., and Gurkin, V. A.,** On the influence of environmental factors on the kinetics of a biosynthesis process, *Pure Appl. Chem (IUPAC)*, 36, 3, 1973.
51. **Andreasen, A. A. and Steier, T. J. B.,** Anaerobic nutrition of *Saccharomyces cerevisiae*. (II). Unsaturated fatty acid requirement in a defined medium, *J. Cell Comp. Physiol.*, 43, 271, 1954.
52. **Lie, S., Haukli, A. D., and Jacobsen, T.,** Oxygen and the fermentation cycle of brewers yeast, Text C (1985), Brewing Industry Research Laboratory, 1985.
53. **Harding, S. and Kirshop, B. H.,** Relative significance of oxygen and other nutrients as fermentation regulations in *Saccharomyces cerevisiae*, *J. Inst. Brew.*, 85(3), 171, 1979.
54. **Babij, T., Moss, F. J., and Ralph, B. J.,** Effects of oxygen and glucose levels on lipid composition of yeast *Candida utilis* grown in continuous culture, *Biotechnol. Bioeng.*, 11, 593, 1969.
55. **Jakobsen, M. and Throne, R. S.,** Oxygen requirements of brewing strains of *Saccharomyces uvarum* (carlsbergensis) — bottom fermentation yeast, *J. Inst. Brew.*, 86, 284, 1980.
56. **Kishore, P. V. and Karanth, N. G.,** Critical influence of dissolved oxygen on glycerol synthesis by an osmophilic yeast *Pichia farinosa*, *Process Biochem.*, 21(5), 160, 1986.
57. **Oura, E.,** Aeration and the biochemical composition of baker's yeast, *Antonie von Leuwenhoek 35 Supplement*, Yeast Symp., G25-6, 25, 1969.
58. **Kono, N. and Terui, G.,** Analysis of the behavior of some industrial microbes toward oxygen. Effect of oxygen concentration upon the growth and hydrolase producing activity of *Bacillus amylosolvens*, *J. Ferment. Technol.*, (Japan) 39, 375, 1961.
59. **Brighton, W. D.,** The dissolved oxygen and respiration rates in pellicle submerged cultures of *Corynebacterium diphteriae*, *J. Appl. Bacteriol.*, 29, 197, 1966.
60. **Paca, J. and Gregr, V.,** Physiological effect of PO_2 on *Candida utilis* in multistage tower fermentor, *4th FEMS Symp., Vienna*, March 28, Abstr. B25, 1977.
61. **Chance, B., Jamieson, D., and Coles, H.,** Energy-linked pyridine nucleotide reduction: inhibitory effects of hyperbaric oxygen *in vitro* and *in vivo*, *Nature, (London)*, 206, 257, 1965.
62. **Yagi, H. and Yoshida, F.,** Oxygen absorption in fermentors. Effect of surfactants, antifoaming agents and sterilized cells, *J. Ferment. Technol.*, 52, 905, 1974.
63. **Vananuvat, E. and Kinsella, J.,** Extraction of protein low in nucleic acid from *Saccharomyces fragilis* grown continuously on crude lactose, *J. Agric. Food Chem.*, 23, 216, 1975.
64. **Akashi, K., Shibai, H., and Hirose, Y.,** Inhibitory effects of carbon dioxide and oxygen in amino acid fermentation, *J. Ferment. Technol.*, 57(4), 317, 1979.
65. **Halász, A. and Szalma-Pfeiffer, I.,** Changes in proteinase A and carboxypeptidase Y activities of *Saccharomyces cerevisiae* at different aeration intensities and glucose concentrations, 16th FEBS Meeting, Moscow, June 25, Abstr. XXII-101, 1984.

66. **Kirshop, B. H.,** Development in beer fermentation, in *Topics in Enzyme and Fermentation Technology,* Vol. 6., Wiseman, A., Ed., Ellis Horowood, Chichester, 1982, 102.
67. **Parker, D., Baker, D. S., Brown, K. G., Tanner, R. D., and Malaney, G. W.,** The effect of carbon dioxide, aeration rate and sodium chloride on the secretion of proteins from growing baker's yeast *(Saccharomyces cerevisiae), J. Biotechnol.,* 2, 337, 1985.
68. **Wöhrer, W., Forstenlehner, L., and Röhr, M.,** Evaluation of the Crabtree effect in different yeast grown in batch and continuous culture, in *Current Developments in Yeast Research,* Stewart, G. G. and Russel, I., Eds., Pergamon Press, New York, 1981.
69. **Wöhrer, N. W. and Röhr, M.,** Regulatory aspects of baker's yeast metabolism in aerobic fed-batch cultures, *Biotechnol. Bioeng..* 23, 567, 1981.
70. **Rickard, P. A. D. and Hewetson, J. W.,** Effects of glucose on the activity and synthesis of fermentative respiratory pathways of *C. utilis, Biotechnol. Bioeng.,* 21, 2337, 1979.
71. **Petrikovich, S. B., Litvinenko, L. A., Kintana, M. E., and Popov, E. J.,** Ultrastructure of *Candida utilis* in continuous culture with various concentrations of carbon source and different feeding methods, *Mikrobiologiya,* 52, 826, 1983.
72. **Luzikov, V. N., Novikova, L. A., Tikhonov, A. N., Yuskovich, A. K., and Zubtov, A. S.,** Disturbances in the proteolysis of mitochondrial translation products under glucose repression of the yeast *Saccharomyces cerevisiae, Biochimiya,* 49, 1571, 1984.
73. **Mohamed, Kasim Mustafa,** SCP Production: Effect of Carbon Source Concentration and Aeration on Protease and RNA-se Activity of Different Yeast Species, Ph.D. thesis, Hungarian Academy of Sciences, Budapest, 1989.
74. **Wöhrer, W. and Röhr, M.,** Investigation on the regulation of aerobic fermentation and glucose ethanol diauxic growth of *Saccharomyces cerevisiae* in a computer-coupled fermentation system, 4th FEMS Symp., Vienna, March 28, Abstr. B2, 1977.
75. **Niederberger, B. P., Eabi, M., and Hüter, R.,** Influence of the general control of amino acid biosynthesis on cell growth and cell viability in *Saccharomyces cerevisiae, J. Gen. Microbiol.,* 129, 2511, 1983.
76. **Saita, M. and Slaughter, J. C.,** Acceleration rate of fermentation by *Saccharomyces* in the presence of ammonium ion, *Enzyme Microbiol. Technol.,* 16, 375, 1984.
77. **Hill, F. F. and Thommel, J.,** Continuous measurement of the ammonium concentrations during the propagation of baker's yeast, *Process Biochem.,* 17(5), 16, 1982.
78. **Kole, M. M., Thompson, B. G., and Gerson, D. F.,** Ammonium concentration control in fed-batch fermentation of *S. cerevisiae, J. Ferment. Technol.,* 63, 121, 1985.
79. **Harju, M., Heikonen, M., Kreula, M., and Linko, M.,** Nutrient supplementation of Swiss cheese whey for the production of feed yeast, *Milchwissenschaft,* 31, 530, 1976.
80. **Maaloe, O. and Kjeldgaarde, F.,** *Control of Macromolecular Synthesis,* Benjamin, W. A. Inc., New York, 1966.
81. **Grick, F. H. C., Barnett, L., Brenner, S., and Watts-Tobbin, R. J.,** General nature of the genetic-code for proteins, *Nature,* 1972, 1227, 1961.
82. **Nadson, S. and Fillipov, P.,** in *The Science of Genetics,* Harper and Row, London, 1966, 221.
83. **Drake, L. M.,** Macromolecular synthesis in temperature sensitive mutants of yeast, *J. Bacteriol.,* 93, 1962, 1967.
84. **Auerbach, Ch.,** *The Science of Genetics,* Harper and Row, London, 1966, 221.
85. **Gradova, N. B. and Robysheva, Z. N.,** Study of mutagenesis in yeasts of *Candida genus* in *Current Developments in Yeast,* Stewart, G. G. and Russell, I., Eds., Pergamon Press, New York, 1981, 205.
86. **Okanishi, H. and Gregory, K. F.** Isolation of mutants of Candida tropicalis with increased methionine content, *Can. J. Microbiol.,* 16, 1139, 1970.
87. **Ken-Ichi, K., Toshio, Y., and Ryoji, K.,** Isolation and characteristics of pool methionine-rich mutants of *Candida* sp., *J. Ferment. Technol.,* 52, 93, 1974.
88. **Young, R. A. and Smith, R. E.,** Production of methionine-excreting mutants of *Streptomyces fradiale, Can. J. Microbiol.,* 21, 587, 1974.
89. **Delgado, J., Herrara, L., and Peréz, M.,** Isolation of auxotrophic mutants of *Candida utilis, Rev. CENIC.,* 7, 1, 1979.
90. **Momose, H. and Gregory, K. F.,** Temperature sensitive mutants of *Saccharomyces cerevisiae* variable in methionine content of their protein, *Appl. Environ. Microbiol.,* 35(4), 641, 1978.
91. **De Zeew, J. R.,** Genetic and environmental control of protein composition, in *Single Cell Protein, I,* Mateles, R. I. and Tannenbaum, R. S., Eds., MIT Press, Cambridge, 1968, 181.
92. **Yanofsky, S. A., Horn, V., and Thorpe, D.,** Protein structure relationship revealed by mutational analysis, *Science,* 146, 1593, 1964.
93. **Kondratyeva, T. F., Linkova, M. A., and Lobacheva, N. A.,** Spontaneous and UV-induced variations in the activity of biomass synthesis in *Candida utilis* haploid and diploid strains, *Mikrobiologiya,* 57, 245, 1988.

94. **Peberdy, J. F.,** Fungal protoplasts: isolation, reversion and fusion, *Am. Rev. Microbiol.*, 33, 31, 1979.
95. **Peberdy, J. F.,** Protoplast fusion. A tool for genetic manipulation and breeding in industrial organisms, *Enzyme Microb. Technol.*, 2, 23, 1980.
96. **Stewart, G. G., Arratt, J. A., Garrison, I., Goring, T., and Hancock, I.,** Studies on the utilization of wort carbohydrates by brewer's yeast strains, *MBAA Tech. Q.*, 16, 1, 1979.
97. **Russell, I. and Stewart, G. G.,** Transformation of maltotriose uptake ability into a haploid strain of *Saccharomyces* sp., *J. Inst. Brew.*, 86, 55, 1980.
98. **Müller, R.,** Die Entstehung entwicklungsfähiger Protoplasten aus Hefezellen und ihre Revision, Wissenschaftliche Filme aus dem Institut f. Mikrobiologie und Experimentelle Therapie der Wissenschaften der DDR, No.T-HF 659, 1966.
99. **Müller, R.,** Contribution to the problem of protoplast fusion, *Acta Fac. Med. Univ. Brun.*, 37, 39, 1970.
100. **Ferenczy, L., Zsolt, J., and Kevei, F.,** Forced heterokaryon formation in auxotrophic *Geotrichum* strains by protoplast fusion, in 3rd Protoplast Symp. on Yeast Protoplasts, Salamanca, Spain, October 2, Abstr. 74, 1972.
101. **Ferenczy, L.,** Microbial protoplast fusion, in *Genetics as a Food in Microbiology,* Glower, S. W. and Hopwood, D. A., Eds., Soc. Gen. Microbiol. Symp. 31, Cambridge University Press, New York, 1981, 1.
102. **Ferenczy, L., Kevei, F., and Zsolt, J.,** Fusion on fungal protoplast, *Nature,* London, 248, 793, 1974.
103. **Ferenczy, L., Kevei, F., and Szegedi, M.,** Increased fusion frequency of *Aspergillus nidulans* protoplasts, *Experentia,* 31, 50, 1975.
104. **Kao, K. N. and Michaluk, M. R.,** A method for high-frequency intergeneric fusion on plant protoplasts, *Planta,* 115, 355, 1975.
105. **Ferenczy, K. and Kevei, F.,** Factors affecting high-frequency fungal protplast fusion, *Experentia,* 32, 1156, 1976.
106. **Weber, H. and Spata, L.,** Characterization of yeast protoplast fusion products, in *Current Developments in Yeast Research,* Stewart, G. G. and Russell, I., Eds., Pergamon Press, New York, 1981, 213.
107. **Senda, M., Takeda, J., Ade, S., and Nakamura, T.,** Induction of cell fusion of plant protoplasts by electrical stimulation, *Plant Cell Physiol.*, 20, 1441, 1979.
108. **Zimmermann, V. and Pilwat, G.,** The relevance of electric field induced changes in the membrane structure basic membrane research and clinical therapeutics and liquors, Poster at the 6th Int. Biophysics Cong., Kyoto, Japan, 1978.
109. **Weber, H., Förster, W., Jacob, H. E., and Berg, H.,** Enhancement of yeast protoplast fusion by electric field effects, in *Current Developments in Yeast Research,* Stewart, G. G. and Russell, I., Eds., Pergamon Press, New York, 1981, 219.
110. **Stewart, G. G., Russell, I., and Panchal, D.,** The genetic manipulation of industrial yeast strains, in *Current Developments in Yeast Research,* Stewart, G. G. and Russell, I., Eds., Pergamon Press, New York, 1981, 71.
111. **Fournier, T., Provost, A., Bourguignon, A., and Heslot, H.,** Recombination after protoplast fusion in the yeast *Candida tropicalis, Arch. Microbiol.*, 115, 143, 1977.
112. **Gunge, N. and Tamaru, A.,** Genetic analysis of products of protoplast fusion in *Saccharomyces cerevisiae, Jpn. J. Genet.*, 53(1), 41, 1978.
113. **Sipiczki, M. and Ferenczy, L.,** Protoplast fusion of *Schizosaccharomyces pombe* auxotrophic mutants of identical mating types, *Mol. Gen. Genet.*, 151, 77, 1977.
114. **de von Brook, M. R., Sierra, M., and de Figueroa, L.,** Intergeneric fusion of yeast protoplasts, in *Current Developments in Yeast Research,* Stewart, G. G. and Russell, I., Eds., Pergamon Press, New York, 1981, 177.
115. **Heslot, M., Massy, A. P., and Fournier, P.,** Process for obtaining hybrid strains of yeasts, U.S. Patent (19) II.4 172, 764.
116. **Forage, A. J. and Righelato, R. C.,** Biomass from carbohydrates, in *Microbial Biomass,* Rose, A. H., Ed., Academic Press, London, 1979, 289.
117. **Rieger, M.,** Untersuchung zur Regulation von Glykolyse und Atmung in *Saccharomyces cerevisiae,* Diss., ETH No. 7264, 1983.
118. **Arreguin, M. M.,** Characterization of the respirative and respiro fermentative glucose metabolism in *Saccharomyces cerevisiae,* Diss. ETH. No. 8089, 1986.
119. **Burnow, S.,** Baker's yeast, in *Economic Microbiology,* Vol. 4, Rose, A. H., Ed., Academic Press, London, 1979, 32.
121. **Halász, A. and Szalma-Pfeiffer, I.,** Dependence of protease activities "A", "B" and inhibitor activities I_A and I_B of *S. cerevisiae* on aeration rate and glucose concentration, 13th Int. Cong. of Bicochemistry, August 25, 1985, Abstr. FR. 599.
122. **Murata, K. and Kimura, A.,** Relationship between glutathion contents and generation times in *Saccharomyces cerevisiae, Agric. Biol. Chem.*, 50, 1055, 1986.

123. **Von Sandmeter, E. P., Herstellung von Backhefe,** Ausganstoffe Melasse und Wasser, *Chem. Rundschau.*, 38, 8, 1985.
124. **Williams, D., Yousefpour, P., and Wellington, E. M. H.,** On-line adaptive control of fed-batch fermentation of *Saccharomyces cerevisiae, Biotechnol. Bioeng.*, 28, 631, 1986.
125. **Williams, D. and Montgomery, P. A.,** Multivariable adaptive control of baker's yeast fermentation, *IEE Proc.*, Vol. 133, Pt.D. No.5, 247, 1986.
126. **Peppler, H. J.,** Production of yeast and yeast products, in *Microbial Technology*, Vol. 1, Peppler, H. J. and Perlman, D., Eds., Academic Press, Orlando, FL, 1979, 157.
127. **Nagadawithana, T. W.,** Yeasts: their role in modified cereal fermentation, in *Advances in Cereal Science and Technology*, Vol. 8, Pomeranz, Y., Ed., St. Paul, MN, 1986, 15.
128. **Crylls, F. S. M., Rennie, S. D., and Kelly, M.,** Process for producing active dried yeast, U.S. Patent 4 188 407, 1980.
129. **Trivedi, N. B., Cooper, E. J., and Bruinsma, B. L.,** Development and applications of quick-rising yeast, *Food Technol.*, 38, 51, 1984.
130. **Beker, M. J., Blumbergs, J. E., Ventina, E. J., and Rapoport, A. I.,** Characteristics of cellular membranes at rehydration of dehydrated yeast *Saccharomyces cerevisiae, Environ. J. Appl. Biotechnol.*, 19, 347, 1984.
131. **Oda, Y., Uno, K., and Ohta, S.,** Selection of yeasts for breadmaking by the frozen-dough method, *Appl. Environ. Microbiol.*, 52, 941, 1986.
132. **Hino, A., Takano, H., and Tanaka, Y.,** New freeze-tolerant yeast for frozen dough preparations, *Cereal Chem.*, 64, 269, 1987.
133. **Ivanov, V. N., Rapoport, A. I., Pindrus, A. A., Saulite, L. A., and Shifruk, T. X.,** The phase specificity of yeast cell damages and damage repair in dehydration-rehydration and freezing-thawing, *Mikrobiologija*, 56, 341, 1987.
134. **Tsutsayeva, A. A., Kazankova, L. N., Balyberdina, L. M., Markova, V. M., Kudokotseva, O. V., and Kadmikova, N. G.,** The effect of cryoconservation conditions on the morphological and functional properties of *Saccharomyces cerevisiae, Mikrobiologija*, 56, 338, 1983.
135. **Jeunehomme, C. R. and Masschlein, C. A.,** Effect represseur du maltose zur la synthese des enzymes respiratoires chez *Saccharomyces carlsbergensis, Cerevisia*, H.v. (2) 82, 1976.
136. **Bonaly, R. and Amri, M. A.,** Nutritive elements and flocculation in two strains of *Saccharomyces uvarum*, 18th Int. EBC. Kong., *Brauindustrie*, 66, 702, 1981.
137. **Rose, A. H. and Keenan, M. H. J.,** Amino acid uptake by *Saccharomyces cerevisiae*, 18th Int. EBC. Kong., *Brauindustrie*, 66, 669, 1981.
138. **Aavenainen, J. and Mäkinen, V.,** The effect of pitching yeast aeration on fermentation and beer flavour, 18th Int. EBC. Kong., *Brauindustrie*, 66, 676, 1981.
139. **Storms, R. K., McNeil, J. B., Khandekar, P. S., Parker, G. A., and Friesen, J. D.,** Chimeric plasmids for cloning of deoxyribonucleic acid sequences in *Saccharomyces cerevisiae, J. Bacteriol.*, 140, 73, 1979.
140. **Tubb, R. S., Brown, A. J. P., Searle, B. A., and Goodey, A. R.,** Development of new techniques for the genetic manipulation of brewing yeast, in *Current Developments in Yeast Research*, Stewart, G. G. and Russell, I., Ed., Pergamon Press, New York, 1981, 75.
141. **Sills, A. M., Panchal, C. J., Russell, I., and Stewart, G. G.,** Genetic manipulation of amylolytic enzyme production by yeasts, *Proc. ALKO Yeast Symp.*, Helsinki, 1983, 8, 209.
142. **Rezessy-Szabó, J., Maráz, A., and Dencsö, L.,** Breeding of brewer's yeast: improvement of flocculation and transfer of killer factor, *Acta Microbiol. Hung.*, 35, 124, 1988.
143. **Hinney, A., Hicks, J. B., and Fink, G. R.,** Transformation of yeast, *Proc. Natl. Acad. Sci. U.S.A.*, 75, 1929, 1978.
144. **Broach, J. R.,** The yeast plasmid 2 μm circle, in The Molecular Biology of the Yeast Saccharomyces-Life Cycle and Inheritance. Strathern, J. N., Johen, E. W., and Broach, J., Eds., Cold Spring Harbor Laboratory, Cold Spring Harbor, New York, 1981.
145. **Panchal, C. J., Russell, I., Sills, A. M., and Stewart, G. G.,** Genetic manipulation of brewing and related yeast strains, *Food Technol.*, 38, 94, 1984.
146. **Russell, I., Jones, R. M., Weston, B. J., and Stewart, G. G.,** Liposome mediated DNA transfer in brewing and related yeast strains, presented at 19th Int. Cong., European Brewing Convention, London, June 5, 1983.
147. **Saddekni, G., Maráz, A., and Deák, T.,** Borélesztö törzsek fiziológiai és technológiai tulajdonságainak felülvizsgálata, *Borgazdaság*, 36(3), 85, 1988.
148. **Houtman, A. C. and du Plessis, C. S.,** Influence du cepage et de la souche de le vure sur la vitesse de fermentation et our la concentration des composants volatiles du vin, *Bull. O.I.V.*, 648, 235, 1985.
149. **Panchal, G. J. and Stewart G. G.,** The effect of osmotic pressure on the production and excretion of ethanol and glycerol by a brewing yeast strain, *J. Inst. Brew.*, 86, 207, 1980.
150. **Moulin, G., Boze, P., and Galzy, P.,** A comparative study of the inhibitory effect of ethanol and substrate on the fermentation rate of the parent and a respiratory deficient mutant, *Biotechnol. Lett.*, 3, 351, 1981.

151. **Hayashida, S. and Ohta, K.**, Formation of high concentrations of alcohol by various yeasts, *J. Inst. Brew.*, 87, 42, 1981.
152. **van Uden, N.**, Ethanol toxicity and ethanol tolerance in yeasts, *Am. Rep. Ferment. Proc.*, 8, 11, 1985.
153. **Jimenez, J. and Benitez, T.**, Characterization of wine yeast for ethanol production, *Appl. Microbiol. Biotechnol.*, 25, 150, 1986.
154. **Saddekni, G., Maráz, A., and Deák, T.**, Breeding of wine yeasts having different physiological properties, Biotechnol. Food Industry Proc. Int. Symp., Budapest, 1988, 133.
155. **Meyrath, J. and Bayer, K.**, Biomass from whey, in *Microbial Biomass*, Rose, A. H., Ed., Academic Press, London, 1979, 208.
156. **Vananuvat, P. and Kinsella, A. J. E.**, Production of yeast protein from crude lactose by *Saccharomyces fragilis*. Batch culture studies, *J. Food. Sci.*, 40, 336, 1975.
157. **Admundson, C. D.**, Increasing the protein content of whey, *Am. Dairy Rev.*, 29, 22, 1967.
158. **Tomisek, J. and Gregr. V.**, Vyroba kvasnicnych bilkovin ze syrovátky, *Kvasny Prum.*, 7, 130, 1961.
159. **Simek, F., Kovács, J., and Sárkány, J.**, Takarmányélesztö elöállítása tejipari melléktermékek felhasználásával, *Tejipar*, 13, 75, 1964.
160. **Anon.**, Immobilized enzyme and fermentation technologies combine to produce baker's yeast from whey, *Food Technol.*, 38, 26, 1984.
161. **Wasserman, A.**, The rapid conversion of whey to yeast, *Dairy Eng.*, 77, 374, 1960.
162. **Chapman, L. P. J.**, Food yeast from whey. New Zealand, *J. Dairy Technol.*, 1, 78, 1966.
163. **Lade, T., Moulin, G., Galzy, P., Joux, J., and Biju-Duval, F.**, Comparison des rendements de croissance sur lactose de quelques *Kluyveromyces* von der Walt, *Lait*, 52, 519, 1972.
164. **Moulin, G. and Galzy, P.**, Une possibilité d'utilization du lactoserum: la production de levure, *Ind. Aliment. Agric.*, 11, 1137, 1976.
165. **Nour-El Dien, H. and Halász, A.**, Attempts to utilize whey for the production of yeast protein. II. Effects of biotin concentration and whey content at constant lactose concentration, *Acta Aliment.*, 11, 11, 1982.
166. **Nour-El Dien, H., Halász, A., and Lengyel, Z.**, Attempts to utilize whey for the production of yeast protein. I. Effect of whey concentration of ammonium sulphate and of phosphate, *Acta Aliment.*, 10, 11, 1981.
167. **Moulin, G., Malige, D., and Galzy, P.**, Etude physiologique de *Kluyveromyces fragilis* consequence pour la production de levure sur lactoserum, *Lait*, 61, 323, 1981.
168. **Bernstein, S., Tzeng, C. H. V., and Sisson, D.**, The commercial fermentation of cheese whey for the production of protein and/or alcohol, *Biotechnol. Bioeng.*, Symp. No. 7. 1, 1977.
169. **Mahmoud, M. M. and Kosikowski, F. V.**, Alcohol and single cell protein production by *Kluyveromyces* in concentrated whey permeates with reduced ash, *J. Dairy Sci.*, 65, 2082, 1982.
170. **Harju, M., Heilkonen, M., Kreula, M., and Linko, M.**, Nutrient supplementation of Swiss cheese whey for the production of feed yeast, *Milchwissenschaft*, 31, 530, 1976.
171. **Davies, R.**, Lactose utilization and hydrolysis in *Saccharomyces fragilis*, *J. Gen. Microbiol.*, 37, 81, 1964.
172. **Spiler, A.**, The biosynthesis of proteins from carbohydrates, Proc. Int. Symp. on Single Cell Protein, Rome, 1973.
173. **Nour-El Dien, H. and Halász, A.**, Attempts to utilize whey for the production of yeast protein. III. Effect of some vital growth factors, *Acta Aliment.*, 11, 125, 1982.
174. **Smith, M. and Bull, A. T.**, Protein and other compositional analyses of *Saccharomyces fragilis* grown coconut waste water, *J. Appl. Bacteriol.*, 41, 97, 1976.
175. **Wasserman, A. E., Horkins, N. J., and Porges, N.**, Whey utilization growth conditions for Saccharomyces fragilis, *Sewage Ind. Wastes*, 30, 913, 1958.
176. **Wasserman, A. E.**, Amino acid and vitamin composition of *Saccharomyces fragilis* grown in whey, *J. Dairy Sci.*, 44, 379, 1961.
177. **Glättli, H. and Blanc, B.**, Gewinnung von Hefebiomasse aus Molke, *Schweiz. Milchwirtsch. Forsch.*, 3, 14, 1974.
178. **Kamel, B. S.**, Utilization of date carbohydrate as substrate in microbial fermentation, *Process Biochem.*, 5, 12, 1979.
179. **Litchfield, J.**, Comparative technical and economical aspect of single cell proteins, *Adv. Appl. Microbiol.*, 35, 21, 1977.
180. **Dimmling, W. and Seipenbusch, R.**, Raw materials for the production of SCP, *Process Biochem.*, 12(3), 9, 1978.
181. **Moo-Yoong, M.**, Economics of SCP production, *Process Biochem.*, 12, 6, 1977.
182. **Cason, D. T., Reid, G. C., and Gatner, E. M. S.**, On the differing rates of fructose and glucose utilization in *Saccharomyces cerevisiae*, *J. Inst. Brew.*, 93, 23, 1987.
183. **Mohamed, Kassim Mustafa and Halász, A.**, The effect of date syrup concentration on growth rate, protein, RNA content and protease activity of *S. cerevisiae*, *C. guilliermondii* and *R. glutinis*, *Acta Aliment.*, 18(2), 177, 1989.

184. **Mohamed, A. M. and Ahmed, A. A.,** Lybian date syrup (Rub-Al Tamr), *J. Food Sci.*, 46, 1162, 1981.
185. **Alberts, B., Bray, D., Lewis, J., Raff, M., Roberts, K., and Watson, J. D.,** *Molecular Biology of the Cell,* Gorland Publishing, New York, 1983.
186. **Tabakoff, B., Eriksson, C. J. P., and von Wartburg, J. P.,** Methionine lowers circulating levels of acetaldehyde after ethanol ingestion, *Alcoholism: Clin. Exp. Res.*, 13(2), 164, 1989.
187. **Buraczewska, L., Zedrowska, T., and Buraczewski, S.,** The rate passage of synthetic lysine and dietary protein from the stomach to the intestine in pigs, 5th Int. Symp. on Amino Acids, Budapest, February 21, 1977, Abstr. p. 13.
188. **Gebhardt, G., Zebrowska, T., Köhler, R.,** An attempt to estimate the endogenous nitrogen and true digestibility of protein and individual amino acids, 5th Int. Symp. on Amino Acids, Budapest, February 21, 1977, Abstr. p. 41.
189. **Komatsu, K., Yamada, T., and Kodaira, R.,** Isolation and characteristics of pool methionine-rich mutants of *Candida* sp., *J. Ferment. Technol.*, 52, 93, 1974.
190. **Lawrence, D., Smith, D., and Bowbury, R.,** Regulation of methionine biosynthesis, in *S. typhimurium:* mutants, resistant to inhibition by analogues of methionine, *Genetics*, 58, 473, 1968.
191. **Greene, R., Su, C., and Holloway, C.,** S-adenosyl-methionine synthese deficient mutants of *E. coli* K-12 with impaired control of methionine biosynthesis, *Biochim. Biophys. Res. Commun.*, 38, 1120, 1970.
192. **Halász, A.,** Biochemical Principles of the Use of Yeast Biomass in Food Production, D.Sc. thesis, Hungarian Academy of Sciences, Budapest, 1988.
193. **Halász, A., Baráth, Á., and Szalma-Pfeiffer, I.,** Versuch zur Herstellung von methionine-reichen Hefemutanten, 5th Int. Symp. Amino Acids, February 21, 1977, Budapest, Abstr. G13.
194. **Halász, A., Baráth, Á., and Szalma-Pfeiffer, I.,** Production of yeast mutants with increased methionine content, *Proc. 5th Int. Spec. Symp. on Yeast,* Keszthely, Hungary, 1977, 125.
195. **Tarka, E. F. and Fields, M. L.,** Yeast mutants containing elevated methionine, *J. Food Sci.*, 51, 1273, 1986.
196. American Type Culture Collection, Catalogue of Strains I., 15th ed., American Type Culture Collection, Rockville, MD.
197. **Muayad, A., Halász, A., Mátrai, B., and Szalma-Pfeiffer, I.,** Efficiency of different mutagenic treatments and the protoplast fusion technique in producing methionine rich mutants from yeasts, *Acta Aliment.*, 12(3), 211, 1983.
198. **Halász, A., Mátrai, B., and Muayad, A.,** S-metabolism of methionine-rich yeasts, *Acta Aliment.*, 18, 36, 1989.
199. **Bowien, B. and Schlegel, H.,** Isolierung and Characterisierung katabolischer Defektmutanten von *Hydrogenomonas* eutropha Stamm H 16., *Arch. Mikrobiol.*, 87, 203, 1972.
200. **Scherr, G. and Rafelson, M.,** The direct isolation of mutants producing increased amounts of metabolites, *J. Appl. Bacteriol.*, 25, 187, 1962.
201. **Colombani, F., Cheresi, H., and de Robinson-Schulmaister, H.,** Biochemical and regulatory effects of methionine analogues in *Saccharomyces cerevisiae, J. Bacteriol.*, 122, 375, 1975.
202. **Mornyska, E., Sawnor-Korszynska, D., Parzenska, A., and Grabski, J.,** Methionine overproduction by *Saccharomycopsis lipolytica, Appl. Environ. Microbiol.*, 32, 125, 1976.
203. **Halász, A., Muayad, A., and Mátrai, B.,** The effect of oxygen absorption rate (aeration rate) on the methionine content and growth rate of some methionine-rich mutants of *Candida guilliermondii 812, Acta Aliment.*, 16(3), 249, 1987.
204. **Benifer, J., Alonso, A., Delgado, J., and Kotyk, A.,** Sulfate transport *in Candida utilis, Folia Microbiol.*, 28, 6, 1983.
205. **Breton, A. and Surdin-Kerjan, Y.,** Sulfate uptake in *Saccharomyces cerevisiae:* biochemical and genetic study, *J. Bacteriol.*, 132, 224, 1977.
206. **Garcia, M., Benitez, J., Delgado, J., and Kotyk, A.,** Isolation of sulfate transport defective mutants of *Candida utilis:* further evidence for a common transport system for sulfate, sulfite and thiosulfate, *Folia Microbiol,* 28, 1, 1983.
207. **Roomans, G., Kuypers, G., Theuvenet, A., and Borst-Pauwals, G. W. F. H.,** Kinetics of sulfate intake by yeast, *Biochem. Biophys. Acta,* 551, 197, 1979.
208. **Alonso, A., Benitez, J., and Diaz, M. A.,** A sulfate, sulfite and thiosulfate incorporating system of *Candida utilis, Folia Microbiol.*, 29, 8, 1984.
209. **Ferenczy, L., Kevei, F., Szegedi, M., Fraknó, A., and Rotik, I.,** Factors affecting high frequency protoplast fusion, *Experimentia,* 32, 1156, 1976.
210. **Laskin, A. I.,** SCP from ethanol, *Biotechnol. Bioeng.*, Symp. No.7, 77, 1977.
211. **Wattew, C. M., Arminger, W. B., Ristroph, D. L., and Humphrey, A. R.,** Production of single cell protein from ethanol by fed-batch process, *Biotechnol. Bioeng.*, 21, 1221, 1979.
212. **Akin, C., Chao, K. C., Lappin, T. A., Rainwater, J., and Wolfe, D. B.,** AMOCO'S aerobic fermentation process for continuous food yeast production, Paper presented at the 2nd Chemical Cong. of the North American Continent, August 26, 1980, Las Vegas, NV, 1980.

213. **Chen, S. L. and Peppler, H. J.**, Single-cell proteins in food applications, *Dev. Ind. Microbiol.*, 19, 79, 1979.
214. **Allen, et al.**, U.S. Patent 4 019 962, 1977.
215. **Annon, I.**, Single-cell protein aus Ethanol, *Chem. Ind.*, 17, 1980.
216. **Christyakova, T. T., Dedyukhina, E. G., and Eroshin, V. K.**, Composition of the biomass of thermotolerant strains of the yeast grown on the ethanol containing medium, *Prikl. Biochim. Mikrobiol.*, 16, 13, 1980.
217. **Sahm, H. and Wagner, F.**, Mikrobielle Verwertung von Methanol: Isolierung und Characterisierung der Hefe *Candida boidinii, Arch. Mikrobiol.*, 24, 135, 1972.
218. **Reuss, M., Gnieser, J., Reng, H. G., and Wagner, F.**, Extended culture of *Candida boidinii* on methanol, *Eur. J. Appl. Microbiol.*, 1, 295, 1975.
219. **Levine, D. W. and Looney, L. L.**, Isolation and characterization of a thermotolerant methanolutilizing yeast, *Appl. Microbiol.*, 26, 982, 1973.
220. **Shannon, L. J. and Stevenson, K. E.**, Growth of fungi and BOD reduction in selected brewery wastes, *J. Food Sci.*, 40, 826, 1975.
221. **Kramarz, M., Lesniak, W., and Ziobrowski, J.**, Production von *Candida pseudotropicalis* aus dem Abfall-saft der Stärkefabrik, *Chem. Mikrobiol. Technol. Lebesm.*, 8, 164, 1984.
222. **Sanchez-Marroquin, A.**, Mixed cultures in the production of single-cell protein from agave juices, *Biotechnol. Bioeng.*, Symp. No. 7., 23, 1977.
223. **Calleja, G. B., Yaguchi, M., Levy-Rick, S., Seguin, J. R. H., Roy, C., and Lusena, C. V.**, Single-cell protein production from potato starch by the yeast *Schwanniomyces alluvius, J. Ferment. Technol.*, 64, 71, 1986.
224. **Augustin, J., Zemek, J., Kockova-Kratochvilova, A., and Kuniak, L.**, Production of α-amylase by yeasts and yeast-like organisms, *Folia Microbiol.*, 23, 353, 1978.
225. **Oteng-Gyang, K., Moulin, G., and Galzy, P.**, Influence of amylase excretion on biomass production by amylolytic yeast, *Acta Microbiol. Acad. Sci. Hung.*, 27, 155, 1980.
226. **Simek, F., Széchényi, Lné, Kárpáti, Gy., and Vas, K.**, Eljárás fehérje-vitamin koncentrátum élesztökkel történö elöállítására állati vagy növényi zsiradékból és készítménnyé alakítására, H.165-381 (C12c 11/08) 1974.
227. **Tan, K. H. and Gill, C. O.**, Effect of culture conditions on batch growth of *Saccharomycopsis lypolytica* on olive oil, *Appl. Microbiol. Biotechnol.*, 20, 201, 1984.

Chapter 5

EXTRACTION OF YEAST PROTEINS, YEAST PROTEIN CONCENTRATES AND ISOLATES

I. INTRODUCTION

In principle the biomass of yeast can be used in different forms. The simplest way can be the utilization of total (intact) cells (e.g., inactive dry yeast). Another possibility is the partial destruction of cell walls and removal of components adversely influencing the end use properties. Finally, extraction of the components, first of all proteins, after disruption of the cell, separation and concentration of proteins and/or other components is also possible (e.g., production of concentrates and isolates).

Ideally microbial cells should be consumable directly as food or food ingredients. Yeast is, e.g., a standard part of the recipe of different bakery products in 2 to 5% quantity (for other examples see the Appendix). However, because of the presence of undesirable physiologically active components, high nucleic acid content and the deleterious effect of cell wall material on protein bioavailability, rupture of cells and extraction of the protein and/or other components is, in most cases, a necessary step in further processing and application.

Cell wall material in unfractioned SCP, also including yeasts, is undesirable not only due to the decreased bioavailability of proteins, but because it may contain antigenic, allergenic agents and factors causing nausea and gastrointestinal disturbances. After cell disruption different separation techniques can be used depending on the component to be produced in relatively pure form.

A. CELL RUPTURE

The principal methods used for rupturing yeast cells are mechanical, disruption and enzymatic digestion of cell walls (see Table 42). Effective disintegration of yeast cell walls has been achieved using mechanical methods.[1-4] Different types of laboratory — or pilot plant scale equipment were described, from simple grinders, bead mills, ball mills, till special vibro-colloid (cavitational) mills[4] and ballistic desintegrators.[5]

Although a lot of results were published on this topic the efficiency of disruption of different mechanical apparatures is less well documented. Zetelaki[6] reported about good effectivity of bead mills in comparsion to other mechanical procedures. Whitworth[7] has found that a homogenizer can be used very efficiently for cell wall disruption. Cavitation effect, especially at lower temperatures, gives the possibility of achieving a very high degree of cell disruption. Concerning commercial apparatures Nevell et al.[8] and Robbins[9] reported that the Manton-Gaulin homogenizer was successfully used on a commercial scale. Similar opinion was expressed in the review paper of Chisti and Moo Young.[10]

In the evaluation of different methods of cell disruption an important step is the determination of integrity of yeast cells. For such purposes the measuring of protein (nitrogen) solubility is widely used. The basis of using this method is that there will be more soluble protein available when the cell wall has been broken and the solvent can enter the cell and solubilize the protein. Microscopy is used under the assumption that it would be possible to see the effect of a cell disruption visually and some sort of subjective measurements might be made. In some cases the release of specific enzymes or other cell components can be measured. The turbidimetric method is also applicable.

Although mechanical methods require large energy expenditures to achieve efficient cell wall rupture and thermal denaturation of the protein during cell breakage is frequent, such procedures are most common in the praxis.

TABLE 42
Methods Used for Disruption of Yeast Cells

Methods	Disadvantages
Mechanical (grinding, milling, homogenization)	High energy consumption and high costs
Freeze-thawing	Energy costs
Sonic oscillation	Special equipment, costs
Hydrolysis with chemicals	High salt concentration in the finished product
Other chemical treatment	Effective only at high concentrations: requires longer period Denaturation and degradation of proteins
Plasmolysis	Less efficient
Autolysis	Slow process
Use of lytic enzymes and thiol reagents	High costs of enzyme preparations. Activation of proteolytic enzymes: significant weight loss
Pressure release	Special high-pressure apparatus

The freeze-thaw procedure includes freezing of the mixture of yeast cells and solvent (e.g., water) at −4°C for 24 h. After freezing the cells are allowed to stand at ambient temperature until thawed. This method is not too effective and is time consuming.

Freeze-pressing of microbial cell suspensions can also be used to disrupt the cells. In the Hughs press[10] the frozen paste of cells is forced through a narrow slit or orifice, either in the presence of an abrasive at temperatures just below zero or without the abrasive at temperatures of about −25°C. In the latter case, phase and consequent volume changes of ice contribute to disruption. In addition, solid shear due to crystalline ice is important. The freeze-pressing yields cell wall membrane preparations that are relatively intact and may be a good method for isolation of membrane-associated enzymes.

The sonic oscillation method is realized in special equipment (ultrasonic disintegrator, oscillation range 11 to 24 kHz). The samples are placed under the oscillator probe more times in short intervals for a total sonification time depending on the equipment (usually 3 to 60 min).

Contradictory data are reported about the effectiveness of the sonic oscillation method. The majority of researchers is of the opinion that effectiveness is lower than in the case of good mechanical apparatuses. Until now ultrasonic equipment is not used on a commercial scale for cell rupture.

The autolysis of yeast is a well-known phenomenon in yeast production and storage.[11] Nevertheless biochemical and biophysical mechanisms of yeast autolysis are still not entirely understood. Most of the problems set by its practical industrial use can be empirically solved. The nature of the compounds released during autolysis pointed out the activities of proteases, mannanases, glucanases, lipases and phospholipases. The action of these enzymes is associated with the changes of cell walls. Electron transmission microscopy revealed a decrease of the thickness of cell wall. A structural loosening of the cell was also observed.[12] Chemical analysis of the cell wall composition at different stages of autolysis agrees with the observations of electronmicroscopy, and also confirm that glucans are constituents of the rigid matrix of yeast cell wall on which the other components are fitted.[13]

Non-polar organic solvents (chloroform, ethylacetate) or inorganic salts activates yeast autolytic enzymes and accelerate autolysis. Autolysis may also be stimulated by heat shock.[14] Shetty and Kinsella[15] have found that low molecular weight thiol compounds are effective for activating endogenous enzymes which degrade cell walls. Yeast cells were exposed to monothioglycerol, 2-mercaptoethanol or dithiothreitol and incubated at 37°C. A progressive release of glucose, indicating the gradual autolysis was observed. Monothioglycerol was most effective. It was also stated, that these treatments markedly facilitated the extractability of proteins from the yeast cell indicating that treatment with thiols was associated with breakage of the cell wall. A big disadvantage of such treatment is degradation of yeast proteins due to the activation of proteolytic enzymes by thiol reagents. The quantity of protein with molecular weight above 100 kDa progressively decreased from 80 to 20% after 10 h incubation.

If the main purpose of the yeast processing is extraction of proteins and production of protein concentrate, the autolysis is not an applicable procedure due to the high losses of protein. Such a procedure is of practical interest if the autolysate itself is used for different purposes or other soluble components will be isolated (see Appendix).

Use of high-concentration alkali or acid results in chemical degradation of cell wall. Treatment of yeast cells with alkali (pH = 11) weakens the cell wall and facilitates the rupture by mechanical methods. This is a common procedure used for the recovery of yeast proteins. However, such treatment has disadvantages if the purpose of the processing is production of proteins for food industry. Alkali treatment may cause hydrolysis of the protein, depolymerization, racemization of component amino acids, beta-elimination reactions and subsequent cross-linking and denaturation.

The pressure released method is also applicable for cell rupture. In a special apparatus the water yeast-cell mixture is pressurized to 1500 psig. After 30 min the chamber is depressurized. After treatment the suspension can be centrifuged and the cell wall particles separated from other components. Plasmolysis is achieved by addition of sodium chloride and/or by raising the temperature of yeast cream to 55 to 58°C. Isopropanol is also used commercially as a yeast plasmolyzing agent.

The enzymatic digestion of the cell walls — as a method of cell wall rupture — is an attractive approach because of its low energy requirement and enhanced efficiency compared to mechanical cell breakage. For such treatment endogenous enzymes of yeast (β-glucanases) can be used or exogenous enzymes obtained from other microorganisms. A review-paper dealing with enzymatic degradation of yeast cell wall was published by Phaff.[16] The important role of glucanases is doubtful. Some authors suggest glucanases to be unique enzymes which can effectively lyse yeast cell walls without any other enzymes taking part in the process. In contrast, several authors reported a similar role to proteinases.[17-19] Kislukhina and Bizulavitchus[20] reported about a preparation of lytic enzymes from the culture of *Bacillus subtilis*. The preparation (liosubtilin) of lysed cells of Gram-positive and Gram-negative bacteria, yeast and fungi. Recently Kalebina et al.[21] isolated a serine proteinase which showed lytic activity against either intact cell or cell wall preparations of *Candida utilis*. The proteinase was isolated from *Bacillus brevis* by affinity chromatography on bacitracin-silochrome and phenylboronate-Sepharose. Both its proteolytic and lytic activities were completely abolished by inhibitors of serine proteases, including phenylmethylsulfonylfluoride and duck ovomucoid. The preparation showed proteolytic activity against casein and the serine protease synthetic substrate Z-Ala-Ala-Leu-$NHC_6H_4NO_2$, as well as yeast cell lytic activity against yeast cell wall preparations and living cells of *Candida utilis*. No glucanase, mannanase or chitinase activity was observed. Yeast lytic enzymes have been isolated on a pilot scale from *Cytophaga* species.[22]

For the control of the action of free enzyme complex on whole yeast the measure of protein release was used. The initial rates of protein release were a linear function of substrate concentration up to 0.15 mg/ml but the conversion over 90 min is not. When the initial rates of protein release minus the control lacking enzyme, were plotted as a function of yeast concentration, the points followed an approximately hyperbolic pattern (Figure 30). The evident complexity of the enzyme mixture and the probable action of its constituents on different bonds does not apply to simple monoenzyme-monosubstrate models. Nevertheless, the kinetics can be characterized by a pseudo-Michaelis constant of maximum rate. The authors compared the activity of lytic enzymes on brewer's yeasts with the activity on baker's yeast. After 90 min the protein release from a brewer's yeast strain was 0.73 mg/

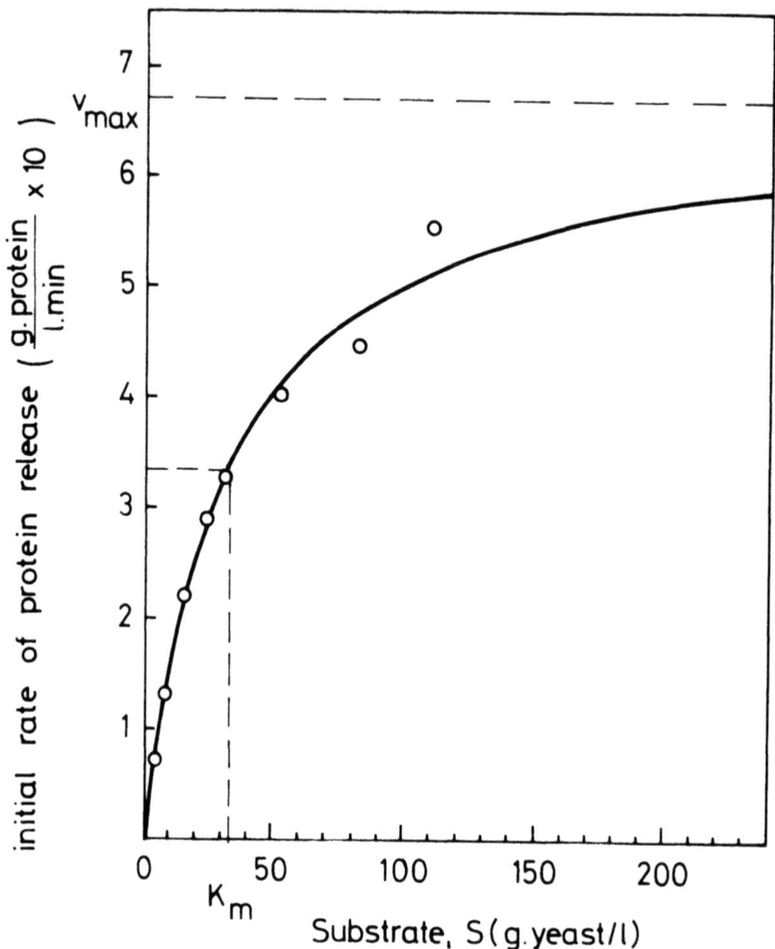

FIGURE 30. The variation of the initial rate of enzymatic protein release, corrected for autolytic release, vs. yeast substrate concentration. (From Asenjo, J. A. and Dunhill, P., *Biotechnol. Bioeng.*, 23, 1045, 1981. With permission.)

ml whereas from baker's yeast it was only 0.04 mg/ml. Comparing the release of reducing groups from brewer's yeast, whole cells with and without enzyme addition, (see Figure 31) showed a linear increase of reducing groups due to the action of enzyme added during a 120 min incubation. This means that degradation of cell wall polysaccharides is still substantial in the enzyme reaction. Cell wall preparations both from brewer's and baker's yeast were produced and the effect of lytic enzymes was investigated measuring the release of reducing groups. After 120 min incubation 0.5 μmol/ml was the concentration of reducing groups in the case of brewer's yeast cell wall and 0.2 μmol/ml for the baker's yeast cell wall preparation.

The lytic enzyme complex was also bound to soluble carbohydrate polymers by the mild periodate technique which leaves no potentially toxic linkage agent in the product. The enzyme bound to Dextran-300 was active on whole yeast cells. The results of these investigations suggest the possibility of elaborating immobilized enzyme technology for enzymatic yeast cell disruption on a commercial scale. Such a solution can significantly reduce the costs of the degradation of yeast cell walls and in the end affect the production of different components of yeast biomass.

A commercial preparation of β-1-3 glucanase from *Basidiomycetes aphyllophoroles* called "Kitalase" was produced by Kumai Chemical Company, Japan. A mixed enzyme

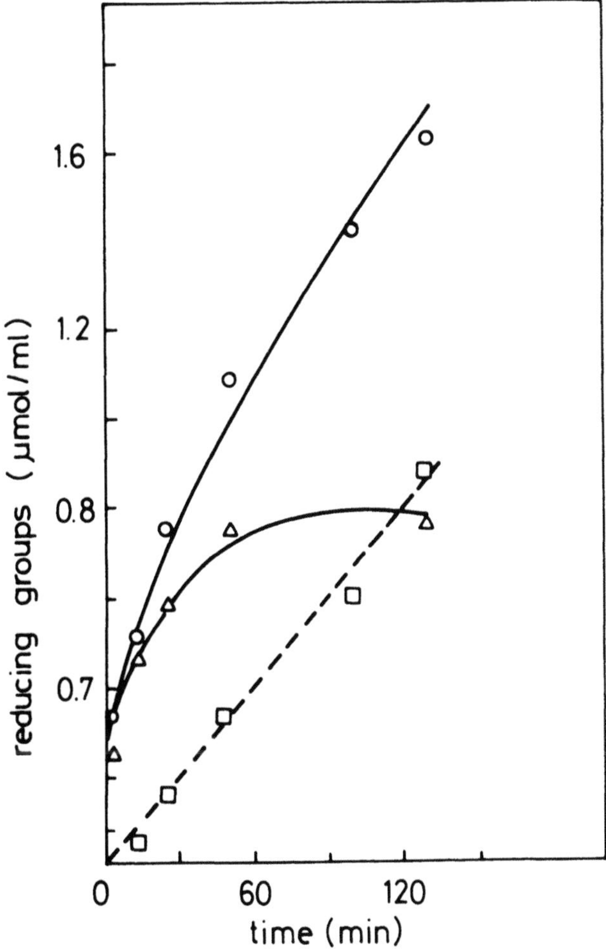

FIGURE 31. The kinetics of release of reducing group from whole yeast cells (brewer's yeast strain) in the presence of lytic enzymes (△△△△). Release due to autolysis: (o) overall release during incubation with enzyme; difference (□) release due to enzyme action. Enzyme concentration 0.15 mg/ml; yeast concentration 33 mg packed weight/ml. (From Asenjo, J. A. and Dunhill, P., *Biotechnol. Bioeng.*, 23, 1045, 1981. With permission.)

preparation containing β-1-3 glucanase, α-1-3-glucanase, xylanase, chitinase and protease activity having the trade name Novozym-234 is marketed by Novo Biolabs, Denmark. The enzyme preparations mentioned above were investigated recently by Ryan and Ward[23] regarding potential application of these enzymes in production of yeast extract. Polysaccharides laminarin and yeast glucan were incubated with lytic enzyme preparation of two *Basidiomycetes* species and Novozym-234. Some of the results are demonstrated in Figures 3 and 4.

Production of glucose, laminaribiose and β-gentiobiose during enzymic hydrolysis of laminarin is illustrated in Figure 32. Laminarin hydrolysis of the enzyme of *Basidomycetes* sp. 04806 was characterized by a rapid rate of glucose production with negligible change in the concentrations of glucose, laminaribiose and β-gentiobiose after 60 min. In contrast, production of glucose was more gradual in the case of the *B. aphyllophoroles* and Novozym-234 enzymes.

In similar experiments enzymatic degradation of yeast glucan was monitored (Figure 33). The most notable feature of this experiment was the early dramatic rise in laminaribiose

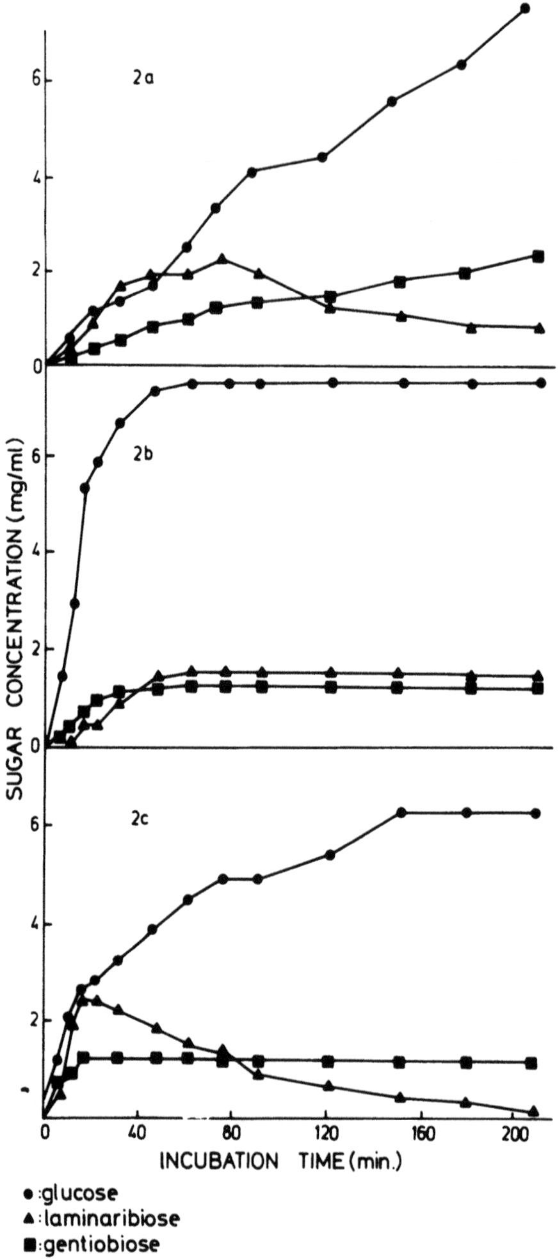

FIGURE 32. Production of glucose and disaccharides during digestion of laminarin with enzyme preparations from *B. aphyllophoroles* (a), *Basidiomycetes sp.* QM 806 (b), and *Novozym-234* (c). (From Ryan, E. and Ward, O. P., *Process Biochem.*, 23(1), 12, 1988. With permission.)

formation when yeast glucan was digested with the β-1-3-glucanase of *B. aphyllophoroles*. Summarizing the results of experiments it was stated that the β-1-3-glucanase of *Basidiomycetes aphyllophoroles* exhibited very high yeast lytic activity. When used in combination with papain in yeast autolysis, 90 to 95% of the yeast dry weight was solubilized. In a comparative study involving three β-1-3-glucanases, laminarinase activity did not correlate well with yeast lytic activity. The possible food applications of enzymatic cell lysis were recently discussed by Andrews and Arsenjo.[60]

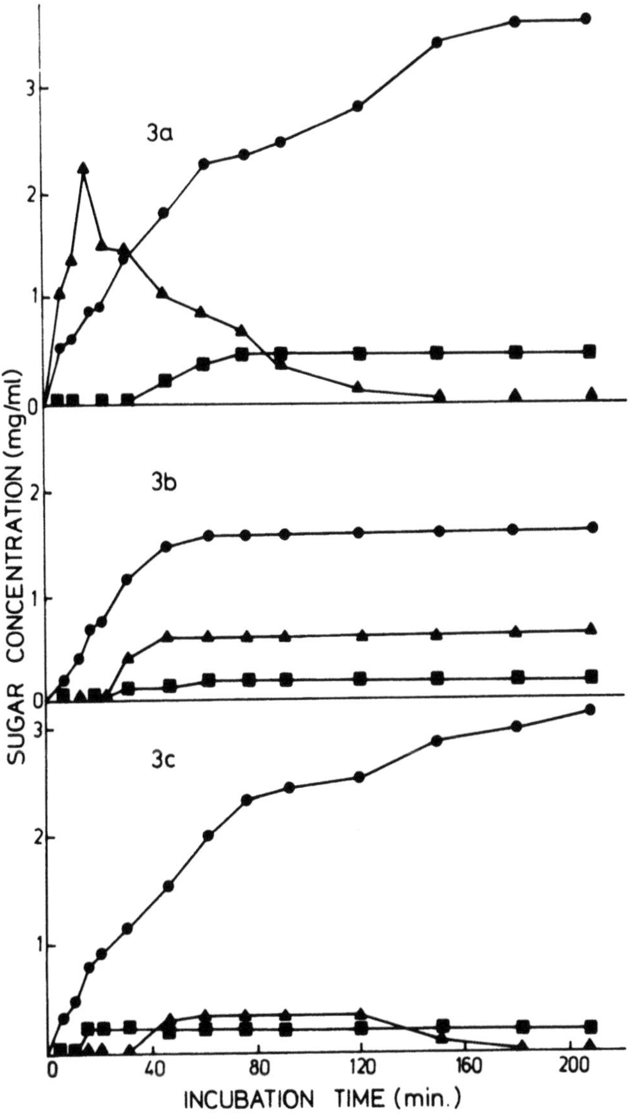

FIGURE 33. Production of glucose and disaccharides during digestion of yeast glucan with enzyme preparations from *B. aphyllophoroles* (a), *Basidiomycetes* sp. QM 806 (b), and *Novozyme-234* (c) (From Ryan, E. and Ward, O. P., *Process Biochem.*, 23(1), 12, 1988. With permission.)

In conclusion regarding cell rupture, it can be stated that for large-scale disruption the mechanical methods of disruption are seemingly the most popular.[10] High pressure homogenizers and bead mills are the most commonly used equipment. The high-pressure homogenizer is best suited for some bacteria and yeasts, while the bead mill appears most suitable for the disintegration of yeast cells and for mycelial organisms and algae. There is enough evidence to show that the mechanical cell disruption methods do not damage most intercellular enzymes and proteins; membrane associated enzymes and multienzyme complexes may be exceptions. Sadkova et al.[27] reported that a combination of organic solvent (methylene

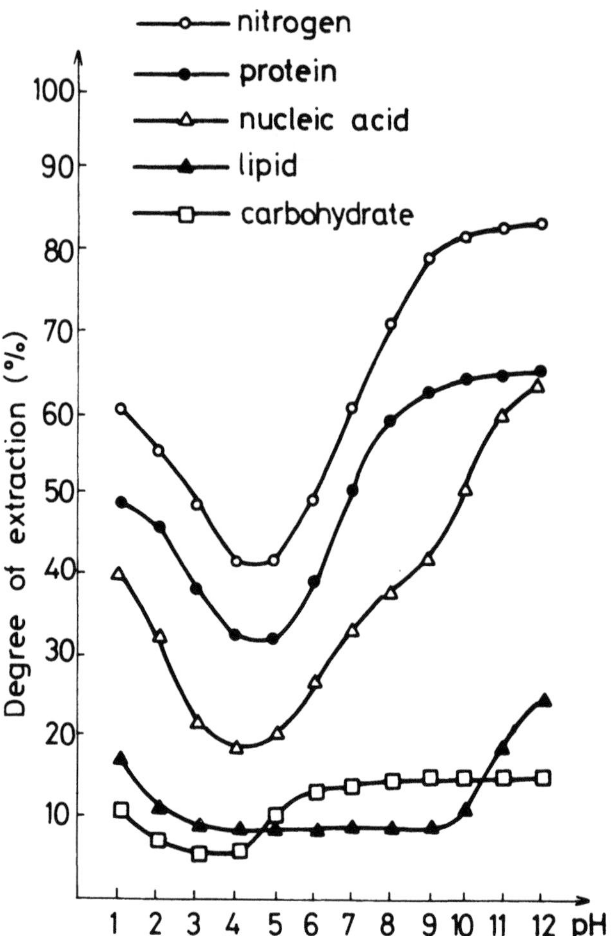

FIGURE 34. Extractability of soluble intracellular components from disintegrated yeast cells depending on pH. (From Sergeev, V. A., Solosenko, V. M., Bezrukov, M. G., and Saporovskaya, M. B., *Acta Biotechnol.*, 4(2), 105, 1984. With permission.)

chloride) and mechanical treatment does not cause any significant change in the properties of protein in disintegrated yeast cell. Other techniques which may have potential large-scale applications are ultrasonication, freeze-pressing and enzymatic lysis. A combination of two or more disruption techniques for disintegration of more resistant microorganisms may have economic advantages.

The resistance of yeast cells to disruption may be dependent on the condition of fermentation. This finding could be favorably used to manipulate the easy disruption of the cell. Studies of disruption kinetics and the influence of cell morphology on kinetics of yeast cell disintegration are needed.

II. EXTRACTION OF PROTEINS

Following the disintegration of the yeast cells using one of the methods described in the previous chapter, the next step in the yeast biomass processing is the extraction of soluble components of yeast cells. The most common procedure is extraction with dilute acid or alkali. The extractability of different components is highly dependent on the pH[24] (see Figure 34). Concerning the proteins, the minimum extractability is observed in the pH region 4 to

TABLE 43
Chemical Composition of the Mechanically Treated Yeast Mass and the Protein Isolate Produced at Different pH Values by Ultrafiltration

	Dry matter composition (%)				
Product	Protein	Nucleic acid	Lipid	Carbohydrate	Insoluble gel phase
Disintegrated yeast mass	46.0	7.0	10.0	18.0	40.0
Extracted at pH = 3.0	52.0	8.3	12.5	3.0	15.0
Extracted at pH = 8.0	60.0	8.5	13.0	4.7	18.0
Extracted at pH = 9.5	64.0	9.5	13.3	4.8	7.5
Extracted at pH = 12.0	66.0	10.5	14.0	4.9	2.5

From Sergeev, V. A., Solosenko, V. M., Bezrukov, M. G., and Saporovskaya, M. B., *Acta Biotechnol.*, 4(2), 105, 1984. With permission.

5. Only 30% of the total protein can be extracted under these conditions. At pH = 3 extractability of protein is significantly higher (40%) and it is important that at such conditions the extractability of nucleic acids is minimal. The extractability of carbohydrates and lipids is also low (see Table 43). If the pH is quite low (pH = 1 to 2) the solubilized amount of protein increases to 50% with a parallel increase of the solubility of nucleic acids, carbohydrates and lipids. Going to the higher pH values (8.0, 9.5, 12.0) after initial rapid increase the extractability of the proteins reaches a maximum (~ 63%). Unfortunately a big increase of the nucleic acid extractability occurs (~ 60%) at the same conditions. The solubilization of carbohydrates does not change practically and over pH = 10 increase of lipid solubilization is observed.

Keeping in mind the dependence of extractability on pH, it is understandable that the common procedure used for extraction of yeast protein is treatment with alkaline solvents (mainly solutions of sodium hydroxide). Extraction with alkaline media at pH = 9 to 12 is the most practical procedure if the main purpose is to achieve a high protein yield. The disadvantage of such a method is the high amount of contaminants, especially nucleic acid, as mentioned earlier (see Table 43 and Figure 34). Several authors reported that for maximizing protein yields and minimizing contaminant nucleic acids, alkali extraction at elevated temperature (80°C) is the most practical procedure.[2,3,25,26]

Extraction of protein from mechanically disrupted yeast cells with concentrated alkali is also possible.[2,25] The high concentration of alkali inactivates the proteolytic enzymes of the yeast, thereby controlling the weight loss due to proteolysis and causes reduction in nucleic acids. On the other side, this procedure has some important disadvantages:

1. Reduction of nutritive (biological) value
2. Formation of lysinoalanine and other amino acid artifacts
3. Partial saponification of lipids and isoelectric coprecipitation of fatty acids with proteins
4. Partial hydrolysis of nucleic acids to useless waste products
5. High viscosity at elevated pH values, leading to troublesome mechanical separation
6. High salt content
7. Large number of extraction, precipitation and washing steps, reducing product yield

To eliminate the drawbacks associated with the conventional methods of protein extraction from yeast a novel method employing succinylation to render the yeast resistant to proteolysis during extraction and nucleic acid reduction was developed by Kinsella and Shetty[35] resp. Shetty and Kinsella.[15] In this process yeast cells are ruptured in aqueous buffer pH = 8.5 and extracted. However, the inclusion of succinic anhydride, and maintenance

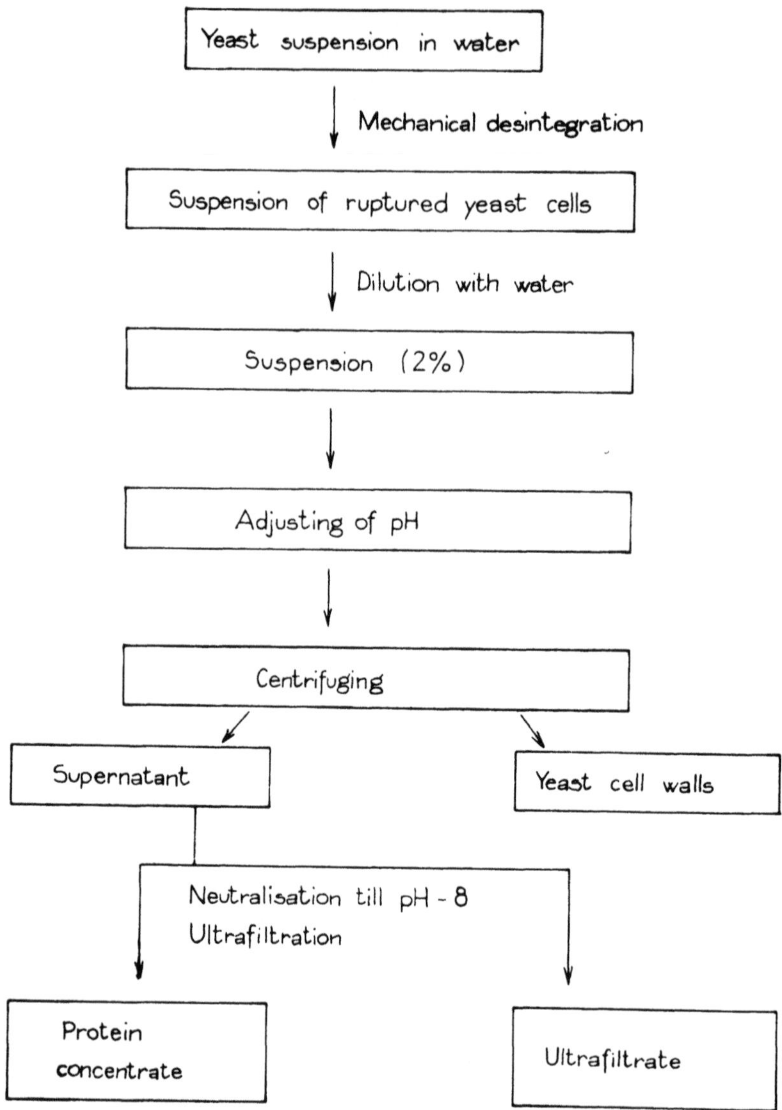

FIGURE 35. Scheme of the production of protein concentrate from yeast mass by ultrafiltration. (From Sergeev, V. A., Solosenko, V. M., Bezrukov, M. G., and Saporovskaya, M. B., *Acta Biotechnol.*, 4(2), 105, 1984. With permission.)

of pH at 8.5 during extraction resulted in a significant increase in protein extractability to 90% of the protein of the yeast cell. This result is very interesting because the pH of extraction represented very mild conditions compared to conventional procedures. Maximum extraction was obtained when 100% of all free amino groups were succinylated. Succinylation eliminated the degradation of proteins by endogeneous proteases and increased yields of intact solubilized protein. However, succinylation also inactivated the ribonuclease thereby impairing the hydrolysis of nucleic acid during extraction. Nevertheless, by precipitating the succinylated protein at pH = 4.5 a protein isolate containing more than 90% protein and less than 1.8% nucleic acid was obtained. The scheme of the procedure is shown in Figure 35.

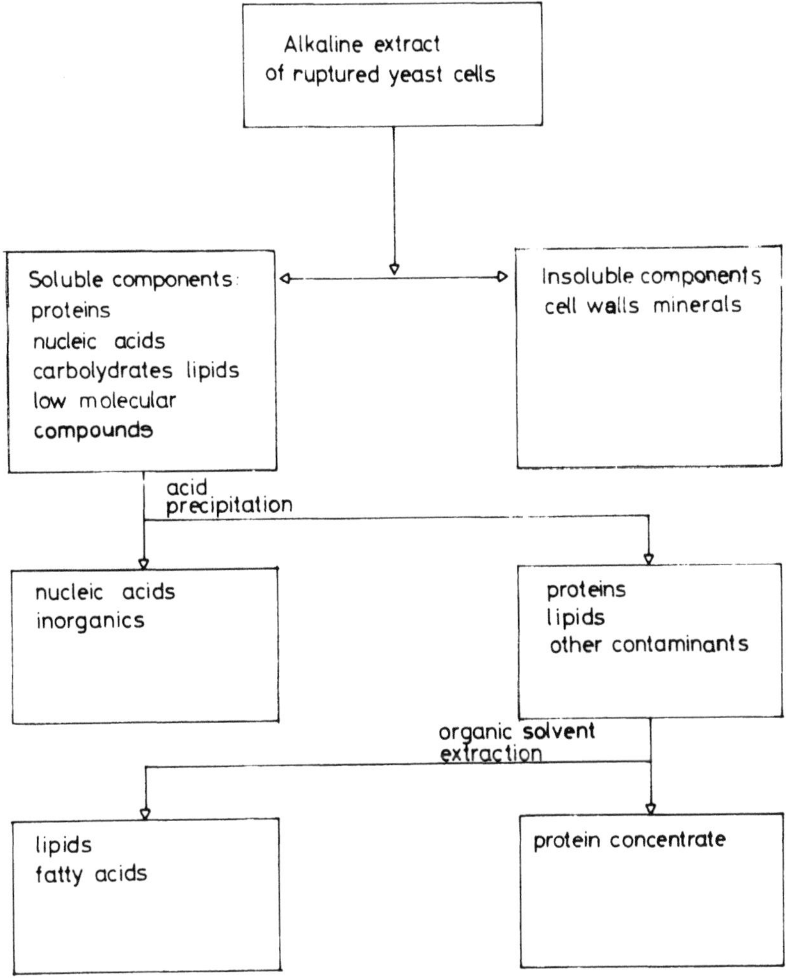

FIGURE 36. Principal steps of the refining of alkaline extract of disintegrated yeast cells.

III. REFINING OF THE EXTRACTED RAW PROTEIN

The alkaline extract of yeast mass is always a multicomponent system containing proteins, nucleic acids, lipids, carbohydrates and low molecular weight components such as minerals, amino acids, vitamins, etc. To produce a product containing a high amount of protein and as few as possible other contaminating components it is necessary to refine the extracted raw protein. The first step in the further processing of extract is precipitation of protein adjusting the pH of the extract to an isoelectric point of the yeast protein (pH ~4) or by heating the extract at 80°C. Washing, or repeated solubilization and precipitation of protein results in the removal of minerals and some low molecular weight soluble components. Use of ultrafiltration is also possible. In Figure 35 a scheme of production of protein concentrate by ultrafiltration is shown according to Sergejev et al.[24] A principal scheme of refining, including lipid removal by organic solvent, is demonstrated on Figure 36. In Figure 37 a scheme of protein isolation including removal and utilization of other components as byproducts is shown. According to the employed process[27] the dry protein concentrate contains about 80 to 90% protein. The cell wall dry matter contains about 20 to 25% of protein and 5 to 10% of lipid substances. The cell wall dry matter can be used for feeding purposes like the dry matter obtained from the soluble phase after separation of protein.

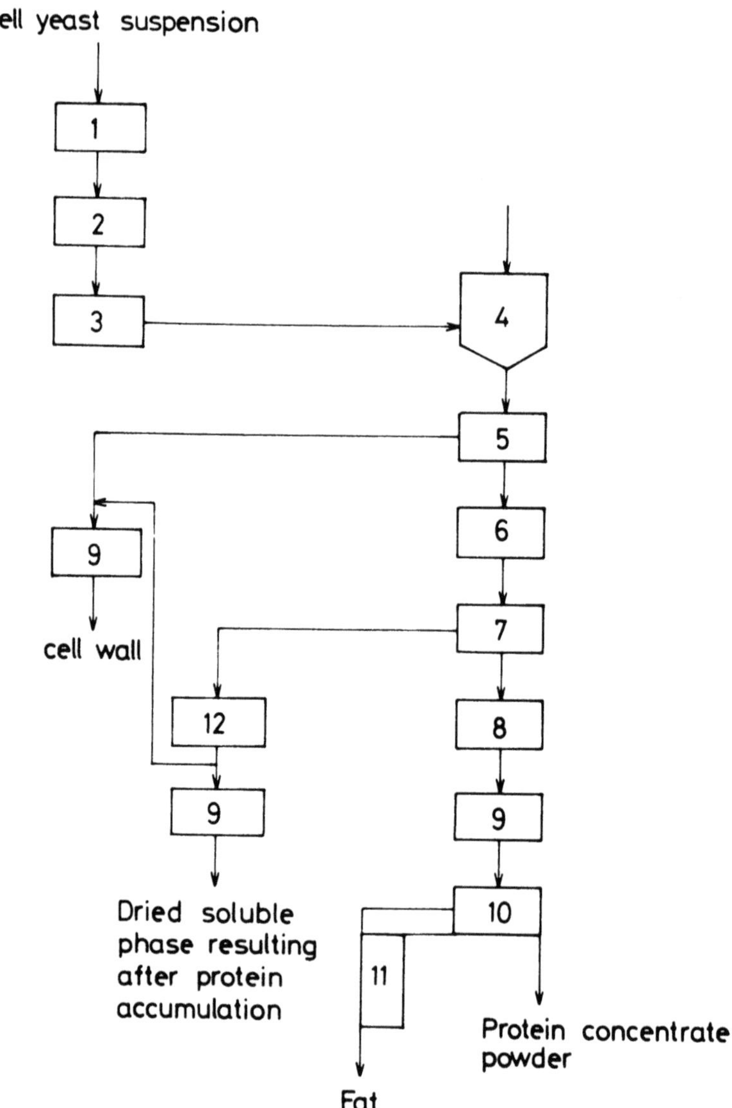

FIGURE 37. Scheme of the production of protein concentrate from yeasts, including the byproducts: 1 — container; 2 — dosing device; 3 — disintegrater; 4 — reactor; 5 — separator; 6 — reactor (removing of nucleic acids and precipitation of proteins); 7 — separation of protein; 8 — treatment of protein (washing, etc.), 9 — spray dryers, 10 — extraction of lipids from protein, 11 — solvent regeneration, 12 — treatment of soluble phase (evaporation). (From Machek, F., Fencl, Z., Beran, K., Behalova, B., Sillinger, V., and Kejmar, R., *Advances in Microbial Engineering, Part 2,* John Wiley & Sons, New York, 1974. With permission.)

This fraction consists mainly of the cell pool substances, such as amino acids, nucleotides, peptides, growth substances, carotenoids and salts. These could be dried together with cell walls and used as fodder. For removing the salts electrodialysis may be used.[28,29] Lipids extracted from the protein concentrate contain fats, fatty acids and a relatively high amount of carotenoids. If the yeast is cultivated on *n*-alkanes, hydrocarbons and odd fatty acids are also present. Depending on the kind of organic solvent and extraction method, lipid substances can be refined and used as edible or technical fat.

By this process 60 to 70% of protein existing in yeast cells can be isolated in the form

TABLE 44
Some Examples of Purification of Enzyme Proteins Extracted from Yeast

Enzyme	Purification step	Ref.
Aldyhide dehydrogenase	Acid and heat treatment, precipitation with ammonium-sulfate	28
	DEAE-Sephadex	
	NAD+-Sepharose 4B	
Phosphoglycerate kinase	Precipitation with ammoniumsulfate treatment	29
	DEAE-cellulose	
	Cibacron blue F3G-A Sepharose 4B	
	Sephadex G-100	
Phosphofructokinase	PEG-precipitation	30
	Cibacron blue affinity partitioning	
	DEAE-cellulose	
	Gelchromatography on Sepharose 4B-CL	
Cytochrome P-450	Subcellular fractionation, solubilization, ammoniumsulfate precipitation, 8-amino-*n*-octyl Sepharose, hydroxy apatite,	31 32
	CM-Sephadex C 50	
	DEAE-Sephacel	
Invertase	Acid and heat treatment, ethanol precipitation, ammoniumsulfate precipitation	33
	SE-Sephadex	
	DEAE-Sephadex	
	Concanavalin A-Sepharose	

of protein concentrate. The rest of the protein remains in the cell wall fraction and in the soluble phase obtained during protein separation.

In the last few years several intercellular enzymes began to be produced industrially. For example, β-galactosidase from *Kluyveromyces fragilis* and *Saccharomyces lactis*, invertase from *Saccharomyces cerevisiae*, glucose-6-phosphate-dehydrogenase from yeast, etc. Some other examples are given in the Appendix. In such cases the fractionation of proteins and purification of enzyme-containing fractions is necessary. Most methods of fractionation and purification require the use of chromatography, usually column chromatography often associated with specific ligand binding (affinity chromatography). Although these separation techniques are widely used in laboratories, an industrial realization is connected with many problems, e.g., zoning edge effect in wide columns and compacting absorbents under their own weight (decreased by using several columns stacked vertically). High flow rates, if so achieved, can cause overheating and other problems, so that careful temperature control is often necessary. Table 44 shows some examples of possible purification processes.

In Figure 38, a purification scheme for components of the microsomal electron transfer system from *Saccharomyces cerevisiae* is demonstrated. For a deeper study of the problems of the isolation and purification of protein and peptide products in the fermentation industry the reader is referred to a recent review of Wiseman et al.[36]

IV. REDUCTION OF NUCLEIC ACID CONTENT OF YEAST PROTEIN

One of the most significant problems associated with consumption of microbial cells or proteins isolated from the cells is the high nucleic acid content. The nucleic acid content ranges from 8 to 25 g nucleic acid/100 g protein in various cells. Most of the nucleic acid is present as RNA. Before yeast protein can be used as a significant source of protein for human consumption, the content of nucleic acids must be reduced so that daily intake of nucleic acid from yeast would not exceed 2 g on a dry weight basis. Higher quantities cause uricemia. Several methods were elaborated for the reduction of nucleic acid content. Main groups of the methods used are summarized in Table 45.

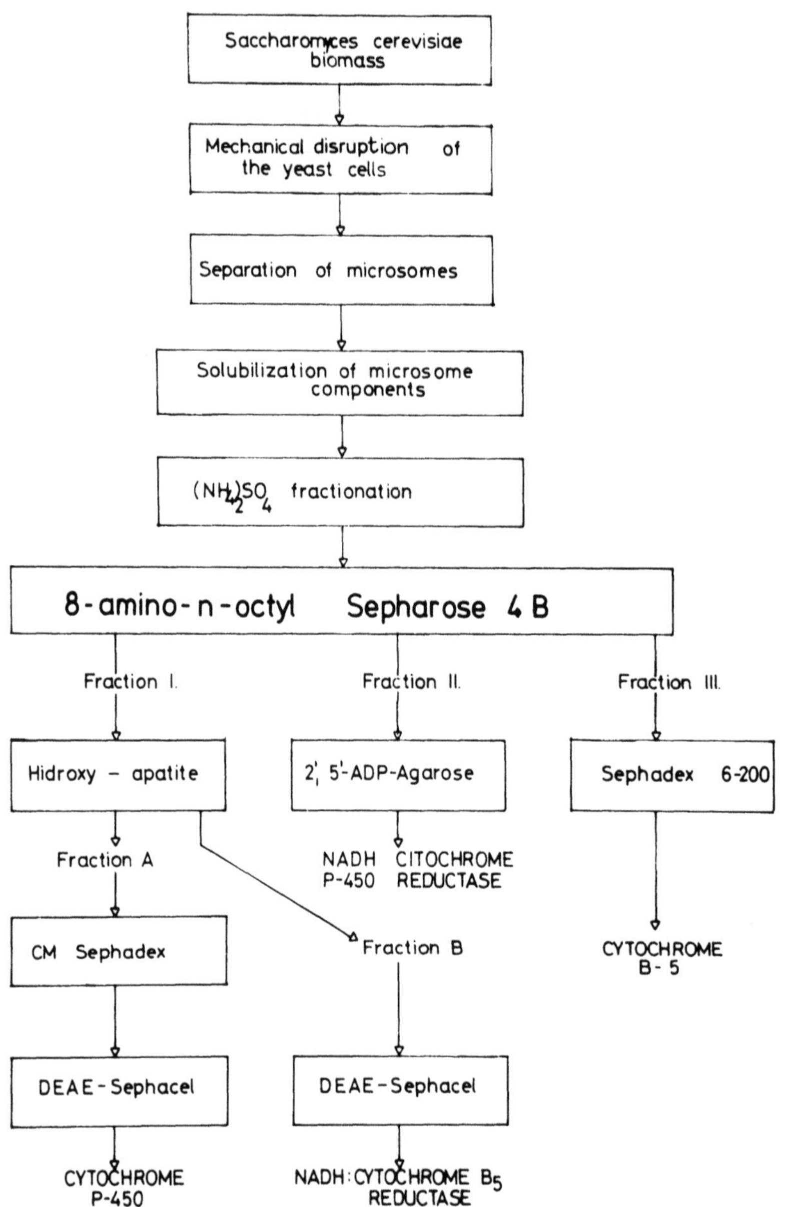

FIGURE 38. Purification scheme for the components of the microsomal electrontransfer system from *Saccharomyces cerevisiae*. (From Wiseman, A., King, D. J., and Winkler, M. A., *Yeast Biology*, Berry, P., Russel, I. R., and Stewart, L. F., Eds., Allen and Unwin Publishers, London, 1987, 433. With permission.)

Nucleic acid content of yeast cells can be reduced by decreasing growth rate, but this is not practical since rapid growth is obligatory in most processes for economic cell production. Few procedures using decreased growth rate are reported.[38,39] Thus, reduction of the nucleic acids present in the yeast cells or protein preparations from yeast cells is necessary if this protein is to be used in reasonable quantities as a source of food proteins.

Alkaline extraction under drastic conditions is a very effective method of reducing nucleic acid content especially at elevated temperatures. The disadvantage of the procedure is — as it was mentioned earlier — the alteration of proteins causing a decrease in nutritive value and functional properties.

TABLE 45
Methods Used for Protein Production with Reduced Nucleic Acid Content

Method	Disadvantages
Decreasing of growth rate during fermentation process	Not practical while for economic cell production rapid growth is needed
Alkaline extraction under drastic conditions (0.2 *M* NaOH followed by heat precipitation of protein at 80°C)	Loss of weight and nitrogen, degradation and denaturation of proteins and formation of new substances because of side reaction
Extraction of proteins from mechanically disrupted cells at slightly alkaline pH and incubating the extract for RNA reduction by making use of endogenous RNase and exogenous RNase	Loss of protein because of proteolysis during incubation process, costs
Treatment with hydrochloric acid	Loss of solid, corrosive nature of the extracting solution
Extraction of nucleic acids from protein with inorganic salts	Loss of protein
Chelating of nucleic acids with cetyl-trimethyl-ammoniumbromide	Costs
Acylation of proteins and alkaline treatment	Costs

With certain strains of yeast the so called heat shock procedure has been very successful in reducing RNA-content. Autolysis is triggered in all cells of the population by momentary exposure to a relatively high temperature (60 to 70°C). The activity of RNase can be increased by this procedure and a rapid hydrolysis of RNA will occur. As RNase in microbial cells is rapidly inactivated at this temperature level, temperature and time of exposure have to be carefully controlled. Generally, a short exposure (maximum 60 s) is used. The heat shock process for reducing RNA content of *Candida utilis* was developed by Otha et al.[40] This consisted of an initial heat shock, at 68°C for 1 to 6 s, of the yeast cells followed by incubation for 2 h around 52°C at a pH of 5 to 6. Hydrolysis of RNA by endogenous RNase and leaking of accumulated hydrolysis products into the suspending medium was observed. The nucleic acid content was reduced from an initial value of 8% to less than 2%. The process is unfortunately associated with increased activity of proteolytic enzymes and with loss of protein.[41]

As it was stated by Lindblom[42] the heat shock process cannot be used for the reduction of the RNA in *Saccharomyces cerevisiae*. Halász,[43] studying the effect of heat shock on the activity of yeast strains *Saccharomyces cerevisiae*, *Candida boidinii*, *R. glutinis* and *C. guilliermondii*, found that the RNase of *Saccharomyces cerevisiae* cannot be activated by heat shock. Heat treatment over 45°C always caused a decrease of RNase activity. In the case of *Rhodotorula glutinis* and *Candida guilliermondii* an activation of RNase was observed at 60 and 68°C, respectively. It was suggested by the author that the heat shock activation is probably associated with a rapid inactivation of a heat sensitive RNase inhibiting protein present in the different strains of yeast in different amounts. On the basis of this hypothesis the difference observed between different yeast strains is understandable.

The addition of carboxylic anions during the heat shock process facilitates the reduction of nucleic acids.[44] Sodium chloride improves the degradation of nucleic acids by endogenous RNase.[42] In a more recent study of heat shock process Bueno et al.[45] reported RNA reduction of different yeast strains. Different results were obtained from each of them. *Candida utilis NRRLY-660* shows its best performance after an 8 s heat shock in the presence of 3% NaCl. For commercial baker's yeast *Saccharomyces cerevisiae* and *Kluyveromyces fragilis L-1930* similar results were obtained with only 1% NaCl. The latter needed longer heat shock periods, e.g., 15 s to give such an RNA reduction. Biomass recovery ranged from 60 to 75%, being higher for *C. utilis* and *K. fragilis* while excessive losses were observed in *S. cerevisiae* cells. No significant protein deterioration was observed in the samples. The amino acid

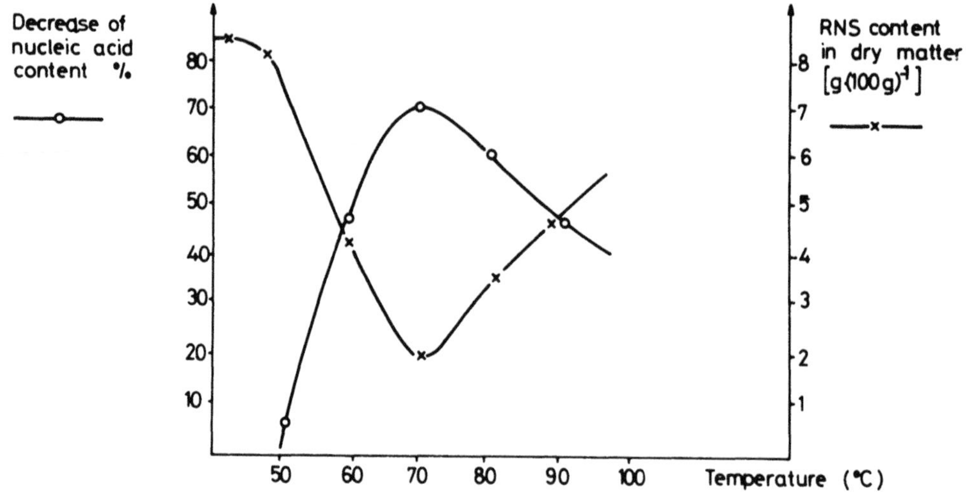

FIGURE 39. The effect of the temperature of heat shock treatment on the nucleic acid reduction in brewer's yeast (incubation at 40°C 1 h and at 50°C 2 h). (From Hálász, A., D.Sc. thesis, Hungarian Academy of Sciences, Budapest, 1988. With permission.)

profile appears to be improved in comparison to the starting material in these strains after RNA reduction. The RNA content of yeast cells treated by heat shock method at 68°C and incubated 2 h at 52°C ranged from 1.81 to 2.68% depending on the yeast strain.

Lindblom and Mogren[46] added 4% NaCl to homogenized yeast cell suspension and observed a remarkable reduction at incubation temperatures between 48 and 62°C and pH 5 to 9. At the lower pH values, protein precipitation occurred during the incubation, hence the pH should be in the neighborhod of 9. Acidic precipitation should be slow so that large flocks of protein are formed. This facilitates separation of the protein with a nucleic acid content of 1 to 3%.

The possible use of the heat shock process for reducing the nucleic acid content of brewer's yeast was studied by Hálász.[43] A heat treatment at 70°C for 60 s followed by incubation at 40°C for 1 h and at 50°C for 2 h have been found to be optimal (see Figure 39). Addition of NaCl increased the activity of the endogenous RNase. The concentration of 4% has been found to be optimal (Figure 40). Experiments made at different pH values showed that incubation at pH = 8 is optimal. Unfortunately, the loss in protein is also the highest at this pH value (Figure 41).

Schuldt[47] patented a method for nucleic acid reduction characterized by the addition of zinc to yeast. This metal acts as a protective factor for RNase during the alkaline treatment for protein extraction. A significant improvement of nucleic acid reduction was achieved using incubation at pH 6 and 60°C. The addition of Zn can be carried out at several stages, i.e., from the fermentor to just before the cell homogenate is made alkaline. At any subsequent point the RNase is not protected.

The use of exogenous enzymes for removal of RNA is also possible. It was reported[47] that exogenous pancreatic RNase could be used for the reduction of nucleic acid content of yeast. According to another patented process[8] of making isolated yeast protein low in nucleic acid, exogenous RNase from malt sprouts is added to the ruptured yeast cell mass. Due to the high costs of exogenous enzyme preparations, procedures based on exogenous RNases are not practical for use in large-scale production.

Although a reduction of nucleic acids is possible using hydrochloride acid treatment, such a procedure is not commonly used in laboratory and large-scale production praxis. In early experiments Trevelyan[50] found that hydrochloric acids (concentration 0.5 to 0.1 *N*)

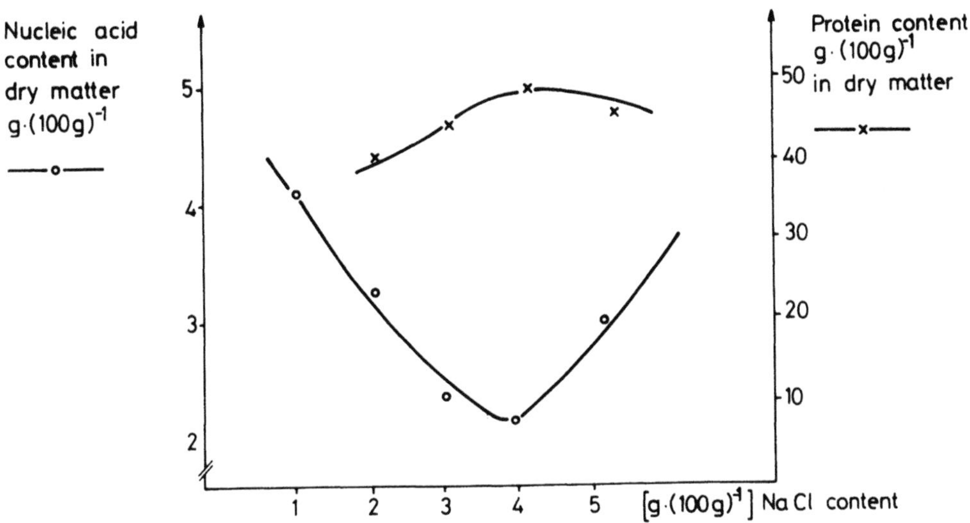

FIGURE 40. The effect of NaCl concentration on the activity of endogenous RNase of brewer yeast. (From Hálász, A., D.Sc. thesis, Hungarian Academy of Sciences, Budapest, 1988. With permission.)

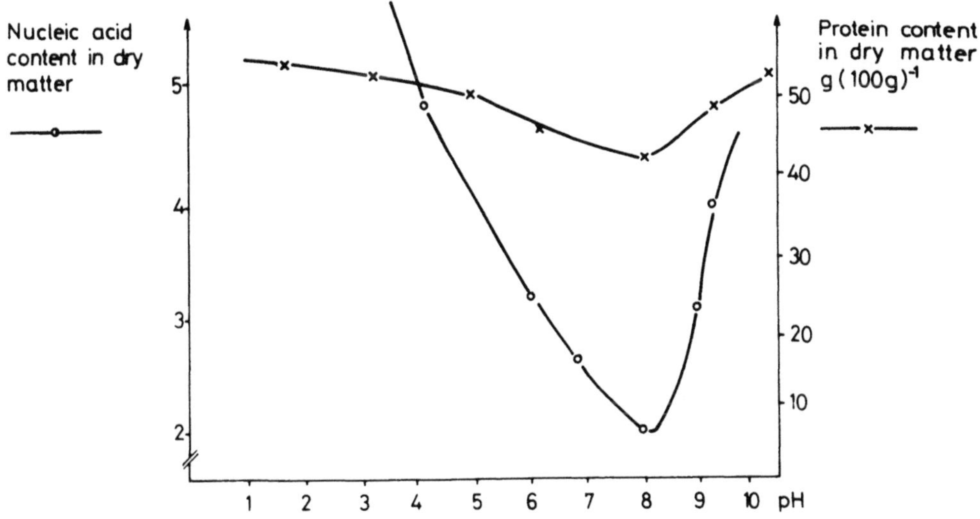

FIGURE 41. The effect of pH on the degradation of nucleic acids by intracellular RNase. (From Hálász, A., D.Sc. thesis, Hungarian Academy of Sciences, Budapest, 1988. With permission.)

extracted RNA at room temperature from pool depleted yeast with a relatively small loss of solids. Further experiments of the same author[49] showed that salt (NaCl) potentiated the action of hydrochloric acid. By operating at a temperature of 60 to 80°C the concentration of acid required could be considerably reduced. Based on these results Trevelyan[49] reported a procedure using sodium chloride containing hydrochloric acid solution and mild heat treatment for reduction of nucleic acids. According to this procedure first the low molecular weight constituents of yeast cell are removed by extraction with water at 70 to 100°C with considerable reduction in nuisance due to odors. A suspension of heat killed yeasts (6% yeast solids) was prepared using 0.5 *M* NaCl solution and the pH of suspension was adjusted to 1.4 with 3.5 *N* hydrochloric acid. With live yeast at 25°C the acid penetrated cell membrane slowly. The initial pH was 0.9 and only after 21 h reached 1.4. After centrifuging the yellow supernatant then contained some 18% of yeast total *N*. At 70°C the membrane was immediately ruptured with the release of coenzymes. In addition, RNA was rapidly degraded to diffusible products. The content of RNA fell in the washed cells to 1.1% after 3 h. At this point the cells had lost 17% of their initial corrected crude protein (CCP); as 15% is to be ascribed to the loss of amino acids and other low molecular weight constituents, the fall in true protein content was small. The protein nucleic acid ratio in the washed cells was 50, considerably above the target value. In conclusion, the author stated that extraction of baker's yeast with sodium chloride solution at pH 1.4 is the most efficient method of reducing RNA content.

Its distinguishing feature is the very small loss of protein from the yeast cells, and its obvious drawback the corrosive nature of the medium, especially at 70°C. Nutritionally valuable constituents can be recovered from the yeast by a preliminary extraction with hot water, a step which reduces odor problems during subsequent extraction of RNA, and also consumption of acid. The spent acid/salt medium is then fairly clean, and possibly could be reused, especially after treatment with an adsorbent for nucleic acid breakdown products. The acid-extracted cells settle rapidly, opening up the prospect of washing by simple decantation before the yeast is neutralized, finally concentrated and dried.

One of the first procedures using an extraction of nucleic acids by salts is reported by Canepa et al.[51] In laboratory-scale experiments they reduced the nucleic acid content by extraction with Na_2HPO_4 in alkaline medium. The investigated yeast strain was *Saccharomyces cerevisiae*. An extraction method using 10% solution of sodium chloride at high temperatures was published by Decker and Dirr.[52] An interesting procedure was developed by Lawford et al.[53] for reducing the nucleic acid content of protein isolates. The protein preparation is solubilized in a citrate-phosphate buffer (pH = 7 at concentration of 0.1 *M*) and acetyl-trimethyl-ammonium bromide is added. The RNA forms an insoluble complex with this reagent and the precipitate can be removed. After treatment the nucleic acid content of the protein isolate is reduced to a value lower than 3%.

Kinsella and Shetty[37] resp. Shetty and Kinsella[17] developed a new method for protein extraction from yeast characterized by succinylation of protein. This procedure resulted in a higher protein yield. In addition, this method was also effective in the reduction of nucleic acid content. The disadvantage of this procedure is that the final product is succinylated protein and cannot be removed under mild conditions. So it is unlikely that succinylated yeast protein can be used as a practical source of dietary protein.

Studies to find a reversible modifying agent were successful and it was demonstrated that citraconic anhydride[54,55] and dimethylmaleic anhydride[56] can be used as such agents. Shetty and Kinsella[57] studied the possiblity of isolation of yeast protein with a reduced nucleic acid level using reversible acylating reagents: maleic anhydride and citraconic anhydride. Brewer's yeast *Saccharomyces carlsbergensis* was used in the experiments. Yeast cells were disrupted using a homogenizer and the pH of disintegrated cell mass was increased to 9.5 by addition of sodium hydroxyde. After centrifuging the protein was precipitated from clear supernatant by three separate methods: isoelectric precipitation (pH 4.5), ammoniumsulfate precipitation and by precipitation with ethanol. All three samples of protein contained high levels of nucleic acids (22 g nucleic acid/100 g protein).

The elution pattern of alkaline extracted yeast proteins from the Sepharose 6B column showed greater adsorbance at 260 nm compared to that at 280 nm in all the fractions, indicating that nucleic acids and proteins are present as nucleoprotein complexes in the disrupted yeast mass. The complexation of the protein and nucleic acids may have occurred during disruption of cell wall and/or the extraction procedure. Nucleoproteins are stabilized by weak noncovalent forces. Ionic linkages occur between the anionic phosphate groups and the cationic groups of the basic amino acids of the proteins, and hydrophobic interactions occur between hydrophic regions of the protein and bases of the nucleic acids. If the basic

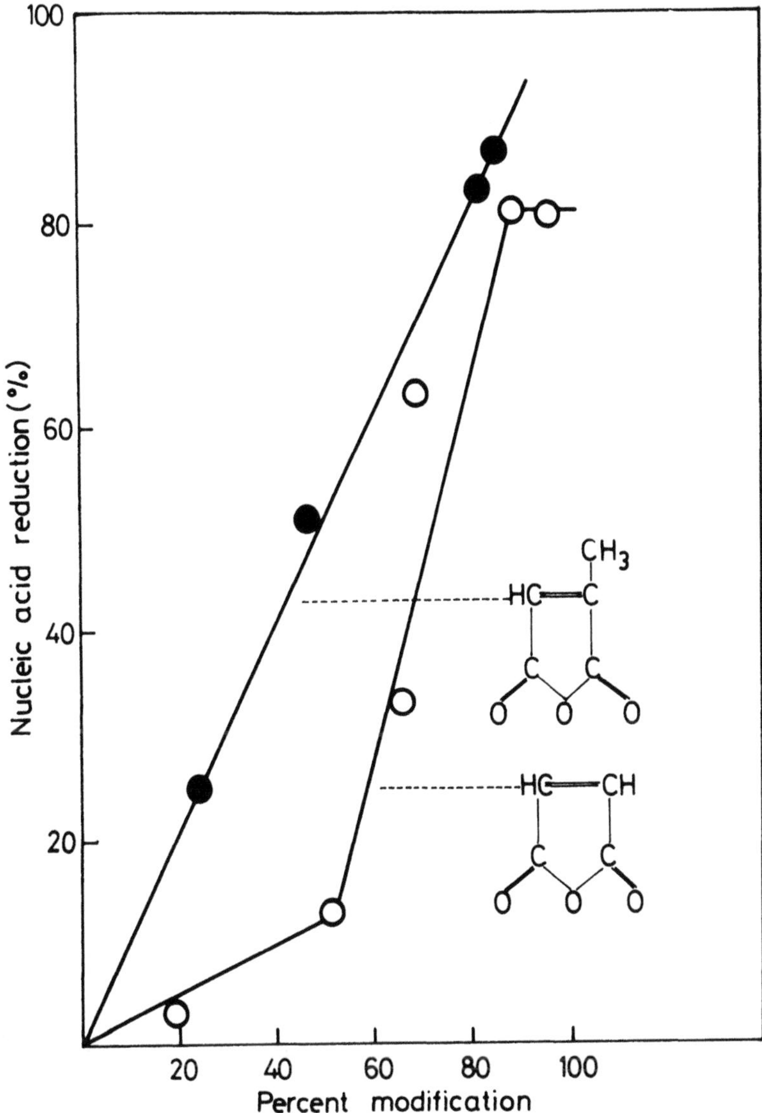

FIGURE 42. Effect of E-NH_2 group acylation on the reduction of nucleic acid content in the yeast protein precipitated at pH 4.2: (o) maleic anhydride; (●) citraconic anhydride. (From Shetty, J. K. and Kinsella, J. E., *Int. J. Pept. Res.*, 18, 18, 1981. With permission.)

amino acid side chains are acylated during maleic anhydride or citraconic anhydride the number of electrostatic bonds will be reduced. In addition, repulsion between anionic phosphate groups and added carboxylic groups will be increased. As a result of these changes the protein-nucleic acid complex will be destabilized and the removal of nucleic acids facilitated. The nucleic acid content of the protein precipitated at pH 4.2 following the modification of the proteins to varying degrees by these anhydrides was progressively decreased (Figure 42). The extent of nucleic acid reduction in yeast protein precipitated at pH 4.2 was proportional to the number of lysine residues modified with anhydrides.

For deacylation of the proteins a 1% (w/v) suspension of acylated protein was dispersed in water. The pH was adjusted to 4.0 and incubated at 30 to 50°C for 1 to 3 h. The integrated scheme of the separation of proteins and nucleic acids from yeast is shown at Figure 43.

Another way to destabilize the nucleoprotein complex in yeast protein preparations with

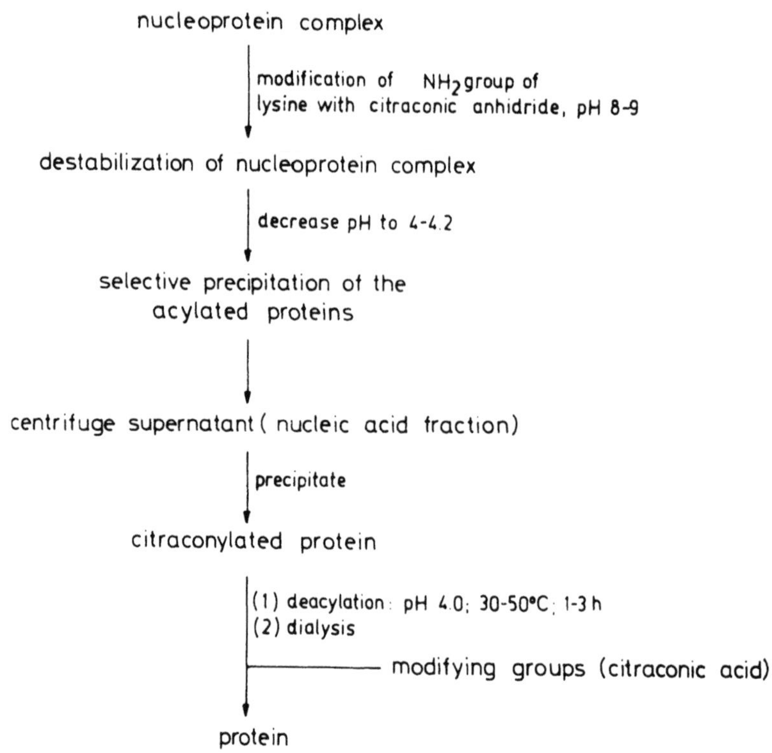

FIGURE 43. Scheme showing the separation of proteins and nucleic acids from a nucleoprotein complex using reversible modifying reagents of free amino groups in the protein. (From Shetty J. K. and Kinsella, J. E., *Int. J. Pept. Res.*, 18, 18, 1981. With permission.)

a chemical method was reported by Chen et al.[58] It was stated that chemical phosphorylation of yeast protein preparation by sodium-trimetaphosphate (STMP) destabilized the nucleoprotein complex. The method involves the addition of STMP at pH 12.0 and incubation for 6 h at room temperature. The modified protein was isoelectrically precipitated at pH 4.2. Most of the RNA after centrifugation remains in the supernatant. The nucleic acid content of the protein isolate was significantly reduced from the original 22% (w/w dry protein basis) to 5%. Some functional properties of the phosphorylated protein are also reported. After protein precipitation, about 90% of the RNA could be precipitated from the supernatant by acidification to pH = 1.5 giving a product containing 50% RNA. The RNA was treated first with 5′ phosphodiesterase, then with 5′ AMP desaminase to produce 5′ IMP and 5′ GMP could be used for preparation of nucleotide flavor enhancers.

The procedure using phosphorylation was also studied by Huang and Kinsella.[59,60] It was found that phosphorylation of 3% of lysil groups of yeast nucleoprotein facilitated preparation of phosphoprotein (72%) with only 2.7% nucleic acid content. It was also stated that some of the functional properties of protein isolate were significantly improved.

Brewer's yeast as a byproduct in brewery for production of preparations for food use not only needs the reduction of nucleic acid content, but the removal of dark color and some bitter substances too. A combined procedure for solving both problems was studied by Halász.[43] Sodium hydroxide and ammonium hydroxide solutions were used in the experiments. The concentration of solutions were 0.5 *N*, 1.0 *N* and 2.0 *N*, respectively; the temperatures of treatment 10, 20 and 30°C, while the time of treatment was 30, 60, and 120 min. Efficient reduction of nucleic acid content was achieved by treatment with NaOH, however, the ammonium hydroxide does not significantly reduce the nucleic acid content of brewer's yeast under conditions mentioned above. A treatment with 0.5 *N* NaOH for 60

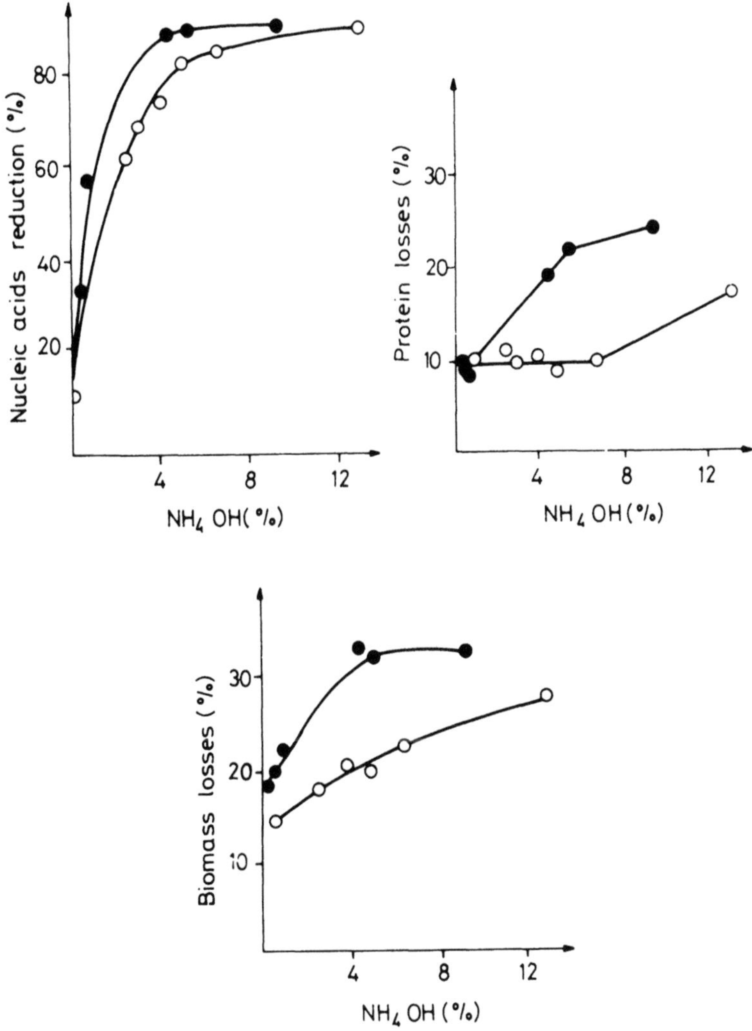

FIGURE 44. Comparison of nucleic acid levels, protein- and biomass losses between the strains of *S. cerevisiae* (○-○-) and *K. fragilis* (●-●-) after 30 min treatment at 65°C. (From Alvarez, R. and Enriquez, A., *Appl. Microbiol. Biotechnol.*, 29, 208, 1988. With permission.)

min at 30°C obtains a product with a nucleic acid content lower than 2% and satisfactory color and taste. As another example of complex treatment a patented procedure of Wagner and Shay[61] should be mentioned. A nucleic acid reduced allergen-free SCP product is obtained by culturing yeast, fungi or bacteria on low sulfate medium, treating the cells with alkali (pH 9.5) then with acid (pH 4.0). Finally the culture is HTST (high temperature, short time), heat shocked, extruded and mealed.

A combined heat treatment and ammonium hydroxide addition was employed in the procedure of Alvarez and Enriquez.[63] Yeast strains *Saccharomyces cerevisiae* and *Kluyveromyces fragilis* were investigated. Yeast suspensions containing 14% dry matter were put in a reactor and treated with ammonium hydroxide of 0.3 to 13.0% concentration (related to yeast dry matter) for 30 min at 45, 50, 60, 65, 70 and 80°C. On the basis of experimental data a 4.5% ammonium hydroxide concentration, 65°C and 30 min incubation temperature and time have been found to be optimal. A low nucleic acid content (less than 2%) was obtained for both strains. Higher losses of proteins and biomass were obtained with *K. fragilis* than with *S. cerevisiae* (see Figure 44). The strain *S. cerevisiae* showed higher resistance to ammonia attack, probably due to more resistant cell walls, as indicated by the lower protein, biomass and nucleic acid losses. In conclusion, the authors stated that the treatment with ammonia shows several advantages in relation to other chemical agents. With a relatively low concentration of ammonium hydroxide (4.5%), a final low nucleic acid content (1.2 resp. 1.4%) can be achieved. In addition, the use of ammonia is more suitable from the point of view of its elimination and/or utilization of residues after treatment. No indication was reported by authors regarding quality of protein with reduced nucleic acid content.

V. NUTRITIVE VALUE AND SAFETY OF YEAST PROTEIN

Nowadays, there is a growing interest in the nutritional quality and safety of foods in general and especially of novel protein sources and foods. From a nutritional point of view the protein quality of foods is interesting. The causes of such situations are complex. The fact that some protein shows different nutritional values for different consumers and the biological value also varies with the age of the consumer offering a new possibility of effective and economical management of proteins; sources of protein can be utilized in the forms in which they represent the highest biological efficacy. Another cause determining the significance of the problem lies in laws and regulations specifying the necessity of ample information. In some countries the biological value of marketed foods must be declared and the registration of new products is bound to be a declaration of protein quality. Analogous problems arise when the task is to maintain the biological value of food where the protein of animal origin is partly replaced by protein of plant resp. microbial origin. Naturally, in the evaluation of the nutritive value of food, other components (carbohydrates, lipids, vitamins, minerals) also play an important role and cannot be neglected. As mentioned in the introduction of this book, a satisfactory energy supply (carbohydrates, lipids) is a prerequisite for efficient protein utilization. Nevertheless, in the framework of this chapter we will discuss only the problems connected with the protein. Readers interested in a more general nutritional evaluation are referred to Chapter 2 (chemical composition of yeast) and to some basic books in food chemistry and nutrition.[64,65]

The ever increasing amount of literature on toxic substances occurring naturally in foods has contributed to the recognition of possible health risks of all types of foods and feeds. Novel protein sources are often derived from groups of organisms which harbor toxin-producing species. Moreover, during their production, the materials are often submitted to treatment by heat, solvents or alkali, each of which may import undesirable, or even toxic properties to the resulting product. It would seem essential, therefore, to demonstrate the safety of all single cell products, including yeast products.

Toxicologists and nutritionists generally dislike detailed instructions concerning the conduct of their studies. Nevertheless several national and international guidelines exist in this field. From our point of view the well known PAG Guideline 6[66] for pre-clinical testing of novel sources of protein could be interesting as well as older and more recent reviews of the subject by different authors.[67-70] Analytical data concerning safety include solvent residues, nucleic acids and nucleotides, trace elements and potential contaminants. However, the most important and major part of the testing program associated with safety of microbial protein is the study of experimental animals and, in the end, human feeding studies too.

A. ESTIMATION OF PROTEIN QUALITY

As it was briefly treated in the introduction of this book the nutritive (biological) value of proteins depends on the relationship between essential amino acids present in the utilizable proportion of the proteins and the essential amino acid requirement of the consumer. Two approaches to the determination of the nutritive value of proteins exist: biological (including

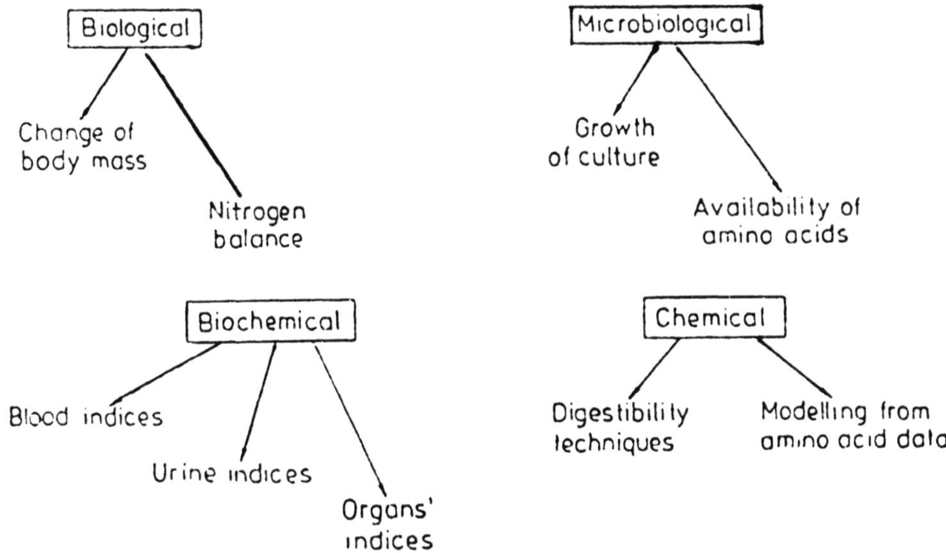

FIGURE 45. Grouping of methods used for evaluating protein quality.

microbiological) and chemical or biochemical (see Figure 45). It is generally accepted to speak about *in vivo* (biological) and *in vitro* (chemical) methods.

The biological methods or bioassay procedures are generally based on measurements of growth or nitrogen retention in experimental animals as a function of protein intake. Clearly, for the evaluation of the protein quality of foods, methods involving the use of humans would be ideal, but cost and time prohibit this. So *in vivo* methods using the rat are well established, although there is debate over which is the most suitable. For reliable accuracy and meaningfulness of the data, several animals must be used per test, and the results must be evaluated statistically. Test conditions must also be standardized. The protein level of the diet is generally kept low (about 10% by weight) so that protein intake remains below requirements. The supply of energy and other nutrients must be adequate. Under conditions mentioned above the growth is slow, the protein is efficiently utilized and the experimental results emphasize differences in nutritive values among proteins and reflect the maximum nutritive value of the tested proteins. In this chapter no attempt has been made to reference the literature completely, as so much published material now exists in the field of protein quality evaluation. Instead, more recent research is discussed, together with comments on problems of interpretation of data for human nutrition. References to reviews of the older literature are given whenever appropriate.

1. Bioassays Using Rats

When protein quality data are to be interpreted in terms of human nutrition, the rat is normally chosen for *in vivo* assay. The advantages and disadvantages of bioassays using laboratory animals are discussed in detail in many reviews.[64,71-74] Here we only note that, in general, man requires less protein for growth and more for maintenance than does the rat. Also the rat's requirements for the sulfur-containing amino acids and lysine are higher than the requirements of man. So the use of the rats may lead to an underestimate of the quality of protein for man. Despite the inadequacies of the rat model, rat bioassays continue to be the principal laboratory technique for protein quality assay. For the numerical characterization of the protein quality the *protein efficiency ratio* (PER) is the most commonly used method because this value is easy to determine. The PER is the weight (g) gained by rats per g protein consumed. The *net protein ratio* is similar to the PER, but also involves a measurement of weight loss of a group of rats fed a protein-free diet.

$$\text{NPR} = \frac{\text{(weight gain)} - \text{(weight loss of protein-free group)}}{\text{protein ingested}}$$

These methods can be improved if values are determined at several different levels of protein in the diet, a plot of change in weight vs. protein intake is prepared, and the slope of this plot is measured on the straight line portion of the response curve (multipoint slope ratio). A second possible modification for improving the method is expressing the values obtained with the test protein as a percentage of those obtained with reference protein (egg, or milk). When both modifications are adopted, the protein values obtained (relative protein values = RPV) are more accurate and more useful in practical situations.

In the nitrogen balance studies the uptake of nitrogen and loss of nitrogen in urine and feces is measured. On the basis of results, the following characteristic values are calculated.

$$\text{Biological value} = \text{BV} - \text{absorbed N from food} - \text{(urinary N} - \text{endogenous urinary N)}$$

$$\text{Net protein utilization} = \text{NPU} = \frac{\text{body N of test group} - \text{body N of group fed non-protein diet}}{\text{N consumed}}$$

$$\text{True digestibility} = \text{TD} = \frac{\text{N intake} - \text{(fecal N} - \text{endogenous fecal N)}}{\text{N intake}}$$

When a bioassay is carried out by making one amino acid nutritionally limiting while all other free amino acids are added to the diet in appropriate amounts, the result indicates the bioavailability of the single limiting amino acid in the test protein.

2. Bioassays with Humans

Methods for the evaluation of protein quality in humans are discussed extensively in the book edited by Bodwell[75] and Lásztity and Hídvégi.[76] Most bioassays for protein quality using humans are based on nitrogen balance studies. As well as the amino acid composition, digestibility of protein is an important factor in determining its protein quality, the digestibility being influenced by its structure or its association with other food components such as lipids. Digestibility of a protein may be measured as *true digestibility* or *apparent digestibility* for which no correction is made for endogenous fecal nitrogen loss on a protein-free diet. Despite the difficulties of protein quality assay using humans, values for true digestibility obtained by different authors on the same foods show little variation.

Commonly, the time needed for nitrogen balance studies in humans is about 14 days. Therefore, there is a tendency to find short-term methods. The basic criticism of nitrogen balance studies is that protein quality tends to be overestimated, as protein is fed in inadequate levels.

3. Bioassays with Microorganisms

Some microorganisms, especially protozoan *Tetrahyema pyriformis*, possess essential amino acid requirements similar to those of humans and the rat. These microorganisms also possess their own proteolytic enzymes. Their growth on test proteins can be used as an index of protein quality (T-PER) or amino acid availability.

Microbiological assays for protein quality have several advantages over animal assays. Although not rapid by chemical standard, they are by those of rat bioassay standards, requiring little space, cost and equipment. The method is highly empirical, the correlation with the results of other bioassays is poor, the assay is not yet at the stage where it could be considered for routine use. Other organisms that have been used in the study of protein quality include *Leuconostoc mesenteroides*, *Streptococcus faecalis* and *Clostridium perfringens*. Microbiological methods at best give a crude ranking order for protein quality, but may be more useful in the future for the determination of the availability of amino acids.

4. Chemical Methods

a. In Vitro Digestibility

All *in vitro* methods for the determination of digestibility of human foods based on the use of enzymes, usually those of the digestive system, and many simulate physiological conditions. In early studies[77,78] one enzyme (pepsin) was used for protein digestion. Saunders et al.[79] studied a number of leaf protein concentrations using a two-enzyme (pepsin and trypsin) technique. These authors obtained a high correlation for apparent digestibility in rats with *in vitro* digestibility obtained by pepsin/trypsin technique. The three-digestive enzyme process (trypsin/chymotrypsin/peptidase) described by Hsu et al.[80,81] introduced a new mode of detection, after beginning of the incubation with enzymes at pH 8 the change in pH after a period of time (usually 10 m) was measured. The results obtained with this *in vitro* method were in good correlation with *in vivo* data in rats. Similarly good correlation was found by Pedersen and Eggum[82] investigating digestibility of different proteins of plant origin and their mixtures with animal proteins. Satterley et al.[83] introduced a four-enzyme digestion technique in which the protease from *Streptomyces griseus* is the fourth digestive enzyme and the basis of detection is also the decrease of pH. The four-enzyme *in vitro* method was studied and compared with *in vivo* data by Rich et al.[83] and Bodwell et al.[84]

Although the pH-decrease technique is a promising rapid method, it is associated with some methodological problems: the buffer capacity of the substance must be measured and taken in consideration in correction and processes of protein digestion proceed *in vivo* in the small intestine at a constant pH 8 value, however, in *in vitro* method the pH is changing.

The methodological problems mentioned above were partly solved by Pedersen and Eggum[85] by the introduction of the pH-stat technique allowing enzyme action at constant pH value. These authors reported that correlation of results obtained with pH-stat technique with *in vivo* data have been improved in comparison with the pH-decrease technique.

A simplified enzymic method for prediction of protein digestibility was published by Salgó et al.[86] A two-digestive enzyme (trypsin/pancreatin) *in vitro* measuring technique was used in the pH-decrease and pH-stat methods, respectively. Parallel to the *in vitro* measurement, true and apparent digestibility was investigated in *in vivo* experiments on rats. It was established from data obtained for 45 test proteins that results obtained with the pH-stat method in a digestion period of 10 min are closely correlated (see Figure 46) with *in vivo* data, digestibility can be well predicted from the *in vitro* data.

The two-enzyme method, using pH-stat technique, gives a similar correlation to that of multienzyme techniques. The costs of the method are less than one third of that of multienzyme tests. In processes accompanied by change in protein quality, the method is suitable for the rapid screening of digestibility.

b. Biochemical Assay

Numerous attempts have been made to use metabolic parameters for predicting protein value. The protein metabolism in skeletal muscles and in liver, muscle growth, cell size or weight, protein concentration in liver, activity of different enzymes, content and relationship of free amino acids in blood plasma, the creatinine excretion in urine should be listed. All these parameters show, more or less, relations to protein intake or protein quality but they are not reliable approaches for evaluating protein quality.[87] It seems that the only exception is the blood urea concentration (BUC) and according to Münchow,[88] the reciprocal blood urea value (RHW on the basis of initials of corresponding words in German). He found a very high correlation between BV resp. NPR and RHW:

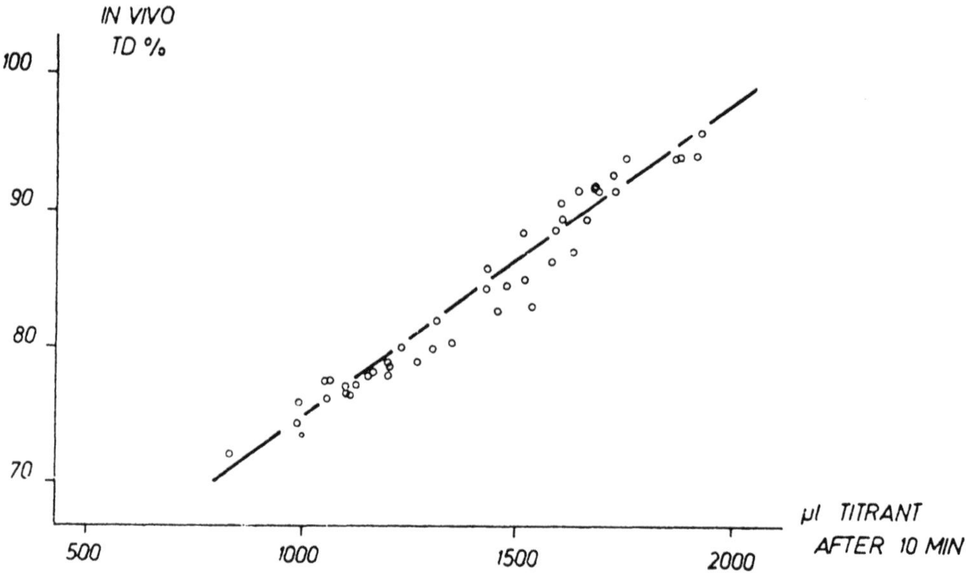

FIGURE 46. Relationship between *in vivo* digestibility and amount of titrant added during the *in vitro* pH-stat digestion procedure.

$$\text{BV} = 0.827 + 0.998 \text{ RHW} \qquad (r = 0.990)$$

$$\text{NPR} = -13.10 + 1.148 \text{ RHW} \qquad (r = 0.997)$$

The reciprocal blood urea value is defined by the following equation:

$$\text{RHW} = \frac{\begin{array}{c}\text{blood urea concentration g/100 ml}\\ \text{with reference protein}\end{array}}{\begin{array}{c}\text{blood urea concentration g/100 ml}\\ \text{with test protein}\end{array}} \cdot \text{BV of reference protein}$$

Several authors[89,90] confirmed the applicability of this assay on humans.

c. *Amino Acid Scoring*

The *in vitro* methods based on amino acid composition serving for determination of nutritive value of proteins are widely used in the practice, despite the aversion of specialists thinking that only results based on biological experiments could be used for such purposes. In contradistinction to the aversion for use of the *in vitro* method evaluating nutritional quality of food and feed proteins, it is beyond all question that they have their own and unreplaceable role in areas of the widely defined food production and consumption. There are at least two kinds of situations where the application of *in vitro* techniques are inevitable:[91]

1. Situations requiring an immediate or fast treatment or observation (i.e., process control), where the time limit rejects the investigation of the *existing* sample by *in vivo* method
2. Cases where the sample does *not exist* at all, and the composition of a food or feed blend is looked for just by computation of the predicted nutritional values in an optimation process

Besides, *in vitro* methods have a true advantage from a financial point of view as well. The whole idea prerequires the existence of reliable, accurately reproducible, relatively simple *in vitro* methods having good correlation with the corresponding *in vivo* method.

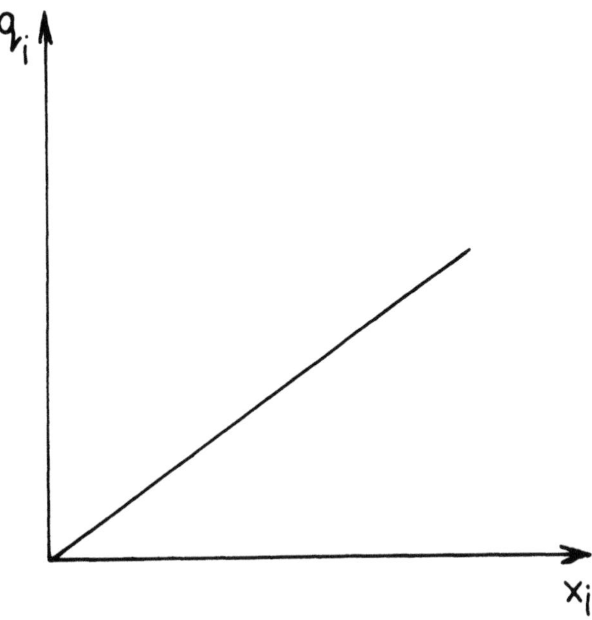

FIGURE 47. Linear model of correlation between concentration of amino acid (a_i) and biological response (amino acid increament = q_i).

Perhaps the most powerful way to investigate the nutritional value by *in vitro* methods is the use of amino acid composition data of the sample applying a mathematical model describing the relationship between this data set and protein nutritional value. These mathematical models are called chemical indices.

Generally speaking, the nutritive value of a protein is characterized by a biological response *R* which is considered to be a function of the amino acid composition of the food or feed

$$R = f(a_1,a_2,a_3,\ldots,a_n)$$

where *R*, the biological response, can be BV (biological value), TD (true digestibility), NPU (net protein utilization) or PER (protein efficiency ratio, or any type of growth response). On the other hand, the a_i values represent the crude amino acid data of the input as determined by chemical methods and usually, no correction is made for availability or any type of variance. In the case of a single variable the a_i will be called the amino acid increment model, which reflects the response of an organism to increasing one amino acid in the diet (q_i). The amino acid increment model can be described by either linear or nonlinear functions. The effect of more amino acids in a diet can be described as either the sum or as the product of the individual response curves. That is, the function describing dependence of nutritive value on amino acid composition can represent summation, product formation or minimun search.

In the following, some of the chemical indices widely used will be shortly discussed. For more detailed information the reader is referred to an excellent review by Hídvégi and Békés.[91]

d. Chemical Indices with Linear Increment Models

These models assume a simple linear dependence of q_i on a_i values, as demonstrated in Figure 47. Such a concept covers the fact that specific interactions between amino acids

TABLE 46
Reference Patterns Proposed by the 1982 Rome FAO/WHO/UNU Committee (92) (g/16 g N)

Essential amino acid	Infant	Child (2—5 years)	Child (10—12 years)	Adult
Histidine	2.59	2.00	1.81	1.60
Isoleucine	4.61	3.01	2.90	1.30
Leucine	9.28	7.01	4.61	1.90
Lysine	6.61	6.21	4.61	1.60
Sulfur-cont.	4.21	2.64	2.30	1.70
Aromatic	7.20	6.61	2.30	1.90
Threonine	4.30	3.60	2.90	0.90
Tryptophan	1.70	1.20	0.90	0.50
Valine	5.50	3.70	2.61	1.30

such as antagonism and synergism, the presence of antinutritive factors and differences in availability of individual amino acids are not included directly in the models.

Historically, the first model expressing protein nutritional quality by amino acid composition was developed in 1946 by Mitchell and Block[92] and is called Chemical Score (CS). Computing the value of CS, the ratios of each essential amino acid of the test protein compared to the quantity of given amino acid in a reference protein (whole egg protein) were calculated and referred to as percentage values. According to Mitchell and Block the nutritive value of a protein or mixture of proteins for any biological function or combination of functions is limited by the relative proportions of the essential amino acids contained in it. This means that the protein quality will be determined by the essential amino acid being present in lowest relative quantity in comparison with reference (egg) protein. This amino acid is called limiting amino acid, and the numerical value of protein quality is the percentage of this amino acid compared with reference protein. In yeast proteins the limiting amino acid is always the methionine. One of the most widely used chemical indices is the FAO/WHO 1973 index. It is based on the same principle as the CS of Mitchell and Block, but the reference protein is a hypothetical one corresponding to the amino acid requirements of man recommended by the FAO/WHO Expert Group.

A further modification of the scoring patterns were proposed by the 1982 Roma FAO/WHO/UNU Committee.[94,95] The Committee suggested four different amino acid scoring patterns for humans considering four different age groups (Table 46). The joint committee also introduced more accurate digestibility factors into the scoring procedure.

Chronologically the last method of this type for protein quality evaluation on the basis of amino acid composition is published by Sarwar.[96] The method is called Available Amino Acid Score (AAAS) and follows the theory of limiting amino acid. Available amino acid ratios are defined as the ratio of mg of available amino acids (mg of total amino acid multiplied by *in vivo* (rat) true amino acid digestibility), in 1 g of test protein to the mg of amino acids in 1 g of reference protein (FAO/WHO/1973 pattern). The smallest ratio multipled by 100 gives the AAAS.

e. Chemical Indices with Saturation Increment Models

In the framework of this book only one index of this type will be discussed characterized by a curve with a breakpoint (Figure 48). The Essential Amino Acid Index (EAAI) was proposed by Oser.[93] The principal novelty of Oser's methods lies upon its mathematics, namely that the EAAI is the geometric mean of the amino acid ratios of tested and reference (egg) proteins. This approach reflects the complexity of the interactions among nutritionally important amino acids, as well as the mutal dependence of amino acids upon each other.

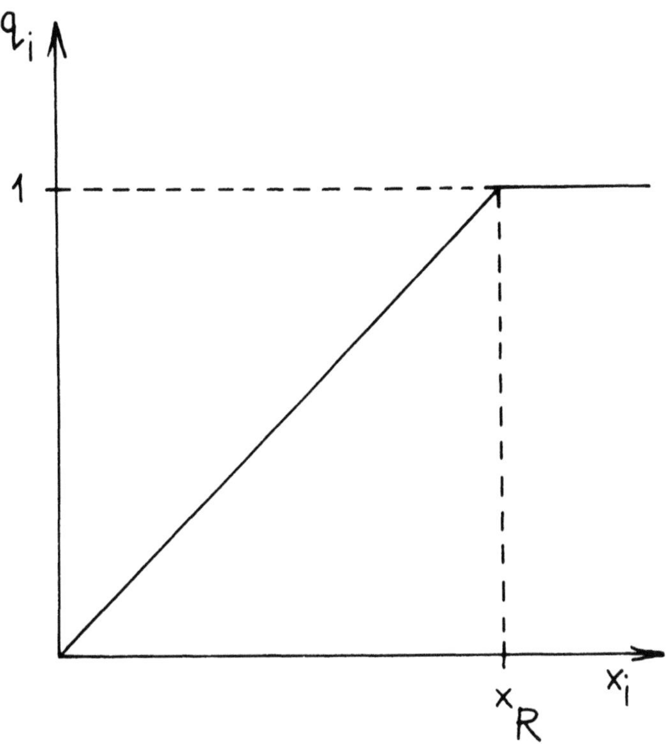

FIGURE 48. Saturation model (with breakpoints) of correlation between concentration of amino acid (a_i) and biological response (amino acid increment = q_i).

In his empirical formula Oser has originally taken into consideration the following essential amino acids, lysine, tryptophan, isoleucine, valine, arginine, threonine, leucine, phenylalanine, histidine and the sum of methionine and cystine. It can be seen that Oser did not consider tyrosine in the formula at first, but later influenced by Mitchell he did include tyrosine in the formula too. It follows from the mathematical character of the EAAI that the nutritive value of a protein, being completely lacking in one or more essential amino acids, becomes zero. In order to avoid the discrepancy between this nature of the formula and the fact that the *in vivo* biological values of proteins, known to have deficiency in one indispensable amino acid, is low but finite which in turn is due to re-use of the products of tissue metabolism (endogenous losses) and coprophagy by the target animal applied in the test, Oser makes the assumption that the minimum egg ratio be 1%. If the concentration of an amino acid in test protein is higher than the same amino acid concentration in egg protein it should be corrected to be equal, not greater, than the standard. Oser has also investigated the correlation between his index and the *in vivo* biological value. He found not only a rather good correlation ($r = 0.85$), but also a numerically similar run of *in vivo* and *in vitro* data.

f. Chemical Indices with Maximum-Type Increment Models

The chemical index with a maximum-type increment model with breakpoint (Figure 49) was first proposed by Korpáczy et al.[97] The index of Korpáczy et al. can be considered an extended version of Oser's index. The modification concerns the following shortcomings of Oser's index: the quantity of nonessential amino acid is not included in the EAA and the negative nutritional effects associated with the excess amino acid intakes were not considered.

The index is a sum of two terms. The first term is practically the EAAI multiplied by

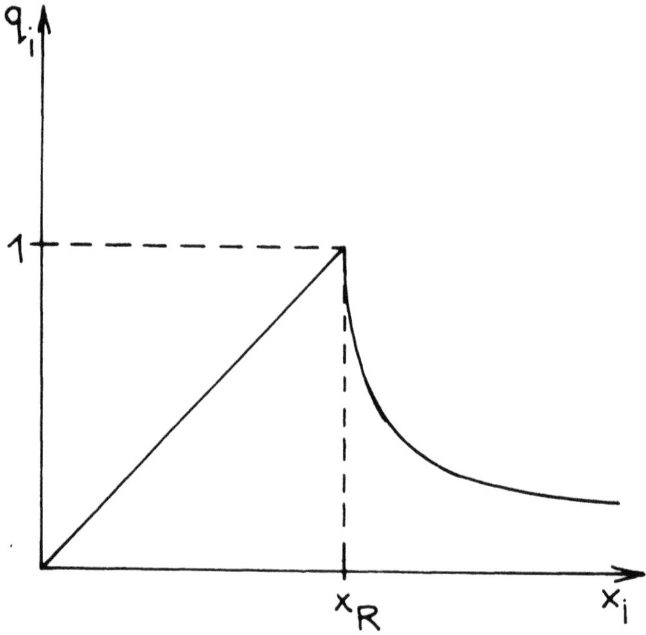

FIGURE 49. Maximum type model (with breakpoint) of correlation between concentration of amino acid (a_i) and biological response (amino acid increment = q_i).

0.75. The excess amino acid intake is taken into consideration by the constraint: if any of the egg ratios is greater than 1.0, its reciprocal must be used. The ratio of the difference between the sums of concentrations of nonessential amino acids in the test protein and the reference protein respectively, and the sum of nonessential amino acids in the egg protein only multiplied by 0.25, gives the second term.

Olesen and Morup[98] resp. Morup and Olesen[99] recommended a new chemical index based on extended experiments on humans and amino acid composition data of proteins used in experiments of Kofrányi,[100] Kofrányi and Jekat[101,102] and Kofrányi et al.[103] The Morup-Olesen index (MOI) was constructed on the assumption that the essential amino acid pattern in terms of nutrial amino acid ratios — not the absolute amounts — has major significance. Therefore, the individual amino acid ratios were formed analogously to those of the FAO/WHO Index but using a different reference pattern. Thus, the variable of the MOI resulted from dividing the content of each of the essential amino acids — expressed in mg/g of total essential amino acids — by that of the reference pattern, which in turn was called provisional reference.

Provisional reference comprised the amino acid pattern of the 35:65 mixture of whole egg and potato protein from Kofrányi's experiments,[103] but with two slight modifications due to the relatively low leucine and tryptophan content of the optimum quality potato-egg mixture. Amino acid imbalances caused by excess intakes were taken into consideration by the following constraint: if any of the amino acid terms (q_i) is greater than 1.0, its reciprocal should be used. In the MOI, the so called Predictive Value (PV) was constructed as follows:

$$PV = 10^{\alpha_0} \prod_{i-1}^{8} q_i^{\alpha_i}$$

where q_i is the amino acid ratio and α_0 and α_i are weighing factors determined by statistical procedures. The essential amino acids considered were isoleucine, leucine, lysine, phenyl-

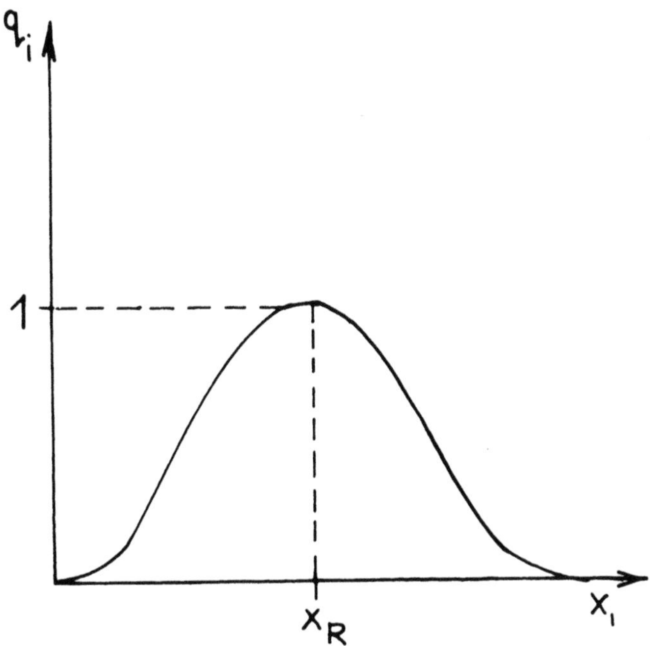

FIGURE 50. Symmetric smooth maximum type model of correlation between concentration of amino acid (a_i) and biological response (amino acid increment = q_i).

alanine + tyrosine, methionine + cystine, threonine, tryptophan and valine. Morup and Olesen eliminated α-exponents, not significantly related to biological value from the equation. From the eight factors three turned out to be zero. Thus, it means that isoleucine, leucine and valine have no individual q_i values in the final formula.

$$MOI = 10^{2.15} \times q_{LYS}^{0.41} \times q_{ARON}^{0.60} \times q_{SULPH}^{0.77} \times q_{THR}^{2.4} \times q_{TRP}^{0.21}$$

The MOI showed good correlation with *in vivo* data. Nutritional value of protein mixtures could be well predicted.

The so called Gaussian Index (GI) was recommended by Békés et al.[104] The reference pattern for the GI is the same as that of the MOI, but a new mathematical function is used for calculation in order to eliminate the breakpoint at the maximum (see Figure 49) and to obtain a symmetrical smooth maximum type-model (see Figure 50).

Following from the nonlinear character of the GI, if it is twice as high, it does not mean that the protein is twice as good. Thus, sometimes for comparative assays the so called Transformed Gaussian Index (TGI) is more useful.[105] This latter was defined as the geometric mean of the individual amino acid terms, multiplied by 100. TGI changes proportionally to the protein quality on a 0 to 100 scale. It has also been proposed by the authors that digestibility could be taken into account by multiplying the TGI by a digestibility factor resulting in TGI Corrected for Digestibility (TGICD).

Sometimes it is important to investigate both quantitative and qualitative aspects in a combined manner. For such comparative purposes, a so called Protein Nutritional Value Index (PNVI) was suggested,[105] which was defined as the geometric mean of TGI and the digestible protein content of the protein tested.

The last derivative of the GI was the so called Extended Gaussian Index[106-109] (EGI) which was defined in the following way: all the α weighing exponents in the GI should be equal to 1 and then a TGI formed. The reference amino acid pattern used in the given EGI should be constructed from the amino acid requirement data of the given consumer (adult, child, infant, etc.). So the EGI is a consumer-specific assay.

g. Assay of the Availability of Amino Acids by Chemical Techniques

The availability of essential amino acids highly influences the biological value of foods. In principle the prerequisite of an accurate estimation of protein quality is the exact knowledge of bioavailability of individual essential amino acids. Nevertheless, in practice it is not common that bioavailability of individual amino acids is determined and taken into consideration while calculating an amino acid score (e.g., FAO/WHO 1973 or FAO/WHO/UNU 1984). In some cases in calculating amino acid scores, the use of determined or literature values for protein digestibility has been suggested. However, the correction for protein digestibility would be of limited applicability because true digestibility of individual amino acids may differ considerably from the digestibility of total protein in the same protein source.[96] Moreover, in the processed food product, digestibility of protein may not be a good predictor of bioavailability of those individual amino acids which are more susceptible to processing loss such as lysine in early Maillard reactions.

Due to the fact that lysine is very often the limiting amino acid in protein sources and that lysine is very susceptible to changes during processing, the chemical tests associated with determination of bioavailability of amino acids are concentrated on lysine. In early research work and practice mainly the fluorodonitrobenzene (FDNB) method was used for determination of bioavailability of lysine. Although some disadvantages of the FDNB method were reported, this method could be a good indicator of processing damage, and there is information based on this method from many laboratories for a wide range of foods.[74] Recently, the dye-binding methods have been modified to estimate available lysine content.[110-112] The dye-binding method is based on the interaction of the azodye Acid Orange 12 with the basic groups of the proteins (lysyl-, histidyl- and arginyl groups). The total amount of free basic group is characterized by dye-binding capacity (DBC). If ϵ-amino group of lysine is blocked with anhydrous propionic acid, the reactive (available) lysine content can be determined specifically by the difference between the two dye-binding capacities (DBC-DBAP), giving the amount of the reactive lysine content (DBL). The main steps of the determination can be seen in Figure 51. Dye-binding methods have the disadvantage of requiring rigid standardization of procedure and establishment of suitable conditions for each class of food material.

VI. NUTRITIVE VALUE OF YEAST PROTEINS

A. DATA BASED ON BIOASSAYS

The greatest part of the bioassays made with yeast proteins are associated with the safety control of these products. Few quantitative data were published in the literature. The review papers dealing with the nutritive value of yeast contain mainly qualitative statements. Kihlberg,[113] in his review paper summarizing the results of bioassays with yeast, makes a statement that mainly feeding tests were made in this field. The digestibility of intact cells is low. Due to the deficiency of sulfur-containing amino acids the BV is relatively low. After supplementation with methionine, however, very high biological values were noted. According to Guzman-Juarez[69] yeast proteins are known to have a lower nutritional quality than other standard proteins such as casein or egg. Calloway[114] reported that the BV of *Saccharomyces cerevisiae* protein ranges from 60 to 90% in rat tests, but the *Torula* and *Candida* species are lower in protein value. About 8 to 9 g of *Torula* nitrogen was adequate to maintain balance in healthy male subjects. The differences in protein value are partly associated with the differences in sulfur-containing amino acids, which are the first limitation

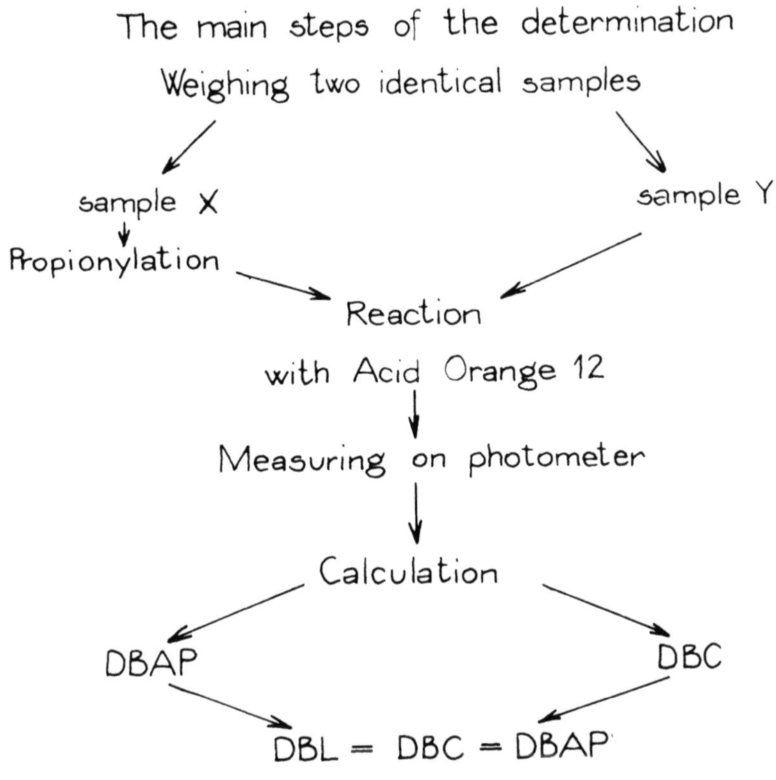

FIGURE 51. Main steps of the determination of reactive (available lysine) using dye-binding method (DBC = total dye-binding capacity, DBL = dye-binding of lysine, DBAP = dye-binding after propionylation of lysine groups). (From Baráth, A. and Hálász, A., *Amino Acid Composition and Biological Value of Cereal Proteins,* Lásztity, R. and Hidvégi, M., Eds., Reidel Publishing, Dordrecht, 1985, 337. With permission.)

in yeast protein all type. Naturally, tests of methionine supplementation have demonstrated improved protein quality. Smith and Palmer[115] in a series of experiments investigated among others six commercial spray-dried yeast samples in rat experiments. NPU, TD and BV were determined. In addition, the effect of supplementation with 0.09 and 0.18% of L-methionine and 0.18% L-methionine + 0.05% L-histidine was studied. The NPU of the yeast samples varied in the range from 47 to 58. The digestibility was high (79 to 93%). The BV changed from 50 to 67. Supplementation with L-methionine significantly increased both the NPU and BV values.

Petrovskij et al.[116] studied the protein quality of yeast protein isolate using male rattlings-weanlings. The following values were measured: PER-1.8, BV-49.1 and NPU-40.3%. The biological value of the isolate amounted to 71% of that of caseine. The method of isolating protein from the yeast biomass does not have any noticeable adverse effect on its biological value.

B. *IN VITRO* DATA OF YEAST PROTEIN QUALITY BASED ON AMINO ACID COMPOSITION

As discussed in the previous part of this book, generally the evaluation of nutritional value of proteins is based on amino acid composition due to the advantages of the *in vitro* methods also discussed earlier. The amino acid composition of yeast has been studied by several early workers and since this time a lot of newer data was published. On the basis of results it is generally accepted that yeast proteins contain a high amount of essential amino

TABLE 47
Protein Content of 8 Yeast Strains Used for Food Yeast Production Grown on Cane Molasses

No.	Strain	Crude protein content (% dry yeast)	No.	Strain	Crude protein content (% dry yeast)
1.	*Torula utilis* SES 6	59.31	5.	*Torula utilis* NRRL-Y-900	57.87
2.	*Torula utilis* NRRL-Y-1082	59.31	6.	*Candida utilis* IFO-0576	60.00
3.	*Torula utilis* NRRL-Y-1084	60.62	7.	*Candida utilis* IAM-4220	60.18
4.	*Torula utilis* NRRL-Y-1427	58.31	8.	*Candida utilis* IAM-4291	60.43

From Kung-Chin, Su, Ming-Chang, Hsie, and Hung Chao Lee, *J. Chin. Agric. Chem. Soc.*, 10, 103, 1973. With permission.

acids, especially a high lysine content is characteristic. The only disadvantage of these proteins is the relatively low content of sulfur-containing amino acids.

Although a lot of data about amino acid composition of yeast proteins is available, there are some facts which cause some problems in the evaluation of these proteins on the basis of these data. Some of the problems may be listed as follows:

1. Lack of the data about tryptophan content (determination of tryptophan needs separate hydrolysis which is not always made by the researchers)
2. Unsatisfactory accuracy of the determination of sulfur-containing amino acids (for this determination a separate procedure is needed)
3. Calculating the protein content the total nitrogen data are not corrected, keeping in mind the nucleic acid content
4 Unsatisfactory indication about the type of yeast resp. yeast product (active yeast, compressed yeast, active dry yeast, instant active dry yeast, inactive yeast, extracted yeast protein, etc.)
5. Neglect of the effect of different growing conditions on yeast composition and the effect of processing (drying, conditions of protein extraction and nucleic acid reduction, etc.) on the amino acids and their bioavailability

In the framework of this chapter only the problems associated with the effect of growing conditions and with the changes of proteins (amino acid side chains) during processing will be discussed.

In an early work King-Chin Su et al.[117] reported on a study of amino acid composition of various yeasts grown on cane molasses under different conditions. Eight yeast strains used for food yeast production were investigated. The protein content of the yeast is shown in Table 47. The amino acid composition is shown in Table 48. No significant differences were observed between three *Candida utilis* species. The amino acid composition of *Torula utilis* yeasts is also very similar with some exceptions. The strain *Torula utilis* NRRL-Y-900 had a significantly higher arginine and alanine content and lower isoleucine content. The cystine and methionine content of strain *T. utilis* NRRL-Y-1082 is very low. It is questionable if this very low value is not associated with an analytical error. To study the effect of growing conditions the *Torula utilis* NRRL-Y-900 strain was grown on cane molasses (with added urea and ammonium sulfate as an additional source of nitrogen) and on synthetic media containing molasses, glucose and magnesium sulfate, potassium hydro-

TABLE 48
Amino Acid Composition of the Yeast Strains Used for Food Yeast Grown on Cane Molasses (g/100 g True Protein)

Amino acid	Strain[a] 1	2	3	4	5	6	7	8
Lysine	8.52	9.07	8.04	8.02	8.95	8.91	8.71	9.17
Histidine	2.03	2.17	2.03	1.86	2.24	2.21	2.15	1.49
Arginine	5.01	4.71	5.10	5.66	4.21	5.47	5.29	4.97
Aspartic acid	8.65	8.45	8.45	9.34	10.57	8.46	8.92	8.68
Threonine	5.26	4.95	5.05	4.91	5.92	5.34	5.23	5.14
Serine	4.57	4.41	4.58	4.43	4.89	4.76	4.67	4.68
Glutamic acid	10.95	10.38	11.13	11.12	11.26	10.67	10.71	10.84
Proline	3.12	3.06	3.03	3.10	2.86	3.05	2.91	3.11
Glycine	4.27	4.20	4.31	4.24	5.05	4.46	4.38	4.34
Alanine	5.28	5.43	5.43	5.93	6.79	5.43	5.34	5.38
Cystine	0.20	0.12	0.45	0.96	0.44	0.73	0.41	0.61
Valine	5.39	5.51	5.39	5.19	5.42	5.75	5.99	5.80
Methionine	0.57	0.12	0.54	1.12	0.75	1.17	0.99	1.17
Isoleucine	4.32	4.42	4.42	4.37	3.76	4.65	4.60	4.65
Leucine	6.34	6.75	6.56	6.65	7.31	7.13	7.10	7.07
Tyrosine	3.99	3.78	4.88	4.57	3.32	3.99	4.04	4.22
Phenylalanine	4.24	4.68	5.05	4.42	4.05	4.10	4.22	4.27
Tryptophan	1.26	1.21	1.18	1.15	1.17	1.20	1.30	1.16

[a] See Table 47.

From Kung-Chin, Su, Ming-Chang, Hsie, and Hung Chao Lee, *J. Chin. Agric. Chem. Soc.*, 10, 103, 1973. With permission.

genphosphate urea and ammonia. In Table 49 the amino acid composition of the yeasts grown on the molasses and synthetic medium is shown. Significant differences were found between the amino acid composition of these two yeasts. In the case of lysine, arginine and glutamic acid the differences are quite big. This fact may probably be explained studying the results reported by Wiemken and Nurse[118] associated with the isolation and characterization of the amino acid pools located within the cytoplasm and vacuoles of the yeast strain *Candida utilis*. The authors demonstrated two distinct amino acid pools in the food yeast *Candida utilis*. Treatment of the cells with basic protein (cytochrome C) under isotonic conditions permeabilized the plasmalemma but left the tonoplast intact. The selective effect on these membranes was indicated by the observation of intact vacuoles but changed contrast of the cytoplasm in the phase contrast microscope and by the free access of a chromogenic substrate to a cytoplasmic enzyme (α-glucoside). Investigation of the released amino acids and their changes indicated a rapidly metabolized amino acid pool locaiized within the cytoplasm. Osmotic shock with water following the treatment with basic protein disrupted the tonoplast, an event which could be followed by phase contrast microscopy. Most of the remaining amino acids were then released. These showed a slow turnover in pulse labeling experiments and a high proportion of some nitrogen-rich amino acids, indicative of a storage function. The ratio of these two pools and also the ratio of metabolically active and storage proteins could influence the total amino acid composition of the yeast cells.

VII. THE SAFETY OF THE YEAST RESP. YEAST PROTEINS

Before any new raw material resp. new food may be used as human food it must be evaluated with respect to the quality of its protein content and its safety for use. This requirement may apply to new varieties of conventional foods where the composition has

TABLE 49
Amino Acid Composition of *Torula utilis* and *Candida utilis* Yeasts Grown at the Different Conditions (g/100 g true protein)

	Yeast strain				
Amino acid	**Torula[a] utilis**	**Torula[a] utilis**	**Candida[b,c] utilis**	**Candida[b,c] utilis**	**Candida[b,c] utilis**
	Medium of growing				
	Cane molasses	**Synthetic (glucose)**	**Calcium ligno sulfate/wood sugars**	**Sulfite waste liquor**	**Ethyl alcohol**
Lysine	8.89	6.31	9.66	9.56	8.73
Histidine	1.41	1.33	2.43	2.44	2.43
Arginine	4.15	8.17	6.54	6.67	7.70
Aspartic acid	10.71	7.65	11.45	11.24	10.55
Threonine	5.87	4.62	6.30	6.00	6.37
Serine	5.07	4.07	5.58	5.59	5.93
Glutamic acid	12.94	16.76	16.71	16.93	21.45
Proline	2.22	1.98	4.08	4.09	3.69
Glycine	5.33	3.28	5.49	5.57	5.23
Alanine	6.68	5.08	7.45	7.84	7.26
Cystine	0.35	0.22	1.25	1.38	1.13
Valine	4.25	4.21	6.72	6.57	6.52
Methionine	0.51	0.96	1.91	1.97	1.50
Isoleucine	4.28	3.58	6.61	6.87	5.69
Leucine	8.15	5.17	9.66	9.66	8.83
Tyrosine	3.01	2.74	4.72	4.73	4.45
Phenylalanine	4.21	2.93	5.44	5.43	5.02
Tryptophan	1.10	0.96	1.38	1.40	1.16
Crude protein	59.62	42.81	40.63[d]	43.63[d]	40.63[d]

[a] Kung-Chin, Su et al.[117]
[b] Saurer, et al.[120]
[c] Recalculated from g/100 g dry yeast data.
[d] True protein.

been genetically changed, but it applies especially to new foods developed by isolation from conventional sources by unusual techniques and to microbial sources such as yeast or yeast protein preparations. Although a prior history of safe use may be taken into account in the evaluation of a protein source proposed for general consumption, this alone is insufficient to preclude adequate preclinical testing by currently available, more objective animal feeding tests.

According to Guideline No. 6 of PAG of UN[64] for the evaluation of safety the following information is needed:

1. Information concerning methods of production
2. Chemical and physical properties
3. Content of microorganisms and their metabolites
4. Toxicological effects on laboratory animals
5. Responses of normal human subjects to limited feeding studies

The information concerning production and especially the control of the SCP process is of particular importance. In other fermentation processes or industries the purpose is to obtain metabolic intermediates, while in the case of yeast (SCP) production the object is that of obtaining biomass. In the production there is some risk of a change in the metabolic activity of the organism when oxygen is exhausted or not adequate. Current processes in industrial production are based on continuous cultures. This presents difficulties in the case of other fermentation processes, especially those using mutants selected for maximal production of a metabolic intermediate, because of risk of back-mutations or of secondary mutations resulting in unfavorable metabolic changes. In the case of yeast production, however, the risk of a mutational change of the culture in the course of a continuous process is minimized, owing to the fact that the selection and stabilization of strains is based on optimal substrate utilization and biomass production.

In light of experience gained at large pilot plants and on an industrial scale, it has been shown that in practice constancy of product over a long period of time can be achieved. It is obvious that both from safety and economic viewpoints, such constancy of the production is essential and has to be routinely monitored. In a round table discussion of experts[121] the group of experts drew attention to the fact that when a SCP process was scaled up manufacturers must be fully aware of the need to monitor the various stages or steps of the process including fermentation, recovery, drying and processing of biomass to ensure the identity of the final product with the one on which extensive testing was originally based. All participants insisted that full disclosure of the process, including the identity of the organism, should be available to the regulatory and legislative authorities.

From the point of view of safety the following chemical analytical data could be of importance: (1) amount of nucleic acids and nucleotides; (2) presence of unusual components such as uneven fatty acids, cell wall polysaccharides, etc.; (3) presence of toxic natural substances and (4) presence of contaminants such as heavy metals, solvent residues, etc.

As it was exhaustively discussed in previous parts of this book, high nucleic acid content of the yeast cells is a factor that perhaps constitutes a universal and major limitation to the use of yeast biomass for feeding or human consumption, unless safe processes for nucleic acid removal prove to be economically competitive.

The final metabolic product formed in man from the purine moiety of the nucleic acids is uric acid (Figure 52). This compound is only slightly soluble at physiologic pHs and if the blood uric acid level is elevated, crystals may form in the joints, causing gout or gouty arthritis. Stones may be deposited in the urinary tract too.

Most animals degrade uric acid to allantoin, a soluble compound which is easily extracted. However, swine is unable to oxidize one of the purines, guanine. Birds do form uric acid, but at the level normally present, this compound serves a useful role in regulating water loss from cloaca.

In man the level of uric acid found in the plasma and urine is affected first by the purine and protein content of the diet. Calloway[114] reported on studies associated with nucleic acid metabolism in man depending on the composition of the diet. A 75 g protein/day was considered a control diet. He found that the average uric acid excretion of men fed a purine-free control diet to be about 350 to 400 mg/day. Their plasma contained 50 μg uric acid/ml. Urinary excretion was less, 300 to 350 mg/day, and plasma uric acid was higher 60μg/ml, when a protein-free diet was fed. Similar values were observed with feeding of 22 to 37 g protein/day. It was suggested by the author that steady-state, endogenous uric acid production of man with marginal levels of protein intake thus appears to be the amount present in urine, 350 mg/day plus about 100 to 300 mg/day degraded by intestinal bacteria.

Urinary uric acid output increases linearly with increased intake of purine-free protein, but plasma uric acid level is unchanged or decreased. At the highest level of protein intake, daily output was 1253 mg and great care had to be taken to maintain urine pH on the alkaline side and the volume high enough to prevent crystallization. Adding 2, 4 and 8 g yeast ribonucleic acid to the diet, urinary uric acid excretion increased by 147 mg/day for each

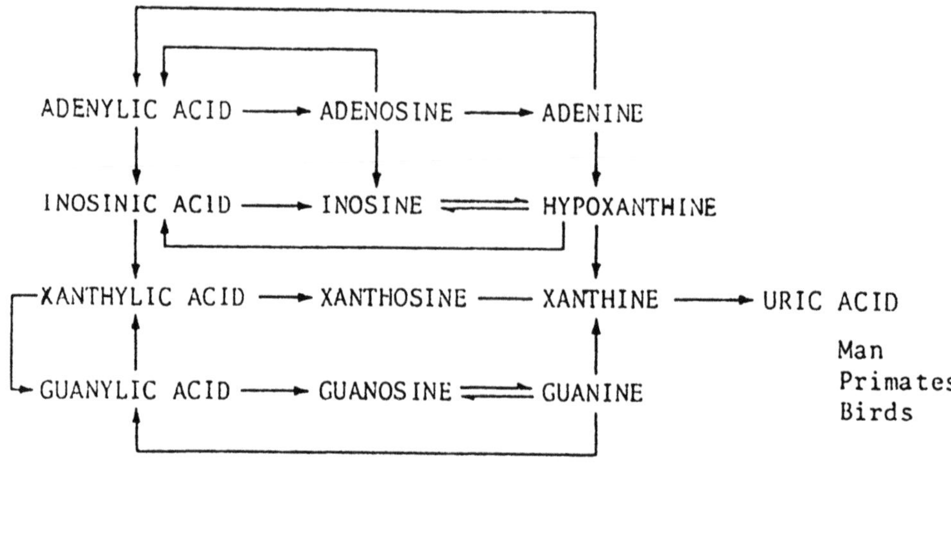

FIGURE 52. Pathways of nucleic acid metabolism.

g of RNA added. With 2 g of RNA in the diet, urinary uric acid quantity was 670 mg/day similar to the excretions reported with normal diets that include moderate amounts of meat and vegetables. Higher levels increased both the urinary excretion and the plasma level of uric acid causing hyperuremia. In conclusion it was stated that in healthy men, dietary protein free of nucleic acid would not be expected to increase risk of gout whereas dietary RNA would do so. Both substances would increase the risk of renal stone formation. It was also noted, that only the amino groups of adenylic acid and guanine would contribute to the body protein pool so 75% of the nitrogen tied up in these compounds would be lost to human utilization.

A detailed analysis of the lipids and carbohydrates of yeasts is needed to detect unusual components present in yeast biomass. Fatty acids with an odd carbon atom number can be metabolized by animals and humans. Recent experiments with broilers fed with a microbial protein product containing large amounts of odd carbon number fatty acids showed little effect on the fatty acid composition of deposit fat on broilers.[122] Contradictory data were reported concerning the effect of special polysaccharides on the cell wall of yeasts. Most of the researchers are of the opinion that cell wall material is undesirable because it reduces the bioavailability of proteins and may contain antigenic allergenic agents and factors causing gastrointestinal disturbances. Calloway[114] reported that *Torula* yeast seems to produce few pathophysiological reactions, except for the dermatologic findings referred to above, which occurred in 12 of 50 young men fed 45 to 135 g yeast/day for 3 to 4 weeks. An antigen prepared from the yeast did not evoke a dermal reaction in any subject, so the response was probably not due to an allergy. In a more recent work Sarwar et al.[119] compared the amino acid composition of three samples of dried food yeast *Candida utilis* grown in three different media: calcium lignosulfate/wood sugars, sulfite waste liquor and ethyl alcohol. Results of the determination of amino acid composition are included in Table 48. No significant differences were observed between the yeasts grown on calcium lignosulfate/wood sugars and sulfite waste liquor, however, the product grown on ethyl alcohol had significantly lower lysine, isoleucine and leucine content and significantly higher arginine and glutamic acid

TABLE 50
Protein and Amino Acid Content in Samples of Yeast Biomass from Six Different Strains Fermented in Cheddar Cheese Whey Permeate at 32°C for Five Days

Strain	1	2	3	4	5	6
Protein (%)	40.25	35.13	37.81	31.25	33.44	30.00
Amino acids			(% w/w)			
Essential						
Threonine	5.19	5.82	5.69	5.80	6.77	6.21
Valine	6.72	6.77	5.95	6.66	6.68	6.57
Methionine	0.86	0.86	1.51	1.50	1.36	1.44
Isoleucine	5.43	5.82	5.78	6.51	5.80	6.07
Leucine	8.21	8.44	9.42	9.63	8.95	9.92
Phenylalanine	4.77	4.69	4.24	5.05	4.38	4.61
Lysine	8.97	9.33	8.66	9.63	8.82	7.94
Nonessential						
Aspartic acid	11.36	10.39	9.72	9.10	8.95	8.37
Glutamic acid	14.85	13.24	16.84	12.96	14.90	15.64
Serine	5.48	5.88	5.66	6.18	7.21	6.45
Glycine	5.68	5.04	4.72	4.66	4.31	4.51
Hystidine	2.93	2.76	2.91	2.27	2.33	2.29
Arginine	5.03	7.73	4.84	6.02	5.38	5.36
Alanine	8.33	7.53	6.11	5.93	6.25	6.39
Proline	4.23	4.34	4.67	4.41	4.14	4.59
Tyrosine	1.88	1.59	2.83	3.62	3.70	3.14
Cystine	0.08	0.07	0.45	0.07	0.07	0.50

Note: 1 — *Saccharomyces fragilis*, 2 — *Kluyveromyces marxianus* var. marxianus ATCC 28 244, 3 — *Candida tropicalis* ATCC 20 401, 4 — *Candida albicans* ATCC 20 402, 5 — *Candida utilis* ATCC 9 226 and 6 — *Candida utilis* ATCC 9 950.

From El-Samragy, Y. A., Chen, J. H., and Zall, R. R., *Process Biochem.*, 23(2), 28, 1988. With permission.

content. It is very interesting that the tendency of amino acid composition change is the same as it was observed in an earlier work of Kung-Chin Su et al.[117] in the case of *Torula utilis* yeast grown on synthetic medium (see Table 49). It seems that natural, more complex media are better from the point of view of lysine content than the synthetic ones.

This view can be supported with the recent report of El-Samragi et al.[120] dealing with the amino acid profile of yeast biomass produced from fermentation of cheddar whey permeate. Protein and amino acid content of six different yeast strains is shown in Table 50. Data presented in Table 50 show that there was no big difference in the concentration of individual amino acids, either essential or nonessential between single cell proteins produced by the different yeasts used to ferment whey permeate. The only exception is the relatively big difference in cystine content of some strains. However, keeping in mind the small amount of this amino acid in yeast protein and also the analytical difficulties in hydrolysis and determination of sulfur-containing amino acids, these differences can be understood.

To demonstrate the effect of different conditions of protein extraction and precipitation on the amino acid composition of protein concentrate in Table 51 based on work of Sergeev et al.[24] amino acid composition of protein concentrate samples is shown. The samples originated from the same yeast biomass, however, were extracted at different pH values. For comparison, the amino acid composition of the yeast cell homogenizate is also shown. Significant differences could be observed both in essential and nonessential amino acids depending on the pH of extraction although some amino acids occur practically in the same amount in different protein concentrates (e.g., tryptophan).

TABLE 51
Amino Acid Composition of Protein Concentrates Prepared From the Same Yeast Biomass by Extraction at Different pH Values (g/100 g protein)

Amino acid	Yeast biomass	Condition of extraction pH 3.0	pH 8.0	pH 9.5	pH 12.0
Lysine	6.79	7.83	9.50	10.04	8.13
Histidine	1.22	1.55	1.50	2.52	3.92
Arginine	2.80	4.60	1.60	4.18	4.02
Aspartic acid	7.06	9.41	10.47	10.46	13.39
Threonine	2.98	4.24	4.76	4.40	4.18
Serine	3.43	4.62	5.77	4.44	5.63
Glutanic acid	26.03	19.35	15.38	19.92	11.32
Proline	4.36	4.45	5.08	3.38	5.86
Glycine	5.48	5.14	4.89	5.90	7.72
Alanine	11.56	8.14	7.69	7.52	10.34
Cystine + Methionine	2.73	2.37	2.62	1.99	0.12
Valine	4.07	4.66	4.65	5.17	5.88
Isoleucine	4.80	4.85	4.60	4.20	3.58
Leucine	9.50	10.45	11.02	7.46	9.30
Tyrosine	2.93	3.53	4.30	3.54	0.62
Phenylalanine	3.56	4.07	5.36	4.09	1.21
Triptophan	0.70	0.74	0.81	0.80	0.83
Not identified	—	—	—	—	3.95

From Sergeev, V. A., Solosenko, V. M., Bezrukov, M. G., and Saporovskaya, M. B., *Acta Biotechnol.*, 4(2), 105, 1984.

Turning to the data reported in literature concerning values of chemical indices of yeast proteins, first some data published by Békés et al.[104] will be mentioned. Investigating brewer's yeast (dried) the following chemical indices were calculated:

FAO/WHO 1972 index	68.0
Korpáczy index	61.5
Morup-Olesen index	58.9
Gaussian index	50.0
Limiting amino acids:	MET, TRP

It was shown by these authors that due to high lysine content yeast protein is an excellent complementing protein for cereal products. In Table 52 MOI for some cereal foods with and without fortification with brewer's yeast are shown. The biological value of white breads especially increased significantly.

Sergeev et al.[24] reported that the FAO/WHO index of yeast protein isolate varied from 32 to 36% due to the low content of sulfur-containing amino acids being limiting amino acids. The *in vitro* digestibility of the isolate was high, practically equal to that of casein. Smith and Palmer[115] investigating six commercial dried yeast samples found that SC varies from 60 to 67. Supplementation with 0.18% L-methionine increased the CS values to 78 to 90.

Presence of natural toxic substances in food yeast strains was until now only reported if it was contaminated with other toxin-producing microorganisms. Solvent contamination or contamination with heavy metals may occur when the production conditions in the large-scale factory are not adequate.

With respect to the microbiological status of novel protein sources, acceptable limits for different types of microorganisms have been proposed by the Protein Advisory Group

TABLE 52
Biological Value of Some Cereal Products Fortified with Brewer's Yeast

Product	PV value	PV value after fortification	Added yeast (%)
Fine white bread	49.7	121.2	9.6
White bread	75.3	111.4	6.4
Buffet bread	80.6	107.8	6.3
Wheat loaf	72.7	114.7	7.1
Brown rye bread	100.9	123.9	1.7
Rye bread	93.2	128.9	2.3
Pastry	92.1	123.1	8.1

From Békés, F., Hidvégi, M., Zsigmond, A., and Lásztity, R., *Acta Aliment.*, 13, 135, 1984. With permission.

of UN. Minimal requirements for SCP and also for yeast biomass which seem to be acceptable to microbiologists are the following ones:[123]

Total aerobic count	10^4/g
Mold spores	10/g
Enterobacteriaceae	absent in 0.1 g
Clostridia group	10^2/g
Lancefield group D streptococci	10^2/g

The standards mentioned may be considered as tight, and products surpassing them by a factor of 10 could still be considered acceptable for use in most animal feeds.

The most suitable species of animals for feeding studies is undoubtedly the rat. Mice are also used but less is known of their nutritional requirements and their size precludes obtaining sufficient blood or urine for examination. Among the nonrodent species, beagle dogs, rhesus monkeys and miniature pigs have been used for short-term but not for chronic (lifecycle) studies. Different types of studies are used as follows: (1) screening test; (2) subchronic (90-day) study; (3) chronic (2 years) feeding study; (4) multigeneration study; (5) teratogenicity study and (6) mutagenicity study.

The most practical way of starting any safety evaluation program is by means of a simple subacute feeding test with rats (10 male and 10 female weanling rats) on a diet with 30% protein content. Generally growth, food intake, hemoglobin content, gross pathology, weight of liver and kidneys is controlled and a microscopy of liver and kidney is used.

If the screening test does not reveal any deleterious effects, it is justified to continue the program with the much more elaborate 90-day (subchronic) study. In spite of the extremely valuable information provided by a well-conducted 90-day study, no indications are obtained of possible real long-term effects like shortening of the lifespan or carcinogenicity. For this purpose it is inevitable to conduct lifetime studies, using the rat as the most suitable animal species.

Although the average lifespan of rats is longer, 2 years is generally considered a sufficiently long experimental period. At this age rats have already developed many aging symptoms, including tumors. The gross inspections, clinical laboratory tests and pathological observations and their frequency in both short- and long-term toxicological studies, are summarized in Tables 53 and 54.

In addition to possible chronic and carcinogenic properties, effects on fertility and lactation performance should also be examined in reproduction studies (multigeneration study). Finally, genetic hazards from food contaminants can now be examined in mammals

TABLE 53
Typical Criteria Used in Toxicological Evaluations: General, Hematology, Pathology (According PAG Guidelines No. 6)

Observation	Frequency	
	Short term	Long term
Physical appearance	daily	daily
Behavior	daily	daily
Body weight	weekly	weekly
Food consumption	weekly	weekly
Hematology		
Hemoglobin		
Hematocrit	0, 4, 8, 12 weeks	1, 3, 6, 9, 12 18, 24 months
Leukocytes		
Platelets		
Reticulocytes		
Gross pathological examination		
Organ (liver, kidney, heart, brain, spleen, gonads, pituitary, adrenals, thyroid) weight	Terminal	Terminal
Histopathology		
20 organs and tissues	Terminal	Terminal
Electron microscopy of liver and kidneys		

TABLE 54
Typical Criteria Used in Toxicological Evaluations: Blood Chemistry, Urine (According PAG Guidelines No. 6)

Observations	Frequency	
	Short term	Long term
Blood chemistry		
Glucose		
Urea-N		
Protein, total		
Albumin/globulin ratio		
Triglycerides	0, 4, 8, 12 weeks	1, 3, 6, 9, 12 18, 24 months
Cholesterol		
SGOT, SGPT		
Alkaline phosphatase		
Uric acid		
Allantoin		
Urine		
Volume, specific gr., pH		
Glucose	0, 4, 8, 12 weeks	1, 3, 6, 9, 12, 18, 24 months
Protein		
Ketone bodies		
Bile		
Occult blood		
Sediment		

by several, relatively simple methods including *in vivo* cytogenesis, host-mediated assay and dominant-lethal assay (mutagenicity study).

Tests for safety and suitability for human consumption, especially for feeding of infants and children, are essential in the development of new foods. The prerequisite to start with human testing is always the preliminary testing with experimental animals as described briefly in the previous part of this chapter. Human testing, as these observations will be termed, will fall into four main categories: (1) acceptability and tolerance tests; (2) growth tests; (3) nitrogen balance measurements and (4) other criteria.

A. CONCLUSIONS

If we are trying to summarize the practical experience of yeast resp. yeast protein nutritional value and safety it can be stated, from the reports published so far, that yeast are the only SCP sources which have been extensively tested in long-term studies. The results of experiments before 1972 were summarized in the report of Kihlberg.[113] About 70 years ago the first experiments were performed with rats kept on diets containing 30 and 40% yeast as a sole source of protein for a period more than one year. Negative effects were observed due to the deficiency in sulfur-containing amino acids and in some cases, deficiency of α-tocopherol. It was realized step by step that the negative effects were not caused by a toxicity of the yeast per se but rather by a dietary deficiency (e.g., liver necrosis, induced particularly by yeast grown on sulfite liquor, could be completely hindered by the addition of a trace amount of selenite or vitamin E to the yeast-based diets). In the late 60s and early 70s the questions of safety of hydrocarbon-cultivated microorganisms were the focus of interest. Extensive experiments have been performed in which gas-oil- and *n*-paraffin-grown yeast were studied mainly as feed components. Most of the published data were satisfactory, no toxic effects were observed. Good digestibility was observed in pigs and chicken fed with yeast. This fact is probably associated with the presence of chitinase and glucanase in the digestive tract.[123]

The views concerning the use of yeast directly as human food are in many cases conflicting. Earlier studies are summarized by Calloway.[114] First of all it must be noted that opposite to the animals which are able to metabolize uric acid and remove the destruction products of RNA from the organism, human organisms cannot metabolize uric acid and physiological disturbances associated with elevated levels of uric acid in blood plasma may occur. Based on the human experiments with graded RNA feeding the safe intake of nucleic acids for adult, healthy people has been estimated to be 2 g/day. That means that without reduction of nucleic acid content of yeast, the use of yeast as human food is very limited (the possible ways of reducing nucleic acid content of yeast were discussed in a previous chapter).

The digestibility of intact yeast cells is rather low due to the poor attachability of cell wall material by enzymes of the digestive tract of men. So a preliminary disruption of the cells, or increase of cell wall permeability is needed (the methods of the cell wall disruption were also discussed in a previous chapter).

Although yeasts seem to produce few pathophysiologic reactions, in a new review Kinsella and Shetty[176] expressed the view that several problems (mainly gastrointestinal upsets, reducing of the digestibility of protein, etc.) emphasize the general need for refining yeast protein for human diets.

The lipid content of yeast varies significantly depending on the conditions of growing. The lipid content in young cultures grown in adequate media is largely phospholipid, containing polyunsaturated fatty acids. Triglycerides accumulate when growth is slowed. From a nutritional point of view the vitamin content of yeast is more important. Yeast is one of the richest sources of the vitamin B group and also provitamins of the D group. The mineral content of yeast biomass is relatively high. Calcium and phosphorus are the main mineral constituents in the group of macroelements and copper and zinc among the microelements.

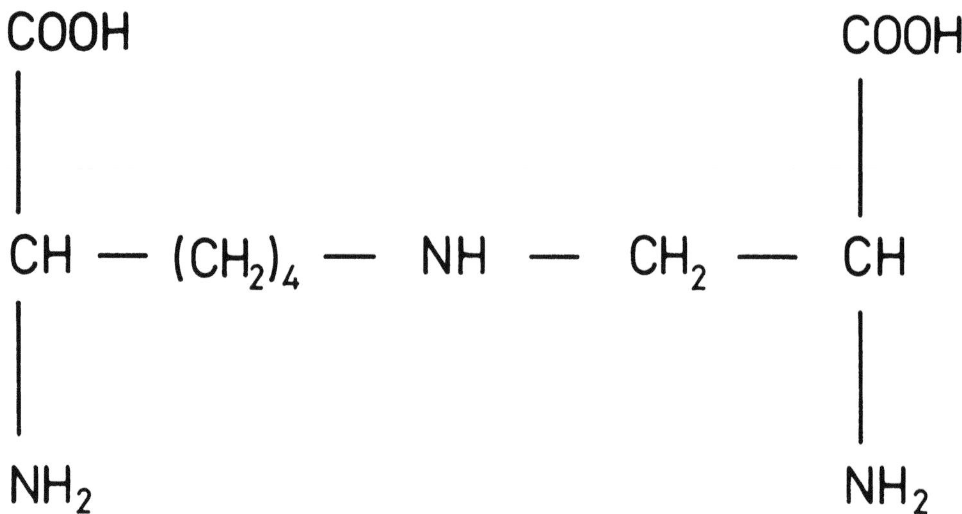

FIGURE 53. α-, β-, and γ-carbolines.

VIII. CHANGES OF PROTEINS DURING EXTRACTION OF PROTEINS AND PROCESSING OF YEAST BIOMASS

Except for direct food use of pressed active intact yeast cells, the yeast biomass is submitted to different treatments and shorter or longer storage. The processing and storage of yeast biomass and/or different protein preparations produced from it causes changes in the protein properties. In many cases the processing has little or no adverse effect on the nutritional and functional properties of proteins. However, some unfavorable reactions can also take place resulting in a decrease in the content of essential amino acids or in the formation of antinutritional and possibly toxic derivatives. In addition, changes of the conformation of proteins may occur adversely influencing the functional properties. In the framework of this chapter we will discuss only the chemical changes causing nutritional or safety problems in the use of yeast proteins. The problems associated with functional properties of these proteins will be treated in Part II.

The effect of processing and storage on food proteins has been studied extensively and a number of general reviews are available.[125-135] In the following the most important changes will be discussed.

A. THE EFFECT OF HEAT TREATMENT

Moderate heat treatment causes denaturation of proteins. Most of the enzymes will be inactivated. A number of proteins become more readily digestible after moderate heat processing. These changes do not affect the nutritive value of proteins (however changes in functional properties may occur).

Thermal treatment at higher temperatures and for a longer time in the absence of any added substance can lead to different changes of amino acid residues, e.g., sterilization at temperatures above 115°C brings about the partial cystine and cysteine residues and the formation of hydrogen sulfide, dimethylsulfide and cysteic acid. Deamidation reactions take place during the heating of proteins at temperatures above 100°C. Amidated residues of glutamine and asparagine will be changed to glutamyl- and asparagyl side chains and ammonia will be released. The changes do not affect the nutritive value of the proteins. Thermal treatment can also cause destruction of tryptophan, especially in the presence of oxygen.

Severe heat treatments applied to proteins may result in the formation of cyclic derivatives, some of which possess a strong mutagenic action. Thus above 200°C tryptophan can be transformed, by cyclization into α-, β- or γ-carbolines (see Figure 53).

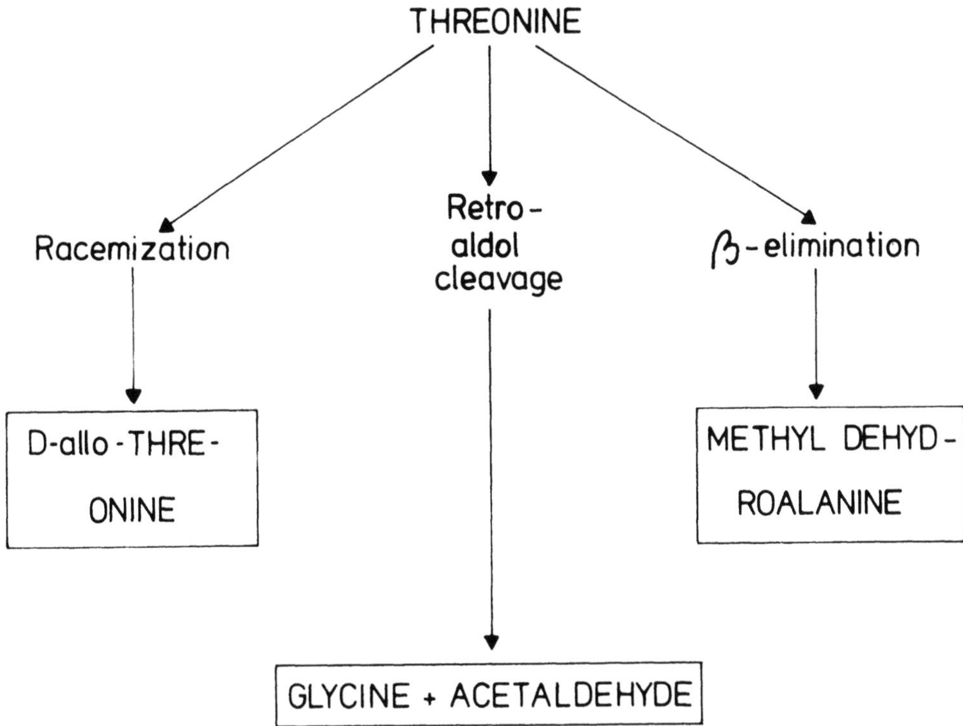

FIGURE 54. Isomerization of amino acid residue in alkaline medium.

B. EFFECTS OF ALKALINE pHs

One of the most typical reactions in alkaline media is the isomerization of amino acid residues. The residues being sensitive to isomerization are: methionine, lysine, cysteine, alanine, phenylalanine, tyrosine, glutamic acid and aspartic acid.

The isomerization involves as a first step dissociation of the hydrogen located at α-carbon atom (see Figure 54) and formation of a carbanion being in equilibrium with a compound without an optically-active carbon atom. After protonation of carbanion randomly D- and L forms are obtained. When the reaction is complete, a racemic mixture of the amino acid residue is formed.

Since most D-amino acids have no nutritional value, racemization of the residue of an essential amino acid reduces its nutritional value. The presence of D-isomers also reduces digestibility of protein, because peptide bonds involving D-residues are less easily hydrolyzed *in vivo* than those containing only L-isomers.

1. Formation of Lysinoalanine, Lanthionine, and Ornithoalanine

Lysinoalanine (LAL) is the name given to the compound E-*N*(DL-2-amino carboxyethyl)-L-lysine (see Figure 55). LAL is composed of a lysine residue whose ϵ-amino group is linked to the methyl group of an alanine residue. When both residues are in the protein chain it is thus an intra- or inter-molecular cross-linkage. It is not a dipeptide, as it does not have a peptide group, and it does not regenerate the two amino acids on acid hydrolysis. There are four possible stereoisomers (LL,LD,DL and DD) of LAL. According to generally accepted views concerning the mechanism of LAL formation, dehydroalanine is formed first in alkaline conditions, then in β-elimination reaction from cysteine or serine residues. The resulting dehydroalanine (DHA) residues are very reactive, combining readily with the ϵ-amino groups of lysine residues, the δ-amino groups of ornithine residues, and the sulfhydryl groups of cystein residues, thereby producing cross-links involving lysinoalanine, ornithoalanine and

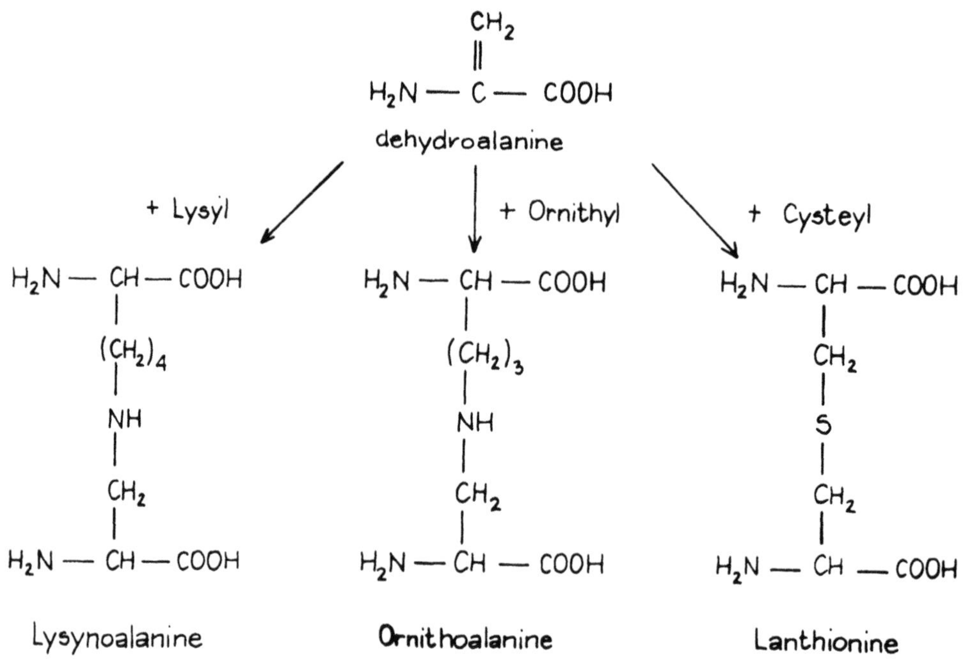

$$CH_2{=}C(NH_2){-}COOH$$

dehydroalanine

+ Lysyl   + Ornithyl   + Cysteyl

$$H_2N{-}CH(COOH){-}(CH_2)_4{-}NH{-}CH_2{-}CH(NH_2){-}COOH$$

Lysynoalanine

$$H_2N{-}CH(COOH){-}(CH_2)_3{-}NH{-}CH_2{-}CH(NH_2){-}COOH$$

Ornithoalanine

$$H_2N{-}CH(COOH){-}CH_2{-}S{-}CH_2{-}CH(NH_2){-}COOH$$

Lanthionine

FIGURE 55. Lysinoalanine.

lanthione, respectively, (Figure 56). A substitution reaction describing the formation of lysinoalanine from serine-phosphate or glycosylserine is also proposed. Methyl dehydroalanine could be formed from threonine and lysinomethylalanine too[179] (see Figure 57).

During formation of LAL, by β-elimination reaction, the H-atom may be added back on either side of the C-atom of dehydroalanine so that the alanine moiety of the newly formed LAL may be present either as the L- or D-isomer. Theoretically, therefore, LL-LAL and LD-LAL should be formed in proteins in equal amounts. The DL and DD isomers may also be formed in strong alkali if lysine racemizes. In the case of substitution reaction whereby lysine combines directly with phosphorylserine without passing through dehydroalanine, LL-isomer will be formed. Unfortunately, analytical methods do not exist to distinguish between the different isomers present in food products. It has been suggested that unusual derivatives may also be formed by condensation between dehydroalanine and other amino acid residues, such as arginine, histidine, tyrosine and tryptophan. The presence of ammonia may prevent the production of these cross-linkages apparently with reacting with DHA to form β-aminoalanine. Cysteine, glucose and sodium bisulfite or hydrosulfite, as well as previous acetylation or succinylation of lysine residues, also decreases the formation of lysinoalanine in proteins during alkaline treaments.[64]

The physical and chemical parameters influencing LAL formation in food proteins were recently reviewed by Hurrel.[124] Because LAL is process induced, much effort has been made to find ways to reduce or eliminate its formation in alkali-treated proteins. Lower temperature of treatment and as low pH as possible decreases the lysinoalanine formation. The amino acid composition and sequence and the three-dimensional structure of the protein also plays an important role. The amount of LAL formed depends on the concentration of lysine and cysteine + serine in the protein chain. Proteins in which lysine and cysteine (or serine) residues are adjacent, or are only separated by one or two other amino acid residues, readily give LAL. Suggested alternative ways in using chemical treatment (acylation, oxidation of cysteine to sulfonic acid) or addition of chemicals (cysteine, glutathione) seem to have little practical relevance to the food industry.

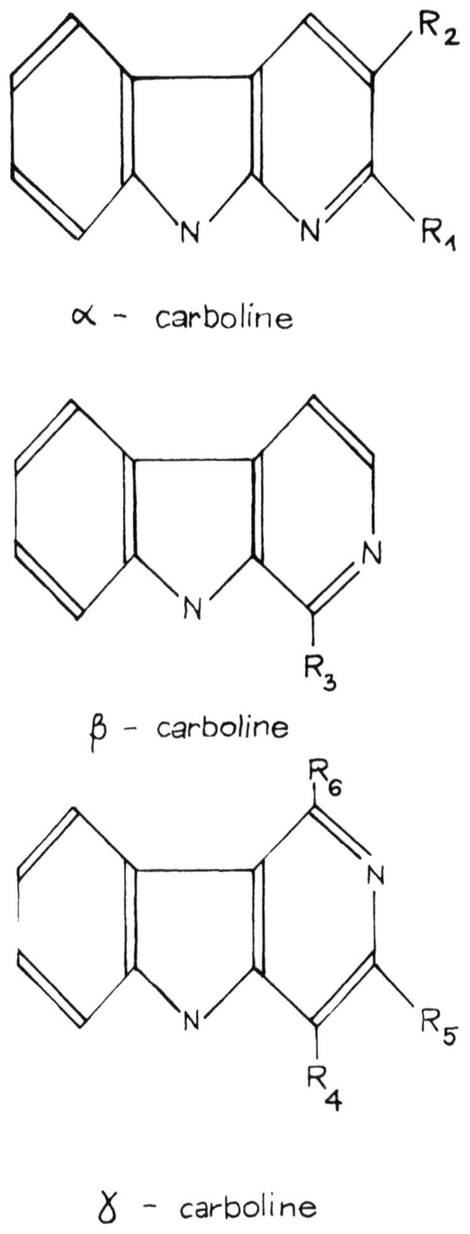

R_2 R_1 N N

α - carboline

N N R_3

β - carboline

R_6 N N R_5 R_4

γ - carboline

FIGURE 56. Reactions of dehydroalanine (DHA).

Until now no data were published about the LAL content of alkali-treated yeast protein preparations. Some informal data measured in our laboratory are shown in the Table 55. Concerning the nutritional value and safety of proteins in which covalent bonds of this kind are formed, there is an agreement among researchers that nutritive value of such proteins is often lower than that of the untreated proteins. In relation to safety (toxicity) of LAL and LAL-containing foods resp. protein preparations the views are sometimes contradictory.

The reduction in nutritive value of alkali-treated protein is due to a combination of factors. The formation of lysinoalanine would be expected to contribute to reduced protein digestibility but would probably have influence on lysine or cysteine bioavailability as the concentrations involved are normally small. It seems that destruction and racemization of

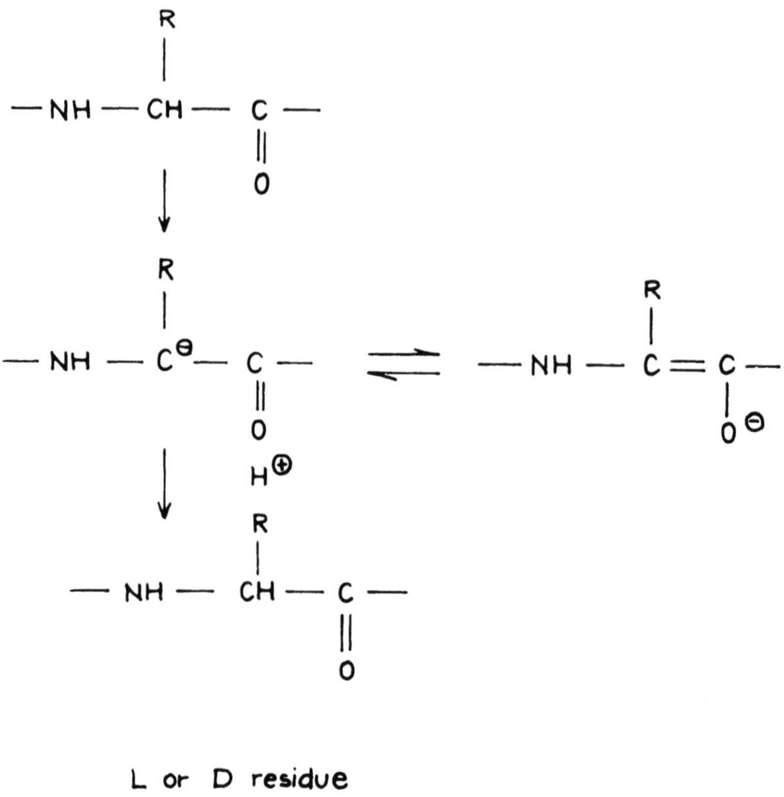

L or D residue

FIGURE 57. Reactions of threonine.

TABLE 55
Composition of Yeast Extracts (Autolysates)

Component	Product	
	Primary[174]	Secondary[175]
Total nitrogen	8—10.5	10.8
NaCl	2.1	8.3
α-amino acids	28.0	26.2
Peptides	21.0	21.3
Glutamic acid	2.5—7.0	9.6
Organic acids	2.0—4.0	3.7
Nucleotides and nucleosides	1.1	9.7
Fat	0.1—0.3	ND
Carbohydrate	12.0—21.0	ND

certain amino acid residues also plays a role in the reduction of bioavailability of some amino acids and reduction of digestibility of alkali-treated protein. Bioavailability studies have shown LAL to be unavailable as a source of lysine to the rat and mouse.

Experiments made with addition of free LAL to the diet of rats showed that free LAL is absorbed by the rat much more readily than the protein-bound LAL. Around 50 to 60% of the administered dose of free LAL is absorbed compared with only 10 to 20% of protein-bound LAL,[125] the rest being excreted in the feces. Protein-bound lysinoalanine is excreted in the feces to the extent of about 50%. Absorbed LAL is largely eliminated in the urine. It has been shown that C^{14}-labeled lysinoalanine is partly catabolized in the kidneys of the rat. The kidney appears capable of releasing lysine from up to 30% of the LAL. The released lysine can be utilized by the kidney cells and also by the liver.[125] These experiments indicate that the rat can utilize LAL to a small extent, which was not evident in the classic bioavailablity studies. Urinary catabolites in the rat are numerous, and some of them are different from those that have been identified in other animal species.

The first observations associated with the toxicity of LAL were published about 20 years ago, based on the experiments with rats fed with soy protein preparation produced with heat treatment in alkaline medium. The renal lesion designated nephrocytomegaly was demonstrated in the rats. It is characterized by increases in both nuclear and cytoplasmic parts of the epithelial cells in the straight part of the proximal tubule. The quantity of nuclear protein increases linearly with nuclear enlargement and enlarged nuclei frequently have 2 to 8 times the normal amount of DNA. The first signs of toxic effect appeared just after the first week of experiments, the first increased nuclei were observed in the fourth week and the maximal changes occurred after 8 to 10 weeks of feeding experiment. The lesion, however, is reversible, after feeding the rats with the LAL free fodder the normal function of kidneys was restored.[126]

The views concerning toxicity of lysinoalanine are recently summarized by Lásztity[127] and Lásztity and Törley.[128] It was stated that concerning the toxicity of free LAL it is generally accepted that a LAL content of 50 to 100 mg/kg is able to cause renal lesions in experimental rats. Extremely high doses (10,000 mg/kg) may result in a total pathological change of kidneys. It was also stated on the basis of long-term experiments that neither carcinogenic nor teratogenic effects were observed.

On the other side, views concerning the toxic effect of protein-bound LAL are often contradictory. The most surprising results were published by the research group of de Groot et al.[129] No pathological changes in the kidneys of rats fed with LAL-containing proteins in long-term studies, were observed. If soy protein treated under severe conditions (pH 12, 4 h, 40°C) was fed, the digestibility of protein decreased. A correlation was found between the NPU-values and the lysinoalanine content of proteins ($r = 0.96$). Only nephrocalcinosis was observed in experimental animals, but increasing the calcium level of diet the nephrocalcinosis diminished.

It was found that a diet containing acid hydrolyzed protein (containing LAL) caused nephrocytomegaly if the LAL concentration reached or surpassed 200 mg/kg. However, unhydrolyzed protein with 2000 to 6000 mg/kg LAL content caused no pathological symptoms.

To explain the contradictory results published by different research workers, a lot of hypotheses were elaborated. First, the question of duration of feeding experiments was discussed. Some authors reported toxic effect of protein-bound LAL in rats used in long-term (1 year) experiments with high levels (10,000 mg/kg) of LAL in the diet. Contradictory results were explained by different conditions of the experiments. It was also noted that the sensitivity of different rat strains could be very different and influenced by other factors too. Among the factors the most important is the protective effect of LAL-free proteins, e.g., native lactalbumin neutralized the toxic effect of LAL-containing soy protein. Similar observations were made in connection with the protective effect of casein.

Keeping in mind the facts mentioned above it is understandable that there is no direct correlation between the absolute quantity of LAL in the diet and the nephrocytomegaly symptoms. The effect of free LAL or oligopeptide-bound LAL is stronger. From the point of view of the evaluation of results, not only the LAL content but the conditions of the production of a protein source are also important.

As mentioned earlier different stereoisomers of LAL may be present in the protein or free LAL preparation. The LD-isomer is about 3 times more active than the DL, 10 times more active than the LL and 30 times more active than the DD. Rats developed cytomegaly in 4 weeks when fed a diet containing 30 mg/kg LD-LAL but needed 300 mg/kg of LL-isomer and 1000 mg/kg of the DD-isomer.

In addition, it can be noted that the LAL-containing proteins included in the experimental diets may contain other biologically active amino acid derivatives too. Dehydroalanine as a precursor of LAL is often present, ornithoalanine also causes nephrocytomegaly, however, due to the low quantity of ornithoalanine the effect of this compound on the results of experiments is probably insignificant.

The data confirming the toxicity of LAL are based on experiments with rats. As mentioned earlier some of the urinary catabolites in the rat are different from those that have been identified in other animal species. The toxic effects of lysinoalanine in rats may therefore be partly due to the formation of these unusual derivatives and lysinoalanine formation thus, may be of concern only in the rat. Cytomegaly has never been provoked in the hamster, rabbit, dog, quail or monkey, although the Swiss mouse is slightly sensitive to high levels of LAL. It is noted that the sensitivity of rats is much higher than the sensitivity of other animals.

The biochemical basis of the high sensitivity of rats is, until now, not clear. Since the mechanism by which LAL induces nephrocytomegaly is still unknown it is impossible to extrapolate to man. On the basis of our knowledge today regarding toxicity of LAL it seems that the small quantity of LAL present in some heat- and/or alkali-treated foods does not mean any hazard in the human nutrition. As infant formula, however, is the sole source of nourishment for young babies, it would still seem wise to ensure that all liquid infant formulae receive the least severe sterilization process possible.

C. OXIDATION OF AMINO ACID RESIDUES

Although the use of oxidizing agents in food processing is limited (hydrogen peroxide for example, is used in the milk industry) the naturally occurring lipid peroxides and their degradation products resulting from lipid oxidation can oxidize the proteins. Amino acid residues can also undergo oxidative modification as a result of photooxidation reactions, irradiation and hot air drying. The amino acid residues most sensitive to oxidative reactions are the sulfur-containing amino acids, tryptophan and to a lesser extent tyrosine and histidine.

As a result of oxidation of cysteine and cystine a lot of derivatives are formed (see Figure 58). Oxidative changes of cysteine and cysteine residues always lead to a decrease of nutritive value of protein. L-cystine mono and disulfoxides and cysteine sulfenic acid are able to replace, at least in part, L-cysteine, however, other derivatives cannot replace L-cysteine.

Methionine residue is also sensitive to oxidation. Not only oxidizing agents but also oxidizing lipids cause oxidative change of methionine. In the presence of light, methionine residues of oxygen and sensitizing dyes (e.g., riboflavin) undergo photooxidation reactions to yield methionine sulfoxide, (see Figure 59). Regarding the nutritive value of oxidized proteins it was found that free or protein-bound methionine sulfoxide may replace methionine in the diet of rats or chicks with an efficiency that varies depending on its configuration (D or L). It appears that methionine sulfoxide is set free from protein during digestion, and is absorbed and then reduced to methionine before being used for protein systhesis. Rats receiving protein-containing methionine sulfoxide show increased levels of methionine sulfoxide in the blood and muscles, which may indicate that the *in vitro* reduction of the sulfoxide is slow. Methionine sulfone cannot replace methionine.

Not only methionine but also tryptophan can undergo photooxidation reactions. Treatment with peracids leads essentially to the formation of β-oxyindolylalanine and/or *N*-formilkynurenine (see Figure 60). From a nutritional viewpoint, kynurenine, formulated or not, cannot replace tryptophan, at least for rats. Also, kynurenine is carcinogenic when injected into animal bladders, and tryptophan degradation products, as a group, inhibit growth of cultured mouse embryonic fibroblasts and exhibit mutagenic activities.

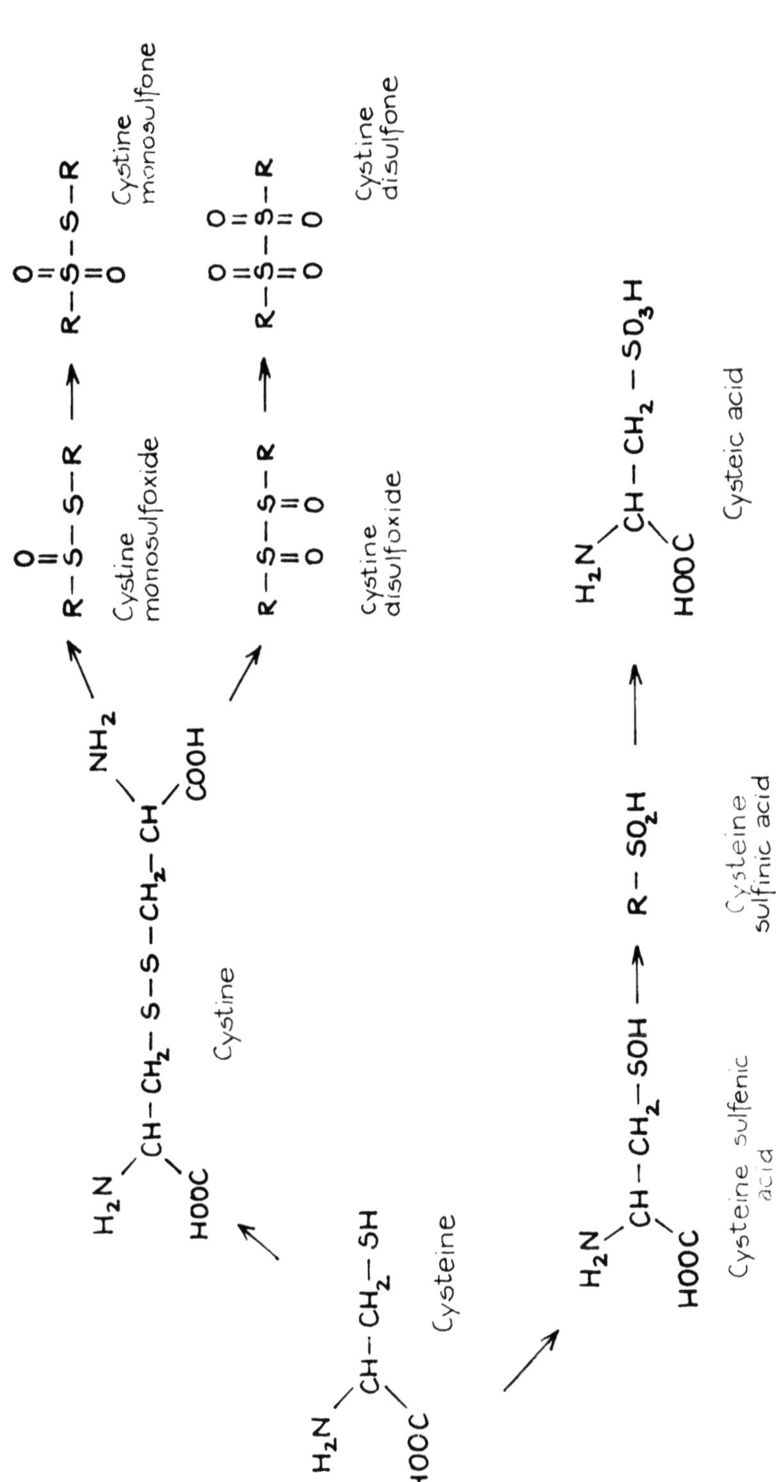

FIGURE 58. Oxidation derivatives of cysteine and cystine.

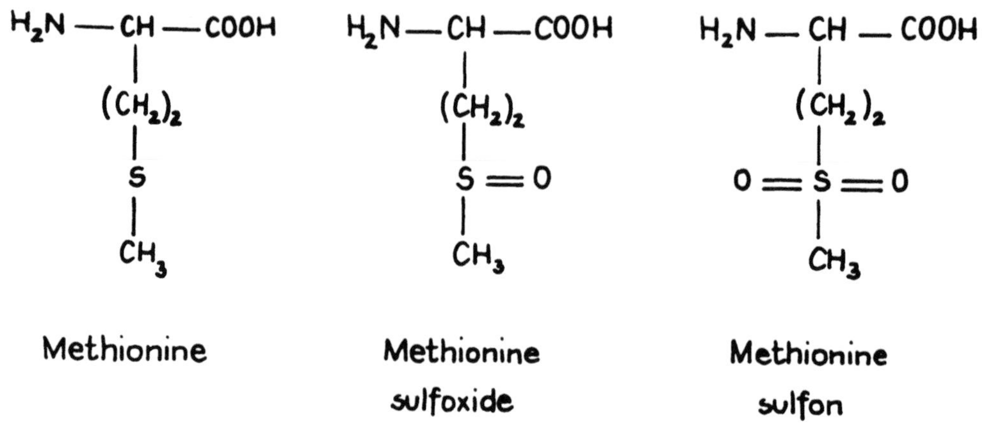

$$H_2N—CH—COOH \quad | \quad (CH_2)_2 \quad | \quad S \quad | \quad CH_3$$

Methionine

$$H_2N—CH—COOH \quad | \quad (CH_2)_2 \quad | \quad S=O \quad | \quad CH_3$$

Methionine sulfoxide

$$H_2N—CH—COOH \quad | \quad (CH_2)_2 \quad | \quad O=S=O \quad | \quad CH_3$$

Methionine sulfon

FIGURE 59. Oxidation of methionine.

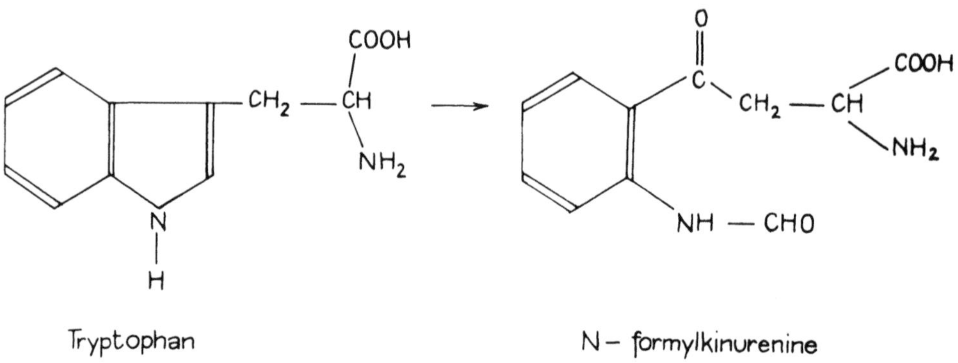

Tryptophan → N- formylkinurenine

FIGURE 60. Photooxidation of tryptophan.

D. PROTEIN-CARBOHYDRATE INTERACTIONS

The reaction of protein with reducing sugars is the major source of nutritional damage to food proteins during processing and storage. As a result of complex multistep reactions brown or black pigments, melanoidins are formed. The reaction called Maillard reaction or nonenzymatic browning was named after the French chemist, Louis Maillard who first described the formation of brown pigments when heating a solution of glucose and glycine. In the foods, the reaction normally occurs between reducing sugars and the amino groups of amino acids or proteins. Most primary amino groups are represented by the ϵ-amino group of lysine and, to a very small extent, the α-amino group of N-terminal amino acids. In addition, most foodstuffs normally contain a certain proportion of free amino acids whose reactions are often more relevant to flavor formation than to losses in nutritive value. The initial step in Maillard reaction is the formation of glycosylamine (see Figure 61). The glycosylamine undergoes the Amadori rearrangement to produce a 1-amino-2-keto sugar (see Figure 62). If ketose is the initial compound in the same way a ketosylamine will be formed and after Heyn's rearrangement an aldoseamine.

During the second step of this complex set of reactions, ketoseamines and aldoseamines evolve into numerous carbonyl and polycarbonyl unsaturated derivatives of which reductones are the best known. The reactions leading to pigments (Figure 63) called melanoidins are not well defined. It is suggested on the basis of numerous studies of the browning process in food[140-142] that there are three main pathways in the advanced Maillard reaction, all leading to production of brown pigments. Two of these pathways begin with the Amadori product

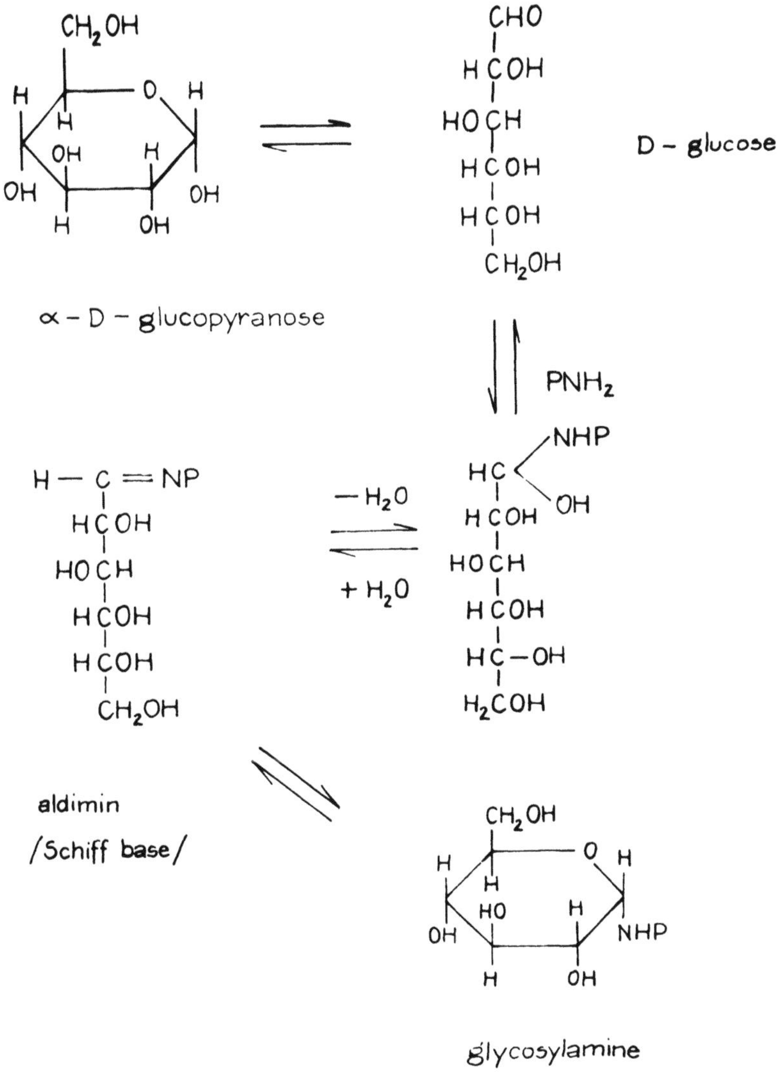

FIGURE 61. Formation of glycosylamine in Maillard reaction.

(Figure 62), the third is the Strecker degradation (Figure 64). This third pathway involves the degradation of free amino acids by the dicarbonyl compounds formed in the first pathway (Figure 63). Amino acids are degraded to the corresponding aldehydes of one carbon atom less with the loss of CO_2 and ammonia.

The nutritional effects of the Maillard reactions are still the object of many investigations. It is evident that advanced Maillard reactions destroy the lysine molecule presumably by reaction with such active intermediate compounds as the dicarbonyls and aldehydes. The loss of the lysine amino groups could be measured if the free lysine amino groups are blocked with FDNB and after hydrolysis of the protein the quantity of FDNB-lysine derivatives is determined. If the nondinitrophenylated protein is hydrolyzed with concentrated hydrochloric acid, Amadori products of lysine such as ϵ-*N*-deoxyfructosyl, release free lysine to the extent of about 50% of the initial substituted lysine. The remainder consists of new derivatives (furosine and pyridosine) which can be determined by ion-exchange chromatography and used as an index of substituted lysine residues and protein damage. Free ϵ-deoxyfructosyl-lysine is well absorbed by the rat. It is not utilized, however, and an average of 64% of the

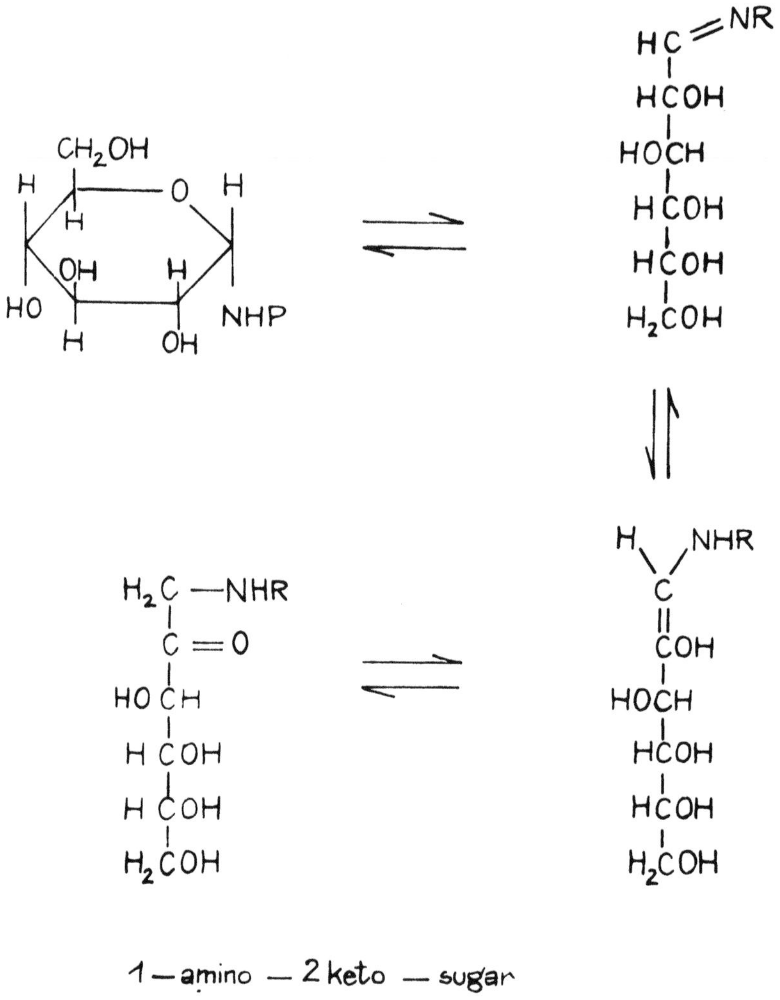

1—amino — 2 keto — sugar

FIGURE 62. Amadori rearrangement of glycosylamine.

ingested dose was excreted unchanged in the urine. Protein-bound ε-deoxyfructosyllysine, on the other hand, was not well digested and only 11% released from the protein, adsorbed and excreted unchanged in the urine. Those units not adsorbed are destroyed by the intestinal flora of the hind-gut, thus explaining the low recovery of only 6% of the ingested ε-deoxyfructosyllysine in the feces. The lysine reacted as Schiff's base (see Figure 61) is bioavailable since it can be set free under the acid conditions prevailing in the stomach.

There occurs a reduction in the biological availability of some other amino acids such as cystine, and even leucine, an amino acid with a chemically inert side chain. Such findings can probably be explained by the formation of profuse enzyme-resistant cross-linkages between various amino acid side chains via the reactive break down products. These cross-linkages reduce the rate of protein digestion, possibly by preventing enzyme penetration or by masking the sites of enzymatic attack.

The melanoidin fraction was almost completely indigestible and about 90% was found in the feces in rat experiments. The possible toxic effects of melanoidins and intermediers of the Maillard reaction are recently the object of many investigations.[145-146] Until now there is no definite answer regarding toxic effect of these products.

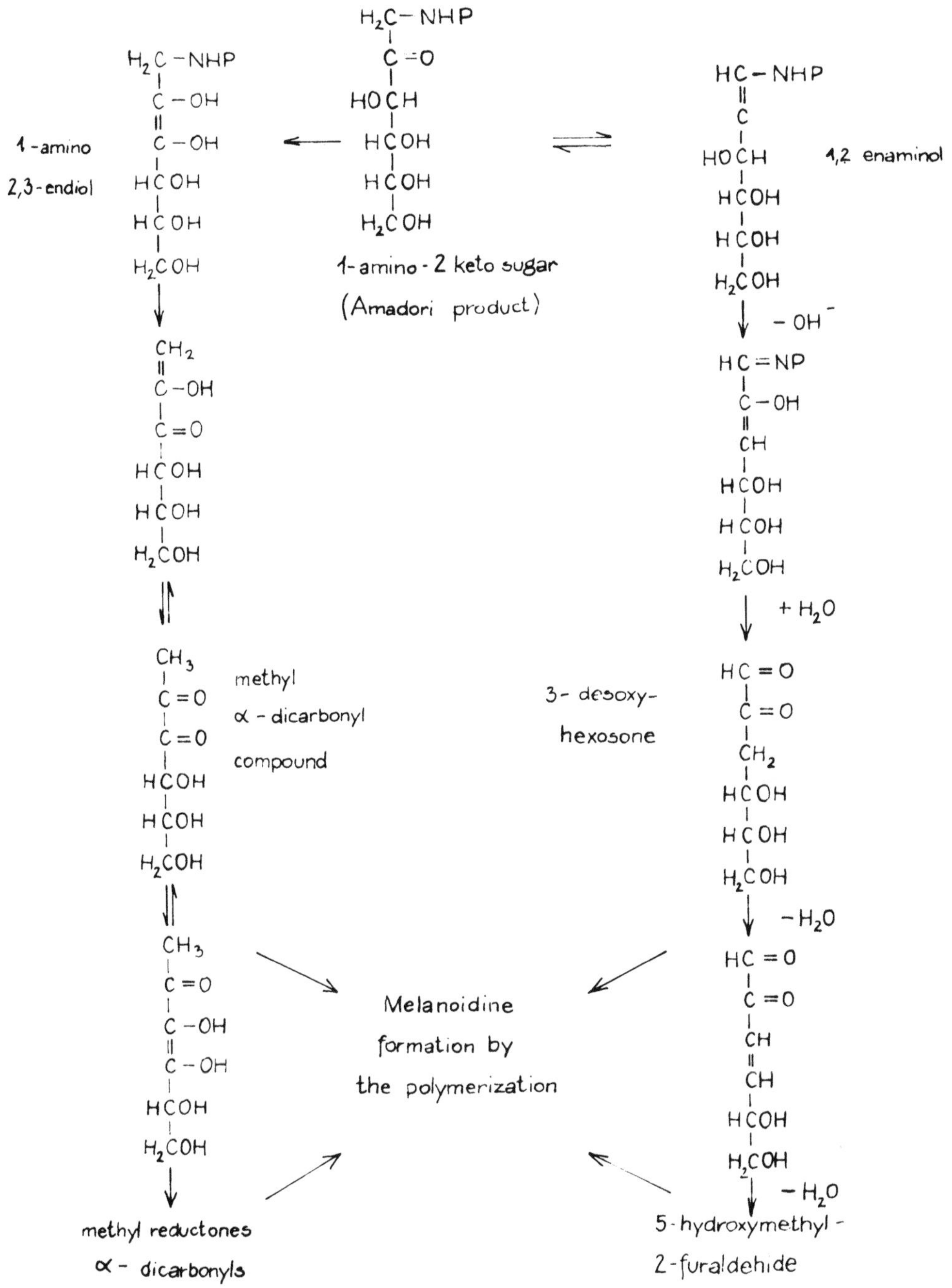

1-amino 2,3-endiol

1-amino-2 keto sugar (Amadori product)

1,2 enaminol

$-OH^-$

$+H_2O$

methyl α-dicarbonyl compound

3-desoxy-hexosone

$-H_2O$

Melanoidine formation by the polymerization

$-H_2O$

methyl reductones α-dicarbonyls

5-hydroxymethyl-2-furaldehide

FIGURE 63. Scheme of melanoidine formation.

E. PROTEIN-LIPID INTERACTIONS

Lipoproteins, consisting of complexes of lipids and proteins, occur widely in living organisms including yeast cells and influence their composition and nutritional value. In the interaction between proteins and lipids theoretically the following bonds may play a role: covalent bonds; ionic bonds; hydrophobic interactions and hydrogen bonds. Covalent bonds rarely occur in lipoproteins as a potential example, the bond between the hydroxyl group of serine residue and the phosphate group of phospholipids could be mentioned. Ionic bonds

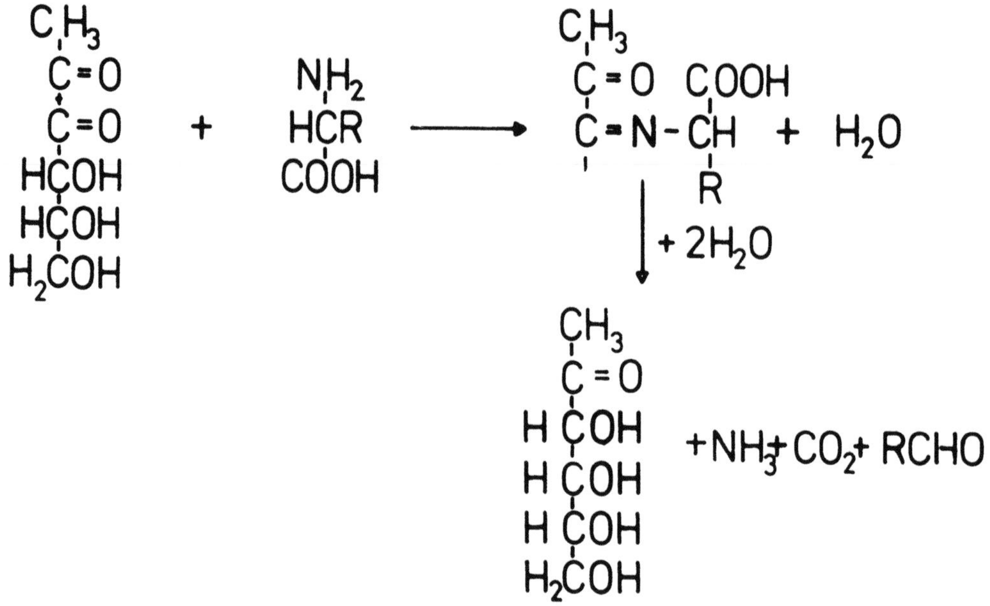

FIGURE 64. Scheme of Strecker degradation.

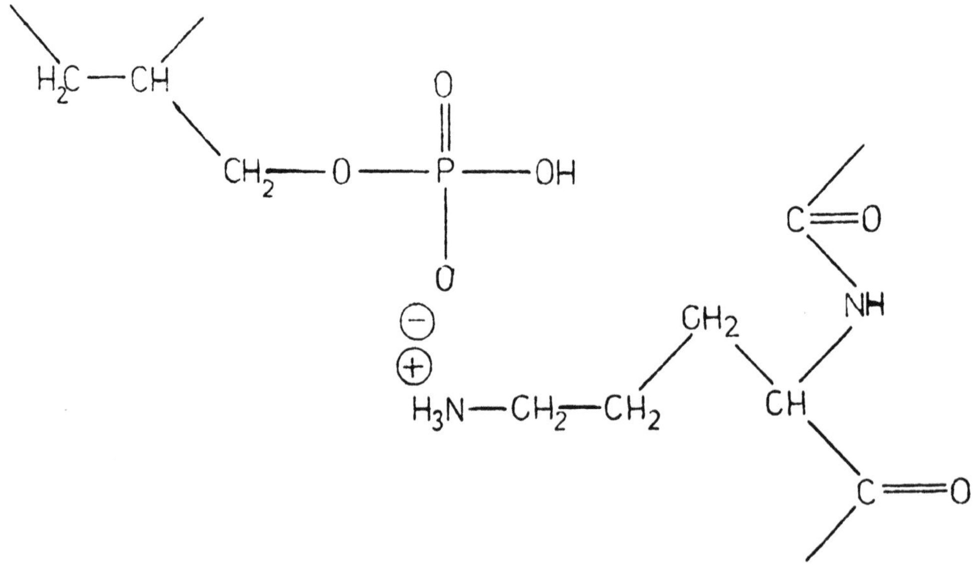

FIGURE 65. Ionic bonds in lipoproteins.

may be formed between phosphate group resp. basic amino group and the corresponding basic or acidic groups of proteins (see Figure 65). In Figure 66 some examples of formation of hydrogen bonds and hydrophobic interactions are demonstrated.

Changes of bonds in natural lipoproteins remove lipids by extraction with organic solvents, in most instances, it does not affect the nutritive value of protein constituents. Most technological operations during extraction, refining and drying of yeast proteins or drying of total yeast biomass are realized in the presence of oxygen. In such conditions formation of lipid oxidation products may occur. The lipid oxidation can be followed by protein-lipid covalent interaction. This interaction was intensively studied in model experiments and also

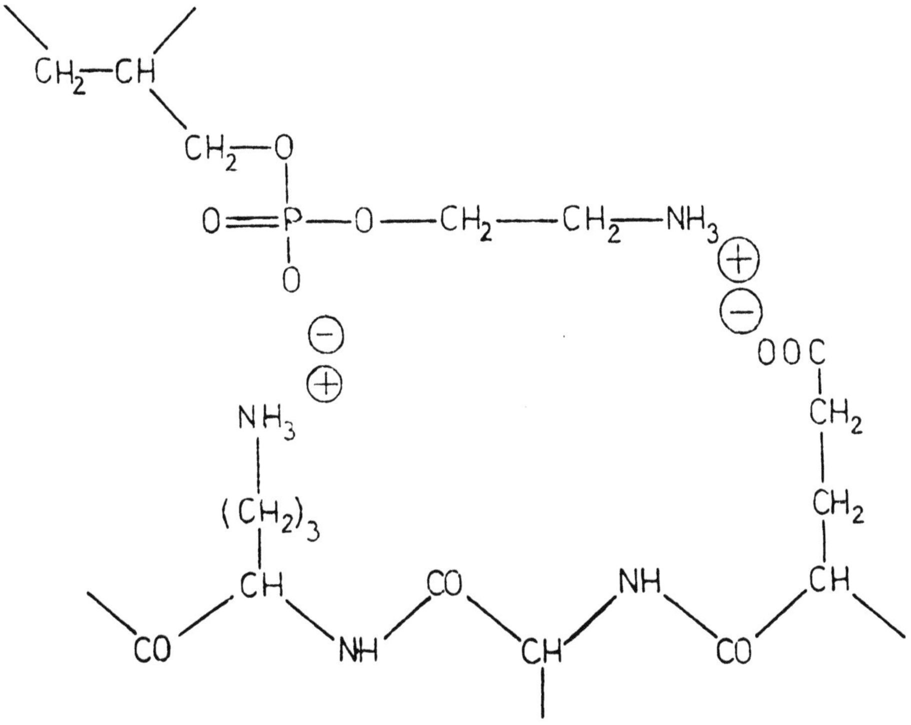

FIGURE 65B (continued)

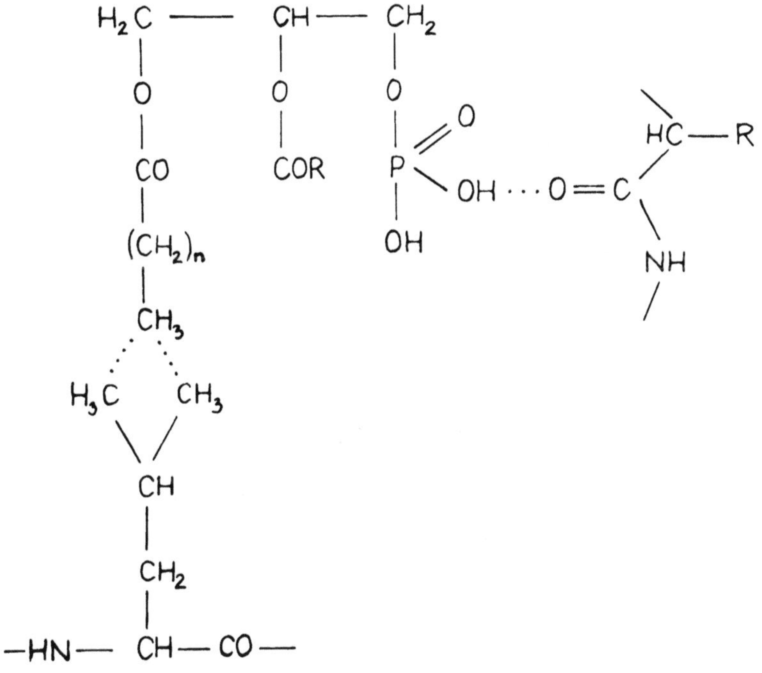

FIGURE 66. Hydrogen bonds and hydrophobic interactions between lipids and proteins.

in some foods. Two types of mechanisms appear to be involved in the covalent binding of peroxidizing lipids to proteins and in the lipid-induced polymerization: free radical reaction[147] and carbonylamine reactions. Free radicals of lipids (L$^{\cdot}$, LO$^{\cdot}$ alkoxi-, LOO$^{\cdot}$ peroxi-) can react with proteins in different ways. By addition a complex radical can be formed:

$$\mathrm{LOO^{\cdot} + Protein\ ^{\cdot}H \rightarrow LOO\ Protein^{\cdot}}$$

This reaction can be followed by measuring polymerization involving cross-linking of protein chains by multifunctional lipid free radicals:

$$\mathrm{^{\cdot}LOO\ Protein + O_2 \rightarrow {}^{\cdot}OOLOO\ Protein}$$

$$\mathrm{^{\cdot}OOLOO\ Protein + Protein \rightarrow Protein\ OOLOO\ Protein}$$

Protein free radicals can form at α-carbon atoms and at the sulfur atoms of cysteine residue. It was found that by increasing water activity the concentration of free lipids decreased. The free protein radicals can form by polymerization of lipid-free protein polymers. The occurrence of such polymerization was confirmed by experiments made with model systems containing water-soluble proteins, enzymes and lipid peroxides. A significant increase of the molecular weight of proteins was observed. Losses were found in methionine, histidine, cysteine and lysine residues of protein. The covalent binding of lipids in the complex was also confirmed.

The carbonylamine reaction is connected with the formation of secondary products of lipid oxidation and their interactions with amino acids or free amino groups of proteins. Aldehyde derivatives resulting from oxidation of unsaturated fatty acids can bind to the amino group of proteins through Schiff's base type reactions.

$$\mathrm{L - C = O + H_2 - Protein \rightarrow L - C = N - Protein}$$

Occurrence of such reactions was confirmed by changes of protein solubility, changes of UV absorption spectra and fluorescence of newly formed compounds. A decrease of available lysine and losses in methionine and histidine were also observed. The interaction of malonaldehyde with proteins is of particular interest. This compound can react with free amino groups of different polypeptide chains. Covalent bonds of the 1-amino-3-iminopropene type are formed (see Figure 67). As a result of the interaction of the products formed in the reaction of proteins and malonaldehyde yellow or brownish pigments may be formed called "age pigments" or "lipofuscines". Another pathway of lipofuscine formation is associated with the reaction of malonaldehyde and free amino groups containing phospholipids (Figure 68).

From a practical standpoint, it is likely that proteinaceous foods undergoing lipid oxidation become organoleptically unacceptable before damage to protein nutritive value has occurred to any significant extent. Regarding the losses of nutritive value it is evident that lysine, methionine, cysteine and tryptophan can all be damaged on reaction of proteins with oxidizing lipids. In terms of relative nutritional importance, however, losses of lysine and methionine would appear to be most important, along with the general reduction in protein digestibility.

IX. YEAST AUTOLYSATES AND THEIR PRODUCTION

The use of yeast extract as a source of nutrients in media for cultivation of fastidious microorganisms has been known for more than 60 years. In the last decades yeast extract is becoming increasingly popular as a food additive. The list of foods where yeast autolysate

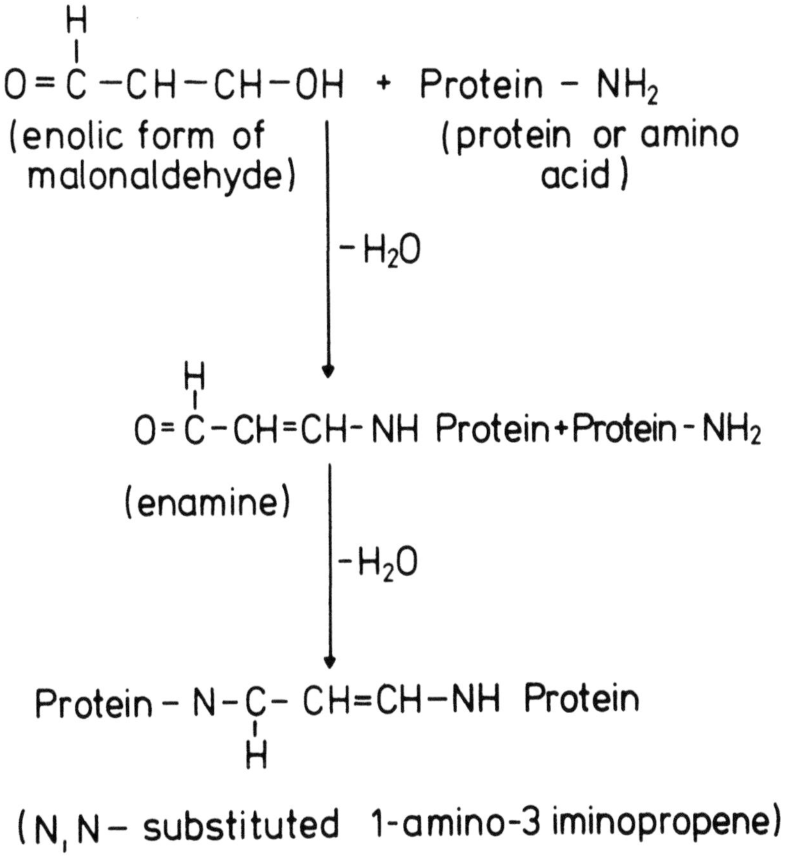

$$O=\underset{|}{\overset{H}{C}}-CH-CH-OH \; + \; Protein-NH_2$$

(enolic form of malonaldehyde) (protein or amino acid)

$$\downarrow -H_2O$$

$$O=\overset{H}{C}-CH=CH-NH\;Protein + Protein-NH_2$$

(enamine)

$$\downarrow -H_2O$$

$$Protein-N-\underset{H}{C}-CH=CH-NH\;Protein$$

(N,N– substituted 1-amino-3 iminopropene)

FIGURE 67. Reaction of malonaldehyde with proteins resp. amino acids.

may be used as an ingredient include such products as meat pastes, meat pie fillings, soups, gravies, sauces, cocktail snacks and savory spreads. Also, the extract is an excellent nutrient source in such fermentation processes as vinegar production. These and other applications are reviewed by several authors.[148,149,159,160]

Yeast autolysis is also an important tool in biochemical research for the extraction and purification of enzymes and coenzymes. The best known example is that of invertase, which is also produced on a commercial basis.

In contrast to the yeast extract manufacturer who wishes to encourage the autolysis of yeast, the brewer and producer of baker's yeast both wish to avoid it. Autolysis of yeast in the brewery can lead to unwanted "off flavors" in the finished beer and to an increased susceptibility to bacterial contamination. Autolysis of baker's yeast can lead to liquefaction and loss in leavening power.

Although, the term "autolysis" was introduced into biological literature more than 100 years ago, and began to be used to designate self-digestion of the cells under the action of their own intercellular enzymes, the intensive research work on the biochemical basis of autolysis only began in the 1960s. Stolpe and Starr,[150] in an earlier review paper, expressed the view that "autolysis should be understood as a lytic event in the cells caused by the action of their own intracellular mureinase (the enzymes that degrade peptidoglycans of the cell wall). On the basis of our knowledge today this definition was incorrect because only bacteria have a mureic type cell wall (the external wall of yeasts consists of some layers containing glucan, mannan, protein, chitin, etc.) and enzymes other than mureinases take part in the autolysis of yeasts.

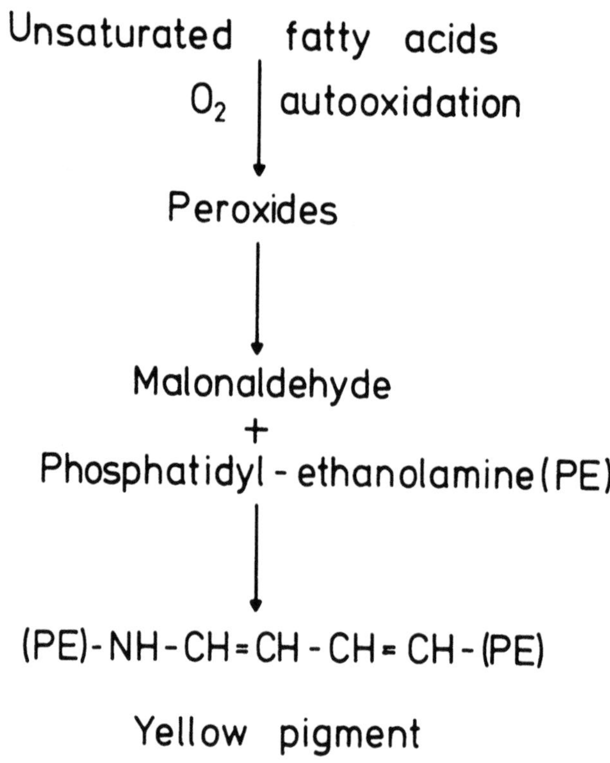

FIGURE 68. Scheme of lipofuscin formation from phospholipids and malonaldehyde.

The development of autolysis in microorganism cells may take place in two ways.[151] The first one beginning with loss of the cell wall under the action of its own hydrolases is characteristic of "exo-type" autolysis, which is observed in most bacteria. The exo-type autolysis may terminate at this stage and not affect other cell components. The second way begins with disturbance of lipoprotein structures in the cell membranes ("endo-type" autolysis, is what fungi, yeasts and certain kinds of bacteria are subject). If denaturation of intracellular hydrolases takes place during disturbance in the cell, the process of autolysis is absent, albeit the cell dies. Hydrolysis of the cell components may begin only following activation of hydrolytic endoenzymes. However, if hydrolase activation takes place within a limited scope of the cell and is genetically controlled, the cell does not die. At various stages of the lifecycle in definite cell sites the need arises for the degradation process to prevail over synthesis.

The regulatory role in living organisms is connected often with so called limited proteolysis. The latter underlies the various physiological processes that take place inside the living cells. New data on the proteolytic system of yeast certify the free cytosol peptidases to participate in cell regulation. The mechanism of cell degradation in yeast during budding was found to involve the autolytic complex, which activates the chitin synthase to lead to synthesis of chitin, forming a membrane and gemmating a daughter cell. Such events also occur in other microorganisms where the balance relationships are moved toward degradation of biopolymers in specific loci. Yet, the cell does not die, since this violation is of a reversible nature. It involves an autolytic process, but autolysis does not occur. A newer definition given by Matile[152] namely that autolysis should be seen as an "irreversible process entailed with cell death occurring as a result of disturbed regulation in the equilibrium work of enzymatic systems toward prevalence of a hydrolytic system" is more accurate than that of

Stolpe and Starr[150] mentioned earlier. Nevertheless, two important characteristics of the yeast autolysis, namely the formation of low-molecular weight products and releasing of these compounds from the intercellular space were not mentioned. The most acceptable view is, according to our opinion, the description of autolysis as a complex process in which hydrolysis of intracellular biopolymers occur under the action of endohydrolases entailed cell death and the formation of low-molecular products releasing from the intracellular space.[151]

Apart from separating the process into the exo- and endo-types, autolysis is also classified into natural- and induced-types, both of which are inherent in absolutely all microorganisms, irrespective of their taxonomy. In periodic cultivation of microorganisms, cell autolysis at the end of the stationary phase is caused by natural aging of the culture. Such autolysis is termed natural, and it is difficult to reveal any single basic cause of its development. The so called induction autolysis of microorganism cells of any age may be artificially caused by means of various physical, chemical or biological agents.

A. INDUCTORS OF MICROORGANISM AUTOLYSIS

On the basis of earlier research Welsh[153] formulated a hypothesis, according to which any disturbance in cell energy supply, irrespective of the cause, induces autolysis. However, it was shown[154] that inhibition of the cell energy supply systems, for example limited ATP supply, do not always result in autolysis but supports anabiosis. An analysis of the few examples of inductors of eukaryotic cell autolysis permit the conclusion that inductors of a varying nature may cause similar disturbances both functional and structural, e.g., the temperature effect and the action of detergents are reduced to changing membrane structural organization. Removal of oxygen from the growing aerobic culture leads to an imbalance between the synthetic and autolytic enzyme systems, i.e., to the same effect produced by lactamic antibiotics. According to Hough and Maddox[149] it appears that in the absence of any other suitable carbon source, the brewer's yeast will adjust its metabolism in order to utilize ribose, although healthy cells do not normally do so. We suppose that although different factors may initiate the autolysis of yeast cells, nevertheless, the most important is the absence of any carbon source. This effect may be enforced with other factors such as the increasing of temperature or removal of oxygen.

Among the *physical inductors* of yeast autolysis first, the temperature conditions can be mentioned. On one side an increase of temperature may be used and on the other side an alternate freezing and melting.[155] In the first case, the development of autolysis is said to be caused by changes in membrane structure, an inevitable process for the autolysis together with some changes in the regulatory system of the cell. In the second case, moreover, ice crystals form to result in mechanical violation of organelle structural integrity. In connection with latter statements it must be noted that the processes occurring during freezing are of a much more complex character.[156] The general scheme of the structural-metabolic changes is shown in Figure 69. The damage of membranes and organelles also depends on the metabolic state of the yeast cells. Experiments made by Kirsop showed that stationary-phase cells, strains grown without free access to oxygen, are generally less sensitive to the freeze damage.[157] The autolysis rate depends on temperature, and the most optimal temperature interval was observed at 45 to 60°C and pH = 5.[155,158]

Osmotic pressure of the medium may effect the yeast cells. Changed osmotic pressure may have two consequences, i.e., complete cell degradation or autoplast formation. A third version of the process is also possible, when in certain sodium chloride and sucrose concentrations it is preceded by plasmolysis, i.e., by inner rearrangements unaccompanied by hydrolysis.

Irradiation by ultraviolet or X-rays which causes deep changes in regulatory and biosynthetical cell processes, but does not impair autolytic capability can also be an inductor of yeast autolysis. Finally, autolysis can be induced by mechanical disintegration — as a

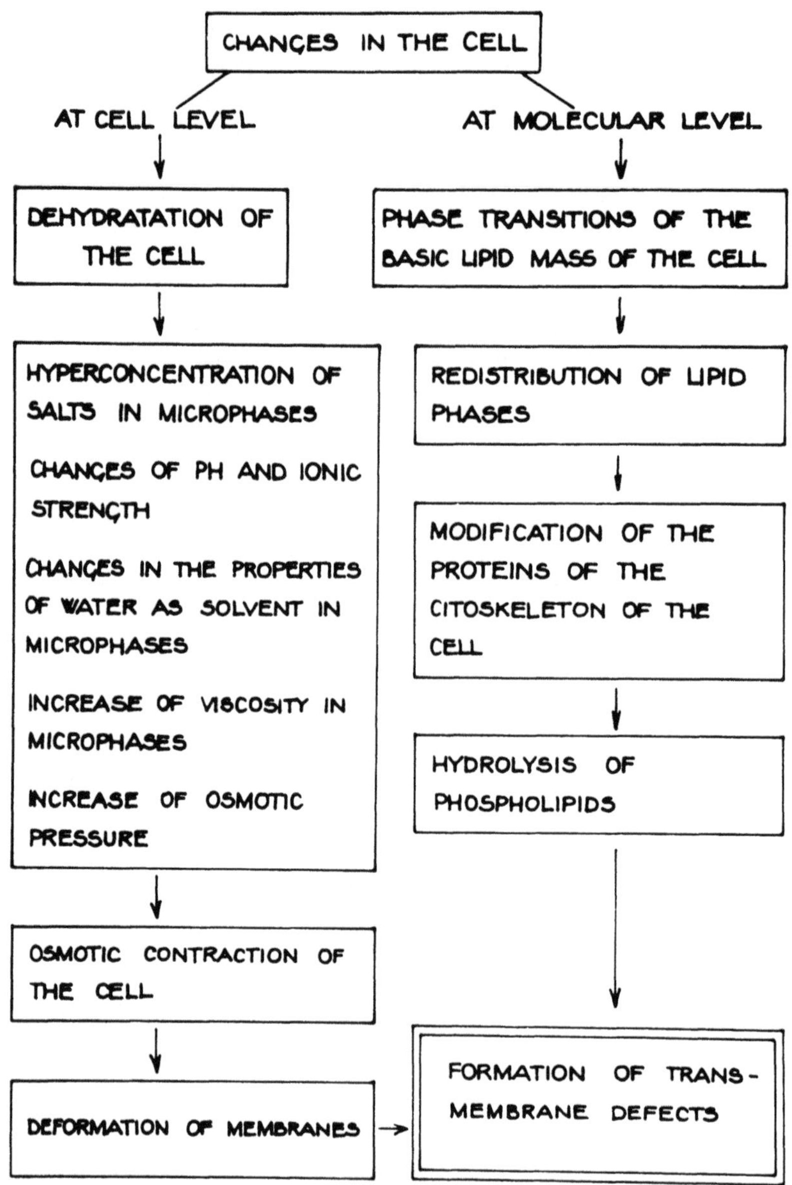

FIGURE 69. General scheme of structural-metabolical changes in the cell in temperature range 0 to − 30°C. (From Belous, A. M., Proc. Third Natl. School Cryobiol. Freeze-Drying, Tsvetkov, T., Ed., Agricultural Academy, Sofia, 1985, G 1-12. With permission.)

physical effect — that leads to the liberation of the autolytic complex enzymes contained in lysozomes, and also those connected with the cell membranes and cell walls.

Among the chemical inductors of autolysis changes of ionic strength, changes of pH and ion composition of medium can be mentioned. All these factors can produce multiple effects leading to disturbed cell metabolic activity. The qualitative ion composition in the incubation medium is highly significant for the autolysis rate. The influence of several metal cations is used in invention[163] and discussed in the paper of Chrenova et al.[164]

Various membranotropic compounds, including detergents,[165] proteins, peptides and amino acids, that violate the structural entity and functional ability of membranes may be used for induction of autolysis. A mixture of the products of biomass autolysis and casein proteolysis is a fine inductor of yeast autolysis, e.g., treatment of cells with basic proteins permeabilize the cell membranes. It was demonstrated in yeasts that these proteins differentially disrupt the membranes of the plasmalemma, but leave those of tonoplast intact.[166] Osmotic shock with water following the treatment with basic proteins disrupts the tonoplast too.

The degrading effect of these agents upon membranes is caused by their ability to solubilize in the membranes, or to change the energetics of hydrophobic interactions at the expense of changed water structure, this leading to changed membrane permeability and leakage of the lysing enzymes. Conversely, some amino acids are inhibitors of autolysis and the inhibitory effect is dependent upon their hydrophobicity. Antibiotics induce autolysis as a result of imbalance between synthetases and depolymerases. Besides, the previously caused imbalance was associated only with cell wall synthesis and at present antibiotics have been proven to effect the membrane.

B. CHARACTERISTICS OF THE PROCESS OF AUTOLYSIS

Autolysis of yeasts is accompanied by characteristic, visually observed changes, including changes in the rheology and color. The presence of at least two stages of autolysis was exemplified by changed morphology of yeast cell studies by electron microscopy. The first stage of autolysis at a submicroscopic level manifests itself in degradation of cell endostructures: turgor is absent, neither the organelles themselves nor their fragments are discernible, all the cell content is uniformly distributed over a space restricted by the cell wall and the cell diameter reduces almost 1.5-fold. Comparing with the starting cell, the cell wall is considerably thickened, but its integrity is not destroyed. The second stage of autolysis involves hydrolysis of intercellular biopolymers, which leads to diffusion of the hydrolysis products into the extracellular medium. However, at pH 5 to 6, the lipids aggregate in the form of large drops and do not pass beyond the cell bounds.

The autolytic changes in the *Saccharomyces cerevisiae* cells induced by starvation, were shown to be distinguished at a submicroscopic level much later than by additive-induced ones (after 2 to 3 days instead of 2 to 3 h, respectively). Contradictory data were published concerning intactness or degradation of mitochondria. Babayan et al.[151] reported on intactness of mitochondria up to the very late stages of autolysis, however, other authors observed these organelles to digest during autolysis.

The number of autolyzing cells in a unit of medium volume does not change, however, many of them assume an irregular form. This signifies that yeast autolysis does not involve clarification of the yeast suspension, whose optical density remains virtually unchanged. The rate of depth of autolysis in bacteria is known to be assessed by a decrease in the optical density of the cell suspension. Since this criterion is unacceptable for yeast, investigators resort to characterizing protein hydrolysis products; the degradation products of other biopolymers may also be measured.

In the process of autolysis the cell wall in most yeasts undergoes only certain structural modifications, however, its completeness is retained, this being observed also at the end of the process, and the cytoplasmic material gradually diffuses into the extracellular space. Autolysis depends on the culture age and the physiological conditions in the cell.[167] Proliferating culture cells autolyze quicker and to a higher degree than cells of the stationary growth stage, for autolytic enzymes are synthesizing most intensively, precisely in the cells of the exponential growth phase. Hence, autolysis occurs when induction by means of a given factor not only fails to lead to degradation of autolytic enzymes, but also when there are optimal conditions for their action.

In natural autolysis at the end of the stationary phase when culture growth is hindered and the culture is considered aged, the cell physiological conditions are such that altogether they promote autolysis. Natural autolysis caused by aging of the microorganism culture has been well-studied in fungi, however, it was reported that autolysis in yeasts differ little from that in other fungi and is called mycolysis.

1. Location and Stages of Autolytic Complex

Generally, it is suggested that the autolytic enzyme system of a living yeast cell is located on the inner side of the cytoplasmic membrane. Some hydrolases may be localized in the periplasma and cell wall. Yeast cell walls were used to isolate the enzyme fixed therein, most of these enzymes proved to belong to the class of hydrolases. The majority of investigators do not dispute that hydrolases are localized in the cell walls, both in prokaryotes and eukaryotes (see also Chapter 2).

There is evidence of the presence of free peptidases in the cytosol of yeast and proteases with pH 8 to 8.5 optimum in yeast nuclei. And still the largest number of hydrolases in yeast are concentrated in lysozymes.[168,169] At present there is no exact understanding of the mechanism of autolysis in yeast. The analysis of available materials permits the conclusion that there is the presence of at least four stages in the autolytic process of yeast.[170] The first step involves a disturbance of supramolecular intracellular structures that violates the spatial apartment of hydrolytic enzymes and their substrates at the expense of any effect upon the lipoprotein membranes of lysozomes and other organelles.

The second step involves interaction of released enzymes with the cytosol inhibitors. Interacting digestion begins through cross proteolysis of inhibitors that specifically inhibits a corresponding enzyme. Inhibitor hydrolysis results in activation of the hydrolytic enzymes themselves. In the third step, activated enzymes interact with intracellular polymer components (substrates), resulting in accumulation of their hydrolysis products in the space restricted by cell wall. In step four, as the molecular mass of hydrolysis products decreases to an extent commensurable with the size of the cell wall parts, the said products diffuse into the extracellular medium. At present, there are still many unstudied and unresolved issues regarding the regulation and physiological role of autolytic enzymes in eukaryotic cells.

2. The Biochemistry of the Process of Autolysis

The biochemical basis of autolysis is not yet completely understood. It is clear that hydrolytic enzymes are responsible for autolytic degradation of most cell polymers, the main role in the process belongs to nucleases, proteases and glucanases. Since yeast autolysis is associated with initial degradation of the cytoplasmic membranes containing protein lipid complexes, it would be logical to assume the involvement is the process of phospholipases. It was reported that in the yeast *Saccharomyces cerevisiae* the degradation of mitochondria is caused by phospholipases A and D during glucose repression.[151] Experiments of Hough and Maddox[149] support the view that in the absence of any other suitable carbon source, the brewing yeast will adjust its metabolism in order to utilize ribose, although healthy cells do not normally do so. In their experiments made with *Saccharomyces carlsbergensis* the yeast was autolyzed at 45°C in distilled water at pH 6.5 with occasional stirring, samples were withdrawn at 1 h intervals for a period of 14 h. Loss of dry weight of the yeast was observed under these conditions, indicating that cell material was appearing in the extract. Among the extracted materials nucleic acids were present and paralleled the nucleic acid content of the decreased yeast cells. The amount of DNA lost by the cell was approximately the same as that gained in the extract, but RNA did not follow this pattern; the amount found in the extract was less than that lost by the yeast. The techniques for measuring RNA in cell and extract were different, however, the nucleotide was measured in the cell, and only the pentose sugar ribose in the extract. If the ribose moiety of RNA was used by the cell the discrepancy would be explained. This explanation has been confirmed by performing autolysis at a temperature less than lethal temperature for the cell (30°C), circumstances where added ribose was rapidly utilized. No comparable utilization of deoxyribose has been shown. From this it was evident that during autolysis RNA is degraded. The pathway probably includes RNase, which produces nucleotides which are themselves attacked by phosphomonoesterases to yield nucleosides and inorganic phosphate. Degradation of nucleosides is then brought about by the nucleosidases, producing ribose and a free base. DNA is probably degraded in a similar fashion. Ribose can be used by the yeast cell through the pentose phosphate pathway and glycolysis.

The release of proteins and amino acids is generally considered to be the most important aspect of yeast autolysis. It is evident that fairly extensive proteases occur outside the cell, showing a release of active enzymes from the cell during autolysis. The proteases of the yeast cell were discussed in detail in Chapter 3. It is interesting that all proteolytic enzymes involved in autolysis appear to be glycoproteins, containing both glucose and mannose residues in various proportions. This suggests that there may be some connection with yeast cell wall synthesis because the wall is substantially a mannan-glucan-protein complex. Synthesis of the wall is thought to take place in spherosomes just beneath the cytoplasmic membrane. The extraction of the proteolytic enzymes by autolyzing yeast means that the enzymes have to pass through this region of wall synthesis.

It is interesting to consider the possible origin of the proteolytic enzymes involved in autolysis. If the yeast protoplasts are prepared by degrading the cell wall with lytic enzymes, the protoplasts that emerge can be readily lysed. From the lysed protoplasts proteolytic enzymes can be isolated. The major difference between these, and those prepared from autolyzing yeast is the lack of bound carbohydrate. Intact vacuoles of yeast contain proteases, nucleases and esterases. Similar vacuoles were permitted to be lysozomes.[149] Subsequently, aminopeptidases and other enzymes of the lytic complex were identified in the enzyme composition of yeast lysozomes and cytosol.[151] A distinctive feature of the lysozome complex is its sets of enzymes which differ in substrate specificity and ensure complete hydrolysis of the respective substrates. At least seven proteases and peptidases take part in protein hydrolysis. The following should be considered as important principles in the action of lysozome enzymatic groups: (1) consecutiveness; (2) absence of narrow specificity and (3) resultant deep splitting of biopolymers to low-molecular products. The resistance of lysozomal enzymes to the action of hydrolases is explained by the linkage with carbohydrate prosthetic groups.[171] It was suggested that the resistance of proteolytic enzymes to self-digestion is allegedly caused by the spatial separation of proteolytic enzymes and their substrates in the cell. Autolytic protein degradation becomes possible only after this situation is violated and biosynthesis is delayed; yet in this case too, a system of inhibitors may prevent autolysis. Every inhibitor is specific to its protease, and inhibition is of a competitive nature. During the mechanical or other disturbances the proteases are released from the vacuoles to combine with corresponding inhibitors. The proteases break away from the inhibitors after the latter are digested to subsequently perform nonselective proteolysis of other yeast proteins. In as much as the inhibitors are separated from their proteases, it would seem that they cannot be regulators of protease levels.

The available data in extracellular enzymes in yeasts are controversial: some of them indicate the presence of proteolytic activity in yeast autolysates, and others the complete absence of proteolytic, nucleolytic and amylolytic activity. The extracellular lysing enzymes found in numerous microorganisms are assumed to be evolutionally related with autolysis.

During autolysis, the release of carbohydrates from the cell into the extract is small compared with that of protein and nucleic acid. The yeast cell wall remains essentially intact, so that living and autolyzed cells cannot readily be distinguished by light microscopy. However, some carbohydrate is released from the cell in the form of glycoprotein and this becomes split, probably by carbohydrates which are released in the same way as proteolytic enzymes. In fact, enzymes degrading both mannan and glucan have been detected in a yeast autolysate, but they have not yet been characterized.[149]

It has already been stressed that proteolysis plays a major role in the phenomenon autolysis. During its growth yeast must have possessed its full complement of proteolytic enzymes. In industry, the temperature of autolysis is usually greater than the death point of the yeast so that no protein synthesis can be assumed to occur under these conditions. To test the hypothesis that no protein synthesis occurs during autolysis, yeast was grown in a synthetic medium with various nitrogen sources, harvested in the logarithmic phase, and the yeast autolyzed with chloroform. The proteolytic activity of the crude, water-soluble extract was then measured at pH 7. The results show that proteolytic activity is present when protein is used as the nitrogen source in the growth medium, but activity is completely lacking when amino acids or ammonium salts alone are used. Even the presence of only a small amount of protein produces a positive effect. Hence, to achieve an efficient autolysis under industrial conditions, it is advisable to use yeast that has been produced on a medium containing protein, such as brewer's wort.

When one is considering a possible mechanism for yeast autolysis, it must be kept in mind that because of the imposed adverse conditions such as high temperature, the yeast will almost certainly be present as a mixture of living, dying and dead cells. The cells will be starved of nutrients and some of them will become moribund, leading to a change in the permeability of the plasma membrane. Intercellular material will be released into the medium. The living cells will respond to the presence of protein in the medium (some of which is derived from the dead cells) and will secrete proteolytic enzymes to utilize it.

3. Autolysis On a Commercial Scale

Carried out under controlled conditions, the autolytic process is induced by holding an agitated slurry of cells at temperatures (40 to 55°C) which produce cidal effects on the cells without affecting their endogenous enzyme activity. Autolysis generally requires 12 to 36 h to attain the desired degree of hydrolysis. If an autolysate is desired the slurry is pasteurized (80 to 90°C), cooled, and either vacuum evaporated to a paste with 70 to 80% solids or spray dried. To produce an extract, the pasteurized slurry is filtered post cooling, then concentrated or spray dried. Extracts may be classified as hydrolysates or plasmolysates. Hydrolysates are prepared under strongly acidic, controlled conditions and are the most efficiently produced extracts.

Plasmolysates are prepared by extracting the cellular material with a plasmolyzing agent.[161] The most popular such agent is sodium chloride, but other plasmolyzing agents are ethylacetate, amyl-acetate, dextrose and ethanol.[162]

In the manufacture of yeast extract yeast cream is obtained either from propagation of yeast grown in molasses or whey, or from extraction of spent brewer's yeast. In the second process, isohumulones, bitter compounds from the hops, are removed by extraction with a mild base. Under adequate conditions this process can be combined with removal of coloring substances.[1] After alkaline treatment the pH is adjusted to 5.5 to 5.7 with phosphoric acid. The process from spent yeast is known as a secondary process.

Whether from spent brewer's yeast or from propagation on molasses, rupture of the cell walls allows the enzymes present to commence autolysis, i.e., hydrolysis of the cell contents; this is an exothermic process. The temperature of hydrolysis used on a commercial scale is around 40 to 50°C. When hydrolysis is complete liquor is separated by centrifugation. Salt and some amino acids may be added for flavoring or preservation. The liquor is then concentrated into pasta or a powder. The autolyzed yeast extracts are defined by the International Hydrolyzed Protein Council both in terms of the method of manufacture and also, to some extent, the composition of the products.[2] The scheme of the processing is shown in Figure 70. The yeast extract is then ready for uses of yeast extract (see Appendix).

The chemical composition of the yeast extracts (autolysates) is of a complex nature and depends on the raw material (yeast), the hydrolysis procedure and the final processing after

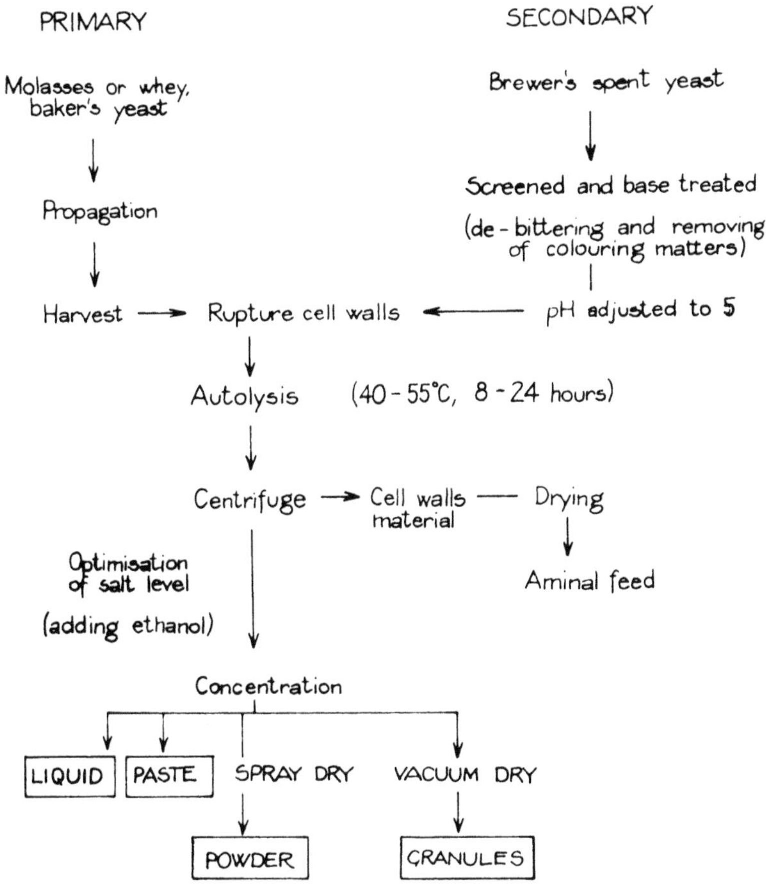

FIGURE 70. Manufacture of autolyzed yeast extract (based on Manley et al.[174] and Halász[175]).

autolysis. A high level of α-amino acids, peptides and nucleotides is characteristic for yeast extracts. Table 56 summarizes data about the macro-composition of the autolysates while, Table 57 gives information about the amino acid composition. Glutamic acid is the amino acid usually present in the largest quantity in hydrolysates. The nucleotide and nucleoside composition of commercial yeast extract is shown in Table 58.

Among other components, the aroma compounds and coloring compounds are the most interesting from a practical point of view. More detailed descriptions of these compounds can be found in the Appendix.

TABLE 56
Lysinoalanine Content of Some Foods

Product	Lysinoalanine content (mg/kg)
Milk protein concentrate	179—377
MCT-Baby food (''Robebi'')	39—103
Roboprotein	38—57
Caseine hydrolysate (MS Chemicals)	1560
Sodium caseinate	387
Baby food (Lida gen)	134
Baby food (Lida lac)	392
Baby food (Sanatogen)	125
Sausages (after 10 min cooking)	50
Condensed milk	700
Canned beef meat	25
Dried milk (defatted)	25
Serum albumin (heat treated at 120°C for 6 h)	8000
Casein (alkali treated)	1000—1500
Sodium caseinate (after heating at 120°C for 6 h)	200—600

Table 57
Amino Acid Composition of Yeast Extracts (g/100 g Product)

	Product	
Amino acid	Primary[174]	Secondary[175]
LYS	1.7	2.80
HIS	0.6	0.82
CYS	—	0.07
ARG	1.3	1.72
ASP	1.4	3.51
THR	1.6	1.44
SER	2.4	1.64
GLU	0.8	5.40
GLY	0.9	1.77
ALA	3.9	2.72
VAL	2.4	1.99
MET	0.6	0.58
ILE	1.8	1.74
LEU	3.2	2.51
TYR	0.6	0.35
PHE	1.8	1.33
TRP	0.7	0.61
PRO	—	0.85

TABLE 58
Nucleotides and Nucleosides of Yeast Extract (Autolysate)

Compound	Product	
	Primary[174]	Secondary[175]
Adenosine monophosphate	—	0.05
Adenosine	0.48	0.55
Adenine	0.50	0.390.39
Cytidine monophosphate	0.40	0.22
Cytidine	0.57	0.61
Cytosine	0.03	0.07
Guanosine monophosphate	0.28	0.41
Guanosine	0.85	0.68
Guanine	0.15	0.08
Uridine monophosphate	0.39	0.22
Uridine	0.33	0.48
Uracyl	0.65	ND
Xanthine	ND	ND
Totals	4.63	3.76

REFERENCES

1. **Dunhill, P. and Lilly, M. D.,** Protein extraction and recovery from microbial cells, in *Single Cell Protein II*, Tannenbaum, S. R. and Wang, D. I., Eds., MIT Press, Cambridge, 1975, 179.
2. **Vananuvat, P. and Kinsella, J. E.,** Extraction of protein, low in nucleic acid, from *Saccharomyces fragilis* grown continuously on crude lactose, *J. Agric. Food. Chem.*, 23, 216, 1975.
3. **Cunningham, S. D., Cater, C. M., Mattil, K. F., and Vanderzant, C.,** Rupture and protein extraction of petroleum grown yeast, *J. Food Sci.*, 40, 732, 1975.
4. **Sadkova, N. J., Valjkovszkij, D. G., Timofejeva, G. I., Kozlova, V. A., Rogoshin, S. V., and Pavlova, Sz. A.,** Study of the effect of conditions of disintegration on the properties of yeast proteins (in Russian), *Prikl. Biokhim. Mikrobiol.*, 11(2), 72, 1975.
5. **Sergeev, A. A., Solosenko, V. M., Bezrukov, M. G., and Saporovskaja, M. B.,** *Acta Biotechnol.*, 4(2), 105, 1984.
6. **Zetelaki, K.,** Disruption of mycelia for enzymes, *Process Biochem.*, 4(12), 19, 1969.
7. **Whitworth, D. A.,** Hydrocarbon fermentation: protein and enzymes solubilization from *C. lipolytica* using an industrial homogenizer, *Biotechnol. Bioeng.*, 16, 1399, 1974.
8. **Nevell, J. A., Robbins, A. L., and Seeley, R. D.,** Manufacture of yeast protein isolate having a reduced nucleic acid content by an alkali process, U.S. Patent 3, 867.555, 1975.
9. **Robbins, E. A., Sucher, R. W., Schuldt, E. H., Jr., Sidoti, D. R., Seeley, R. D., and Nevell, J. A.,** Yeast protein isolate with reduced nucleic acid content and process of making same, U.S. Patent 3, 887, 431, 1975.
10. **Chisti, Y. and Murray, M. Y.,** Disruption of microbial cells for intracellular products, *Enzymol. Microbiol. Technol.*, 8, 194, 1986.
11. **Arnold, W. N.,** *Autolysis in Yeast Cell Envelopes: Biochemistry, Biophysics and Ultrastructure*, Vol. 2, Arnold, W. N., Ed., CRC Press, Boca Raton, FL, 65, 1981.
12. **Charpentier, C., Nguyen, V. L., Benaly, R., and Feullat, M.,** Alteration of cell wall structure in *Saccharomyces cerevisiae* and *Saccharomyces bayanus* autolysis, *Appl. Microbiol. Biotechnol.*, 24, 405, 1986.
13. **Bacon, J. S. D.,** Nature and disposition of polysaccharides within the envelope, in *Autolysis in Yeast Cell Envelopes: Biochemistry, Biophysics and Ultrastructure*, Vol. 1, Arnold, W. N., Ed., CRC Press, Boca Raton, FL, 65, 1981.
14. **Johnson, J. C.,** Yeasts for Food and Other Purposes, Noyes Data Corp., Park Ridge, NJ, 1977.
15. **Shetty, K. J. and Kinsella, J. E.,** Effect of thiol reagents on rupture of yeast cells, *Biotechnol. Bioeng.*, 20, 46, 1978.

16. **Phaff, H. J.,** Enzymatic yeast cell wall degradation, in *Advances in Chemistry,* Feeney, R. C. and Whitaker, J. R., Eds., American Chemical Society, Washington, D.C., 1977, 244.
17. **Obata, T., Iwata, H., and Namba, Y.,** Proteolytic enzymes from *Oerskovia* sp. SK lysing viable yeast cells, *Agric. Biol. Chem.,* 41, 2387, 1977.
18. **Funatsu, M., Oh, H., Aizono, J., and Shimoda, T.,** Protease of *Arthrobacter lutens,* properties and function on lysis of viable yeast cells, *Agric. Biol. Chem.,* 42, 1975, 1978.
19. **Farkas, V.,** Biosynthesis of cell walls of fungi, *Microbiol. Rev.,* 43, 117, 1979.
20. **Kislukhina, O. V. and Bizulavitchus, G. S.,** Study of the properties of the preparation of lytic enzymes from the culture of *Bacillus subtilis* (in Russian), *Prikl. Biochim. Mikrobiol.,* 13, 55, 1977.
21. **Kalebina, T. S., Rudenskaya, G. N., Selyakh, I. O., Khodova, O. M., Chestukhina, G. G., Stepanov, M. S., and Kulaev, I. S.,** Serine proteinase from *Bacillus brevis* lytic action on intact yeast cells, *Appl. Microbiol. Biotechnol.,* 28, 531, 1988.
22. **Asenjo, J. A. and Dunhill, P.,** The isolation of lytic enzymes from *Cytophaga* and their application to the rupture of yeast cells, *Biotechnol. Bioeng.,* 23, 1045, 1981.
23. **Ryan, E. and Ward, O. P.,** The application of lytic enzymes from *Basidiomycetes aphillophoroles* in production of yeast extract, *Process Biochem.,* 23(1), 12, 1988.
24. **Sergeev, V. A., Solosenko, V. M., Bezrukov, M. G., and Saporovskaya, M. B.,** Vergleichscharakteristik von Isolaten der Gesamteiweisse der Hefe in Abhängigkeit von den Bedingungen ihrer Abscheidung, *Acta Biotechnol.,* 4(2), 105, 1984.
25. **Lindblom, M. A.,** The influence of alkali and heat treatment on yeast protein, *Biotechnol. Bioeng.,* 16, 1495, 1974.
26. **Hedenskog, G. and Mogren, H.,** Some methods for processing of single cell protein, *Biotechnol. Bioeng.,* 15, 129, 1973.
27. **Machek, R., Fencl, Z., Beran, K., Behalova, B., Sillinger, V., and Kejmar, I.,** Production of native protein from yeasts, in *Advances in Microbial Engineering, Part 2,* Sykita, B., Prokop, A., and Novák, M., Eds., John Wiley & Sons, New York, 1974, 977.
28. **Lopez-Leiva, M.,** The use of electrodialysis in food processing. Part 1. Some theoretical concepts, *Lebensm. Wiss. Technol,* 21, 119, 1988.
29. **Lopez-Leiva, M.,** The use of electrodialysis in food processing. Part 2. Review of practical applications, *Lebensm. Wiss. Technol.,* 21, 177, 1988.
30. **Bostian, K. A. and Betts, G. F.,** Rapid purification and properties of potassium-activated aldehyde dehydrogenase from *Saccharomyces cerevisiae, Biochem. J.,* 173, 773, 1978.
31. **Kulbe, K. D. and Schuer, R.,** Large scale preparation of phosphoglycerate kinase from *Saccharomyces cerevisiae* using Cibacron Blue-Sepharose 4B affinity chromatography, *Ann. Biochem.,* 93, 46, 1979.
32. **Kopperschlager, G. and Johansson, G.,** Affinity partitioning polymer bound Cibacron Blue F 36-A for rapid large-scale purification of phosphofructokinase from baker's yeast, *Ann. Biochem.,* 124, 117, 1982.
33. **Azari, M. R. and Wiseman, A.,** Purification and characterization of the cytochrome P-448 component of a benzo(a) pyrene hydroxylase from *Saccharomyces cerevisiae, Ann. biochem.,* 122, 129, 1982.
34. **King, D. J., Azari, M. R., and Wiseman, A.,** Studies on the properties of highly purified cytochrome-448 and its dependent activity benzo(a) pyrene hydroxylase from *Saccharomyces cerevisiae, Xenobiotica,* 14, 187, 1984.
35. **Neumann, N. P. and Lampen, S. O.,** Purification and properties of yeast investase, *Biochemistry,* 6, 468, 1967.
36. **Wiseman, A., King, D. J., and Winkler, M. A.,** The isolation and purification of protein and peptide products, in *Yeast Biology,* Berry, P., Russel, I.R., and Stewart, L. F., Eds., Allen and Unwin Publishers, London, 1987, 433.
37. **Kinsella, J. E. and Shetty, K. J.,** Chemical modification for the improvement in functional properties of vegetable and yeast proteins, Proc. ACS Symp., Pour El, A., Ed., ACS Publications, American Chemical Society, Washington, D.C., 1978.
38. **Moebus, O. and Teuber, M.,** Herstellung von Single Cell Protein mit *S. cerevisiae* in einer Tauchstrahlbegasungsanlage, *Kiel. Milchwirtsch. Forschungber.,* 31, 297, 1979.
39. **Moebus, O., Kiesbye, P., and Teuber, M.,** Verbesserung der Technologie der Herstellung von Einzellerprotein aus hochbelasteten Lösungen der Molkreien, Versuche zur technischen Produktion von ribonucleinsaurearmer *Saccharomyces cerevisiae, Kiel. Milchwirtsch. Forschungber.,* 29, 131, 1977.
40. **Ohta, S., Maul, S., Sinskey, A. J., and Tannenbaum, S. R.,** Characterization of heat-shock process for reduction of the nucleic acid content of *Candida utilis, Appl. Microbiol.,* 22, 415, 1971.
41. **Rosales, F.,** Yeast protein source for human nutrition, *Acta Microbiol. Hung.,* 31, 159, 1984.
42. **Lindblom, M.,** Properties of intracellular ribonuclease utilized for RNA reduction in disintegrated cells of *Saccharomyces cerevisiae, Biotechnol. Bioeng.,* 19, 199, 1979.
43. **Halász, A.,** Biochemical and Biotechnological Principles of the Use of Yeast Biomass in Food Industry (in Hungarian), D. Sc. thesis, Hungarian Academy of Sciences, Budapest, 1988.

44. **Sinskey, A. J. and Tannenbaum, S. R.,** Removal of nucleic acids in SCP, in *Single Cell Protein II*, Tannenbaum, S. R. and Wang, D. I., Eds., MIT Press Cambridge, 1975, 158.
45. **Bueno, G. E., Otero, M. A., Klibansky, M., and Gonzales, A. G.,** Nucleic acid reduction from yeast. Activation of intracellular RNase, *Acta Biotechnol.*, 5(1), 91, 1985.
46. **Lindblom, N. and Mogren, H. L.,** Process for preparing a protein concentrate from a microbial cell mass and the protein content thus obtained, *British Patent*, 1, 474, 313, 1977.
47. **Schuldt, E. H.,** Process for RNA reduction in yeast involving the addition of Zn to yeast, *U.S. Patent*, 4, 135.000, 1979.
48. **Castro, A. C., Sinskey, A. J., and Tannenbaum, S. R.,** Reduction of nucleic acid content in *Candida* yeast cells by bovine pancreatic ribonuclease — a treatment, *Appl. Microbiol.*, 22, 422, 1971.
49. **Trevelyan, W. E.,** Processing yeast to reduce nucleic acid content. Induction of intracellular RNase action by heat shock procedure, and an efficient chemical method based on extraction of RNA by salt solutions at low pH, *J. Sci. Food Agric.*, 29, 141, 1978.
50. **Trevelyan, W. E.,** Chemical methods for the reduction of purine content of yeast, a form of single cell protein, *J. Sci. Food Agric.*, 27, 225, 1976.
51. **Canepa, A., Pieber, M., Romero, C., and Toha, J.,** A method for large reduction of the nucleic acid content of yeast, *Biotechnol. Bioeng.*, 14, 173, 1972.
52. **Decker, P. and Dirr, V.,** Non protein nitrogen of yeast. II. Composition of purine fraction and extraction of nucleic acids, *Biochem. Z.*, 316, 348, 1974.
53. **Landord, G., Klingerman, A., and Williams, T.,** Production of high quality edible protein from *Candida* yeast grown in continuous culture, *Biotechnol. Bioeng.*, 21, 1163, 1979.
54. **Brineger, A. C. and Kinsella, J. E.,** Reversible modification of lysine in soybean protein using citraconic anhydride. Characterization of physical and chemical changes in soy protein isolate, the 7S globulin and lipoxygenase, *J. Agric. Food Chem.*, 28, 818, 1980.
55. **Brineger, A. C. and Kinsella, J. E.,** *Int. J. Pept. Protein Res.*, 18, 18, 1981.
56. **Palacian, E., Lopez-Rivas, A., Pinter-Toro, J. A., and Hernandez, F.,** *Mol. Cell. Biochem.*, 36, 163, 1981.
57. **Shetty, J. K. and Kinsella, J. E.,** Isolation of yeast protein with reduced nucleic acid level using reversible acylating reagents: some properties of the isolated protein, *J. Agric. Food Chem.*, 30, 1166, 1982.
58. **Chen, S. H., Chen, H. J., and Sung, H. Y.,** Studies on the protein isolates and nucleic acids from brewer's yeast, *J. Chin. Agric. Chem. Soc.*, 23, 318, 1986.
59. **Huang, Y. T. and Kinsella, J. E.,** Functional properties of phosphorylated yeast protein: solubility, water holding capacity and viscosity, *J. Agric. Food Chem.*, 34, 670, 1986.
60. **Huang, Y. T. and Kinsella, J. E.,** Phosphorylation of yeast protein: reduction of ribonucleic acid and isolation of yeast protein concentrate, *Biotechnol. Bioeng.*, 28, 1690, 1986.
61. **Wegner, E. H. and Shay, L. K.,** Protein product of reduced nucleic acid content and low allergenicity, U.S. Patent 460, 986, 1986.
62. **Andrews, B. A. and Ansenjo, J. A.,** Enzymatic lysis and disruption of microbial cells, *Trends Biotechnol.*, 5, 273, 1987.
63. **Alvarez, R. and Enriquez, A.,** Nucleic acid reduction in yeast, *Appl. Microbiol. Biotechnol.*, 29, 208, 1988.
64. **Fennema, O. R., Ed.,** *Food Chemistry*, 2nd Ed., Marcel Dekker, New York, 1985.
65. **Tannenbaum, S. R., Ed.,** *Nutritional and Safety Aspects of Food Processing*, Marcel Dekker, New York, 1979.
66. Protein Advisory Group of the United Nations, System Guideline No. 6. for Preclinical Testing of Novel Source of Protein, March 13, Rome, 1972.
67. **Oser, B. L.,** The safety evaluation of new sources of protein for man, in *Evaluation of Novel, Protein Products*, Proc. Symp., Stockholm, 1968, Pergamon Press, Oxford, 1970.
60. **de Groot, A. P.,** Minimal tests necessary to evaluate the nutritional qualities and the safety of SCP, in *Single Cell Protein*, Davis, P., Ed., Academic Press, London, 1974, 75.
69. **Guzman-Juarez, M.,** Yeast protein, in *Developments in Food Proteins-2*, Hudson, B. J. F., Ed., Applied Science Publishers, London, 1983, 263.
70. **Schilingmann, M., Faust, U., and Scharf, V.,** Bacterial proteins, in *Developments in Food Proteins-3*, Hudson, B. J. F., Ed., Applied Science Publishers, London, 1984, 139.
71. **Porter, I. W. G. and Rolls, B. A., Eds.,** *Protein in Human Nutrition*, Academic Press, New York, 1973.
72. **Friedman, M., Ed.,** *Protein Nutritional Quality of Foods and Feeds*, Vol. 1 and Vol. 2, Marcel Dekker, New York, 1975.
73. **Brodwell, C. E.,** Problems associated with the development and application of rapid methods of assessing protein quality, *Nutr. Rep. Int.*, 16, 163, 1977.
74. **Walker, A. F.,** The estimation of protein quality, in *Developments in Food Proteins-2*, Hudson, B. J. F., Ed., Applied Science Publishers, London, 1983, 293.

75. **Bodwell, C. E., Ed.,** *Evaluation of Proteins for Humans,* AVI Publishers, Westport, CT, 1977.
76. **Lásztity, R. and Hidvégi, M., Eds.,** *Amino Acid Composition and Biological Value of Cereal Proteins,* Reidel Publishing, Dordrecht, 1985.
77. **Buchanan, R. A.,** *In vitro* and *in vivo* methods of measuring nutritive value of leaf protein concentrates, *Br. J. Nutr.,* 23, 533, 1969.
78. **Magan, J. A., Lorenz, K., and Onayemi, O.,** Digestive acceptibility of proteins as measured by the initial rate of proteolysis, *J. Food Sci.,* 3, 173, 1973.
79. **Saunders, R. M., Connor, M. A., Booth, A. N., Bischoff, E. M., and Koehler, G. O.,** Measurement of digestibility of alfalfa concentrates by *in vivo* and *in vitro* methods, *J. Nutr.,* 103, 530, 1973.
80. **Hsu, H. W., Vavak, D. L., Satterlee, C. D., and Miller, D. A.,** A multienzyme technique for estimating protein digestibility, *J. Food Sci.,* 42, 1269, 1977.
81. **Hsu, H. W., Sutton, N. E., Banjo, M. O., Satterlee, L. D., and Kendrick, J. G.,** The C-PER and T-PER assay for protein quality, *Food Technol.,* 32(12), 69, 1978.
82. **Pedersen, B. and Eggum, B. O.,** Prediction of protein digestibility by *in vitro* procedures based on two multi-enzyme systems, *J. Anim. Physiol. Anim. Nutr.,* 45, 190, 1981.
83. **Rich, N., Satterlee, L. D., and Smith, J. L.,** A comparison on *in vivo* apparent protein digestibility in man or rat to *in vitro* protein digestibility as determined by using human and rat pancreatins and commercially available proteins, *Nutr. Rep. Int.,* 21, 285, 1980.
84. **Bodwell, C. E., Adkins, J. S., and Hopkins, D. T.,** *Proteins Quality in Humans: Assessment and In Vitro Estimation,* AVI Publishing, Westport, CT, 1981.
85. **Pedersen, B. and Eggum, B. C.,** Prediction of protein digestibility by an *in vitro* enzymatic pH-stat procedure, *J. Anim. Physiol. Anim. Nutr.,* 49, 265, 1983.
86. **Salgó, A., Ganzler, K., and Jécsai, I.,** Simple enzymic methods for prediction of plant protein digestibility, in *Amino Acid Composition and Biological Value of Cereal Proteins,* Lásztity, R. and Hidvégi, M., Eds., Reidel Publishing, Dordrecht, 1985, 311.
87. **Walger-Kunze, B.,** *In vivo* methods in the evaluation of the nutritional quality of cereal proteins, in *Amino Acid Composition and Biological Value of Cereal Proteins,* Lásztity, R. and Hidvégi, M., Eds., Reidel Publishing, Dordrecht, 1985, 131.
88. **Münchow, A. and Bergner, H.,** Untersuchungen zur Proteinbewertung von Futtermitteln, 2. Mitt.: Die Harnstoffkonzenration in Blut von Ratte und Schwein in Abhängigkeit vom biologischen Wert des gefütterten Nahrungsproteins, *Arch. Tiernaehr.,* 17(3), 141, 1967.
89. **Bodwell, C. E., Kyle, E. M., Schuster, E. M., Vanghan, D. A., Wornack, M., Ahrens, R. A., and Hackler, L. R.,** Biochemical indices in humans of protein nutritive value. III. Fasting plasma urea nitrogen and urinary metabolites as a low protein intake level, *Nutr. Rep. Int.,* 19(5), 703, 1979.
90. **Tayler, Y. S. M., Scrimshaw, N. S., and Xoung, V. R.,** The relationship between serum urea levels and dietary nitrogen utilization in young man, *Br. J. Nutr.,* 32, 107, 1974.
91. **Hidvégi, M. and Békés, F.,** Mathematical modeling of protein nutritional quality from amino acid composition, in *Amino Acid Composition and Biological Value of Cereal Proteins,* Lásztity, R. and Hidvégi, M., Eds., Reidel Publishing, Dordrecht, 1985, 205.
92. **Mitchell, H. H. and Block, R. J.,** Some relationships between the amino acid contents of proteins and their nutritive value for the rat, *J. Biol. Chem.,* 163, 599, 1946.
93. **Oser, B. L.,** Method for integrating essential amino acid content in the nutritional evaluation of protein, *J. Am. Diet. Assoc.,* 27, 396, 1951.
94. FAO/WHO/UNU: *Protein and Energy Requirements,* FAO, Rome, 1984.
95. **Pellet, P. L.,** Amino acid scoring systems and their role in the estimation of the protein quality of cereals, in *Amino Acid Composition and Biological Value of Cereal Proteins,* Lásztity, R. and Hidvégi, M., Eds., Reidel Publishing, Dordrecht, 1985, 183.
96. **Sarwar, G.,** Available amino acid score method for protein quality evaluation, in *Amino Acid Composition and Biological Value of Cereal Proteins,* Lásztity, R. and Hidvégi, M., Eds., Reidel Publishing, Dordrecht, 1985, 305.
97. **Korpáczy, J., Lindner, K., and Varga, K.,** Recent contribution to the composition of foods. VII. Improved method for the calculation of the biological value of food proteins (in Hungarian), *Élelmiszervizs galati Közl.,* 7, 11, 1961.
98. **Olesen, E. S. and Morup, I. K.,** Prediction of protein value in human nutrition, new index based on revised evaluation of amino acid pattern, in Abstr. of the Xth Int. Cong. Nutr., Kyoto, Japan, 1975.
99. **Morup, I. K. and Olesen, E. S.,** New method for prediction of protein value from essential amino acid pattern, *Nutr. Rep. Int.,* 13, 355, 1976.
100. **Kofrányi, E.,** Evaluation of the traditional hypotheses on the biological value of proteins, *Nutr. Rep. Int.,* 7, 45, 1973.
101. **Korfányi, E. and Jakab, F.,** Zur Bestimmung der biologischen Wertigkeit von Nahrungsproteinen. VIII. Die Wertigkeit gemischter Proteine, *Z. Physiol. Chem.,* 335, 174, 1964.

102. **Kofrányi, E. and Jakab, F.,** Zur Bestimmung der biologischen Wertigkeit von Nahrungsproteinen. X. Vergleich der Bausteinanalysen mit dem Minimalbedarf gemischter Proteine für Menschen, *Z. Physiol. Chem.*, 338, 159, 1964.
103. **Kofrányi, E., Jakab, F., and Müller-Wecker, H.,** The determination of the biological value of dietary proteins. XVI. The minimum protein requirement of humans tested with mixtures of whole egg plus potato and maize plus beans, *Z. Physiol. Chem.*, 351, 1485, 1970.
104. **Békés, F., Hidvégi, M., Zsigmond, A., and Lásztity, R.,** Studies on the evaluation of the *in vitro* biological value of food proteins, *Acta Aliment.*, 13, 135, 1984.
105. **Békés, F., Hidvégi, M., Zsigmond, A., and Lásztity, R.,** A novel mathematical method for determining *in vitro* biological value of proteins and its application for non-linear optimization of the nutritional quality of food and feed formulas, in *Developments in Food Science, Vol. 5B, Progress in Cereal Chemistry and Technology,* Holas, J. and Kratochvil, J., Eds., Elsevier, Amsterdam, 1983, 1213.
106. **Békés, F., Lásztity, R., and Hidvégi, M.,** Evaluation of food proteins on the basis of chemical indices. II. Off-line amino acid analyzer-computer system and its application in the food industry (in Hungarian) *Élelmez. Ipar,* 36, 402, 1982.
107. **Hidvégi, M.,** Modeling of Protein Nutritional Quality From Amino Acid Data (in Hungarian), Ph.D. thesis, Technical University, Budapest, 1983.
108. **Hidvégi, M.,** New methods of protein evaluation for grain and feed products (in Hunagrian) *Gabonaipar,* 30, 46, 1983.
109. **Hidvégi, M., Örsi, F., and Békés, F.,** Equation predict BV, TD and NPU from amino acid analysis, in *Cereals Healthful Food for All,* Steller, W. Ed., C. Kersting, St. Augustin, 1984, 106.
110. **Walker, A. F.,** Determination of protein and reactive lysine in leaf protein concentrates by dye-binding, *Br. J. Nutr.*, 42, 455, 1979.
111. **Baráth, A. and Halász, A.,** Determination of reactive lysine by dye-binding, in *Amino Acid Composition and Biological Value of Cereal Proteins,* Lásztity, R. and Hidvégi, M., Eds., Reidel Publishing, Dordrecht, 1985, 337.
112. **Munck, L.,** Optimization of lysine composition in plant breeding programmes and in feed technology by the dye-binding analysis, in *Amino Acid Composition and Biological Value of Cereal Proteins,* Lásztity, R. and Hidvégi, M., Eds., Reidel Publishing, Dordrecht, 1985, 325.
113. **Kihlberg, R.,** The microbe as a source of food, in *Annual Review of Microbiology,* 1972, 427.
114. **Calloway, D. H.,** The place of SCP in Man's Diet, in *Single Cell Protein,* Davis, P. Ed., Academic Press, New York, 1974, 129.
115. **Smith, R. H. and Palmer, R.,** A chemical and nutritional evaluation of yeasts and bacteria as dietary protein sources for rats and pigs, *J. Sci. Food Agric.*, 27, 763, 1976.
116. **Petrovszkij, K. S., Suchanov, B. P., and Rogoshin, S. V.,** Investigation of the biological value of yeast isolate (in Russian), *Vopr. Pitan.*, 1978, No. 3., 48.
117. **Kung-Chin Su, Ming-Chang Hsie, and Hung Chao Lee,** Amino acid composition of various yeasts prepared from cane molasses, *J. Chin. Agric. Chem. Soc.*, 10, 103, 1973.
118. **Wiemken, A. and Nurse, P.,** Isolation and characterization of the amino acid pools located within the cytoplasm and vacuoles of *Candida utilis, Planta,* 109, 293, 1973.
119. **Sarwar, G., Shah, B. G., Mongeau, R., and Hopfner, K.,** Nucleic acid, fiber and nutrient composition of inactive dried food yeast products, *J. Food Sci.*, 50, 353, 1985.
120. **El-Samragy, Y. A., Chen, J. H., and Zall, R. R.,** Amino acid and mineral profile of yeast biomass produced from fermentation of cheddar whey permeate, *Process Biochem.*, 23(2), 29, 1988.
121. **Giacobbe, F., Griesmer, G. I., Littlehailes, J., Peri, P., Senej, J., and Solomons, G.,** Control of SCP processes, in *Single Cell Protein,* Davis, P. Ed., Academic Press, London, 1974, 195.
122. **Lusky, K., Stoyke, M., Göbel, R., Busch, A., and Ackermann, H. I.,** Untersuchungen zum Einfluss von mikrobiellen Eiweiss auf Kohlenwasserstoffbasis (fermosin) mit difinierter Fettsäuerzusammensetzung auf den Stoffwechsel und die Fettzusammensetzung von landwirtschaftlichen Nutztieren. 1. Mitt. Einfluss von fermosin auf die Zusammensetzung von Broilerdepotfett, *Die Nahrung,* 32(6), 627, 1988.
113. **Hibino, S. and Terashimsa, H.,** Enzymatic digestion of yeast in some animals, in *Single Cell Protein,* Davis, P. Ed., Academic Press, London, 1974, 114.
124. **Hurrel, R. F.,** Reactions of food proteins during processing and storage and their nutritional consequences, in *Developments in Food Proteins-3,* Hudson, B. J. F., Ed., Elsevier Applied Science Publisher, London, 1984, 213.
125. **Carpenter, K. J.,** Damage to lysine in food processing: its measurement and its significance, *Nutr. Abstr.*, 43, 423, 1973.
126. **Mauron, J.,** Influence of industrial and household handling on food protein quality, in *Protein and Amino Acid Function, Vol. 2.*, Bigwood, E. J., Ed., Pergamon Press, Oxford, 1972, 417.
127. **Cheftel, J. C.,** Chemical and nutritional modifications of food proteins due to processing and storage, in *Food Proteins,* Whitaker, J. R. and Tannenbaum, S. R., Eds., AVI Publishing Westport, CT, 1977, 401.

128. **Hoyem, T. and Kvala, O.,** *Physical, Chemical and Biological Change in Food Caused by Thermal Processing,* Applied Science Publishers, London, 1977.
129. **Bender, A. E.,** *Food Processing and Nutrition,* Academic Press, London, 1978.
130. **Priestley, R. J., Ed.,** *Effects of Heating on Foodstuffs,* Applied Science Publishers, London, 1979.
131. **Hurrel, R. F.,** Interaction of food components during processing, in *Food and Health: Science and Technology,* Birch, G. G. and Parker, K. J., Eds., Applied Science Publishers, London, 1980, 369.
132. **Feeney, R. E.,** Overview on the chemical deteriorative changes of proteins and their consequences, in *Chemical Determination of Proteins,* Whitaker, J. R. and Fujimaki, M., Eds., American Chemical Society, Washington, D.C., 1980, 1.
133. **Finot, P. A.,** Nutritional and metabolic aspects of protein modification during food processing, in *Modification of Proteins,* Feeney, R. E. and Whitaker, J., Eds., American Chemical Society Washington, D.C., 1982, 91.
134. **Tannenbaum, S. R., Ed.,** *Nutritional and Safety Aspects of Food Processing,* Marcel Dekker, New York, 1979.
135. **Cheftel, J. C., Cuq, J. L., and Lorient, D.,** Amino acids, peptides and proteins, in *Food Chemistry, 2nd ed.,* Fennema, O. R., Ed., Marcel Dekker, New York, 1985, 245.
136. **Finot, P. A., Bujard, E., and Arnot, G.** *Protein Crosslinking, Nutritional and Medical Consequences,* Friedman, M., Ed., Plenum Press, New York, 1979, 51.
137. **Feron, V. J., Van Beek, L., Slump, P., and Beems, R. B.,** in Biochemical Aspects of Protein Foods, Adler, N., et al., Eds., Federation of European Biochemical Societies, 44-Symposium A 3, 1977, 139.
138. **Lásztity, R.,** *Principles of Food Biochemistry,* (in Hungarian), Mezögazdasági Kiadó, Budapest, 1981.
139. **Lásztity, R. and Törley, D.,** *Food Chemistry,* Vol. 2 (in Hungarian) Mezögazdasági Kiadó, Budapest, (in press).
140. **Dworschák, E., Örsi, F., Zsigmond, A., Trézl, L., and Rusznák, I.,** Factors influencing the formation of lysinoalanine in alkali treated proteins, *Nahrung,* 25, 441, 1981.
141. **Dworschák, E.,** Nonenzymic browning and its effect on protein nutrition, *CRC Crit. Rev. Food Sci. Nutr.,* 13, 1, 1980.
142. **Eriksson, C., Ed.,** *Maillard Reactions in Food: Chemical, Physiological and Technological Aspects,* Pergamon Press, Oxford, 1981.
143. **Örsi, F. and Hollósy, I.,** Thermogravimetrische Untersuchung der Maillard Reaktion, *Ernährung,* (Nutrition), 7, 387, 1983.
144. **Lásztity, R., Örsi, F., and Békés, F.,** The role of protein-lipid and protein-carbohydrate interactions in cereal processing (in Serbian), *Zito Hleb.,* 12, 51, 1985.
145. **Aeschbacher, H. V.,** Possible cancer risk of dietary heat reaction products, in *Proc. Interdisciplinary Conf. Natural Toxicants Food,* University of Zürich, Ed., 1986, 112.
146. **O'Brien, J. M., Morissey, P. A., and Flynn, A.,** The effect of Maillard reaction products on mineral homeostasis in the rat, in *Proc. Interdisciplinary Conf. Natural Toxicants Food,* University of Zürich, Eds., 1986, 214.
147. **Schaich, V. M.,** Free radical imitation in proteins and amino acids by ionizing and ultraviolet radiations and lipid oxidation. Part I. Ionizing radiation. Part II. Ultraviolet radiation and photolysis. Part III. Free Radical transfer from oxidizing lipids, *CRC Crit. Rev. Food Sci. Nutr.,* 13, 89, 1980; 13, 131, 1980; 13, 189, 1980.
148. **Dziezak, J. D.,** Yeast and Yeast derivatives: definitions, characteristics and processing, *Food Technol.,* 41, 104, 1987.
149. **Hough, J. S. and Maddox, I. S.,** Yeast autolysis, *Process Biochem.,* 6, 50, 1971.
150. **Stolpe, H. and Starr, M. P.,** Yeast autolysis, *Ann. Rev. Microbiol.,* 19, 79, 1965.
151. **Babayan, T. L. and Bezrukov, M. G.,** Autolysis of yeasts, *Acta Biotechnol.,* 5(2), 129, 1985.
152. **Matile, P.,** *The Lytic Compartment of Plant Cells,* Springer Verlag, New York, 1975.
153. **Welsch, M. J.,** Lysine agents of microbial origin, *J. Gen. Microbiol.,* 18, 491, 1958.
154. **Duda, V. I., Pronin, S. V., El-Registan, G. I., Kaprelyants, A. S., and Mitushina, L. L.,** Formation of dormant refractile cells in *Bacillus cereus* under effect of an autoregulatory factor (in Russian), *Mikrobiologija,* 51(1), 77, 1982.
155. **Ohmija, K. and Sato, Y.,** Promotion of autolysis in *Lactobacilli, Agric. Biol. Chem.,* 39, 585, 1975.
156. **Belous, A. M.,** Molecular and intracellular mechanisms of membrane cryodamage, in Cryobiol. Freeze-Drying, Proc. Third Natl. School Cryobiol. Freeze-Drying, Tsvetkov, T., Ed., Agricultural Academy, Central Laboratory for Cryobiology and Freeze-drying, Sofia, 1987, 27.
157. **Kirshop, B.,** A comparison of the effects of different preservation methods on yeast survival and stability, in Cryobiol. Freeze-Drying, Proc. Third Natl. School Cryobiol. Freeze-drying, Tsvetkov, T., Ed., Agricultural Academy, Central Laboratory for Cryobiology and Freeze-drying, Sofia, 1985, G 1-12.
158. **Halász, A., Hajós, Gy., and Teleky-Vámossy, Gy.,** Use of yeast autolysates in production of flavouring preparations, Research Report No. 8, Central Research Institute of Food Industry, Budapest, 1987.

159. **Dziezak, J. D.,** Yeast and yeast derivatives: applications, *Food Technol.*, 41(2), 122, 1987.
160. **Weir, G. S. D.,** Protein hydrolysates as flavouring, in *Development in Proteins-4*, Hudson, B. J. F., Ed., Elsevier Applied Science Publishers, New York, 1986, 180.
161. **Maltz, M. A., Ed.,** Biochemical methods of protein production, in Protein Food Supplements-Recent Advances, Noyes Data, Park Ridge, NJ, 1981, 28.
162. **Johnson, J. C., Ed.,** Other food additives: yeast and related products, in Food Additives — Recent Developments, Noyes Data, Park Ridge, NJ, 1983, 377.
163. Improvement of yeast autolysis by addition of metallic cations, USSR Inventors Certificate, N 141135000, 1979.
164. **Chrenova, N. M., Bezrukov, M. G., Kogan, A. S., and Sergeev, W. A.,** Autolyse der Hefe *Saccharomyces cerevisiae*, induziert durch Metallkationen, 1. Mitt. Metalle der II. Gruppe, *Die Nahrung*, 25, 837, 1981.
165. **Miozzeri, G. F., Niederberger, P., and Hütter, R.,** Permeabilization of microorganisms by Triton-X-100, *Anal. Biochem.*, 90, 220, 1978.
166. **Wiemken, A. and Nurse, P.,** Isolation and characterization of the amino-acid pools located within the cytoplasm and vacuoles of *Candida utilis*, *Planta*, 109, 293, 1973.
167. **Brown, W. C. and Cuhel, R. L.,** *J. Gen. Microbiol.*, 91, 429, 1975.
168. **Reyes, F., Lahor, R., and Val Moreno, A.,** *J. Gen. Microbiol.*, 347, 1981.
169. **Kusunose, M., Nakanish, T., Minamiura, N., and Yamamoto, I.,** Yeast cell wall bound proteolytic enzymes, *Agric. Biol. Chem.*, 44, 2779, 1980.
170. **Babayan, T. L. and Lator, V. K.,** Autolysis of microorganisms, in Thesis of the 3rd All-union Conf. Amino Acids, Erevan, U.S.S.R., April 23, 1984, 110.
171. **Pokrovsky, A. A. and Tutelyan, V. A.,** *Lysosomes*, Nauka Publishing House, Moscow, 1977.
172. **Halász, A.,** Biochemical Principles of the Use of Yeast Biomass in Food Industry, D.Sc. thesis, Hungarian Academy of Sciences, Budapest, 1988.
173. International Hydrolyzed Protein Council, Comments with additional data on SCOCS tentative evaluation of the healths aspects of protein hydrolysates as food ingredients — Report 37b, Washington, D.C., 1977.
174. **Weir, G. S. D.,** Protein hydrolysates as flavourings, in *Developments in Food Proteins-4*, Hudson, B. J. F., Ed., Elsevier Applied Science Publisher, New York, 1986, 175.
175. **Halász, A.,** Waste free technology in brewery, Research Report No. 7, Central Research Institute of Food Industry, Budapest, Hungary, 1987.
176. **Kinsella, J. E.,** Relationships between structure and functional properties of food proteins, in *Food Proteins*, Fox, P. F. and Condon, J. J., Eds., Applied Science Publishers, 1982, 51.
177. **Sherr, B., Lee, C. M., and Jelesciewicz, C.,** Absorption and metabolism of lysine Maillard products in relation to utilization of L-lysine, *J. Agric. Food Chem.*, 37, 119, 1989.
178. **Friedman, M. and Pearce, V. N.,** Copper II and Cobalt II affinities of LL-and LD-lysinoalanine diastereoisomers: implication for food safety and nutrition, *J. Agric. Food. Chem.*, 37, 123, 1989.
179. **Kruse, B. and Steinhart, H.,** Evidence for lysinomethylalanine formation in model systems, *J. Agric. Food Chem.*, 37, 304, 1989.

Part II

Chapter 6

INTRODUCTION

The first records of the use of yeast are concerned with the production of a type of acid beer called "boozah" in 6000 BC Egypt. The processes for producing wine and beer and leaven bread probably developed in parallel over the next few thousand years.[1] In the time of Romans beer was a common drink among the inhabitants of Gaul, Spain and Germany. The Britons welcomed the production of beer by Romans and adopted the beer as their national drink.

The people of South Asia have for thousands of years stored food by fermentation methods. For example, rice wine, sugar cane wine, pickles, fish sauces and different soy foods are produced in such a way.[2,3] A lot of other foods (cultured dairy products, fermented sausages, etc.) also contain microorganisms including yeast. So microbial cells have been consumed for a long time by humans, in some cases in quite high quantities.[4]

Kaffir beer, for example, which is a traditional beverage in Africa, is still produced by the Bantu and is almost a complete diet due to its balance of yeast and grain.[5] The nature of yeast has long remained unknown. It is generally agreed that it was the work of Luis Pasteur, published in his "Études sur Vin" in 1866 which established beyond doubt the role of yeasts in the fermentation of sugar to alcohol and that yeast was a living organism.

The technology of microbial cell production on a commercial scale for food and feed has developed only within the past 100 years. In ancient times people recovered top fermenting yeast of the genus *Saccharomyces* from the production of fermented beverages such as beer, and used these yeasts as a leavening agent in baked goods. Preparation of compressed brewer's yeast began in the late 18th century in Great Britain, Netherlands and Germany. By 1868 Fleischmann had founded the compressed yeast industry in the U.S. and concurrent improvements in the biochemistry of fermentation and technical engineering greatly aided the development of the modern yeast factory. Specified technology with continuous aeration and separation with centrifuge was developed for baker's yeast production. Modern technology for microbial cell production dates from World War I. Large scale fermentors with effective aeration were developed and newer continuous processes were used. Newer and newer raw materials (molasses, sulfite liquor, hydrocarbons, whey, starchy wastes from potato processing, date syrup, agave juice, etc.) were detected and used. The number of yeast strains and species grown on different raw materials was also enlarged.

In fermented foods and beverages, the consumption of microbes is incidental. The conscious growth of microbes for human consumption started in Germany in about 1910 with brewer's yeast, which was separated from the beer, washed and dried. During and following World War II *Candida utilis* yeast was grown on wood sugar containing sulfite liquors, although the main usage of microbial biomass is at present in feeds, good possibilities exist for the increase of food applications based on a history of yeasts as foods, the recognition of their safety and developed technology of production.

Four species of yeasts are presently used in foods. *Saccharomyces cerevisiae, Saccharomyces uvarum* (formerly *Saccharomyces carlsbergensis*), *Candida utilis* (*Torula utilis*), and *Kluyveromyces fragilis* (formerly *Saccharomyces fragilis*). *Saccharomyces uvarum* and *Saccharomyces cerevisiae,* the yeasts of the brewing industry, are byproducts (secondary products) of the production of beer. Other yeasts are grown specially for food resp. feed use and are often called *primary* yeasts. *Torula utilis* yeast is grown on sulfite liquor; and *Kluyveromyces fragilis* on cheese whey. Yeast may also be grown on hydrocarbons. In spite of the high expectations of the 1960s and early 1970s, however, the two operating plants have been closed, and the other plants never started production. This is due largely to the rising costs of petroleum products and supposed safety problems.[6-8] Based on detailed food

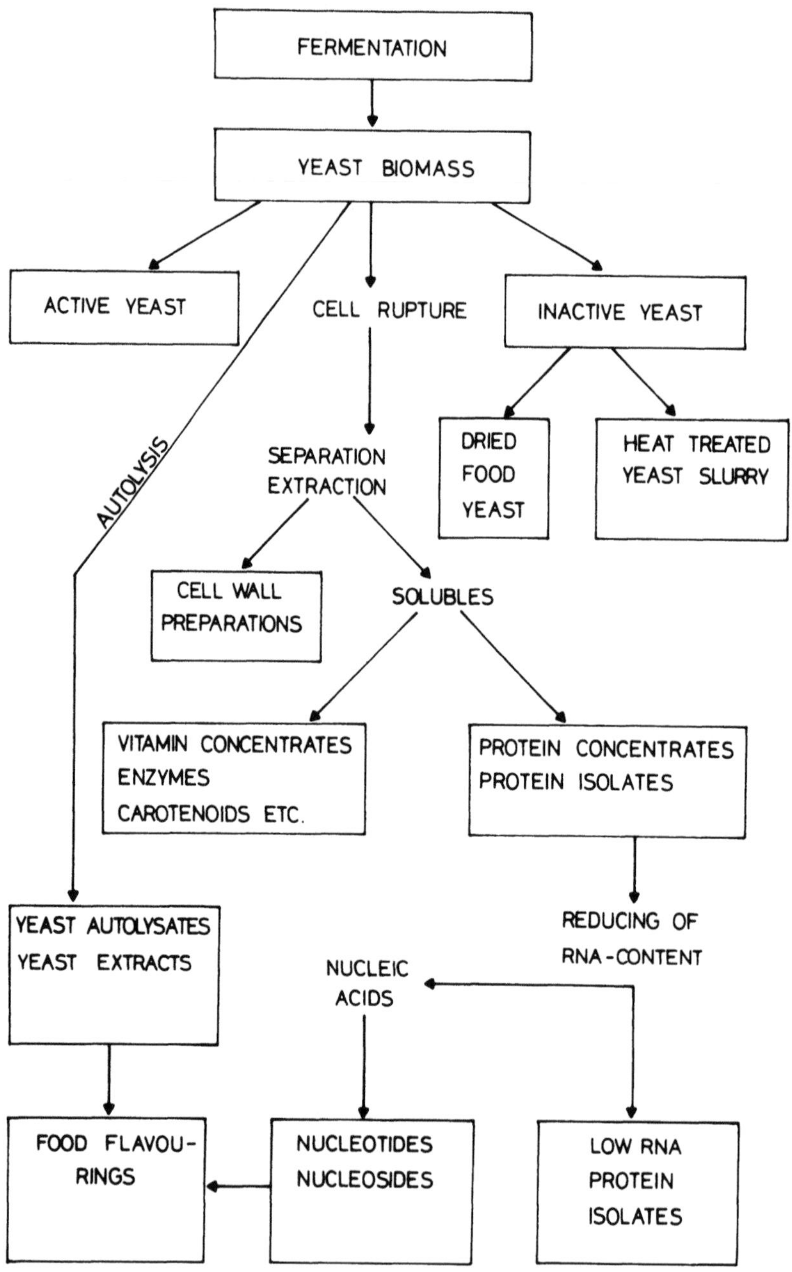

FIGURE 71. Main products of yeast biomass processing potentially useful in food production.

analysis it was acknowledged that yeast is a rich source of proteins, essential minerals and one of the richest sources of the B vitamin group.

Although fermentation and leavening account for the most common usages of yeast in food production, the application of yeast and yeast derivatives for nutritional fortification (as protein and vitamin source) of processed foods and also for refining their flavor profile has a growing tendency.[1,3,5,9,10] In Figure 71 the main products of yeast biomass processing potentially applicable in food production are summarized. The simplest possibility is the use of inactive whole yeast cells. Due to the high nucleic acid and cell wall content, the use of whole yeast cells is limited, therefore an extracted and purified protein preparation has advantages as a food ingredient. The cell wall preparation (yeast glycan) can also be used at first as an emulsifier.[10,12] Yeast autolysates are applicable as food ingredients or as a rich source of natural flavor,[10,13] (yeast autolysates sit in the area of the flavor spectrum between meat extracts and hydrolyzed animal or vegetable protein). Conditions of autolysis can be changed to allow production of yeast extracts with different characteristics. Extracts produced by digestion of heat-treated yeast with commercial enzyme preparations have also been described. The heat-treatment inactivates the enzymes which normally participate in autolysis so that the breakdown products formed are controlled by the type of enzymes used. As a final stage, fermentation of extract with a culture of lactic acid bacteria is possible resulting in a product with modified characteristics.

In the framework of Part II of this book the possible applications of whole yeast cells, protein isolates and autolysates (extracts) as food ingredients will be reviewed. The production of yeast-based flavors, the use of native yeast for production of commercial enzyme preparations and some other applications of lower interest in food production will be shortly treated in the Appendix.

REFERENCES

1. **Hobson, J. C.,** Yeasts, *Food Flavourings, Ingredients Packag. Process.*, 7(1), 41, 1985.
2. **Kozaki, M.,** Lactic acid bacterial flora of fermented foods in Southeast Asia, Ann. Rep. Intl. Center of Coop. Res. and Dev. in Microbial Eng., Vol. 1, 1978, 363.
3. **Reed, G.,** Use of microbial cultures: yeast products, *Food Technol.*, 32, 89, 1981.
4. **Litchfield, J. H.,** Microbial cells on your menu, *CHEMTECH*, 8, 218, 1978.
5. **Schmidt, G.,** Using yeast: bubbling over with yeast, *Food Flavourings, Ingredients, Packag. Process.*, 9(5), 25, 1987.
6. **Kamazawa, M.,** The production of yeast from n-paraffin, in *Single Cell Protein II*, Tannenbaum, S. R. and Wang, D. I. C., Eds., MIT Press, Cambridge, 1975.
7. **Knecht, R., Präve, P., Seipenbusch, R., and Sukatch, D. A.,** Microbiology and biotechnology of SCP produced from n-paraffin, *Process Biochem.*, 12(4), 11, 1977.
8. **Mauron, J.,** Haben Einzeller Proteine noch eine Zukunft, in *Problems in Nutrition and Food Sciences No. 5, Problems with New Foods*, Harmer, R., Auerswald, W., Zöllner, N., Blanc, B., and Brandstetter, B. M., Eds., Wilhelm Maudrich Verlag, Vienna, 1978, 125.
9. **Tusé, D.,** Single-cell protein: current status and future prospects, *CRC Crit. Rev. Food Sci. Nutr.*, Vol. 19, Issue 4, 1984, 273.
10. **Dziezak, J. D.,** Yeast and yeast derivatives: applications, *Food Technol.*, 41(2), 122, 1987.
11. **Seeley, R. D.,** Functional properties of SCP are a key to potential markets, *Food Prod. Dev.*, 9(9), 46, 1975.
12. **McCormick, R. D.,** Baker's yeast — world's oldest food — is newest source of protein and other ingredients, *Food. Prod. Dev.*, 7(6), 17, 1973.
13. **Halász, A.,** Biochemical and Biotechnological Principles of the Use of Yeast Biomass in Food Industry, D.Sc. thesis, Hungarian Academy of Sciences, Budapest, 1988.

Chapter 7

FOOD USES OF WHOLE YEAST CELLS

I. ACTIVE YEAST

Active yeast is widely used in the baking industry and different fermentation industries where their leavening activity and ability to ferment different raw materials (first of all their carbohydrate content) to improve the storability, taste and flavor is utilized. Theoretically, active whole yeast cells can also be directly used as human food or an ingredient in food. Nevertheless, in practice exclusively inactivated yeast biomass is used for food processing if the purpose of application is enrichment in nutrients and improvement of flavor or other important characteristics from the point of view of quality.

This fact is first of all connected with the short storability of compressed yeast and high costs of mild drying process suitable for production of active dried yeast cells. Brewer's yeast slurry could be stored even at a lower temperature (4.4°C) only for a short time without significant changes.[2] In foods containing carbohydrates undesirable fermentation may occur or the physical properties of food (e.g., dough) will change adversely under the effect of enzymes of active yeast.[3] In Figure 72 changes of the wheat flour dough consistency caused by addition of partially active yeast are demonstrated. The changes may be explained with the remaining proteolytic activity of yeast added due to the inadquate heat treatment during drying. These observations indicate that by determination of the parameters of the drying process, the heat inactivation injury, and repair of yeast cells must be specifically controlled.[4,5] Although the use of active yeast in traditional fermentation technologies such as bakery production, wine making, beer production, ethanol production, etc. lies out of the topics discussed in detail in the framework of this book, some newer specific applications of active yeast influencing the nutritional properties of the products will be shortly discussed in the following.

The possibility of simultaneous activation of compressed yeast and a starter containing *Lactobacillus fermenti-23* based on the symbiosis of these microorganisms, was studied[34] using *Saccharomyces cerevisiae-14* as yeast (50 to 60 min fermentation capacity), a lactic starter (LS) of 23 to 250 acidity and grade II wheat flour with average baking properties. LS ratios used were 1:0.25, 1:0.75, 1:1.25, 1:2.5 and 1:5 and the activation temperature was 32°C. The yeast activation mixture consisted of flour, (5 to 6 kg) water, (11 to 12 kg) yeast (1.3 to 1.5 kg) and LS (1.5 to 3.2 kg).

Results showed that pressed yeast could be activated using nutrient medium containing LS enriched with yeast autolysate, a yeast-LS ratio 1:1 to 1:3 and pH 4.5 to 4.6. The fermentation capacity of yeast improved, the semi-manufactured products were more intensively fermented and the quality of the finished products was improved.

A soft cheese (yeast cheese) was made by Kang et al.[35] using a lactic acid bacteria starter and then adding a mixed yeast culture (*Saccharomyces fragilis* + *Debaromyces hansenii*) before renneting. The resulting yeast cheese had a soft texture and a slightly alcoholic flavor. Compared with a similar cheese made with lactic acid bacteria only, the yeast cheese had less acidity, a higher count of lactic acid bacteria, ripened more quickly and was less susceptible to mold contamination. After ripening for 15 days, the yeast cheese contained 47.5% moisture, 3.31% total N, 1.13% soluble N and 0.032% ethanol.

In another study,[36] red pepper was mashed with mixtures of starter cultures containing various yeast strains: (1) *Saccharomyces rouxii* + *Torula etchellii*; (2) *Saccharomyces rouxii* + *Torulopsis versatilis* and (3) *Torula versatilis* + *Torula etchellii*. A control with no added starter was used. Enzyme activity and various chemical parameters were determined

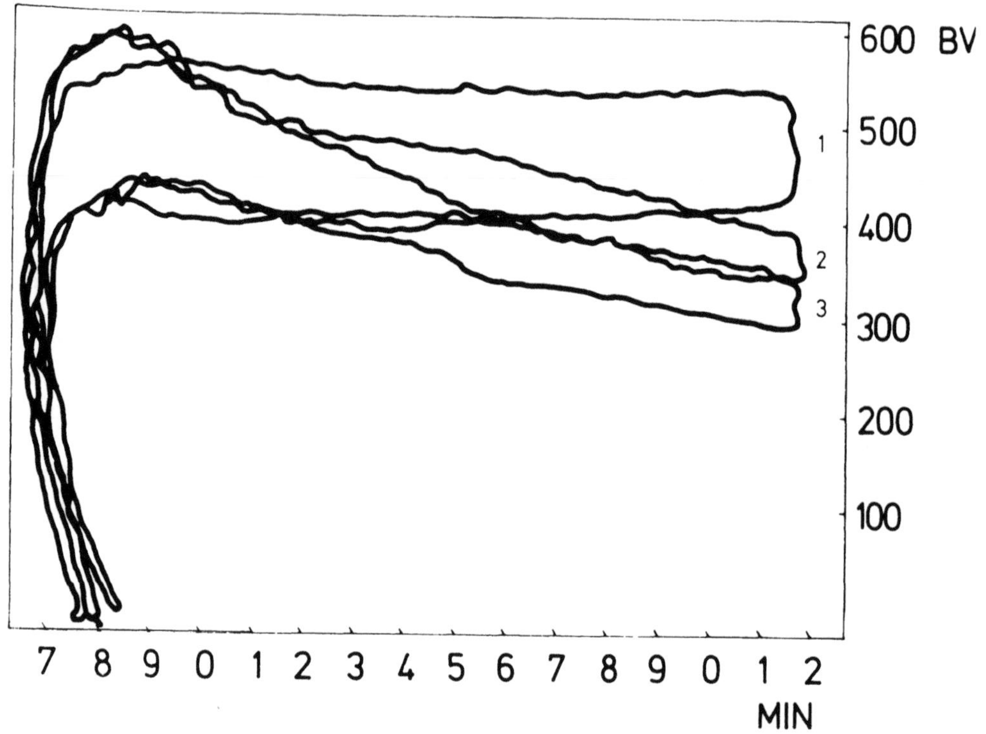

FIGURE 72. The effect of the addition of partially active dry yeast on the consistency of wheat flour dough (1 — control, 2 — 3% yeast added, 3 — 6% yeast added).

during 180 days aging. Mixture 3 had the best liquefying and saccharogenic amylase activities. Ethanol content was highest using mixture 2 (2.9% after 10 days of aging), but levels decreased steadily on aging. Reducing sugar content increased initially especially using mixture 3 (16.5% after 10 days of fermentation). Amino N increased dramatically in the first 10 days, and then more slowly up to 90 days (220 mg% was the highest value using mixture 3). All Ko-chu-jang made using starters had superior organoleptic (taste, flavor, color) properties comparing to the control. Mixture 3 was preferred.

The effect of the addition of active *Saccharomyces cerevisiae* and/or *Torulopsis* yeasts on composition and quality of soy sauce mash (moromi; used in Japan) was studied by Noda et al.[47] Growing of yeasts was not substantially inhibited and composition and overall sensory properties have been improved. In another report the *Torulopsis* yeast was found to be the best from the point of view of flavor of moromi.[52] In the framework of improvement of soy sauce production using active yeast addition, the effect of salt concentration on yeasts was investigated.[59] Salt tolerant yeasts *Saccharomyces rouxii* and *Torulopsis versatilis* and salt sensitive yeast, *Saccharoymces sake* were cultured in media with and without NaCl. Salt tolerant yeasts had a higher cell wall content than sensitive yeasts and the presence of NaCl in the media decreased the mannan content of cell walls of both types. The chitin content in cell wall of tolerant yeast was reduced by NaCl in the medium, that of the sensitive species was low. Protein in cell walls was increased in both types by NaCl in the media.

Metabolization of lactic and acetic acid by yeast and yeast growth during cheese production with starter containing lactic acid bacteria and yeasts was investigated by Petona et al.[62] *Debaromyces hansenii* was added to ewes's milk at the same time as the cheese starter for manufacture of Pecorino romano cheese according to traditional methods. Analysis of samples showed that L (+) lactic acid concentration decreased in cheese during ripening especially at higher yeast content. Production of free fatty acids in the cheeses was higher and that of acetic acid was lower in the presence of added *Debaromyces hansenii*. The yeast developed well in the cheese and resulted in a more rapid proteolysis and overall ripening; the organoleptic score was not accepted by the presence of added *Debaromyces hansenii*. The possibility that *Debaromyces hansenii* might inhibit clostridial spores is suggested.

II. FOOD USES OF INACTIVE DRIED WHOLE YEAST CELLS

The use of yeast as a primary foodstuff for human consumption was first suggested by Max Delbrück, a scientist working in Germany in the early part of this century. Up until then, yeast had been consumed, but only because it formed an integral part of fermented food and beverages.

Numerous feeding experiments carried out between 1910 and 1940, showed that primary grown yeast was a valuable source of proteins and vitamins, especially B group vitamins. Brewer's and baking yeast were first of all investigated (*Saccharomyces cerevisiae* species grown on molasses, *Saccharomyces carlsbergensis*) but other species such as *Candida utilis, Candida tropicalis* grown on paper pulp mill waste and later *Kluyveromyces fragilis* grown on whey were also studied.[6,7] Brewer's yeast is generated at a rate of 0.5 lb yeast solids per barrel as a byproduct of beer production of which only a minor amount is recycled through the brewing process.[8] Following the centrifugal separation of the yeasts from the beer, the yeast solids are concentrated to 10 to 15% and pasteurized. Most of the solids are either dried as is or blended together with the spent grain, then dried and sold to the feed industry.[5] The term "debittered brewer's dried yeast" is applied to the yeast that has been washed with an alkaline solution to remove the surface-adsorbed, bitter tasting principles, that is hops resins and tannins. The final product is sold to the health food industry.[9,10]

Primary grown yeasts are those organisms that are aerobically grown on a variety of substrates specifically for use in the human food and feed industries. These products are designated with different terms including dried yeast, dried inactive yeast, microbial protein, dried Torula yeast, and single cell protein (SCP). The U.S. Food and Drug Administration (FDA) permits the usage of dried yeast from *Saccharomyces cerevisiae* and *Kluyveromyces fragilis,* and dried torula yeast from *Candida utilis* in human food, provided that the total folic acid content of the yeast is not greater than 0.4 mg/g yeast (approximately 0.008 mg of pteroyl glutamic acid per gram of yeast[11]). Dried yeast derived from *Saccharomyces cerevisiae* is also granted prior sanction for optional use in bakery products under food standards, provided the total inactive dried yeast content does not exceed 2 parts for each 100 parts by weight of flour used.[12] *Saccharomyces cerevisiae* strains propagated aerobically on molasses give a raw material of constant high quality. It is this consistency of starting material which allows a wide range of goods to be produced to very high specifications.[13]

Simplest of all is dried food yeast, produced by passing a yeast cream (500 to 600 g yeast/1) over steam-heated rollers. This process kills and dries the yeast to a fine, light-colored flake material which is called "flake yeast". This material can also be milled to a powder which is marketed in U.S. as processed yeast powder (PYP).

The macaroni products prepared from durum wheat are foods which are economical, easy to prepare, shelf stable, and can be served in many different ways. The major nutritional limitation is the low lysine content of wheat protein. A selective addition of a protein rich in lysine (e.g., yeast protein) improves the nutritive value of such products. Since the products are extruded, additives can be easily blended into a formula: yeast protein has a real possibility to be used in pasta production. The yeast was blended with semolina to produce high protein pasta.

Products containing 9% torula yeast and 91% semolina had a protein content of 18.0% and the PER value was almost doubled over conventional pasta.[14] The cooked weight was slightly lower, the cooking loss was unreported and the product retained a slight meaty

TABLE 59
Biological Value (Morup-Olesen Index) of Different Bakery Products and Quick-Frozen Doughs Enriched with Dried Yeast

Product	Morup-Olesen index						
		Dried yeast product added					
	Control	1	2	3	4	5	6
Extra white wheat bread	49.7	101.3	105.5	114.0	112.6	73.6	121.2
White wheat bread	75.3	82.2	106.8	109.7	108.9	76.5	111.4
Rye bread	100.9	104.6	110.8	118.4	116.8	102.2	123.9
Rye-wheat bread	93.2	104.6	122.5	126.2	125.2	114.8	128.9
Quick-frozen dough:							
with water	61.8	99.6	104.0	113.5	110.0	77.9	118.1
with milk	92.4	109.6	114.7	123.4	120.4	102.3	122.0
with 1 egg/kg	93.0	108.8	114.0	120.7	119.8	103.3	121.3
with 2 eggs/kg	93.4	109.6	114.7	120.7	120.3	102.2	121.8
with 3 eggs/kg	94.7	110.0	114.9	119.0	120.3	101.8	121.8
with 4 eggs/kg	95.3	108.0	113.0	119.5	118.0	100.4	119.4
with 5 eggs/kg	95.9	108.5	115.0	123.4	118.5	102.8	119.9
Fatty cake dough	90.4	109.2	113.2	122.0	121.5	100.9	123.1
Butter dough	92.1	107.1	112.4	119.6	120.4	98.1	120.8
with mashed potato	59.1	112.6	113.5	122.3	120.9	86.1	126.4

1 — MAGGI yeast powder; 2 — Brewer's yeast dried in fluid bed; 3 — Brewer's yeast dried with geizir dryer; 4 — Brewer's yeast dried with band dryer; 5 — Brewer's yeast liophilized and 6 — Brewer's yeast spray dried.

TABLE 60
Characteristics of White Wheat Breads Enriched with Inactive Dried Brewer's Yeast

No.	Volume (cm^3)	Density (g/cm^3)	Rheological properties (penetrometer units)			Yeast added (%)
			Total compressibility	Plastic compressibility	Elastic deformation	
1	932	0.3917	2.0	0.4	1.6	0.0
2	1090	0.3250	2.7	0.7	2.1	1.5
3	1145	0.3185	3.0	0.9	2.1	3.0
4	1120	0.3210	2.9	0.9	2.0	4.5
5	1000	0.3596	2.2	0.6	1.6	6.0

flavor. An elastic texture was reported as well as good stability to heating. No report of method of flavor evaluation was given.

Recently Breen et al.[15] and Halász[10] reported the use of various protein sources in pasta, including yeasts. Although addition of inactive dry yeast increases the protein content and nutritional value of the product, the organoleptic quality of product is not satisfactory. The darker color and yeasty taste is not desirable. A lot of bakery products, biscuits (cakes) and quick-frozen doughs could be prepared with the addition of inactive dried yeast.[6] An increase of the nutritive value was observed in all cases as expressed with chemical index calculated according Morup and Olesen. In Table 59 the nutritional value of 19 different bakery products (and quick-frozen doughs) are shown enriched with different inactive dried yeast products. Table 60 shows some characteristics of white wheat breads prepared by the addition of different quantities of inactive dried brewer's yeast. Although a slight increase in volume was observed a decrease of elastic properties of crumb occurred.[10] The protein content increased with the yeast addition and higher amounts of lysine were measured. As a consequence the nutritive value increased too. Some spray-dried yeast preparations caused a decrease of volume probably due to the remaining proteolytic activity of the preparation.

Addition of flour improvers (e.g, ascorbic acid) eliminated this disadvantage.

The use of single cell protein in the form of dried inactive yeast as a supplement has been investigated in numerous studies.[16-20] The species most commonly used were *Candida utilis* (torula yeast), *Saccharomyces cerevisiae* and *Saccharomyces fragilis*. In a study, Ching-Ming Lin et al.[21] investigated the sensory and nutritional properties of wheat breads supplemented with an inactive dried mass of *Torula* yeast. The *Torula* yeast used in the study was a commercial product grown on ethanol (Torutein-10, Amoco Foods Company). With increasing amounts of Torula yeast flour (TYF), the height of the ends of loaves decreased sharply. In some trials an unsymmetrical shape occurred in breads containing 10 and 12% of TYF. Crust color of all breads supplemented with TYF was darker than that of control bread (0% TYF). Interior color of supplemented breads was golden-yellow due to the color of TYF and also to the heat-accelerated Maillard reaction. The higher protein content (and lysine content of yeast protein) of TYF breads resulted in greater browning. The grain of breads containing up to 8% TYF was typical of standard quality wheat bread. However, as levels of TYF were increased above 8%, air cells became larger and the grain became coarser, breads were heavier with a compact consistency at the bottom, and loaves broke apart readily. Breads containing 10 to 12% TYF were unacceptable both from the standpoint of texture and flavor. The dough failed to develop elastic properties when kneaded and did not rise properly in proofing and baking. This was probably due to reduced gluten development resulting in failure to retain gas, which in turn caused the heavy loaf with poor volume.

The acceptability of breads supplemented with TYF was also investigated. The scores given by judges were relatively low. When test bread was identified as high-protein bread that could be the main source of protein in the diet of some people, the scores were significantly improved. The scores were higher than for unidentified bread but not as high as when butter and jelly were available to mask the flavor. To mask the TYF flavor experiments were made with whole wheat flour, honey and molasses. The average results of panel scores showed that judges significantly preferred the flavor of breads containing whole wheat flour, honey or molasses compared to basic TYF bread. In conclusion it was stated by the authors[21] that although TYF was found to depress loaf volume, bread made with 8% TYF had acceptable volume, flavor and moistness. Panelists gave higher ratings to bread supplemented with TYF if the yeasty flavor was masked by butter, jelly, honey or molasses. The addition of 8% TYF increased the protein content from 12.42% for all wheat flour bread to 14.22% for supplemented bread. Amino acid analysis revealed a marked increase in amino acid content especially lysine, the limiting amino acid in wheat flour. The PER values increased from 1.31 to 2.28 for supplemented bread. No indication was reported concerning elevated nucleic acid content of the product.

In another experiment commercially prepared baker's yeast single cell protein (B-SCP) was studied in bread dough. Hard red spring wheat flour was supplemented at four levels (0, 3, 6 and 12%) with B-SCP. The composition of the flour and B-SCP is given in Table 61. Five treatments were investigated at each level of B-SCP substitution: no additive; an oxidant-50 ppm-potassium bromate/potassium iodate (3:1); a conditioner — 0.5% sodium stearoyl-lactylate (SSL); 2% sodium chloride and a 1.25 min microwave treatment of the dry B-SCP. The bread formula was: flour, 100%; yeast, 3.0%; salt, 1.5%; sugar, 5.0%; and water (24°C), 3.0% higher than farinograph absorption. The bread was prepared by a straith dough method, fermented for 60 min at 30°C and 85% relative humidity, degassed with a dough sheeter scaled to 150 g, molded, proofed for 35 min at 30°C and 85% relative humidity, and baked at 218°C for 20 min. Physical properties and pH of the dough, the loaf volume and the organoleptic characteristics of the final product were measured. As it is shown in Tables 62 and 63 the addition of B-SCP increased the water absorption of the flour — B-SCP mixture, decreased the stability of dough and also the extensibility. B-SCP

TABLE 61
Composition (%) of Hard Red Spring Wheat Flour and Yeast Single Cell Protein

Component	Hard red spring wheat flour	Yeast single cell protein
Moisture	14.0	3.5
Protein	14.0	75.0
Fat	2.0	9.5
Ash	0.5	1.8
Carbohydrate	69.5	10.2

From Volpe, T. A. and Zabik, M. E., *Cereal Chem.*, 58(5), 441, 1981. With permission.

TABLE 62
Farinograph Data[a] for Doughs Prepared from Flour Supplemented with 0, 3, 6, and 12% Single Cell Yeast Protein (SCP) Under Varied Treatments

Farinograph measure	Treatment[b]	Level of SCP substitution (%)			
		0	3	6	12
Absorption (%)	None	65.6	69.7	75.8	92.0
	Oxidant	66.6	69.4	74.3	83.9
	Salt	63.9	68.3	74.2	82.3
	Conditioner	63.4	68.5	73.4	84.3
	Heat	—	69.3	72.5	73.3
Arrival time (min)	None	1.9	1.5	1.5	2.3
	Oxidant	5.1	2.0	1.5	1.6
	Salt	3.1	2.1	1.8	2.5
	Conditioner	1.6	1.8	1.8	2.0
	Heat	—	3.0	3.0	3.1
Peak time (min)	None	6.1	3.0	2.5	2.9
	Oxidant	9.1	7.1	2.0	2.1
	Salt	15.0	15.0	4.0	5.3
	Conditioner	3.0	3.3	3.3	2.8
	Heat	—	5.6	5.1	5.3
Stability (min)	None	13.0	12.5	3.0	1.0
	Oxidant	9.1	12.5	2.0	0.8
	Salt	15.0	15.0	13.1	8.5
	Conditioner	13.3	10.3	2.0	2.1
	Heat	—	7.0	5.6	4.5

[a] Average of three replications.
[b] None = control; oxidant = $KBrO_3/KIO_3$(3:1), 50 ppm; salt = NaCl, 2%; conditioner = sodium stearoyl lactylate, 0.5%; heat = microwave heat on dry SCP, 1.25 min.

From Volpe, T. A. and Zabik, M. E., *Cereal Chem.*, 58(5), 441, 1981. With permission.

had a buffering effect on the pH of the dough during fermenting and proofing. As the level of added B-SCP increased a slight increase of the final pH of the dough was observed from 5.18 to 5.45. Addition of oxidant, conditioner and salt to the dough or heat treatment of B-SCP had no effect on the final dough pH. Bread with 3% B-SCP had loaf volumes similar to those of the controls. Higher B-SCP levels yielded loaves with reduced volumes. The

TABLE 63
Means[a] of Measurements of Extensibility (mm) and Resistance to Extension (BU) for Doughs Made with Flour Supplemented with 0, 3, 6, and 12% Single Cell Yeast Proteins (SCP) Under Varied Treatments

Treatment[b]	Time of measurement (min)	Level of SCP substitution (%) 0 mm	0 BU	3 mm	3 BU	6 mm	6 BU	12 mm	12 BU
None	45	237	570	235	426	211	316	137	307
	90	251	601	225	484	201	384	115	325
	135	256	700	210	465	193	416	103	315
Oxidant	45	240	846	245	616	207	516	132	490
	90	195	1001	217	787	181	660	121	524
	135	171	995	200	907	170	752	108	524
Salt	45	295	762	278	654	230	495	170	315
	90	275	937	257	724	216	579	142	341
	135	233	925	217	791	203	604	141	355
Conditioner	45	316	640	258	449	231	399	106	261
	90	295	667	263	490	207	421	126	270
	135	305	750	245	512	187	416	146	270
Heat	45			267	494	230	286	155	266
	90			247	504	208	336	142	285
	135			243	526	196	375	131	289

[a] Average of four replications.
[b] None = control; oxidant = $KBrO_3/KIO_3$(3:1), 50 ppm; salt = NaCl, 2%; conditioner = sodium stearoyl lactylate 0.5%; heat = microwave heat on dry SCP, 1.25 min.

From Volpe, T. A. and Zabile, M. E., *Cereal Chem.*, 50(5), 441, 1981. With permission.

volume of loaves with 12% B-SCP was only 265 cm^3 in comparison to 473 cm^3 average volume of the control. Treatment of the 6% B-SCP dough did not help to improve volume. The use of conditioner, oxidant, or salt had no improving effect on breads containing 12% B-SCP. Heat treatment of the B-SCP before incorporation into the bread had no volume improving effect at any level of B-SCP.

Proximate analysis of the bread crumb showed that the protein content of the bread increased from 11.0 to 13.33% as the B-SCP addition increased. The moisture content of the breads ranged from 39.26 to 45.49%. At the high moisture levels (increased B-SCP content) the crumb was very gummy and wet. Sensory evaluation showed that the most acceptable product was the 3% B-SCP bread treated with SSL. The crumb color and grain of the 3% B-SCP-SSL bread was very good, yielding an acceptable product. A short review about the production and various ways of utilization of food yeasts (*Kluyveromyces lactis Dombrowslu* and *Kluyveromyces fragilis*) grown on deproteinized whey with lactose as a main carbon source in France is given by Guerviére[30] and Marzolf.[31]

Stewart and Guilliland[24] and Bostian et al.[25] reported the use of yeast-whey proteins to improve the nutritional value of snack foods. A pilot plant system has been developed that utilizes a fermentation process for converting cottage cheese whey into a usable food product. The process involves treating whey with the yeast *Kluyveromyces fragilis* which reduces the lactose level and produces single cell protein. After drying the yeast-whey material contains abut 73% single cell protein and can be effectively utilized as an ingredient in the manufacturing of bakery products and other foods. One important use of the recovered yeast-whey protein material is in the processing of snack foods. Snack foods are composed essentially of wheat flour and other sources of carbohydrates, but are generally lacking in protein. By supplementing such products with protein-rich ingredients a high protein snack food can be produced. The authors developed four different cookie and cracker formulas. The ingredients in a basic vanilla wafer were altered to create a recipe with cookies containing approximately 15% yeast whey protein material. The same recipe was further altered by incorporating cocoa to produce chocolate wafer. Oatmeal cookies and cheese-type crackers were also prepared by supplementing with yeast-whey protein. Yeast-whey protein was prepared not only in dried form but also in pressed form. The latter was unworkable in the recipes because it produced an undesirable flavor as well as poor dough consistency and baked cookie texture. However, the dry material provided an acceptable flavor and cookie texture. The addition of various ingredients was effective in improving the flavor and texture of the cookies. An increase in the sugar, vanilla, and shortening seemed to conceal the underlying flavor of the yeast-whey protein material. Cookie structure was improved by the addition of egg and shortening, and a more workable dough was obtained by decreasing the amount of milk. The experimental vanilla wafers contained less fat, but almost 4 times as much protein as the original wafer. At the 35% level of yeast-whey protein the cheese-type crackers were highly acceptable and had an excellent aroma and flavor showing no trace of the underlying flavor of the material. Of all the experimental cookies the oatmeal products were the most acceptable. They were very crispy, had good flavor, and an appetizing aroma and appearance. By decreasing the sugar content and adding apple pie spice, the cookie flavor was greatly enhanced. Approximately 20% yeast-whey protein material was optimal in the recipe. The experimental cookies contained more fat and 3 times as much protein as the commercial products.

The use of extrusion cooking as a method of food fabrication has increased during the past few years. This increase is due to the growing demand for snack-type convenience foods and the advantages extrusion cooking offers over other production methods. The starch based extruded products generally also contain proteins. The quantity and quality of added protein greatly influences the extrusion properties of starch and also the properties of the extrudates. Yeast protein as a high quality protein can also be used in the production of extruded foods.[26]

Development of a lactic yeast preparation (Protibel) by Bel Industries as a means of utilizing whey from cheese production was reported by Kusachi.[29] Biomass of yeast *Kluyveromyces lactis* and *Kluyveromyces fragilis* was produced on whey and the dried preparation (Protibel) was used as a supplement in bread making. Addition of Protibel to bread dough at the 2.5% level decreased the expansion without affecting its adhesion. When added to pies and snack foods, lactic yeast can reduce the bench time before cooking by more than 80%.

An interesting report was published by Smolyak et al.[32] reporting about the evaluation of bakery products enriched with yeasts containing carotene. A new carotenoid-producing yeast (*Phaffie rhodozyma*) was described by Phaff[33] and some possible uses of this yeast were discussed.

The possible use of yeast biomass in pizza formulation was studied by Kamel et al.[37] Pizzas were prepared using nine different levels and concentrations of ground beef, mozarella cheese, textured vegetable protein (TVP), SCP and imitation cheese, provolone and cheddar cheese. Proximate amino acid, fatty acid and sterol analysis were carried out on samples from each treatment. Protein content ranged from 12.15 (with cheese addition) to 16.88% (in the beef + 50% SCP pizza). Fat content was the lowest in pizzas with TVP. The addition of SCP caused significant increase in lysine, arginine and threonine.

Kann et al.[53] reported the application of debittered brewer's yeast in breadmaking. The nutritional value of breads was improved without significant changes in the bread quality. Possibilities of enrichment of local bread with dried brewer's yeast were investigated by Alian et al.[55,56] Debittered, dried brewer's yeast was added at 0.2 and 5% levels, to wheat flours (70 and 90% extraction) and the BV of the flour and breads was investigated. The

BV increased from 31.09 to 76.11 if 5% fried brewer's yeast was added to wheat flour of 93% extraction. The BV changed from 43.31 to 76.15 by addition of 5% of dried yeast. Addition of dried brewer's yeast to Balady bread resulted in no significant differences in appreciation. In white bread, however, the addition of dried yeast caused considerable differences in taste. The three white breads (control, 2 and 5% yeast addition) could be easily distinguished from each other.

A new accelerated method for pastry production was devised and tested by Asenova et al.[57] It is based on preparation of an emulsion formed from the dough components and on the addition of a biological preparation from brewer's yeast. The fermented emulsion is prepared in an ultrasound dispersing device. It contains 25% of the total flour amount, all other dough ingredients and the improver (brewer's yeast preparation). The improver accelerates biochèmical and microbial processes in dough and yields a product with better organoleptic properties. Uniform dispersion of the emulsion in dough results in better structure and physical properties and yields bakery products with large volume. The addition of fat emulsion leads to higher dough resistance, to oxidation and to better digestibility. The new method is effective only if the biological preparation is applied.

A dried yeast protein product (Provesteen-T) was produced by a subsidiary of Philips Petroleum Corporation. It was reported[58] that this product could be used in production of egg noodles, peanut butter cookies and potato snacks.

Chemical properties of Ras cheese (an Egyptian specialty cheese) during ripening as affected by the addition of inactive dry yeast and yeast autolysate were investigated by Hagrass et al.[63] Dried yeast (DY) was added at 1, 2, 3 and 4% and autolysate (YA) 0.25, 0.5 or 1% to the curd before hopping. Addition of DY up to 1% or YA up to 0.5% appeared to increase the waterholding capacity of the cheese. Yeast addition has no apparent influence on pH of fresh or ripening cheese. Total volatile fatty acids, formol N (as % of the total N) soluble tyrosine and soluble tryptophan concentration increased with DY up to 1% or YA up to 0.5%. It is concluded that ripening of Ras cheese may be accelerated and its quality enhanced by addition of 1% DY or 0.5% AY, the latter having the greater effect. A cocoa alternative was made from yeast and carob.[60] The new product called CoCoNo (Coors Food Products) comprises a yeast product made from spent brewer's yeast, combined with carob. The product was successfully used to partially replace cocoa in brownies served in schools.

The effect of substitution of meat in sausages or meat balls with (1) brewer's grain, (2) brewer's yeast and (3) distillers stillage was investigated by Junnila et al.[41] The addition of brewer's grain impaired the quality of both meat products, the degree of impairment increased with an increasing level of addition; these adverse effects are thought to be due to the lack of gelling or emulsifying capacity of brewer's grain. Overall acceptability of samples with brewer's yeast was relatively good; sausages containing 1% yeast were rated higher than control. An off-flavor was observed in products with 10% yeast addition.

Studies on use of Protibel lactic yeast preparation as an ingredient in finely comminuted heat-preserved pate were conducted by Pinel.[42] The yeast was added at levels of 0.5, 1.2 or 5.0 % in the absence of other functional additives, used in combination with sodium caseinate (0.8% caseinate + 0.4% yeast, 0.5% caseinate + 0.1% yeast or 1.0% caseinate + 0.5, 1.0 or 2.0% yeast). The yeast was also used in the presence or absence of sodium glutamate, or substituted for a part of the lean meat. The sensory properties, rheological properties and stability (fat and jelly separation) of the pate were evaluated. No effect of levels up to 2% yeast (with or without sodium glutamate or caseinate) on the characteristics studied were observed. Addition of the yeast at levels of 2 to 4% reduced jelly and fat separation and gave a firmer texture, but impaired an undesirable off-flavor.

It was reported[43] that yeast biomass has an antioxidative activity positively influencing the shelf-life of some food systems. Two processed meat systems (an extended meat pattie held under frozen conditions and wieners containing sodium nitrite held at refrigerated temperature) were evaluated to determine whether the presence of yeast influenced microbial levels.

Samples were frozen for 65 days and then evaluated for a total plate count. There were no significant differences in microbial die off "between the control and the experimental patties". The results show that the addition of 2.5% primary grown *Candida utilis* did not decrease freezer die off.

Other tests indicate that the shelf-life of oil-containing products may actually be extended because of a natural antioxidant-like property formed in the yeast. Examples considered are the effect of the specially processed primary grown yeast, Torutein-94, and commercial antioxidants on the oxidation of palm oil and its effect on the peroxide value of salad dressing.

Long term (18 months) storage experiments[44] with ground beef patties containing 8% TVP and 25% yeast SCP showed that frozen products can be stored for only 6 months without loss of high quality characteristics. The temperature of storage of frozen patties was −10 to −30°C. Studies on effects of storage on protein quality of (1) commercial ground beef, (2) ground beef + SCP, (3) ground beef + TVP, (4) ground beef + TVP + SCP, (5) commercial beef patties + TVP, (6) meat patties with TVP + SCP, (7) SCP, (8) TVP and (9) skim milk powder were performed by Butrum.[46] Storage for 6 month at −20°C had no effect on relative nutritive value (RNR) determined by *Tetrahymena* assays or PER of samples 3, 4 or 8, but resulted in a decreased RNR and PER in skim milk powder. Storage at 0°C for 6 months did not change the PER or RNR of TVP, but reduced the protein nutritional value of SCP and skim milk powder. Commercial beef patties supplemented with TVP, 6, and 8 showed no significant loss in PER and RNR after 3 months at −10, −20 or −30°C, but significant losses in their characteristics after 6 months. There was little further loss in protein nutritional value during subsequent storage for up to 18 months PER was highly significantly correlated with RNR. Storage of beef for 6 months at −20°C caused losses in contents of essential amino acids. Little change in amino acids was observed in samples SCP 7, 8 or 9 stored at −20, 0°C or room temperature for 6 months. A study of the possibility of adding alcohol-fermentation derived yeast to luncheon meat-type products was reported by Oluski et al.[64] A preparation derived from brewer's yeast as an emulsifying agent was used in the manufacturing of luncheon meat-type products. Trials were conducted on samples made with sodium caseinate, with brewer's yeast, or with 50:50 or 75:25 blends of these materials. The results showed that use of blends in the above ratios gave results similar to those achieved with sodium caseinate alone.

Nutritional and technological properties of a protein preparation made from brewer's yeast and possibilities of their use in meat products were studied by Modic et al.[65] *Saccharomyces cerevisiae* and *Saccharomyces carlsbergensis* yeast milk was deactivated at 70 to 80°C. Bitter substances were removed by mixing with 2% sodium chloride solution, separating, washing and finally spray drying with hot air (250°C). The products were added to meat products and an improving of sensory properties was observed.

The effect of the inactivated yeast, soy meal and wheat gluten on the stability of the meat-fat-water system was investigated by Stamenkovic et al.[66] The control dispersion system was prepared in the ratios veal meat-adipose tissue-water 1:1:1 and 1:2:1. In the experimental dispersion systems 2 and 4% of meat were substituted by the protein preparations. The protein preparations used were the following ones: (1) Beviprot, a protein preparation produced from distiller's yeast by drying; (2) toasted soy meal and (3) wheat gluten. The protein content of preparations was 45.5, 46.5 and 70.4% resp. Sodium chloride (3 g/100 g) was added to the meat-fat-water mixture. The homogenized mixtures were heat treated in aluminum cans at 120°C for 35 min. After cooling and storage for 24 h the loss of water and fat was measured. Addition of protein preparations decreased the cooking loss of water

TABLE 64
Organoleptic Quality of Liver Pastes Containing Brewer's Yeast and/or Brewer's Yeast Autolysate

Sample	Taste (max. score 60)	Flavor (max. score 20)	Texture (max. score 20)	Total scores (max. score 100)
Control	41	13	13	67
2% autolysate added	46	18	14	78
2% brewer's yeast + 2% autolysate added	45	17	17	79
After 1 month storage				
Control	39	15	12	66
2% autolysate added	37	15	13	65
2% brewer's yeast + 2% autolysate added	44	17	16	77
After 3 month storage				
Control	37	17	12	66
2% autolysate added	28	15	11	54
2% brewer's yeast + 2% autolysate added	54	18	17	89

and fat in all samples. Soy flour and wheat gluten was more effective than the dried yeast product.

Liver paste and a chicken meat based luncheon meat-type product was produced in pilot plant experiments by Halász et al.[67] with the addition of inactivated brewer's yeast and/or brewer's yeast autolysate. The following formula was used for liver paste:

- Pork liver — 290 g
- Pork meat — 150 g
- Beef meat — 130 g
- Fat — 170 g
- Milk — 85 cm^3
- Rice flour — 66 g
- Salt mixture (with nitrite) — 12 g
- Seasonings
- Water — 95 cm^3
- Yeast autolysate — 2% (dry weight basis)
- Brewer's yeast — 2% (dry weight basis)

The luncheon meat contained chicken meat, flour, egg seasonings and 3% of yeast autolysates.

The comminuted mass of ingredients was filled into cans and sterilized. The products were organoleptically tested for taste, flavor and consistency immediately after production and after 1 and 3 months storage. Some results are summarized in Table 64. The results show a slight improvement in the organoleptic quality in the case of supplemented samples. However, after storage the samples prepared with the addition of 2% yeast autolysate had a lower organoleptic quality. Simultaneous addition of inactive brewer's yeast and yeast autolysate improved the flavor and texture of the liver paste. Addition of 3% yeast autolysate to the luncheon meat-type product does not significantly change the organoleptic properties. A Japanese patent describes production of a beverage based on single cell protein from *Torula* yeast in which odorous and astringent components are removed by hot water (60 to 150°C) or aqueous alcohol (>30%) extraction to yield a residue and is milled to crush the cell membrane and accelerate proteinase action (e.g., papain).

A vitamin-containing beverage with improved organoleptic and prophylactic properties is manufactured by Skvortsova and Vorontsova[40] by addition of citric acid to pH 5.0 to 5.5, prefermentation of yeast in 2 to 1 sugar-mountain ash fruit solution with Amylorhizin P10 × (37 to 38°C for 7 to 8 h, then 6 to 8°C for 15 to 16 h), and clarification. Then sugar syrup (10 to 15 g/l), black currant essence (0.3 to 0.5 mg/l) and the prefermented (3 to 7 generations) *Saccharomyces carlsbergensis* (water content 74 to 75%) 10 to 10.5% in solution is mixed. Fermentation produces the improved product.

The use of whole yeast in food processing is very often limited due to undesired flavor and taste of dried yeast. To eliminate the unpleasant taste and aroma different procedures were elaborated and patented. According to a Japanese patent[27] substances having unpleasant taste and aroma can be effectively removed from proteinaceous materials contained in single cell microorganisms, especially yeasts and bacteria, by extraction with a solvent mixture containing a 1 to 3 C aliphatic alcohol. The resultant product is suitable for human consumption. Another procedure for modifying the flavor of yeast was patented in the U.S.[28] This method improves the color and flavor of a yeast product derived from a yeast fermentation process which yields a slurry of fresh yeast cells.

The procedure mentioned above includes oxygenating the slurry with an oxygen-containing gas and heat treatment. In carrying out this method it has been found that a slurry of fresh yeast cells is necessary to obtain the desired effects in flavor and color of the yeast. Yeast cells which have been spray dried and reslurried will not benefit from this treatment method. Although this invented process is applicable to any slurry of fresh yeast cells, a preferred yeast is *Candida utilis*. The treatment of yeast slurry resulted in a product free of yeast's slightly astringent taste and the product was very mild, slightly sweet, slightly meaty and slightly mushroom like. The color changed from light pink to creamy, and the slurry was much less foamy.

Alkali treatment at 60°C is recommended as a simple and cheap treatment for improving nutritional safety of SCP.[38] A treatment of proteinaceous materials such as commercial spray-dried yeast, soy flour and their mixtures with anhydrous ammonia resulted in the improvement of flavor.[48] Increase of water solubility of the nucleic acid present was also observed. The yeasts that have food application are preferred e.g., *Candida utilis* and *Saccharomyces cerevisiae*. A process for reducing the adsorption of a flavoring agent into plant or single cell protein was patented in the U.K.[45] According to the inventor the treatment involves use of oils or fats or their mixtures with emulsifier before or concurrently with the addition of flavor. In such a way the amount of flavoring materials (and/or spices) adsorbed by protein will need lower and smaller quantities for aromatization.

III. FUNCTIONAL PROPERTIES OF INACTIVE DRIED WHOLE YEAST

The functional properties of protein preparations (concentrates and isolates) were intensively studied and different practical methods were developed for their control (see Chapter 3). However, our knowledge concerning functionality of whole yeast in food systems is relatively poor. Although such properties as color, flavor, and antioxidant effect are always controlled other properties (e.g., colloidal properties of dried yeast) have been neglected. The dry whole yeast cells are insoluble both in water and oil, nevertheless, their surface properties and size may influence the structure of a food system supplemented with inactive dried yeast. In a review paper Cooney et al.[49] summarize the possible interaction of dried yeast cells with other components of an emulsion as follows:

1. When the oil droplet diameter is greater than the diameter of yeast particle, the individual yeast particles could stabilize the emulsion by stacking along the phase boundary.
2. When the yeast particle diameter is greater than the oil droplet, the latter may adhere to the yeast, or the oil can wet the yeast particle surface, depending on the relative tension. This in turn, may associate further to form a larger unit consisting of oil phase and yeast.
3. The stacking of phases, lamellar or cage structures that are more organized or regular than core (2) could result, owing to surface energy and geometry.

Based on these theoretical possibilities it can be stated that the particle size and particle size distribution of dried yeast would be critical to its behavior in food systems. Naturally such properties as surface hydrophobicity of yeast cell, the eventual electric charge of surface (depending on ionizable groups or presence of adsorbed cations) and other interacting groups of the surface may influence the yeast-yeast or yeast-water and yeast-oil interactions. Finally, it is possible that the emulsion-stabilizing effect of yeast is a nonspecific effect of thickening of the continuous phase in the food system.

The hydration of dried whole yeast is slow as reported by Conney et al.[48] investigating a dried *Candida utilis* product. The slow rate is probably due to the particle clumping and surface characteristics. Alkaline pH increased the amount of hydration. In contrast to the low level of interactions with water, a relatively high capacity of yeast for hydrophobic substances was observed as demonstrated by good oil emulsifying properties of yeast.

Whole yeast, or unseparated broken cells can be processed into texturized products by extrusion.[50,51] Texturizing of mixtures of yeasts and other proteins is also possible.

Solubility, water and fat binding, emulsifying and foaming properties of brewer's yeast dried with drum dryer were determined in different environmental conditions by Koivurinta et al.[54] The pH solubility profiles indicated a high degree of denaturation, no isoelectric points were recorded. The protein solubility of brewer's yeast was 20% of the total N on a wide range from pH 3 to 8. Temperatures of 60 and 90°C or 2% NaCl had little effect on water-oil emulsions, the emulsion capacity being 12 ml/g. The foams obtained with brewer's yeast suspension (15%) were almost comparable to whole egg foams and they could be used in cookies to replace up to 25% of egg white on a dry matter basis without serious negative effects on the organoleptic properties of the products. Functional properties of inactivated dried whole brewer's yeast (debittered with dilute sodium hydroxides) were studied by Halász et al.[67] Mixtures of dried brewer's yeast, dried *Rhodotorula glutinis* yeast, wheat flour (ash content 0.80%) and soy isolate (Purina 500E) were also tested. Some results of experiments are demonstrated in Tables 65-67. Inactivated *Rhodotorula glutinis* yeast was found to have the highest water-binding capacity and baker's yeast the lowest. The oil-binding capacity of Baker's yeast, brewer's yeast and *Rhodotorula glutinis* is similar. In this case the alkaline treated brewer's yeast also had the lowest fat-binding capacity. Texture evaluation of spun yeast protein fibers was studied by Kawasaki.[69] A baker's yeast SCP extract was spun into fibers with or without the addition of carrageenan, chopped into short pieces and gelled with an equal amount of egg white by boiling for 20 min. Incorporation of spun fiber into egg white made the latter more elastic. Fiber with carrageenan required only 70 to 80% of the compression force for 70% height reduction as shown by the model food without carrageenan. Texture profile analysis parameters were calculated for the model food meat sausages; the yeast protein egg white model had similar or superior properties to commercial foods.

TABLE 65
Water Binding of Yeast + Wheat Flour, and Yeast + Purina Samples

No. of sample	Sample	Water binding g H_2O/g sample
1.	90% baker's yeast + 10% Purina 500 E	9.86 ± 0.05
2.	95% baker's yeast + 5% Purina 500 E	9.84 ± 0.06
3.	98% baker's yeast + 2% Purina 500 E	9.69 ± 0.06
4.	90% baker's yeast + 10% wheat flour	8.67 ± 0.05
5.	95% baker's yeast + 5% wheat flour	8.52 ± 0.16
6.	98% baker's yeast + 2% wheat flour	8.50 ± 0.03
7.	90% *R. glutinis* + 10% Purina 500 E	16.47 ± 0.11
8.	95% *R. glutinis* + 5% Purina 500 E	15.95 ± 0.09
9.	98% *R. glutinis* + 2% Purina 500 E	15.68 ± 0.08
10.	90% *R. glutinis* + 10% wheat flour	10.44 ± 0.10
11.	95% *R. glutinis* + 5% wheat flour	14.44 ± 0.10
12.	98% *R. glutinis* + 2% wheat flour	16.54 ± 0.18
13.	90% water treated brewer's yeast + 10% Purina 500 E	12.64 ± 0.18
14.	95% water treated brewer's yeast + 5% Purina 500 E	12.25 ± 0.10
15.	98% water treated brewer's yeast + 2% Purina 500 E	10.82 ± 0.23
16.	90% water treated brewer's yeast + 10% wheat flour	11.58 ± 0.08
17.	95% water treated brewer's yeast + 5% wheat flour	11.76 ± 0.11
18.	98% water treated brewer's yeast + 2% wheat flour	11.95 ± 0.16
19.	90% basic treated brewer's yeast + 10% Purina 500 E	14.61 ± 0.16
20.	95% basic treated brewer's yeast + 5% Purina 500 E	13.61 ± 0.18
21.	98% basic treated brewer's yeast + 2% Purina 500 E	12.60 ± 0.16
22.	90% basic treated brewer's yeast + 10% wheat flour	12.12 ± 0.14
23.	95% basic treated brewer's yeast + 5% wheat flour	12.60 ± 0.13
24.	98% basic treated brewer's yeast + 2% wheat flour	13.42 ± 0.18

TABLE 66
Oil Binding of Yeast + Wheat Flour and Yeast + Purina Samples

No. of sample	Sample	Oil binding (%)
1.	90% baker's yeast + 10% Purina 500 E	219.08 ± 4.42
2.	95% baker's yeast + 5% Purina 500 E	219.59 ± 9.33
3.	98% baker's yeast + 2% Purina 500 E	221.66 ± 3.28
4.	90% baker's yeast + 10% wheat flour	222.94 ± 8.17
5.	95% baker's yeast + 5% wheat flour	208.10 ± 10.80
6.	98% baker's yeast + 2% wheat flour	214.69 ± 9.84
7.	90% *R. glutinis* + 10% Purina 500 E	265.85 ± 1.80
8.	95% *R. glutinis* + 5% Purina 500 E	269.87 ± 7.26
9.	98% *R. glutinis* + 2% Purina 500 E	281.68 ± 4.03
10.	90% *R. glutinis* + 10% wheat flour	264.06 ± 6.47
11.	85% *R. glutinis* + 5% wheat flour	267.95 ± 8.63
12.	98% *R. glutinis* + 2% wheat flour	283.10 ± 0.87
13.	90% water treated brewer's yeast + 10% Purina 500 E	211.75 ± 9.08
14.	95% water treated brewer's yeast + 5% Purina 500 E	217.06 ± 10.35
15.	98% water treated brewer's yeast + 2% Purina 500 E	220.83 ± 9.76
16.	90% water treated brewer's yeast + 10% wheat flour	215.44 ± 7.71
17.	95% water treated brewer's yeast + 5% wheat flour	221.93 ± 9.09
18.	98% water treated brewer's yeast + 2% wheat flour	212.17 ± 13.01
19.	90% basic treated brewer's yeast + 10% Purina 500 E	86.11 ± 8.17
20.	95% basic treated brewer's yeast + 5% Purina 500 E	78.58 ± 1.47
21.	98% basic treated brewer's yeast + 2% Purina 500 E	75.61 ± 1.88
22.	90% basic treated brewer's yeast + 10% wheat flour	73.68 ± 4.52
23.	95% basic treated brewer's yeast + 5% wheat flour	81.04 ± 6.37
24.	98% basic treated brewer's yeast + 2% wheat flour	79.90 ± 0.66

TABLE 67
Emulsion Activity Index and Emulsion Stability Index of Yeast + Wheat Flour and Yeast + Purina Samples

	EAI and EST $y = a.x^b$ (m = 10)				
No. of Sample[a]	r (P = 1%)	a	b	EAI m (m^2/g)	EST (min)
1.	0.9794	0.3852	−0.2904	0.2765 13.57	1.65
2.	0.9651	0.3530	−0.2800	0.2434 11.94	1.65
3.	0.9739	0.4259	−0.3027	0.3133 15.37	1.61
4.	0.9618	0.3608	−0.2861	0.2558 12.55	1.34
5.	0.9665	0.3794	−0.2756	0.2608 12.80	1.68
6.	0.9469	0.3857	−0.2763	0.2658 13.04	1.68
7.	0.8418	0.3754	−0.0563	0.0615 3.02	1.52
8.	0.9739	0.3866	−0.0784	0.0867 4.25	2.09
9.	0.9424	0.3492	−0.0542	0.0552 2.71	2.15
10.	0.9629	0.2967	−0.0501	0.0435 2.13	1.52
11.	0.6804	0.3100	−0.0329	0.0303 1.49	2.20
12.	0.7865	0.3625	−0.0523	0.0554 2.72	2.81
13.	0.9600	0.1632	−0.0535	0.0255 1.25	2.15
14.	0.9572	0.2090	−0.1297	0.0746 3.66	1.97
15.	0.8780	0.1690	−0.0648	0.0517 1.56	2.32
16.	0.9350	0.1751	−0.0884	0.0440 2.16	2.07
17.	0.8989	0.1866	−0.0725	0.0388 1.90	2.11
18.	0.9362	0.1975	−0.0577	0.0332 1.63	2.15
19.	0.8762	0.1216	−0.0436	0.0145 0.71	1.96
20.	0.9899	0.0921	−0.1293	0.0309 1.52	1.80
21.	0.9199	0.0620	−0.2141	0.0328 1.61	1.67
22.	0.8210	0.0777	−0.0538	0.0114 0.56	1.97
23.	0.9841	0.0873	−0.0403	0.0097 0.46	1.95
24.	0.8209	0.0799	−0.0466	0.0102 0.50	2.00

[a] See Tables 6 and 7.

REFERENCES

1. **Smith, J. L. and Palumbo, S. A.,** Microorganisms as food additives, *J. Food Prot.*, 44, 936, 1981.
2. **Steckley, J. D., Grieve, D. G., McLeod, G. K., and Moran, E. T., Jr.,** Brewer's yeast slurry. I. Composition as affected by length of storage, temperature, and chemical treatment, *J. Dairy Sci.*, 62, 941, 1979.
3. **Halász, A.,** Use of new protein source in bakery products. Central Research Institute of Food Industry, Report No. 11, Budapest, 1979.
4. **Graumlich, T. R. and Stevenson, K. E.,** Recovery of thermally injured *Saccharomyces cerevisiae*: effects of media and storage conditions, *J. Food Sci.*, 43, 1865, 1978.
5. **Torregiani, D. and Toledo, R. T.,** Influence of sugars on heat inactivation, injury and repair of *Saccharomyces cerevisiae, J. Food Sci.*, 51, 211, 1986.
6. **Halász, A., Baráth, Á., and Mátrai, B.,** Yeast as a human protein source, *Acta Aliment.*, 17, 374, 1988.
7. **Halász, A., Mietsch, F., Baráth, Á., and Kassim, M.,** Potential use of SCP in food and their functional properties, in *Functional Properties of Food Proteins*, Lásztity R. and Ember-Kárpáti, M., Eds., MÉTE Publishing, Budapest, 1988, 122.
8. **Peppler, H. J.,** Production of yeast and yeast products, in *Microbial Technology*, Vol. 1., Peppler, H. J. and Perlman, D., Eds., Academic Press, Orlando, FL, 1979, 157.
9. **Dziezak, J. D.,** Yeast and yeast derivatives: definitions, characteristics, and processing, *Food Technol.*, 41(2), 104, 1987.
10. **Halász, A.,** Biochemical Principles of the use of Yeast Biomass in Food Production (in Hungarian), D.Sc. thesis, Hungarian Academy of Sciences, Budapest, 1988.
11. FDA, Dried yeasts, Code of Federal Regulations, Title 21, 172.896. April ed., Food and Drug Administration, Washington, D.C., 1986.
12. **Rotschild, L., Jr., Ed.,** Dried yeasts, in *Food Chemical News Guide*, Food Chemical News, Washington, D.C., 1984, 176.4.
13. **Schmidt, G.,** Bubbling over with yeast, *Food Flavourings, Ingredients, Packaging and Processing*, 9(5), 25, 1987.
14. **McCormick, R. D.,** Improved flour and nutritional enhancement for pasta products, *Food Prod. Dev.*, 9(6), 11, 1975.
15. **Breene, M. D., Bonasik, O. J., and Walsh, D. E.,** Use of various protein sources in pasta, *Macaroni J.*, February 1977, p. 26.
16. **Yanez, E., Wulf, H., Ballester, D., Fernandez, N., Gattas, W., and Monckeberg, F.,** Nutritive value and baking properties of bread supplementation with *Candida utilis, J. Sci. Food Agric.*, 24, 519, 1972.
17. **Young, V. R. and Scrimshaw, N. S.,** Clinical studies on the nutritional value of single-cell proteins, in *Single Cell Protein II*, MIT Press, Cambridge, 1975.
18. **Eroshin, V. K., Rakhmankulova, R. G., and Beletskaya, N. M.,** Method of production of bread, USSR. Patent 563, 952, 1977.
19. **Kiovurinta, J.,** The use of pekilo biomass as a flour, meat substitute, *Kenia-Kemi*, 5, 632, 1978.
20. **Lindblom, M.,** Bread making properties of yeast protein concentrates, *Lebensm. Wiss. Technol.*, 10, 341, 1977.
21. **Ching-Ming Lin, J., Chastain, M. F., and Strength, D. R.,** Sensory and nutritional evaluation of wheat bread supplemented with single cell protein from Torula yeast *(Candida utilis), J. Food Sci.*, 51, 647, 1986.
22. **Volpe, T. A. and Zabik, M. E.,** Single-cell protein substitution in a bread system: rheological properties and product quality, *Cereal Chem.*, 58(5), 441, 1981.
23. **Tsen, C. C., Weber, J. L., and Eyestone, W.,** Evaluation of distiller's dried grain flour as a bread ingredient, *Cereal Chem.*, 60, 295, 1983.
24. **Stewart, C. F. and Gilliand, S. E.,** Utilizing yeast-whey proteins to improve the nutritional value of snack foods, American Dairy Review, Manufactured Milk Products Supplement, 30A, 1979.
25. **Bostian, M., Smith, N., Gilliard, S. E., and Stewart, C. F.,** Snack foods provide natural target for supplementation with yeast-whey protein, *Food Prod. Dev.*, 12(9), 68, 70, 1978.
26. **Lai, C. S., Davis, B., and Hoseney, R. C.,** The effect of yeast protein concentrate and some of its components on starch extrusion, *Cereal Chem.*, 62, 293, 1985.
27. Removing of unpleasant taste and aroma from proteinaceous materials, Japanese Examined Patent, 5618181, 1981.
28. Method for modifying the flavour of yeast, U.S. Patent 4.166.135, 1979.
29. **Kusochi, M.,** Use of lactic yeasts in confectionery and bread, *New Food Ind.*, 23(5), 24, 1981.
30. **de la Gueriviére, J. F.,** An example of biotechnology: production and improved utilization of lactic yeasts, *Ind. Aliment. Agricoles*, 98(5), 395, 1980.
31. **Marzolf, J.,** Whey yeast, *Med. Nutr.*, 17(3), 217, 1981.

32. **Smolyak, V. I., Sobakar, L. V., Malyuk, E. V., and Portnaya, E. S.,** Evaluation of bakery products enriched with yeasts containing carotene, Ratsionalnoe Pitanie, Resp. Mezhved. Sb., No. 15, 96, 1980.
33. **Phaff, H. J.,** Ecology of yeasts with actual and potential value in biotechnology, *Microb. Ecol.*, 12, 31, 1986.
34. **Dudikova, G. N., Shin, A. P., Vitavskaya, A. N., and Matveev, A. I.,** Activation of pressed yeasts, *Khlebopek. Konditer. Prom.*, No. 12, 28, 1979.
35. **Kang, K. H., Lee, K. H., Choe, Y. J., and Lee, J. Y.,** Studies of the making of yeast cheese with *Debaromyces hansenii* and *Saccharomyces fragilis*, *Korean J. Anim. Sci.*, 22(2), 101, 1980.
36. **Lee, T. S., Yang, K. J., Park, Y. J., and Yu, J. H.,** Studies on the brewing of Kochujang (red pepper paste) with the addition of mixed cultures of yeast strains, *Korean J. Food Sci. Technol.*, 12(4), 313, 1980.
37. **Kamel, B., Kramer, A., Sheppard, A. J., and Neukirk, D. R.,** Amino acid, fatty acid, cholesterol and other sterols analysis of different pizza formulations, *J. Food Qual.*, 2(2), 123, 1979.
38. **Otero, M. A. and Cabello, A.,** Increasing digestibility of yeast by chemical treatment, Cuba Azucar, April-June 13, 1979.
39. Dar-Nippon Seifo Co. Ltd., Protein beverage, Japanese Examined Patent 5.411 379, 1979.
40. **Skvortsova, R. I. and Vorontsova, N. L.,** Vitamin beverage, *Otkrytiya Izobret*, 3, 98, 1986.
41. **Junilla, M., Koivurinta, J., Kurkela, R., and Koivistoinen, P.,** Functional properties of brewer's grain, brewer's yeast and distillers stillage in food systems. II. Application to sausages and meat balls, *Fleischwirtschaft*, 61(7), 1024, 1049, 1981.
42. **Pinel, M.,** Functional properties of lactic yeasts: application in meat products, *Viandes Prod. Carnes*, 2(6), 27, 1981.
43. **Schackelford, J. R. and Murray, D. G.,** Studies show no shelf-life loss from usual food yeasts in products, *Food Prod. Dev.*, 14(9), 40, 1980.
44. **Anon.,** Storage conditions, nature of protein source affect shelf-life of extended meat products, *Food Prod. Dev.*, 14(3), 38, 1980.
45. **Firmenich, S. A.,** Process for the aromatization of protein materials, British Patent 1579 297a, 1980.
46. **Butrum, R. R.,** Effect of storage on the protein quality of meat analogs and their mixtures by rat and *Tetrahymena pyriformis* growth and amino acid analysis, Dissertation Abstr. Int., B 39 (2) 641, Order No. 78-11946, 1978, 112.
47. **Noda, Y., Inoue, H., Kusuda, H., Oba, K., and Nakano, M.,** Control of fermentation in soy sauce mash (moromi). V. Effectiveness of adding yeasts and lactic acid bacteria on an individual scale, *J. Jpn. Soy Sauce Res. Inst.*, 8(3), 108, 1982.
48. **Ridgeway, J. A.,** Treatment of proteinaceous materials with anhydrous ammonia gas, U.S. Patent 4291063, 1981.
49. **Cooney, C. L., Chokyun Rha, and Tannenbaum, S. R.,** Single cell protein, economics, and utilization in foods, *Adv. Food Res.*, 26, 1, 1980.
50. **Tannenbaum, S. R.,** Texturizing process for single-cell protein, U.S. Patent 3, 845.222, 1974.
51. **Tannenbaum, S. R.,** Texturizing process for single-cell protein containing protein mixtures, U.S. Patent 3, 925.562, 1975.
52. **Tsuboi, A. and Takeda, R.,** Effect of *Torulopsis* on fermented flavour of moromi Ann. Rep. Kagawa Prefectural Fermentation Food Exp. Station, No. 75, 11, 1982.
53. **Kann, A. G., Kask, K. A., Annusver, K. Kh., and Rand, T. I.,** Possible applications for brewing by-products, *Tallinna Polütech. Inst. Toim.*, No. 537, 21, 1982.
54. **Koivurinta, J., Junnila, M., and Koivistoinen, P.,** Functional properties of brewer's yeast and distiller's stillage in food systems, *Lebensm. Wiss. Technol.*, 13, 118, 1980.
55. **Alian, A., El-Akher, M. A., Abdon, I., and Aid, N.,** Enrichment of local bread with dried brewery yeast. I. Chemical and biological evaluation of dried brewery yeast and flour, *Egypt. J. Food Sci.*, 11 (1/2) 23, 1983.
56. **Alian, A., El-Akher, M. A., Abdou, J., and Aid, N.,** Enrichment of local bread with dried brewery yeast. II. Physical properties of dough and biological and sensory evaluation of bread, *Egypt. J. Food Sci.*, 11, (1/2), 31, 1983.
57. **Asenova, E. K., Alenova, D. Zh., and Nusupkulova, A. N.,** New technological procedure for bakery product manufacture, *Khlebopek. Konditer. Prom.*, No. 8, 29, 1984.
58. **Anon.,** Dried yeast protein enhances flavour and nutritional value, *Confect. Manuf. Marketing*, 23(10), 18, 1986.
59. **Higashi, K., Yamamoto, Y., and Yoshi, H.,** Study of salt tolerance of miso and soy-sauce yeast. I. Effect of sodium chloride on cell wall content and chemical composition of the cell walls, *J. Jpn. Soc. Food Sci. Technol.*, 31(4), 225, 1984.
60. **Anon.,** Healthy cocoa alternative made from yeast and carob, *Food Eng.*, 56(3), 105, 1984.
61. **Akin, C. and Ridgeway, J. A.,** Heat treatment for live yeast pastes, U.S. Patent U.S. 4472439, 1984.
62. **Petana, P., Fatichenti, F., Farris, G. A., Mocrout, G., Lodi, R., Todesco, R., and Cecchi, L.,** Metabolisation of lactic — and acetic acids in Pecorino Romano cheese made with a combined starter of lactic acid bacteria and yeast, *Lait*, 64, 380, 1984.

63. **Hagrass, H. E. A., Sultan, N. E., and Hammad, Y. A.,** Chemical properties of Rads cheese during ripening as affected by the addition of inactive dry yeast and yeast autolysate, *Egypt. J. Dairy Sci.*, 12(1), 55, 1984.
64. **Oluski, V., Popovic, M., Oluski, T., and Pribis, V.,** Study of the possibility of adding alcohol fermentation-derived yeast to luncheon meat type products, *Technol. Mesa,* 22(3), 71, 1981.
65. **Modic, P., Polic, M., and Trumic, Z.,** Nutritional and technological properties of a protein preparation made from brewer's yeast, *Technol. Mesa,* 19 (7/8), 194, 1978.
66. **Stamenkovic, T., Perunovic, M., Hromis, A., and Maric, L. J.,** Effect of the inactivated yeast, soy meal and wheat gluten on the production and thermostability of systems: meat-fat-water, *Technol. Mesa,* 24(3), 70, 1982.
67. **Halász, A., Mietsch, F., and Pfeiffer-Szalma, I.,** Use of brewer's yeast in food industry, Research Report of Central Research Institute of Food Industry, No. 11, 1987.
68. **Halász, A., Mietsch, F., Baráth, Á., and Kassim, M.,** Potential use of SCP in food and their functional properties, in *Functional Properties of Food Proteins,* Lásztity, R. and Ember-Kárpáti, M., Eds., MÉTE Publishing Budapest, 1988, 122.
69. **Kawasaki, S.,** Texture evaluation of spun yeast protein fibres, *J. Jpn. Soc. Food Sci. Technol.*, 31(2), 72, 1984.

Chapter 8

USE OF YEAST PROTEIN PREPARATIONS IN FOOD INDUSTRY

As mentioned in the previous chapter the yeast proteins can be used in different forms in the food industry. The simplest possibility is the direct use of inactive whole yeast cells. Due to the high nucleic acid- and cell wall content, the use of whole yeast cells is limited, therefore an extracted and purified protein preparation (isolate or concentrate) has advantages as a food ingredient. Schlingmann et al.[1] in a review paper differentiates four types of microbial proteins (isolates):

1. Bioprotein Type I — insoluble denatured protein isolate with high-water, fat-binding and retention capacity
2. Bioprotein Type II — partially hydrolyzed protein isolate with improved solubility, foaming and emulsifying properties
3. Bioprotein Type III — water soluble at neutral and acid pH; low molecular weight protein which can be easily shipped, emulsification properties lead to accelerated resorption in intestine
4. Bioprotein Type IV — partially hydrolyzed protein isolate, fully suspendible in aqueous solution: emulsifying and viscosity improving properties

The classification mentioned above can be extended with protein preparations containing chemically modified (phosphorylated, acetylated, succinylated, etc.) proteins having improved functional properties and slightly changed nutritional value due to the different digestibility (availability) of chemically modified amino acid residues.

In the framework of this chapter we will first give a short review of general problems connected with functional properties of proteins including main methods of their measurement. The second part of the chapter will discuss the practical production of concentrates and isolates, their properties and applications in different branches of food industry.

I. FUNCTIONAL PROPERTIES OF YEAST PROTEINS

A. GENERAL ASPECTS

Successful utilization of SCP as human food ultimately will depend on developing palatable products from this ingredient. Until the day when human protein needs exceed supplies, men probably will not eat unpalatable proteins. Even today in areas of protein deficiency, some malnourished people will not consume protein-rich foods foreign to their tastes.[2]

The steps in producing SCP for human food start with growing microorganisms in a fermentation process. The technology of fermentation has been intensively developed in the last decades and techniques for yeast biomass production economically on available substrates are currently under intense study.[3] However, the area of post fermentation treatment holds the key to direct human consumption of microbial protein. A detailed review of earlier research work (until 1976) connected with functional properties of proteins in food was given by Kinsella.[4] Special aspects of yeast protein functionality are shortly reviewed by Guzman-Juarez.[5]

If we are speaking about the functional properties of proteins the biochemist is always thinking about the biological function of proteins on the function-structure relations. In the eyes of the food scientist protein is the most important nutrient but it is also a compound influencing from one side the physical, organoleptic properties of food products and from the other side the methods of processing the whole technological process. So in food pro-

TABLE 68
Typical Technofunctional Properties of Protein Preparations

No.	Property	Mode of action	Food system
1.	Hydration, water absorption, water binding	Water binding via hydrogen bonds	Meat, meat products, bakery
2.	Solubility	Solubilization of proteins	Beverages
3.	Viscosity	Thickening, water binding (immobilization)	Soups, sauces, dressings
4.	Gel forming	Formation of protein matrix	Meats, curd, cheese, bakery products
5.	Emulsifying	Formation and stabilization of emulsions	Cream, dressings, meat products, soups, cakes, baby foods
6.	Foaming	Formation of stable films, immobilization of gases	Foam cakes
7.	Cohesivity, adhesivity	Protein as adhesive material	Meats, meat products, bakery products, etc.
8.	Elasticity	Hydrophobic binding, S-S-bonds in gels	Meats, bakery products
9.	Fat absorption	Binding of free fat	Meat products, cakes
10.	Flavor absorption	Adsorption, absorption	All foods
11.	Phase changes during heating	e.g., denaturation	Meat products

duction the words functional properties of protein are strongly connected with the operations included in the food processing and also with the physical (eventually other) properties of the end product.

There are two approaches in the definition of the term functional properties. The more general definition given by El Pour[6] is the following: ''any property of food or food ingredients except its nutritional ones which affects its utilization''. The second approach is connected with the growing use of plant resp. animal protein preparations in fabrication of food products. A great variety of protein preparations and the properties they possess necessitate the determination of precise technological procedures that are to be applied in the particular products in order to utilize their potential functionality. This is an indispensable condition for obtaining the final products with appropriate desirable properties to be accepted by consumers. Such an approach requires deep knowledge of properties of the particular protein preparations and their impact on the properties and quality of the product to which they are incorporated. All these properties of protein preparation are called functional properties. In other words, functional properties are those properties of the protein preparation which affect the behavior of the product to which it is added during processing, storage and consumption. After this short review of definitions we continue with the most important functional properties and finally with the discussion of the methods of their measurement.

B. FUNCTIONAL PROPERTIES OF PROTEIN PREPARATIONS

In Tables 68 and 69 two propositions are shown concerning classification of functional properties of food protein preparations. It is clear from the two tables that properties of the protein preparations which affect their functionality and thereby direction of their utilization comprise first the physicochemical properties and second, sensory properties. In the framework of this chapter physicochemical properties will mainly be treated while the dominant attribute that affects functionality of proteins in food systems seems to be the intrinsic physicochemical properties of proteins. Physicochemical basis of protein functionality originates first from the properties of the proteins themselves, and is also affected by other components (carbohydrates, lipids, etc.) as well as by conditions of the processing (temperature, pH, ionic strength, etc.) and others. The physical behavior of protein is determined by its amino acid composition and amino acid sequence coded in its genes, its molecular

TABLE 69
General Classes of Technofunctional Properties of Proteins Important for Food Applications

General property	Specific functional term
Organoleptic	Color, flavor, odor, texture mouth-feel
Binding (hydration)	Solubility, dispersibility, water absorption, water-holding capacity, swelling, viscosity, gelation, dough formation, fat absorption
Texturizability	Fiber spinning, extrudability, plasticity, etc.
Surface properties	Emulsifying properties (emulsifying activity and emulsion stability, emulsifying capacity) foaming properties (aeration, whipping, foam expansion and foam stability, protein-lipid film formation)

size, the conformation of the protein molecule, the charge of the protein and its distribution, the quality, quantity and distribution of the secondary intra- and intermolecular bonds including quaternary structure. Conformation of protein appears to be critical in governing their functionality in food systems.

As it is known there are two basic types of proteins from the point of view of conformation: (1) globular proteins and (2) asymmetric fibrillar (lamellar) proteins. The molecular weight of globular proteins is low, the structure is compact with a relatively high amount of helical structure. More polar groups are oriented toward the surface. This favors solubility and hydration. The unfolded elongated conformation is desirable for gel forming, foam stability, formation of fiber-like structures, etc. After this general overview we continue with the specific functional terms.

1. Hydration, Water Absorption, Water Binding

Most foods are hydrated solid systems, and the physicochemical, rheological behavior of proteins and the other constituents of the food is strongly influenced not only by the presence of water but also by the activity of water. Furthermore, dry protein concentrates or isolates must be hydrated when used. For these reasons, the hydration or rehydration properties of food proteins are of great practical interest.

Protein preparations used in food production always contain numerous polar side chains. The dry protein depending on the relative humidity of air interacts with water molecules and the polar groups will be hydrated with a monomolecular layer. At higher relative humidity values multilayer adsorption may occur and depending on the colloidal structure of protein preparation, liquid water may condensate in the capillaries (see Figure 73). Immersing the protein preparation in liquid water phase additional water absorption occurs due to the swelling (osmotic water absorption). This type of water absorption could be much higher than the first type of hydration. Depending on the structure of protein molecules and their colloidal properties the swelling could be limited and a cohesive swollen mass will be formed, or unlimited and we obtain a protein solution (Figure 73). Which of two types of hydration process is needed that depends on the properties of end product? It should be mentioned that the ability to absorb water by the given preparation depends not only on the origin and concentration of protein ingredient but also, to a large extent, on the content and nature of other nonprotein compounds, methods of processing, including pH, heat treatment and others.

A lot of environmental factors such as protein concentration, pH, temperature, time, ionic strength and presence of other components affect the forces underlying protein-protein and protein-water interactions. Most functional properties are determined by the interrelations among these forces. Total water absorption increases with increasing protein concentration. Changes of pH alter the ability of proteins to associate with water. This fact is associated with changes of ionization and as a consequence with changes of the magnitude of the net

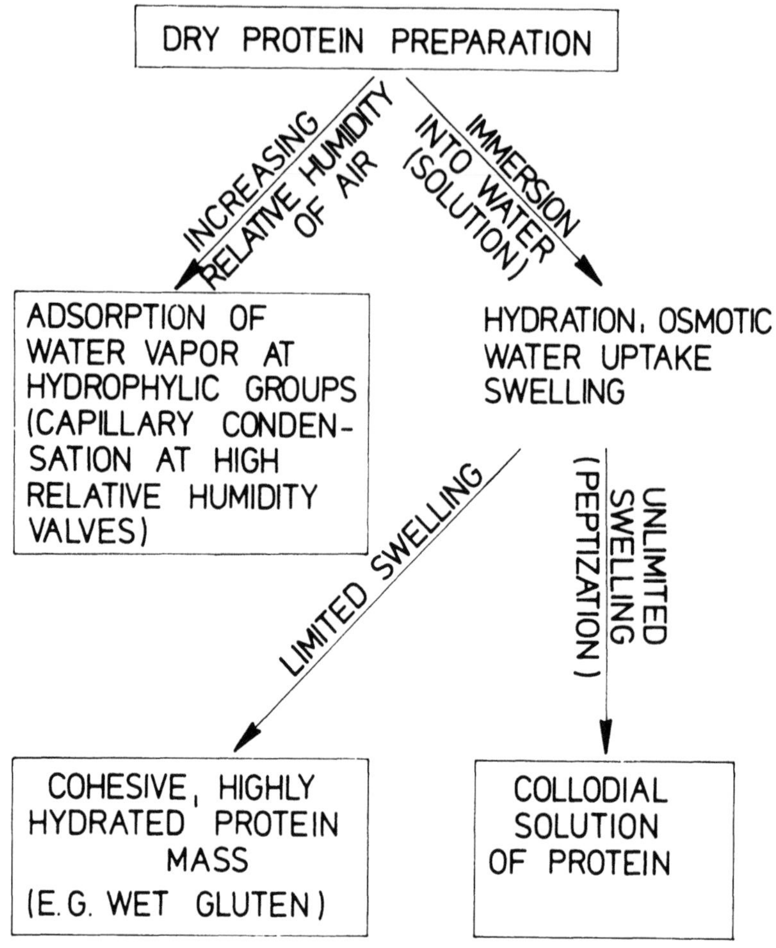

FIGURE 73. Possible ways of hydration of proteins.

charge on the protein molecule altering the attractive and repulsive interactive forces of the molecules. At the isoelectric point the protein-protein interactions are maximal and the associated shrunken proteins exhibit minimal hydration and swelling.

Water binding by proteins generally decreases as the temperature is raised because of decreased hydrogen bonding. Denaturation and aggregation occur during heating and the latter may reduce protein surface area and the availability of polar groups for water binding. However, heating may be advantageous during hydration and swelling of dry protein preparations causing dissociation and unfolding of the compact molecules with resulting improvement in water binding.

The quality and quantity of ions present in the system has a significant effect on the water absorption, swelling and solubility of proteins. At low salt concentrations generally protein hydration increases. At higher ionic strength water-salt interactions predominate over water-protein interactions and the level of hydration of the protein molecule may decrease.

2. Solubility

From a practical standpoint, data on solubility characteristics are very useful for determining optimum condition for the extraction and purification of proteins from natural sources and for the separation of protein fractions. Solubility behavior under various conditions also provides a good index of the potential application of proteins. It is assumed that the solubility of proteins is one of the properties which affects their functionality and application. This results from the fact that high solubility of protein preparation is usually linked to good emulsifying and foaming properties. Good solubility is also needed in dried instant beverage products. In another group of products solubility is not so important in the evaluation of usefulness of the protein preparation. In meat processing favorable technological effects are also produced by the use of protein preparation with low solubility. The solubility of proteins also depends on different factors such as pH, temperature, presence of other compounds, etc.

The main advantage of initial solubility is that it permits rapid and extensive dispersion of protein molecules or particles. This leads to a finely dispersed colloidal system with homogenous macroscopic structure and a smooth texture. Also, initial solubility facilitates protein diffusion to air/water and oil/water interfaces, thus improving surface activity.

3. Viscosity

The viscosity and consistency of protein-containing systems are important functional properties in fluid foods such as beverages, soups, sauces and creams. Knowledge of flow properties of protein dispersions is also of practical significance in optimizing operations, such as pumping, mixing, heating, cooling and spray drying, that involve mass and/or heat transfers. Viscosity of the system plays a significant role in stabilization of some types of emulsions and foams too.

The effect of protein preparation on the viscosity depends on a number of factors. One of the most important factors is the structure (conformation) of the proteins. The viscosity of the solutions of globular proteins increases slowly with the increase of concentration in accordance with Einstein's equation

$$\eta = \eta_0(1 + 2.5\phi)$$

where: η = viscosity of the solution; η_o = viscosity of solvent and Φ = partial volume of solved protein. In the case of fibrillar protein solutions the increase of viscosity is rapid especially if the degree of polimerization is high

$$\eta = P\eta_0^n$$

where: η = viscosity of solution; η_o = viscosity of the solvent; P = degree of polymerization and n = constant specific to protein.

One of the main single factors influencing the viscosity behavior of protein fluids is the apparent diameter of the dispersed molecules or particles. This diameter depends first on the molar mass, size, volume, structure and asymmetry and electric charge of the protein molecule. Environmental factors such as pH, ionic strength and temperature can influence these characteristics through unfolding. Finally, protein-solvent and protein-protein interactions influence the hydrodynamic hydration sphere surrounding the molecule and determine the size of aggregates. The protein solutions often have a nonNewtonian character (or pseudoplastic character) showing a decrease of apparent viscosity with increased shear stress.

4. Gel Forming

Gelation is a very important functional property of several proteins. It plays a major role in the preparation of many foods including various dairy products (e.g., cheese, yogurt, etc.), coagulated egg white, gelatin gels, various heated comminuted meat or fish products, soybean protein gels, texturized proteins, doughs, etc. Protein gelation is utilized not only for the formation of solid viscoelastic gels, but also for improved water absorption, thickening, particle binding (adhesion), and emulsion foam-stabilizing effects.

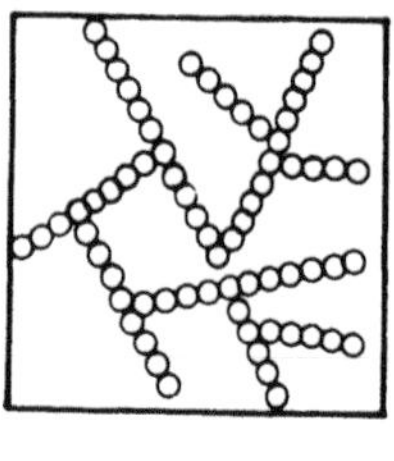
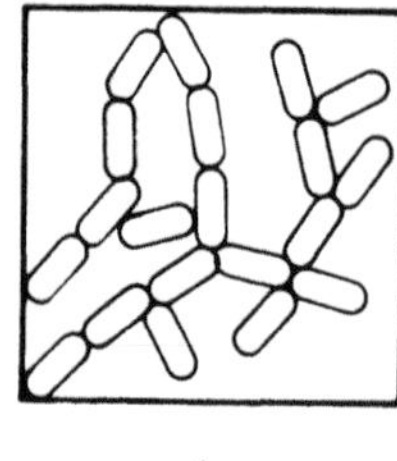
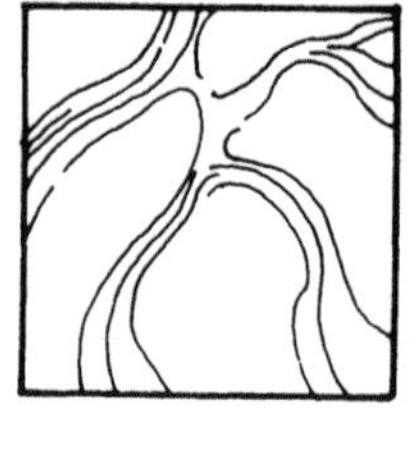
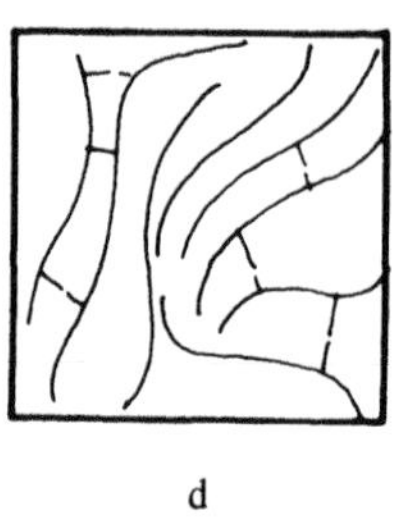

a b c d

FIGURE 74. Different types of gel formation processes; a — globular particles; b — anisodimensional particles; c — fibrillar macromolecules and d — cross-linked linear molecules.

As it is known, the gels are colloidal systems which contain minimally two phases, have an ordered network of solid phase, have immobilized liquid phase and have quasi-solid mechanical characteristics. There are two typical ways of formation of gels: (1) from solution by interactions of protein molecules resulting in a three-dimensional network and (2) from dry protein mass (xerogel) by hydration and swelling. Although it is possible to form gels from concentrated solutions of globular proteins (see Figure 74), the unfolded linear protein molecules form stronger gels and at a lower protein concentration, which type of proteins will be used depends on the product to be fabricated. Cross-linking of the linear molecules contributes to the formation of mechanically stronger gels.

Practical conditions for the gelation of various proteins are fairly well established, although optimization with respect to environmental factors, pretreatments of the protein, use of protein mixtures, etc. is not easily attained. In several cases a thermal treatment is a requisite of gelation. Subsequent cooling may be necessary. Gelation may be affected by addition of salts. Other proteins, however, may gel without heating.

The mechanism and interactions underlying the formation of the three-dimensional protein networks characteristic of gels are not fully understood. As mentioned earlier globular proteins can form gels only at higher concentrations and these gels have a very low strength. So it is understandable that practically all studies point to the necessity of denaturation of globular proteins and unfolding prior to the step of ordered protein-protein interaction and aggregation. This explains, for instance, why some proteins denatured previously form gel without heat treatment. In some cases a reduction of disulfude bonds may result in unfolding of globular molecules. The formation of the protein network is considered to result from a balance between protein-protein and protein-water interaction and between attractive and repulsive forces of polypeptide chains. Among the attractive forces the disulfide bonds, the ionic (electrostatic) interactions, the hydrogen bonds and the hydrophobic interactions can be mentioned. Their relative contribution may vary with the nature of the protein, the environmental conditions, and the various steps in the gelation process. Electrostatic repulsions (especially at pH values far from isoelectric point) and protein-water interactions tend to keep polypeptide chains apart.

Dissociation and/or unfolding of protein molecules generally increases the exposure of reactive groups, especially the hydrophobic groups of globular proteins. Protein-protein interactions are therefore favored and are usually the main cause of subsequent aggregation. Proteins with a high molar mass and a high percentage of hydrophobic amino acids, therefore, tend to establish strong networks. Hydrophobic interactions are enhanced at high temperatures; formation of hydrogen bonds is favored during cooling. Heating may also unmask internal thiol groups and promote formation or interchange of disulfide bonds. The presence of a large number of thiol and disulfide groups strengthens the intermolecular network and tends to make the gel thermally irreversible. Nakai et al.[8] investigating eight food proteins in connection with thickening, heat coagulation and gelation found a correlation between the heat coagulation, number of thiol groups and surface hydrophobicity. The proposed equation is the following: HC(heat coagulation) $=$ constant -2.495 SH $+$ 0.009Se where

Se = exposable hydrophobicity. Bikbov et al.[9] also found that the geling ability of individual globulins and their monomeric forms have been found to correlate with the effective mean hydrophobicity according to Tanford and Bigelow.

Some kinds of proteins can form gels when heated together (cogelation) if they are thermodynamically compatible.[10] Investigation of blends of proteins in aqueous medium showed that such a blend may be homogenous or heterogeneous depending on the production conditions and its composition. Since the phase state of the blend determines its properties and choice of the processing method as well as the protein products to be obtained, a study on the compatibility of proteins and a search for methods to control the phase state of protein mixtures are of great importance. Polyakov et al.[10] investigating seventeen systems containing water and two different proteins (water-protein A-protein B system) found for sixteen systems conditions, under which they separate in the two phases with a predominant concentration of each of the proteins in one of the phases. Thus, the limited thermodynamic compatibility of proteins can be claimed to feature a common phenomenon, when a complete miscibility is rather infrequent. Proteins can also form gels through interactions with polysaccharide geling agents.[11,12] Nonspecific ionic interactions between positively charged gelatin and negatively charged alginates and pectates give gels with high melting points. Generally, any process which intends to give rise to a solid fabricated foodstuff in principle can be broken down into the following three major stages (of course they may be combined in the practice):

1. The production of a multicomponent liquid system (it can be either a true solution or a colloidal solution or a coarse disperse system)
2. The shaping of the liquid system
3. The converting of the liquid system into the gel state which provides the retention of the product shape

5. Emulsifying Properties

Many food products are emulsions (milk, cream, ice cream, butter, processed cheese, mayonnaise, comminuted meats containing larger amount of fats, etc.), and proteins often play a major role in stabilizing these colloidal systems.

An emulsion is a dispersed system consisting of two immiscible or sparingly soluble liquids generally termed (in food industry) the oil phase and the water phase. The prerequisite of a stable emulsion is always the presence of a third component separating the two phase by formation of an interface layer at the surface of two phases. The third component is generally called emulsifier. The stability of the emulsion droplets greatly depends on the properties of the emulsifier used.[13-15] From these properties the surface activity, the mechanical strength of the surface layer of emulsifier and the viscosity of the continuous liquid phase are the most important.[16] Proteins as amphiphatic molecules decrease the surface tension at the interface (which is advantageous from the point of view of formation of emulsion) and by interacting with each other form network structures of films that stabilize emulsion droplets. Generally, the increasing effect on viscosity of continuous phase is also a stabilizing factor.[16-20] Recently much attention has been paid to the surface hydrophobicity of proteins. It has long been recognized that many properties of proteins can be related to the proportion of the amino acids that have nonpolar side chains. The correlation between surface hydrophobicity, foaming and emulsification was recently reviewed by Mitchell.[8] Kato and Nakai[21] reported good correlation between the surface hydrophobicity of 28 different protein samples and their emulsifying activity. Voutsinas et al.[22] also found that emulsifying properties of proteins could be well predicted solely on the basis of surface hydrophobicity but not on the basis of solubility level. However, the emulsion stability and fat-binding capacity of proteins studied could be explained and more accurately predicted using surface hydrophobicity and solubility together.

Unlike synthetic surfactants and some proteins of animal and plant origin that are intensively studied in terms of emulsion stability and interfacial activity,[20,23-28] detailed information about the emulsifying activity of yeast proteins under controlled conditions is limited. There is a need to study the relative emulsifying activities of different microbial proteins under standard conditions by altering variables, e.g., protein concentration, oil: water ratio, temperature, pH, ionic strength and energy input per unit volume. In investigating the formation and stabilization of emulsions by proteins, first the surface properties of proteins must be studied. The soluble proteins are able to diffuse toward, and adsorb at the oil-water interface. It is generally assumed that once a portion of a protein has contacted the interface, the nonpolar amino acid residues orient toward the nonaqueous phase, the free energy of the system decreases, and the remainder of the protein adsorbs spontaneously. During adsorption most proteins unfold extensively and if a large surface is available, spread to a monomolecular layer. Globular proteins with a stable structure are poor emulsifying agents unless they can be unfolded by prior treatment without loss of solubility (e.g., reduction of disulfide bonds).[29] Adsorbed protein films that are thick, highly hydrated, and electrically charged probably result in optimal emulsion stability due to the mechanical stability of the film. The stability of the film depends on the binding forces between the molecules in the film. In light of this, it would seem reasonable that a mixture of two proteins with opposite net charges might have superior emulsifying properties to either protein alone.

6. Foaming Properties

Food foams are usually dispersions of gas bubbles in a continuous liquid or semi-solid phase that contains a soluble surfactant. Many food products are foams from a colloid chemical point of view (e.g., whipped cream, whipped toppings, souffles, some of confectionery products, etc.). The gas phase of the foams is generally air (in some cases carbon dioxide, nitrogen oxide) and the liquid phase an aqueous solution (or suspension) of proteins. Similarly to the emulsions the prerequisite of the formation of stable foams is an adsorption of proteins at the air/water interface and formation of a mechanically stable film. For such purposes soluble proteins are needed which are able to interact one with other. A good correlation was found between the hydrophobicity of proteins and foaming properties.[8,20] Although many studies stress the importance of high protein solubility as a prerequisite to good foaming properties, it also appears that insoluble protein particles can play a beneficial role in stabilizing foams, probably by increasing viscosity of liquid phase. Although foam overrun is generally not great at the protein isoelectric point, foam stability is often quite good.

Salts can also affect the solubility, viscosity, unfolding and aggregation of proteins, and this can alter foaming properties. Sodium chloride often reduces foam stability, and calcium ions may improve stability by forming bridges between carboxyl groups of the protein. Sucrose and other sugars often depress foam volume but improve foam stability. This effect is probably associated with the increase of viscosity in liquid phase.

The presence of contaminating lipids and especially low molecular weight surface active agents seriously impair the foaming performance of proteins. Moderate heat treatment prior to foam formation are found to improve foaming properties of several proteins. These heat treatments cause overrun to increase, but decrease stability. Because many foams have very large interfacial areas, they are often unstable. There are three main destabilizing mechanisms: (1) drainage of lamella liquid due to gravity, pressure differences, and/or evaporation; drainage is lessened if the bulk liquid phase is viscous and the surface viscosity of the adsorbed protein film is high, (2) gas diffusion from small to large bubbles; such disproportionation results from solubility of the gas in the aqueous phase, (3) rupture of the liquid lemellae separating gas bubbles; such ruptures result in an increase in bubble size through coalescence, and ultimately lead to a collapse of the foam.

7. Cohesivity, Elasticity

The ability of proteins to interact with other components of the food system in such a way that it will be able to hold together all the ingredients in solid or semi-solid form is an important property. This property plays an important role, e.g., in the formulation of many foods such as sausages, cheeses, meat analogs, etc. The salt-soluble proteins of meats are important in binding the meat components together, and also in the structure and texture of processed meats. The adhesive or binding capacity of these salt-soluble proteins is exemplified by the ability to firmly bind chunks and pieces of ground meat and deboned homogenates together to form meat rolls and loaves suitable for slicing, following heating under pressure.

The binding conditions are also important in extended meat products containing nonmeat proteins and/or other nonmeat proteins and/or other nonmeat protein components. The added protein preparation does not impair (or only slightly decrease) the binding properties of meat proteins. The polysaccharides included in the formula of fabricated foods must have such groups which can interact with proteins to form bonds, strengthening the cohesivity of the system.

The elasticity of protein systems was extensively studied in wheat gluten and wheat flour dough. The important role of high molecular weight storage proteins (glutenins), disulfide bonds and thiol-disulfide interactions was demonstrated.[34] Because of the potential of fortifying breads with proteins some practical research was done on measuring the effects of added proteins on dough formation and dough quality, and determining methods to minimize some resultant deleterious effects. The elucidation of the changes in elasticity at a molecular level needs further research.

8. Fat Absorption

The ability of protein preparations to bind and retain fats is an important property especially for such applications as meat replacers and extenders, principally because it enhances flavor retention and reputedly improves mouth feel. Although mechanism of fat absorption has not been fully explained, absorption of fat is attributed mainly to the physical entrapment of oil. In support of this a correlation coefficient 0.95 between bulk density and fat absorption by alfalfa leaf protein was found. Chemical modification of protein which increases bulk density concomitantly enhances fat absorption.[4]

9. Flavor Binding

The flavor of a food results from a very low concentration of volatile compounds near its surface, and these concentrations are dependent on a partition equilibrium between the bulk of the food and its ''headspace''. Adsorption at the surface of food or penetration into the food interior by diffusion are the characteristic processes of flavor binding. The binding can be reversible (physical adsorption) or irreversible chemical adsorption via covalent or electrostatic linkages. As it was observed in model experiments, addition of proteins lowered the concentration of volatile compounds in the ''headspace.''[30] Concerning the mechanism of binding of volatile compounds it can be stated that polar compounds, such as alcohols, are bound via hydrogen linkages. In the binding of low-molecular-weight volatile compounds the hydrophobic interactions with nonpolar amino acid residues predominate. In some cases covalent linkage is also possible. For example, binding of aldehydes or ketones to amino groups and binding of amines to carboxyl groups are usually irreverisble, although reversible Schiff-basis may form between carbonyl volatiles and α- or ϵ-amino groups of proteins and amino acids. Irreversible fixation is more likely to occur with volatiles of higher molecular weight.[31,32] Binding of volatiles by proteins is possible only when binding sites are available. The number of binding sites depend on the amino acid composition and conformation of the proteins. Protein-protein interactions in the system decrease the flavor binding. Any factor-modifying protein conformation influences the binding of volatile compounds. Water enhances the binding of polar volatiles but barely affects that of apolar compounds. Increasing the protein concentration decreases the binding of volatiles due to the higher degree of protein-protein interactions. In the solutions the availability of polar or apolar amino acid residues for the binding of volatiles is influenced by a number of factors; the pH of solution and presence of salts may influence the protein conformation. Reagents that tend to dissociate proteins or to open up disulfide bonds usually improve the binding of volatiles. Protein denaturation by heat generally results in increased binding of volatiles. Voilley et al.[33] based on experiments made with some volatile compounds (acetone, ethylacetate, diacetyl, 2-propanol and *n*-hexanol) concluded that three factors are the most important in aroma retention:

1. Water content — the retention of volatiles depends largely on the initial water concentration.
2. Characteristics of the volatile — the retention is related to volatility and diffusivity of aroma compounds.
3. Characteristics of the substrate — the chemical reactions between protein and carbohydrate on ϵ-amino groups limit the aroma retention.

10. Texturization, Phase Changes During Heating

Proteins constitute the basis of structure and texture in several foods, whether these come from living tissues or from fabricated substances (cheese, sausage emulsions, doughs, etc.). The ability of proteins to form ordered macrostructures (fibrils, films, sheets expanded three-dimensional networks, etc.) is of particular importance in fabrication; a lot of new protein foods including simulated or "synthetic" food products. The special processes resulting in an ordered structure of "amorphous" protein mass or a mixture containing proteins is called *texturization*. The most known processes are thermal coagulation and film formation, fiber formation (meat analogs), thermoplastic extrusion and freeze texturization.

The first three methods are widely known so we will now discuss only the freeze texturization. In recent years there has been increased practical interest in using freezing as a method of creating an aligned fibrous structure in an amorphous protein slurry. The current aim of processes inducing fiber formation by freezing is to fabricate a product which simulates some of the textural characteristics of meat, especially the fibrous nature of its proteins myosin and actin which are arranged in a parallel direction. The general freeze texturization process entails freezing an aqueous protein solution in such a way as to control and align the ice crystal growth. Until a eutectic point is reached only the water freezes, forming distinct and separate ice crystals. Cooling only one surface of the protein system will generally cause the crystals to grow aligned and perpendicular to that surface. As the ice crystals grow from the surface as "spears" into the slurry they force the proteinaceous material out of the space occupied by the advancing crystals and draw water molecules out of the slurry to bind onto the ice crystal surface. This action centrates the proteinaceous material in the interstitial spaces between the ice crystals and the branches of each crystal. Upon the melting point of the ice, the proteinaceous material remains compacted in the former interstitial spaces, and long, thin, parallel wids are formed where the crystals existed. In extracted meat or poultry proteins the freezing and frozen storage will partially set the fibrous structure. In most cases the fiber set must be stabilized by heat. The characteristics of the fibrous protein structure may be influenced by rate of cooling resp. by the ratio of the rate of nucleation and rate of crystal growth.

Heat treatment is one of the most commonly used procedures in food processing. As a consequence of heat treatment denaturation of proteins may occur. As it is known protein denaturation is any modification in conformation (secondary, tertiary or quaternary) not accompanied by the rupture of peptide bonds involved in primary structure. Denaturation is an elaborate phenomenon during which new conformations appear, although often intermediary and short-living. The ultimate step in denaturation might be a totally unfolded polypeptide structure. The effects of denaturation are numerous such as decreased solubility, altered water-binding capacity, increased susceptibility to attack by proteases, etc. It is understandable that such changes influence the functional properties of proteins and must be kept in mind by evaluation of these proteins from the point of view of their functionality in food systems.

The susceptibility of proteins to denaturation by heat depends on many factors such as the nature of protein, protein concentration, water activity, pH, ionic strength and the kind of ions present. Although extensive dehydration of proteins even by mild methods such as freeze drying, may cause some protein denaturation, it is well known that proteins are more resistant to heat denaturation when dry than when wet. Thus, the presence of water facilitates denaturation.

C. METHODS OF THE EVALUATION OF FUNCTIONAL PROPERTIES

Generally, methods of evaluation of functional properties of protein preparations may be divided into the following three groups: (1) measurement of general functional properties (ingredient test); (2) examination on model systems (model test) and (3) pilot processing of the product required (utility test). The first type tests (measuring the water absorption, fat absorption, viscosity, emulsifying activity, etc.) are the simplest. The main disadvantage of these methods is the lack of knowledge of the exact correlations between characteristics of end product and the measured data. The most complex information could be obtained by a utility test, but this is the most expensive and time consuming method. The model test seems at this time to be the most suitable. Compared to the utility test, the model test is inexpensive, uses simple equipment, has high resolution and tests for properties of multiple utility.

1. Hydration Properties

Four groups of methods are commonly used for the practical determination of the water absorption and the water-holding capacity of protein ingredients:[35,36]

1. The *relative humidity method* (or equilibrium moisture content method) measures the amount of water absorbed at a given water activity. This method is useful for assessing the hygroscopicity of a protein powder and the risk of caking phenomena, with the resulting deterioration of flow properties and solubility.
2. The *swelling method* uses a device (Baumann apparatus) consisting of a graduated capillary tube attached to a sintered glass filter. The protein powder is placed on the filter and allowed to spontaneously absorb water located in the capillary tube beneath the filter. Both the rate and the extent of hydration can be determined.
3. *Excess water methods* involve exposing the protein sample to water in excess of that the protein can bind, followed by filtration or application of mild centrifugal or compression force to separate the excess water from retained water. This method is applicable to poorly soluble proteins (is widely used for measurement of water absorption of flour or vital gluten), and a correction is necessary to compensate for the loss of soluble protein.
4. In the *water saturation method,* the amount of water needed to establish a state of protein saturation (maximum water retention as determined by centrifugation) is determined.

2. Solubility

The basic principle of the different methods used in practice is the same. The protein sample is dispersed in water (buffer, salt solution, etc.), after adjusting the pH and temperature the suspension is stirred for a given time. After centrifuging the nitrogen (protein) content of the supernatant is determined and compared with the total nitrogen content of the protein preparation. In many countries the measuring of the nitrogen solubility (nitrogen solubility index = NSI) is standardized. The dependence of the solubility on pH is both of practical and theoretical importance. The solubility of protein preparation is measured at different pH and a solubility curve may be constructed; protein solubility profile (PSP).

3. Viscosity

Different types of viscosimeters may be used for the measurement of viscosity from simple capillary viscosimeters until well-equipped, computerized rotational viscosimeters can be used. Due to the nonNewtonian character of protein solution for exact characterization viscosimeters with the ability to change the shear should be used and consistency curves should be constructed. Otherwise, only an apparent viscosity may be determined.

4. Emulsifying Properties

The common feature of all the methods used in practice is the production of an oil in water emulsion using mechanical treatment of a mixture of water, protein preparation and edible oil. In special cases other components (e.g., salts, etc.) may be added to the system. For determining the emulsifying capacity two methods may be used. In a simpler procedure after preparing an initial oil in water emulsion in a mixer, additional oil is added until the break of the emulsion is observed. The other way is adding an excess of oil and centrifugation and separation of emulsion and free oil.

The determination of the stability of emulsion may be realized using longer storage and observation. More often accelerated procedures are used with heat treatment and centrifugation of the emulsion. Good results were reported[3] using an accelerated combined centrifugation-conductivity measurement method.

Several methods are available for making emulsions. The ultraturrax apparatus, colloid mill, liquid whistle and value homogenizer are commonly used[38] but of these the value homogenizer is the only instrument that controls and monitors the energy input.[39] Hence, the value homogenizer may be the method of choice for preparing standard emulsions of constant "initial" surface area for comparing the emulsion stabilizing properties of proteins.

Haque and Kinsella[40] recommend the value homogenization method under controlled conditions as a useful standard method for comparing the emulsifying activity of different food proteins. To characterize food emulsions from a theoretical point of view the determination of droplet size and size distribution (and hence of the total interfacial area) is necessary. This can be done by microscopy, light scattering, centrifugal sedimentation, or the use of such devices as the coulter counter (passage of droplets through orifices of known sizes). Measurement of the amount of protein absorbed at the interface (= surface concentration) requires prior separation of the oil droplets (in the case of o/w emulsions), for instance, by repeated centrifugation and washing to remove loosely bound protein. Protein concentrations of a few milligrams per milliliter emulsion generally lead to a few milligrams protein absorbed per square meter. When the dispersed phase volume fraction is large and the droplet size is small, more protein is needed to obtain an adsorbed protein film that is an effective emulsion stabilizer.

5. Whipping Properties

Nakamura and Sato[41] used a simple apparatus (See Figure 75) for measuring whipping properties of yeast protein isolates. Protein solutions were charged into a glass cylinder (2.5 × 50 cm) and the cylinder was filled with foams by passing air through a glass filter. The air flow was monitored with a flow meter. The height of foam volume was measured at every one minute. Foam volume was defined by the following equation:

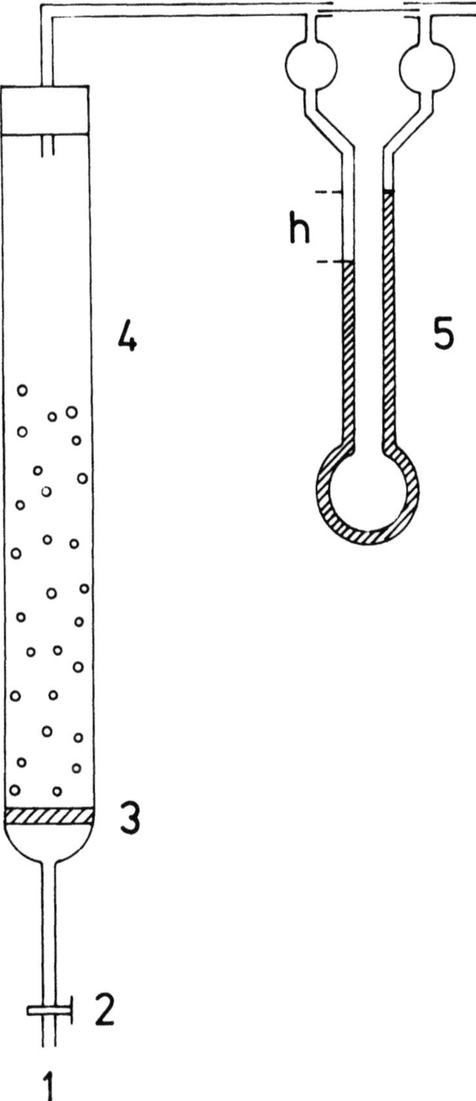

FIGURE 75. Apparatus measuring whipping properties, constructed by Nakamura and Sato: 1 — air pump; 2 — cock; 3 — glass filter; 4 — glass cylinder; 5 — flow-meter and 6 — height. (From Namamura, R. and Sato, Y., *Agric. Biol. Chem.*, 28, 530, 1964. With permission.)

$$\text{foam volume} = \frac{\text{average foam volume (ml/min)}}{\text{protein solution charged (ml)}}$$

After stopping air flow, the cylinder, which was filled with foams, was inverted over a glass funnel on a 10 ml messcylinder. Drainage of liquid from the foams was collected for 15 min and then the volume of the drainage was measured. Foam stability was defined by the following equation:

$$\text{foam stability} = 100 - \frac{V_0 - V}{V_0}$$

where V was the volume of drainage of liquid for 15 min after stopping air flow, and V_o was the volume of the drainage of liquid for 15 min. After stopping air flow no foams remained in the cylinder. Recently, a critical review of the methods of measuring foaming properties of proteins was published by Gassman et al.[44] and a new method of determination was proposed.

A variety of methods have been proposed to produce and characterize protein foams. Reports in literature principally describe three procedures for determining foaming capacity of proteins: whipping, shaking or sparging.[42,43] One important difference between these methods is the amount of protein required for foam production. The amount of protein ranges from 3 to 40% for whipping, around 1% for shaking horizontal graduated cylinders containing the protein dispersion, and ranges from 0.01 to 2% for spraying of gas through perforated disks into the protein dispersions.[42] Another difference between the methods used for determining foaming properties of proteins is the manner in which the foam is formed. While shear forces are involved in all three methods, they are of greatest importance in whipping, the most commonly used method. During whipping, the amount of gas incorporated steadily increases until a maximum volume is obtained. This maximum volume depends on the whipping agent used, the composition and the conditions of the system and on the level of energy supplied. Considering the equipment available for performing the foaming operation on a commercial scale, the following characterization based on the type of equipment and operational conditions has been proposed by the author: (1) batch operation at atmospheric pressure; (2) batch operation using compressed air; (3) continuous operation using compressed air and static dispersors; and (4) continuous operation, using compressed air and rotating dispersers.[45]

Due to the limitations of the amount of energy and/or gas available for whipping, in most cases laboratory equipment for aeration are not comparable to the industrial practice. Therefore, the foaming properties determined in the laboratory from foam volume, foam density and from volume of liquid drained off during a specific time do not necessarily reflect the true performance of protein-based whipping agents. Gassmann et al.[44] expressed some doubts that methods used in GDR for such purposes[46] are adequate to predict the properties of proteins under commercial scale production.[46-50]

Gassmann et al.[44] describe a useful method for whipping protein dispersions with an electric motor driven whisk, the rotation speed of which can be steplessly adjusted. For overcoming viscous friction with constant rotation speed and thus obtaining the optimum degree of aeration the energy output is electronically controlled. Whipping properties of nine protein preparations were measured and characterized by the following data: foaming capacity expressed as foam volume in percentage of starting-volume of liquid-phase and foam stability defined by the expression. Foam volume immediately after whipping shows liquid volume drained off, registered 60 min afterwards whipping and related to 100 ml of foam; foam density measured by weighing a known volume of the foam and firmness of foam measured by special penetrometer expressed in percentage of maximal penetration; the firmness was measured immediately and 60 min after whipping.

The use of the method in evaluation of different protein preparations serving as whipping agents permits determination of such conditions which are needed for preparation foams corresponding to the purpose and function of foams in produced food system.

A new method was established to estimate the foaming properties of proteins from the conductivity of foams using a simple apparatus consisting of a glass column with the conductivity cell.[60] A close correlation was observed between the initial conductivity of foams and the foam volume of 11 native proteins, suggesting that the initial conductivity of foams can be used as a measure of foaming power. In addition a close correlation was obtained between the foam stability determined from changes in the conductivity and foam volume with the time of 11 heated, denatured native proteins, suggesting that foam stability can also be estimated from changes in the conductivity of foams.

A review paper dealing with the whipping properties of different proteins was published by Muschiolik and Schmandke.[43] It was reported that in the last decade very intensive research work was done in this field. A lot of different protein preparations of animal-, plant- and microbial origin (including yeast proteins) was investigated. Concerning the methodology of characterization of whipping properties of protein preparations the following methods were used:

1. Characterization of protein solutions — measurement of surface tension;[51] study of the changes of surface tension in time[52] and measuring the volume of the foam[53]
2. Characterization of the foam formed — determination of the drained liquid phase after a given time (foam stability);[54] determination of the time interval until the first drop of drained liquid phase is formed;[55] measurement of the firmness of foam by penetrometer;[56] measurement of the rheological properties of foam by Brookfield-Viscosimeter;[57] control of consistency changes of foam by measuring the energy requirement charges of mixing machine[58] and study of the desiccation properties of foam.[59]

II. PRODUCTION OF YEAST PROTEIN PREPARATIONS

In Chapter 5 of Part I the general procedure of the production of yeast protein concentrates and isolates was reviewed. In the framework of this part of the book production of some protein preparations on a commercial or laboratory scale will be shortly reviewed including yeast protein containing mixtures and coprecipitates.

A process for recovery of protein for human nutrition was patented in GDR[61] in which 50% of the yeast protein is not subjected to several alkali treatment. After mechanical degradation of the yeast cells, the aqueous extract (which contains about 50% of the total protein) is treated with the acids of salts and surfactants to cause decomposition of nucleic acid by the endogenous nucleases. Nucleic acid is eliminated from the insolubule yeast protein by conventional alkaline extraction. The yeast protein is then mixed with fresh skim milk or casein solution and precipitated as coprecipitate. A yeast-based product is produced according to a French patent[62] by mixing 50 parts of plasmolyzed yeasts (*Kluyveromyces fragilis* + *Kluyveromyces lactis* concentrated to 20% TS) with 50 parts of crystallized ultrafiltration permeate containing 50% TS and 85% lactose in dry matter. The dried product contains 5% moisture, 30% protein, 9.5% minerals and 1.0% calcium. Recommended major uses are meat products, sauces, soups, pate and dietetic products.

Brewer's yeast (*Saccharomyces carlsbergensis*), after debittering with alkali solution and washing with water, was inactivated at 80°C and spray dried. The product, a yellowish to pinkish-grey powder, was investigated as a potential emulsifier in meat products, bakery products and processed cheese by Turubatovic et al.[63] Three processes for isolation of food-grade protein from baker's yeast were compared in another study:[64] (1) autolysis followed by high pressure homogenization and precipitation; (2) autolysis during homogenization, followed by selective precipitation and (3) high pressure homogenization followed by selective precipitation. Efficiency and cost of the three processes were compared, together with composition and quality of the products. High yields may be achieved with high homogenization pressure at low temperature and subsequent adjustment of pH to 10.5. Only slight alkali damage to the proteins occurs. Process 2 shows no clear advantage over the other two processes. Process 1 is the most interesting, although it has highest cost/wt protein isolate. Approximately 65% of the yeast protein can be recovered as a 90% protein isolate, with a very low RNA content. The residue from the first separation step is a RNA-rich fraction, which could be used as a raw material for RNA production.

A special process[65] is described for the removal of lipids from the protein globulin fraction of baker's yeast. Polysaccharides (gum arabic, arobinogalactan) were used for

TABLE 70
Comparing of the Amino Acid Composition of Whole Cell Protein and Protein Isolate from *Hansenula* Yeast (mol/100 mol amino acid)

Amino acids	Protein isolate	Whole cell
Arg	4.4	4.8
Lys	7.5	6.5
His	3.2	2.5
Phe	5.5	5.3
Tyr	4.0	3.4
Leu	8.5	7.2
Ile	5.7	5.1
Met	0.3	1.0
Val	7.4	7.0
Cys	0.1	Trace
Ala	8.6	10.2
Gly	7.5	8.8
Pro	4.6	5.2
Glu	9.2	8.7
Ser	6.6	6.4
Thr	5.8	5.8
Asp	11.4	9.5

From Sato, Y. and Hayakawa, M., *Lebensm. Wiss. Technol.*, 12, 41, 1979. With permission.

purification of globulin fraction of baker's yeast protein. Amino acid composition of the whole cell protein and protein isolate extracted with 0.5 *N* NaOH was determined by Sato and Hayakawa.[73] The results are summarized in Table 70. The amino acid pattern of whole cells of *Hansenula* yeast were similar to that of protein isolate. However, the amino acid contents of isolate were slightly higher than the whole cells excluding Met, Ala, Gly and Pro. This is consistent with the report of Vananuvat and Kinsella,[70] who found that the concentration of most amino acids in the proteins precipitated with acid or heating from the alkaline extracts of *Saccharomyces fragilis* was similar to that of whole cells even though there might be slight degradation of amino acids by alkali during extraction. These data show that major proteins contained in whole cells might be extracted with little losses.

A patented ion-exchange method is described by Levic et al.[74] for removing RNA from soluble homogenates of microorganisms. It was successfully applied to different prokaryotic and eukaryotic microorganisms (*Candida utilis, Saccharomyces cerevisiae, Saccharomyces carlsbergensis, Zymomonas mobilis*). The protein obtained is functionally enhanced and has a maximum RNA:protein ratio (w/w) of 0.03. Protein recoveries were 59 to 74%. The resin can be regenerated and the process is rapid.

Chemical phosphorylation of yeast protein by sodium trimetaphosphate (STMP) destabilized the nucleoprotein complex and facilitated the separation of protein and nucleic acid.[77] The method involves the addition of STMP at pH 12, and incubation for 6 h at room temperature. The modified protein was isoelectrically precipitated at pH 4.2. Most of the RNA remains in the supernatant. After protein precipitation about 90% of the RNA could be precipitated from the supernatant by acidification to pH 1.5 giving a product containing 50% RNA.

Studies of defatting of baker's yeast protein extracts by extraction with supercritical CO_2 are reported.[78] Pressures in the range 11.5 to 13.3 MPa and temperature of approximately 40°C were used. Under these conditions supercritical CO_2 failed to extract lipids; addition of ethanol as a selective component gave excellent lipid elimination, the residual lipid content being reduced to 1%, (i.e., comparable performance to conventional lipid elimination with isopropanol). Protein denaturation was much less than with isopropanol extraction, protein solubility at pH 10.5 being over 95%. Overall, results were better for isolates prepared from fresh yeast than for those prepared from freeze-dried yeast.

A method is described by Tolstoguzov[79] for purifying and concentrating protein solutions. The method, membraneless osmosis, is efficient and relatively cheap and is particularly applicable to food protein isolation on an industrial scale. Its recognized uses in concentration of skimmed milk proteins, legume seed and oilseed storage globulins, proteins of green leaves and gelatin are described. The use of membraneless osmosis to purify baker's yeast proteins, lipids and nucleic acids is also recognized as an application of the method and is discussed.

Extraction and recovery of protein from brewer's yeast *Saccharomyces uvarum* was investigated by Cheng and Chang.[84] Optimum conditions for extracting protein from broken *Saccharomyces uvarum* cells were pH 12, 30 min at room temperature which resulted in 85 to 87% protein extractability. The extracted proteins were best recovered by adjusting pH to 3.5 and centrifuging.

Protein content was increased from 45 to 54 to 56% and RNA decreased from 7.3 to 3.2%. In the protein isolate recovered the protein content was further increased to 75% while RNA remained the same as in the concentrate. A method for the preparation of protein isolate from yeast and other raw materials is patented in FRG.[80]

The protein source, e.g., cereals, faba beans, oilseeds, *microbial proteins* is extracted for solubilizing its protein and forming a protein solution with an aqueous food-type salt solution of defined ionic strength. The protein concentration in the resulting solution is increased, its ionic strength being largely kept constant. The resulting solution is diluted to ionic strength 0.2 for forming individual protein particles in the aqueous phase, at least partly in protein micelle form. These particles are allowed to settle to form an amorphous, sticky, gelatinous glutin-like protein micelle mass. The final product can be used in traditional applications of protein isolates and can also be made into protein fibers for use as a meat analog, and to proteins in foods containing protein binders.

Another process for production of protein for human nutrition is described in a GDR patent.[81] According to the process the properties of yeast protein are modified in a manner permitting separation of protein from cells and cell fragments by means of conventional separating equipment. After mechanical disintegration, lipids are extracted with an aqueous/alcoholic solution, and nucleic acids are separated after alkaline hydrolysis. The alkaline protein solution is mixed with fresh skim milk, yeast protein and milk protein being coprecipitated. The yeast protein so prepared is especially suitable for preparation of textured products for food use. Removal of nucleic acids from yeast nucleoprotein complexes by sulfitolysis was described by Damodaran.[85] Treatment of yeast nucleoprotein with sodium sulfite followed by sodium tetrathionate caused destabilization and dissociation of nucleoprotein complexes. Subsequent precipitation of protein at pH 4.2 resulted in a protein preparation with low levels of nucleic acids. A good correlation between the extent of nucleic acid removal and disulfide content of the sample was also observed.

The effect of sodium hydroxide concentration on the protein extraction from *Saccharomyces cerevisiae* cells which had been disintegrated in a vibrogen milk was studied by Asano et al.[92] The greatest part (86.3%) of the total *N* was extracted with 0.4% NaOH. The extracted fraction consisted of a dry matter basis of approximately 70% protein and 9.7% RNA. Amino acid analysis indicated that the protein was rich in aspartic acid, glutamic acid, lysine, methionine (?) and histidine. The extracted fraction was characterized by acetic-acid-urea PAGE. Results indicated the presence of more than 15 components.

Extraction of protein from *Candida tropicalis* and *Candida utilis* grown by a batch method on cassava hydrolysate by a combination of mechanical and mild chemical (0.4% NaOH) treatment and precipitated by 1 *N* HCl at pH 4.0 to 4.5, yielded about 71 and 68% protein resp. with nucleic acid concentration reduced by 75 to 80% (to 1.2%) in the protein concentrates of both organisms. The amino acid profiles were comparable to that of whole egg except methionine and tryptophan being the limiting amino acids.[93]

An improved method is recommended for producing functional yeast protein according to the following patented procedure comprising: (1) slurrying whole yeast cells for 5 to 30 min in a dilute alkaline solution having a pH 8.5 to 10.0 at a temperature of 85 to 95°C to extract nucleotidic materials and undesirable flavor and color bodies; (2) separation of the yeast cells from the alkaline extract; (3) reslurrying the separated yeast cells, preferably for 15 to 45 min in a solution of 0.1—0.3 *N* sodium hydroxide or potassium hydroxide at a temperature of 85 to 100°C to extract proteins; (4) neutralizing the slurry; (5) separating the undigested cell residue from the supernatant containing the extracted protein; (6) acidifying the supernatant at a pH 3.5 to 4.5 to precipitate proteins; (7) separating the precipitated proteins from the mother liquor solution containing the soluble proteins; (8) neutralizing the mother liquor solution to a pH 6.7 to 7.2 and (9) drying (spray drying) the mother liquor solution to yield a yeast protein product.[95]

The acid-precipitated protein can also be neutralized and dried. As a result of the procedure described above two protein products can be obtained from yeast: an acid-precipitated yeast protein isolate and a well whippable yeast protein. The isolate contains about 88 to 89% protein, the whippable product has high ash content (24.8%) and lower protein content (32.8%). The ash content could be reduced by dialysis. The yeast whippable protein is soluble at all acid pH values and the yeast protein isolate can be completely dissolved at pH 3.3. These properties demonstrate their potential for use in making acidic protein beverages. No data are given by the inventor about the nucleic acid content of protein preparations.

The optimal parameters of production of protein isolate from *Candida utilis* yeast (grown on sulfite waste liquor) were studied by Achor et al.[96] Alkaline and acid treatment were used and compared. The results indicated that an increase in concentration, time, temperature and pH of the alkaline treatments tend to increase yield of protein isolates and that optimum conditions for maximizing the yield of protein isolate were to heat a 4% (w/v) suspension of disrupted *Candida utilis* cells in NaOH solution for 31.9 min at 81.9°C and pH 11.7. The predicted yield of isolate there is 51%.

Although the isolates from the alkaline extractions were mainly proteinaceous, their actual amount (quantities) of protein was lower than for the cellular residue after acid treatments. The differences between total protein contents within the acid treatments were slight but the values for HCl were consistently lower. Possibly these lower values reflect better extraction of nucleic acid nitrogen with this acid compared with perchloric acid and trichloroacetic acid. From a nutritional standpoint, it is likely that the alkaline isolates would be more readily digested compared to the acid treated products containing cellular debris.

The nucleic acid content of isolates obtained by extracting disrupted *Candida utilis* cells with NaOH solution ranged from 0.43 to 11.97%, whereas those for the acid treated residues varied from 2.37 to 6.90%. The nucleic acid content of untreated disrupted *Candida utilis* cells was 12.1%. The most severe alkaline treatment yielded products with the lowest nucleic acid content. Among the acids HCl was most effective in reducing the nucleic acid content of the cells. For the acid treatment, heating low yeast concentrations at low pH and elevated time and temperature resulted in lower nucleic acid content. The predicted conditions for maximum removal of nucleic acids from alkaline protein isolate was to heat an 8.3% (w/v) disrupted cell in NaOH solution at 92.6°C and pH 11 for 32.3 min. Under these conditions the residual nucleic acid content would be zero but from consideration of subsequent data on damage to proteins as a result of alkaline extractions, it would seem that the aforementioned conditions to effect complete removal of nucleic acids would be too damaging to the proteins.

The acidic treatments seemed most detrimental to the ε-amino groups of lysine resulting in decreased availability of lysine. Increase in time, temperature and pH of treatment resulted in decreased available lysine in the isolates from alkali-treated material.

For the acid treatments, increased time and temperature but decreased pH seemed to decrease the availability of lysine. The best conditions for maximum retention of available lysine were to heat an 8.6% (w/v) cell in NaOH solution at 58.3°C and pH 8.6 for 15.2 min. However, under these conditions the removal of nucleic acids is unsatisfactory. Aqueous acid solutions catalyzed reactions between carbonyl and amino groups also involving the ε-amino group of lysine. Among acids HCl was the optimal from the point of view of lower loss of lysine.

Generally, it can be stated that major problems associated with alkaline treatment are the racemization of amino acids and formation of lysinoalanine (for details see Chapter 5, Part I), whereas with the acid extraction the destruction of lysine.

From the facts mentioned above, the difficulty in making recommendations for an extraction procedure becomes readily evident. However, from the various treatments it appears that NaOH extraction of an 8% suspension of disrupted *Candida utilis* cells for 30 min at 100°C and pH 10.0 would offer the best compromise between residual nucleic acid and damage to amino acids. From the point of view of reduction of protein damage consideration should be given to the heat-shock activation of endogenous nucleases to reduce the levels of nucleic acids in single cell protein.

Procedure for the production of a high-quality edible protein from *Candida* yeast grown in continuous culture is reported by Lawford et al.[98] A specific feature of the yeast protein production is addition of an organic cation (cetyltrimethyl ammonium bromide = cetavlon) forming an insoluble complex with RNA. A disadvantage of the procedure is the loss of protein due to the formation of RNA-cetavlon protein clot.

Phosphorylated yeast protein preparation were produced by Huang and Kinsella.[99] Phosphorus oxychloride was used for phosphorylation and approximately 30% of the ε-lysine amino groups were phosphorylated. No data were presented about influence of phosphorylation on the nutritional value of yeast protein isolates.

The procedure of Roshkova et al.[100] results in a low RNA protein isolate, a polysaccharide-protein complex (cell wall material) and a liquid yeast fraction which can be concentrated and dried. The isolates contained 65 to 70% protein, 1.0 to 1.5% nucleic acids, 10 to 12% of polysaccharides and 10 to 12% of ash and other substances. No data were published about the yield of isolate and losses of protein. The amino acid composition data (very low concentration or absence of sulfur-containing amino acids) suggest that the removal of RNA and alkaline extraction caused damage of proteins.

Technology of enzymatic modification of proteins from unconventional sources is discussed by Bednarski.[102] *Candida utilis* and *Saccharomyces fragilis* yeast biomass was disintegrated in the experiments, defatted, mixed with water, pH 8.0 insoluble residues removed, dialyzed, pasteurized and preserved with 0.05% sodium benzoate. Then a 2% protein solution was hydrolyzed, resynthesis carried out with α-chymotrypsin with concentrated hydrolysates. Yield was 47.87% after resynthesis and no RNA content was found in the modified yeast protein. Admixtures of whey or blood proteins improved biological value from 1.47 to 2.62. Advantageous modifications in physical properties and chemical properties were also observed.

III. FUNCTIONAL PROPERTIES OF YEAST CONCENTRATES AND ISOLATES

Some general statements concerning the functional properties of yeast proteins are summarized in the review papers of Schlingmann,[1] Guzman-Juarez,[5] Kinsella[4] and Cooney et al.[66] The solubility of yeast SCP, especially at pHs common in food, is somethig lower than

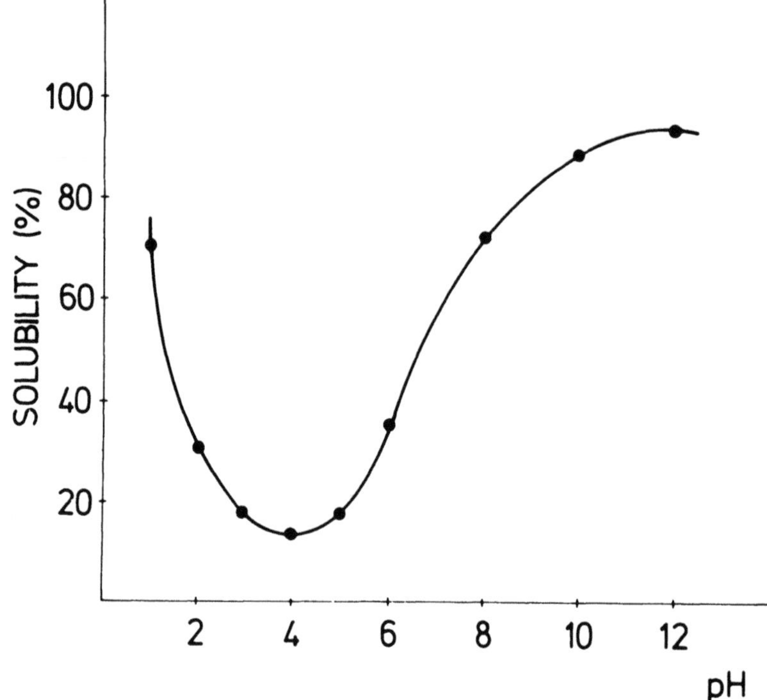

FIGURE 76. Typical solubility profile of food yeast proteins.

the other proteins such as soy. Although water and salt soluble proteins or the major part of native yeast proteins, the procedure of extraction, precipitation and drying may change these properties and increase the insoluble part of proteins. Heat treatment and alkali treatment at higher temperatures decreases the solubility. A typical sclubility profile of yeast protein isolates is shown at Figure 76. As it is seen the lowest solubility is observed at pH 4 to 4.5, and the best solubility may be obtained at slightly alkaline pHs. Water absorption of yeast protein concentrates and isolates vary in the range from 3.0 to 4.0 ml g^{-1} samples. These values are comparable to or less than the water-holding capacity of unheated soybean proteins.

However, the values are lower than those for heated soybean precipitates. The emulsifying characteistics of yeast protein preparations vary in a wide range depending on the procedure of production. Yeast proteins (alkali extracted and acid precipitated) showed similar behavior to soy isolates in water and sodium chloride solutions (4% w/v).

Heat treatment of the emulsions to test for stability did not change the emulsifying characteristics of yeast proteins, whereas these of soy isolates were increased by a further 20%. The presence of sodium chloride had a detrimental effect upon both the activity and stability of soy isolate emulsions. No such effect was detected in yeast isolates under similar conditions.

Heat coagulation of yeast proteins generally begins at 60°C. A chemical coagulation by acids and/or salts is also possible. The network structure is far less developed in isoelectric or calcium coagulates of yeast proteins than that in soybean protein isolate. This fact is attributed to the lower average molecular weight of yeast proteins, electrostatic repulsion and rapid rate of coagulation.

Structure potential and texturability of SCP (without chemical modification) is evaluated to be lower than those of soy proteins, however, these properties could be substantially improved by a combination of yeast proteins with compatible polysaccharides.[67,68] After this general review in the following some special, more detailed reports and information in this field will be reviewed.

A report was published by Vananuvat and Kinsella[70] on the functional properties of yeast proteins. Yeast cells were grown on crude lactose and ruptured by homogenization in a blender containing glass beads. The protein was extracted with 0.4% sodium hydroxide solution or water and the proteins were precipitated at pH 4 and 26°C, or by heating to 80°C at pH 6.[71] The proteins were characterized by emulsifying activity by blending a 7% aqueous suspension of the protein with vegetable oil, centrifuging at 20°C and determining the amount of free and emulsified oil produced. Different yeast protein preparations showed an emulsifying activity ranging from 51 to 59% compared to the value of 46% for commercial soy protein isolate. The protein extracted with alkali and precipitated by heat treatment (80°C) had no emulsification ability. A short review about the use of microbial proteins in emulsification was recently published by Morr.[69]

Functional properties of yeast protein isolate obtained from different yeast were studied by Hassan et al.[72] Compressed, freeze-dried, and hot-air dried yeast (*Saccharomyces cerevisiae*) were used for protein isolate production after alkaline extraction. The crude protein content of isolates was 87.55, 88.88 and 87.32% and the total nucleic acid only 2.26%. Solubility profiles of the samples were similar and foam volume of the product was significantly lower. No significant differences were observed in emulsifying capacity. The solubility profile of protein isolated from *Hansenula* yeast was investigated by Sato and Hayakawa.[73] The protein was very soluble in an aqueous solution of different pH except a low solubility range at pH 4 to 5.

Viscoelastic properties of concentrated solutions of yeast protein and some other protein preparation were investigated by Kawasaki and Nomura.[103] The purpose of the studies was to clarify fundamental rheological properties of yeast proteins, soybean protein and casein solutions and their applicability to the theory of linear viscoelasticity. Results were as follows: the viscoelastic behavior of each protein sample was found to be applicable to the method of reduced variables, and the master curves of storage modulus G' were obtained over a wide range of time scale. The master curves were characteristic for each protein, resembling each other among the same protein source. Yeast protein and soybean protein were optimal to form a gel-like structure, suggesting the formation of cross-links. However, casein behaved very similarly to an uncross-linked amorphus polymer. Their differences were especially apparent at the terminal zone.

Solubility, water-holding capacity, and viscosity of phosphorylated yeast protein was intensively investigated by Huang and Kinsella.[99] It was reported that phosphorylated yeast proteins were very wettable and hydrated easily. At all pHs studied (5.0 to 8.0), the water-holding capacity (WHC) of yeast protein was enhanced following phosphorylation. This could be attributed to the increased hydration of the added phosphoryl group, which can bind five to six water molecules per group and the loosening of the protein structure resulting from the electrostatic repulsion between the negatively charged phosphoryl groups and anionic carboxyl groups. Above pH 6.0 a semi-transparent gel-like pellet was observed when the phosphorylated yeast protein dispersions were centrifuged. Possibly, the gel-like network structure was formed during or following phosphorylation of yeast protiens and water was entrapped in the network, resulting in a substantial increase in WHC. The addition of salt caused a marked decrease in the amount of bound water.

The solubility profile of phosphorylated yeast protein showed an improvement in solubility in the pH range of 5.0 to 7.0. The phosphorylated protein dispersed very rapidly and formed a clear stable suspension. Upon centrifugation much of the phosphorylated protein sedimented to form a gel-like pellet in the tube, and only a fraction (45%) of the protein remained in true solution. Calcium in excess precipitated more than 90% of the soluble proteins probably due to the formation of protein-calcium complexes. Formation of network structure by cross-linking of the proteins during phosphorylation is also possible.

The apparent viscosity of phosphorylated yeast protein dispersions increased progressively with increasing protein concentration. This increase is probably due to the protein-protein interactions in more concentrated solutions. Phosphorylated yeast protein dispersions exhibited shear thinning over a range of shear rates, having pseudoplastic properties similarly to some other proteins (e.g., egg albumin, soy protein, casein and whey protein, alkali-treated yeast protein concentrate, etc.). The apparent viscosity of phosphorylated yeast protein increased nearly four-fold when the pH was increased from 5.0 to 8.0. This may be related to the increased hydration with increasing ionization of the phosphoryl groups and possibly greater electrostatic repulsion between molecules.

Ionic strength has a marked effect on the flow properties of protein investigated. The apparent viscosity was reduced by 60% in medium containing sodium chloride. Increased temperatures resulted in decreased viscosity by destabilizing both protein-protein and protein-water interactions. No increase in viscosity nor coagulation (gelation) was observed even at high temperatures.

In conclusion, it was stated by authors[99] that some functional properties of yeast proteins, especially WHC and apparent viscosity were substantially improved by phosphorylation. The increases in solubility, WHC and viscosity were presumably due to the increased hydration of the added phosphoryl group and the loosening of the protein structure resulting from the electrostatic repulsions between the negatively charged groups. The surface active properties of yeast proteins were also improved by phosphorylation, indicating that chemical phosphorylation while reducing nucleic acids is a feasible method for preparing functional proteins from yeast.

Whipping property (foam volume and foam stability) of the protein, extracted with 0.5 *N* sodium hydroxide from *Hansenula* yeast grown on methanol, was measured under various conditions and the effect of additives such as sodium chloride, calcium chloride, sucrose, methylcellulose and polyethyleneglycol on its whipping properties were examined by Sato and Hayakava.[73] The spray dried cells were pretreated with 1 *N* hydrochloric acid in boiling water for 1 min. Protein was extracted with 0.5 *N* sodium hydroxide at 40°C for 3 h. Protein was precipitated adjusting the pH of the extract to 4.5. The precipitate was dissolved at pH 12 and dialyzed and the purified precipitate was finally freeze-dried. This preparation was used for further experiments.

The method of Nakamura and Sato[41] was used for the determination of whipping property. The protein preparation was very soluble at various pH values excluding the lower solubility range between pH 4 to 5. Foam volume of the protein was not changed and foam stability was enhanced with an increase of protein concentration. Foam volume of 1% protein aqueous solution with or without salts (sodium chloride or calcium chloride) showed a pH dependence with maximum at pH 4.5 and 7 and minimum at pH 6. Although foam stability of the protein solution was poor at various pH, it was increased by the addition of salts in alkaline region.

The effect of viscous materials was not remarkable but sucrose enhanced foam volume at every pH and stability in acidic region. By the simultaneous addition of sucrose and salt, both foam volume and foam stability increased in the whole range of the pH tested. Little content of lipids (3 to 4%) in the protein isolate might cause high foam volume in the vicinity of pH 4.5 or there might be an adequate amount of soluble protein to form foams even at a lower solubility range of the protein.

Comparison of functional properties of protein isolate from yeasts grown on cassava hydrolysate[93] with those of soya protein concentrate showed very good wettability, whippability and emulsion capacity but poor emulsion stability. Yeast protein concentrates were about 8 times more wettable and 2 to 3 times more whippable and emulsifiable than soy protein. Water swelling and water-binding indices were lower than those of soy protein. Potential applications include substitution for egg white in confectionery and dessert, as well as in meat and high-protein food products.

Common methods of recovery of yeast proteins and reduction of nucleic acid content were evaluated by Asano and Shibasaki.[75] It was stated that previous methods resulted low yield and impairment of desirable functional characteristics. Emulsifying properties of yeast protein isolates prepared by three different methods were compared with those of soy protein and a caseinate at various pH ionic strength and lecithin contents. Protein isolated by an improved method elaborated by authors had an emulsifying capacity far superior to those of the soy protein and caseinate, particularly at extreme pH. Isolates made by previous methods were comparable in emulsifying properties with the other protein.

Preparation of yeast protein isolates with low nucleic acid content and improved functionality was developed by chemical phosphorylation.[76] Phosphorylation of water-soluble proteins from baker's yeast was performed by incubation with sodium trimetaphosphate at pH 11.5 at 40°C for 6 h, followed by isoelectric precipitation at pH 4. Nucleic acid content (w/w dry protein basis) was reduced from 33 to 36 to 2 to 3.5%, and functional properties of the protein (solubility and emulsifying capacity) were improved.

A comestible whipped protein product produced by treating yeast cells for RNA reduction followed by treating RNA-reduced cells with relatively HTST heat-shock and whipping is described by Shay.[82] The product has good whipping properties and forms a highly stable foam, a property important in the preparation of light, fluffy foams with minimal weight of ingredients. Various protein products obtained from *Saccharomyces uvarum* cells were investigated for protein solubility, WHC, emulsifying and whipping properties and compared to soybean concentrate STA-PRO by Cheng and Chang.[83] Results revealed that acetic-acid and HCl used in the acid precipitation did not cause much difference, while functional properties of the products at different pHs showed significant differences especially when acid and ethanol were used together to precipitate the protein. The heating during spray drying had little effect on protein denaturation, but a great change in surface structure was noticed which was visualized by scanning electron microscopy and resulted in a decreased WHC. Among the samples investigated the protein isolate obtained at pH 8 and adding ethanol to 20% exhibited the highest WHC and best emulsifying properties which were comparable to those of STA-PRO.

The emulsifying properties of *Candida utilis* grown on *n*-paraffin, acetic acid and sulphite liquor were studied in comparison with those of soya protein.[86] The emulsion stability (ES) and emulsifying capacity of protein isolates prepared from these yeasts by alkaline extraction were superior to those of soya protein, especially in the slightly acidic region. The ES was in order of 3>2>1> soy protein. These results were presumed to depend mainly on the extent of partial hydrolysis caused during protein preparation. Hydrolysate obtained by direct partial hydrolysis of damaged cells using HCl indicated a high ES level in the slightly acidic region, regardless of the removal of the insolbule fractions. This simple one-step process was significant from the standpoint of actual utilization of single cell proteins.

Gel-forming proteins (geling agents) may be produced from microbial cells after cell rupture by homogenization and alkali treatment.[94,95] The gel-forming substance is almost odorless and tasteless and is easily geled by the addition of water and heat.

A. TEXTURIZATION OF YEAST PROTEINS

The application of yeast proteins in food processing may need their texturization. So it is understandable that intensive research work was done in this field. The general questions of texturization were treated in the previous part of this chapter. Here some reports about texturization experiments with yeast protein isolates will be shortly reviewed.

A process for texturizing proteinaceous materials is patented by Akin and Darrington.[87] According the patented procedure an aqueous slurry of single-cell protein is combined with a texturizing agent such as gelatin and heated to 30 to 130°C. The slurry is then cooked to form a foam which is frozen, dehydrated and heat treated.

Texturized protein isolate was produced and described by Schmandke et al.[88] Physi-

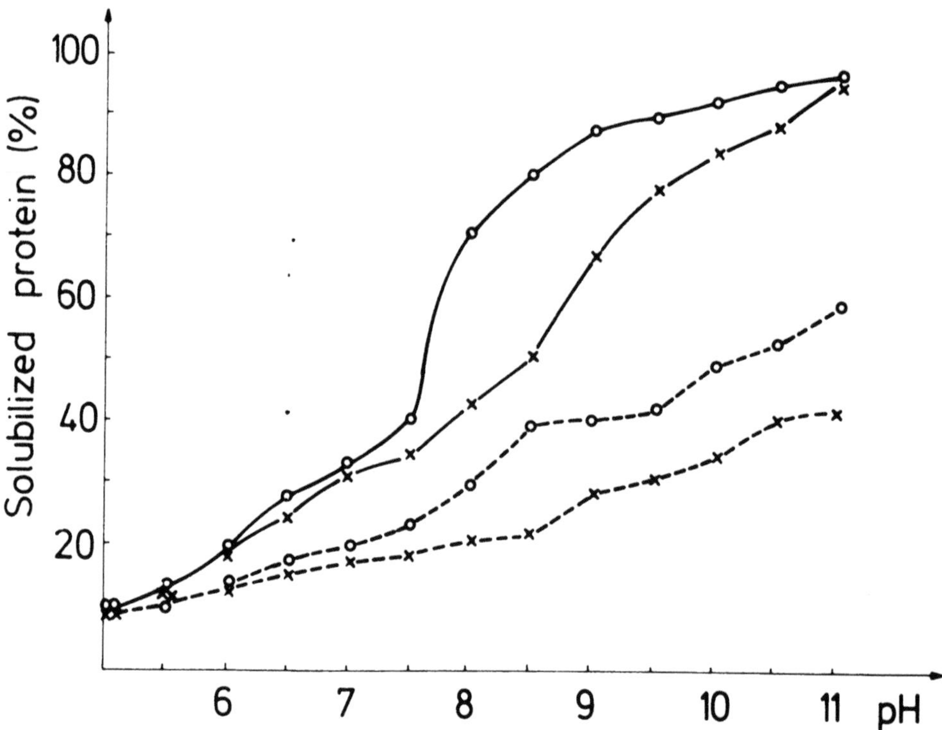

FIGURE 77. Solubility of ethanol-yeast protein isolate in water (——) and in 5% sodium chloride solution (- - -) without (xxx) and with (OOO) β-mercaptoethanol depending on pH. (From Schmandke, H. and Schmidt, G., *Nahrung*, 22(7), 643, 1978. With permission.)

cochemical aspects of texturization and fiber formation from globular proteins is treated by Shen and Morr.[89]

According to another patent[90] stretched fibers or fibrils from single-cell dopes are prepared by extruding the dope into a moving, coagulating bath wherein the fluid motion of the bath causes the extrudate to gradually stretch. The resultant stretched fibers are useful as textured food products. Shaped, cohesive, textured microbial protein is produced by passage of biomass dough through a sheeter or bar former to give a laminated bar. The process is particularly useful for the production of textured meat analogs from single cell protein.[91]

Solubility of ethanol yeast protein and corresponding spun-protein fibers in dependence of pH was studied by Schmandke and Schimdt.[97] The following materials were used for the experiments: (1) Freeze-dried ethanol-yeast protein isolate with a protein content of 71.6%, produced with alkaline extraction of disrupted yeast cell biomass; (2) sulfuric acid precipitated casein with 84.1% protein content; (3) mixed-fibers produced from a casein-ethanol yeast protein isolate mixture (1:1) solubilized with 1.8% NaOH to obtain a solution with 19% protein content; (4) fibers from ethanol yeast protein isolate produced from a solution of protein in 1.8% NaOH having a protein concentration of 15%. The fibers were additionally modified with dialdehyde starch and treated with sodium chloride solution and heated. The effect of beta-mercaptoethanol on the solubility of the fibers was also studied. Some results of these investigations are summarized as follows (see also Figures 77 to 79). An increase of the solubility of ethanol yeast protein isolate in water and 5% sodium chloride solution resp. was observed in the presence of β-mercaptoethanol in pH range 7.5 to 10.5 and 7.5 to 11.0 resp. The addition of sodium chloride to solvent decreases the solubility. According to the Figure 78 the solubility of a mixed (1:1) fiber produced from ethanol yeast

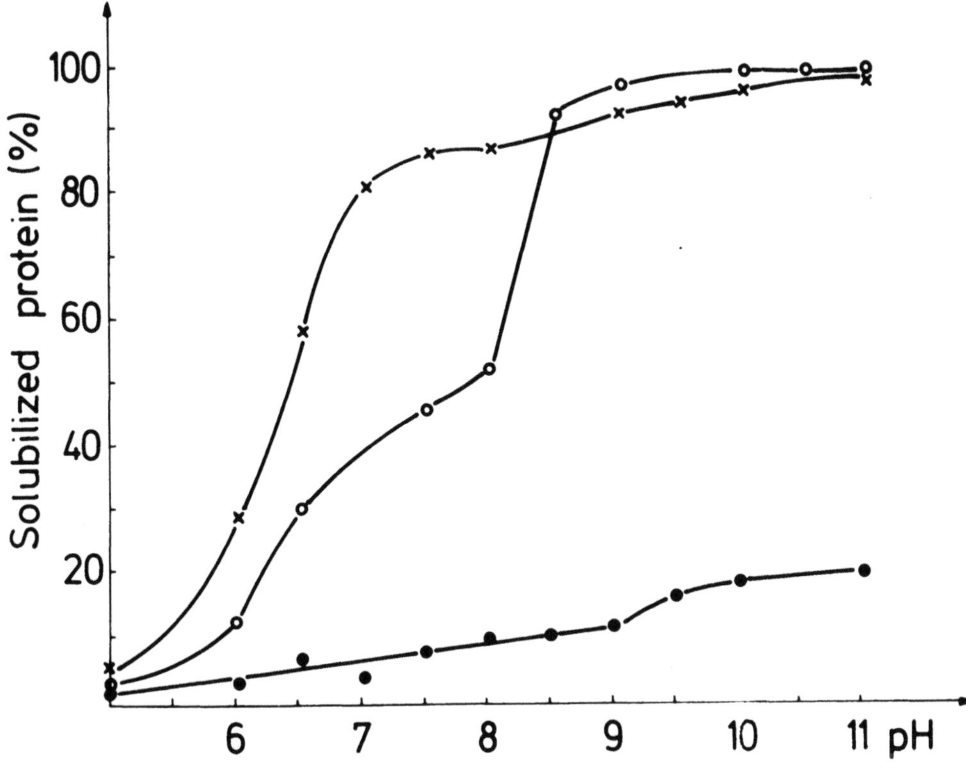

FIGURE 78. Influence of dialdehyde starch and heat treatment in sodium chloride solution on the solubility of a mixed (1:1) ethanol yeast protein isolate and casein fibers: x — x without treatment; O –– O heat treated in sodium chloride solution and O — O a modified with dialdehyde starch. (From Schmandke, H. and Schmidt, G., *Nahrung*, 22(7), 643, 1978. With permission.)

protein isolate and casein decreased significantly due to heat treatment in sodium chloride solution in the pH range 6.0 to 8.0 — casein fibers have a much lower solubility. The same tendencies in solubility were observed in the case of fibers produced exclusively from ethanol-yeast protein isolate. However, in this case the effect of the heat treatment in sodium chloride solution is much greater. The solubility of fibers modified with dialdehyde starch is similar. Sodium chloride addition to the solvent decreases the solubility. The addition of β-mercaptoethanol (a reagent splitting the disulfide bonds) increases the solubility in the pH range 6.0 to 8.0.

The viscosity of SCP prepared from commercial baker's yeast and the potentiality for forming a continuous and uniform filament were studied by Kawasaki and Nomura.[101] Several kinds of additives were added to SCP at various levels to investigate their effects on flow behavior and spinnability. Tensile strength of the resulted filaments was measured. Under the experimental conditions 1 to 3% of carboxy-methyl-cellulose (CMC) or carrageenan improved the spinnability and streetchability of SCP with an increase in apparent viscosity.

This increase in apparent viscosity might reflect an increase in the intermolecular bonding resulting from the addition of hydrocolloids. However, at the higher levels of addition, samples failed to yield filament. This failure is perhaps mainly due to the mechanical aspect of the viscosity rather than the chemical nature of the mixtures. At the same concentration CMC was a more desirable additive than carrageenan for making SCP filament. On the other hand, addition of 35% casein was needed to obtain similar results for the SCP-CMC filament and the SCP-casein mixtures and with increasing addition, the stretchability and tensile strength of the filaments were improved.

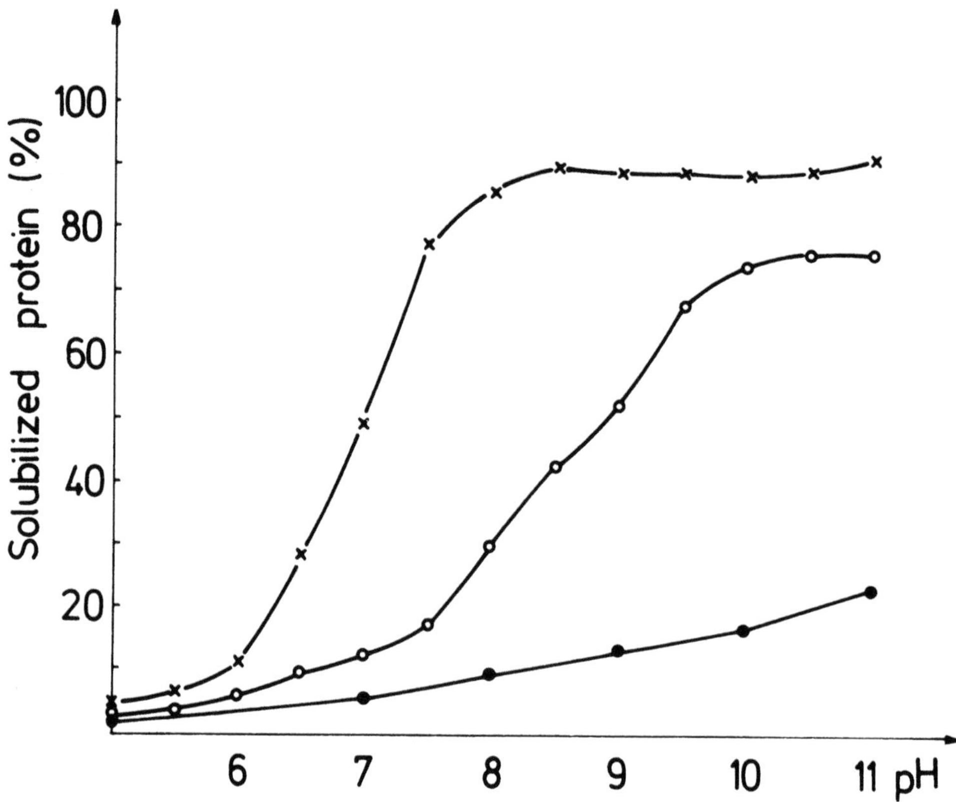

FIGURE 79. Influence of dialdehyde starch and heat treatment in sodium chloride solution on the solubility of ethanol-yeast protein isolate fibers: x — x without treatment; O — O heat treated in sodium chloride solution and O — O modified with dialdehyde starch. (From Schmandke, H. and Schmidt, G., *Nahrung,* 22(7), 643, 1978. With permission.)

B. FOOD APPLICATION OF YEAST PROTEIN ISOLATES AND CONCENTRATES

Although the use of yeast protein concentrates and isolates in food processing is a possibility recommended by a lot of scientific papers and has advantages from a nutritional point of view, the application of these protein preparations is concentrated on meat products, meat and milk analogs. This fact may be explained with a relatively high cost of yeast protein isolate production. So only the food products having a higher price may be supplemented with isolate in an economical way. In lower priced foods the addition of whole yeast is more economical despite the limitations caused by RNA content and cell wall content of the inactive dried yeast. In the following some applications of yeast protein isolates and concentrates will be treated.

The strategy of the use of meat extenders is discussed in the paper of Sofronova et al.[104] and Vancikova and Simunek.[105] According to authors mentioned above the main task that arises in the course of development of combined meat products lies in the design of protein compositions whose biological value would not be inferior to that common to proteins contained by meat or to that common to an ideal protein. The authors discuss two varieties of combined meat products. The first variety based on the replacement principle consists of the utilization of the amino acid reserves of meat proteins, thereby enabling the introduction of certain amounts of less valuable proteins into meat products causing no significant change in biological value. The second variety is based on the design of preliminary developed protein composition in which the amino acid content is balanced at the expense of mutual protein supplementation. The strategy described by the authors is satisfactory from the point of view of nutrition, however, a study or prediction of functional properties and palatability of products must also be kept in mind.

A series of reports dealing with the application of a brewer's yeast based protein preparation (Belvit-S) in meat products was published.[106-112] In Yugoslavia (and many other countries) food legislation permits the addition of 3% (in Hungary, e.g., 2%) protein of nonmeat origin into canned meat products and smoked products. A brewer's yeast-based protein preparation (Belvit-S) has good possibilities to be used for such purposes instead of with blood plasma protein due to high costs of manufacturing of dried blood plasma protein preparations. Experiments with model meat systems showed that Belvit-S and mixtures (70:30, 60:40, 50:50) of Belvit-S and dried blood plasma resulted in heat stable emulsions regardless of the method of thermal processing. Stability and organoleptic properties of meat paste containing added Belvit-S and various ratios of water to fat were also studied. The products investigated were prepared from the following raw materials: heat meat, hog skin, liver, fat, water and spices. The emulsion was homogenized and emulsified in a cutter at 45°C ± 2°C. Samples were then placed in aluminium foil containers and heated at 118°C for 35 min. Seven samples with fat/water ratios ranging from 1:0.52 to 1:1.91 were prepared. The pastes were studied organoleptically and chemically. The best ratio was between 1:1.33 to 1:1.91. A 1:1 ratio was still acceptable. The best water/fat ratio was found to be 1:1.59.

In another experiment paste was made from solidified roasted fat (46 kg), water (24 kg), pork heads (8 kg), pork liver (12 kg), skin suspension (4 kg), salt and spices (3 kg). Additionally, 1.0, 1.5, 2.0 and 2.5% Belvit-S was made. Fat and water do not separate with 2% Belvit-S addition, and color and consistency are best when such an addition was made. The best processing temperature is 47°C.

Pilot plant experiments were made using different protein preparation mixtures (Belvit-S, dried blood plasma and dried whole blood) in the amount of 2% calculated on the meat paste produced from veal meat, water (in form of ice and 25% amount calculated on meat quantity) and fat (adipose tissue, 30% meat) in a high speed cutter. The samples were placed in tin cans and heat treated at 80°C for 60 min and at 118°C for 40 min. The heat stability and organoleptic properties of products were determined. The loss of water and fat was measured, and the color, consistency and flavor was evaluated organoleptically. The best results were obtained using dried blood plasma or a mixture (60:40 and 50:50) of Belvit-S and dried blood plasma. Products with the addition of Belvit-S in an amount of 2% were acceptable but small losses of water and fat were observed. Combinations of Belvit-S and dried whole blood or whole blood alone resulted in products with unsatisfactory heat stability.

The color of the products with dried blood plasma and Belvit-S had a light red color (sometimes with slight greyish attenuation) while products with whole dried blood or mixtures containing whole blood were brownish or greyish. The consistency of all experimental products was satisfactory, no significant differences between normally prepared commercial products and experimental products were observed. The flavor of products was acceptable except for whole dried blood containing products having an off-flavor.

Sodium caseinate and Belvit-S were also used for production of frankfurter-type sausages.[110] It was stated that the two additives are equally suitable for sausages and similar products. Additions of 1 to 2% were made to the meat emulsion and the effects on consistency, weight loss during cooking and smoking, and organoleptic properties were studied. It was found that Belvit-S addition to frankfurters is advantageous and markedly improves the organoleptic properties.

A possibility of using Belvit-S as a component of spice mixtures was also studied.[107] Three types of spice mixtures were investigated: (1) oleoresins with sodium chloride as a carrier; (2) natural spices and (3) mixtures of oleoresins in oil as a carrier. All mixtures received Belvit-S as 100, 300 and 500 g/100 kg of spice mixtures. The preparations were used for frankfurters and "extra"-type sausages. The Belvit-S allows reduction of levels of spices in the meat, it does not affect the flavor of the product.

One of the possible uses of yeast protein isolates in the milk industry might be production of cheese analogs or cheese-like products. From the point of view of the technology of such types of products the coagulation of yeast protein isolates, the effect of heat treatment and coagulants on the microstructure, WHC, hardness, springiness, adhesiveness and cohesiveness of the curds is of interest. Yeast protein can be aggregated either by calcium or at the isoelectric pont. Studies of the microstructure and mechanical properties of yeast protein curds were reported by Tsintsadze et al.[113]

Candida lipolytica yeast was used for preparation of yeast protein isolate. After alkaline extraction and isoelectric precipitation the protein was resuspended and dialyzed. Four different curds were prepared: (1) the protein was coagulated by adjusting pH to 4.0; (2) $CaCl_2$ was added to protein solution for coagulation; (3) the protein solution was heated at 80°C for 30 min, cooled to 25°C and isoelectrically coagulated and (4) the protein solution was heated at 80°C for 30 min, cooled to 25°C and precipitated by the addition of $CaCl_2$.

The optimum calcium concentration for the coagulation of yeast protein concentrate was found to be 13 to 16 m*M* $CaCl_2$ for unheated yeast protein and 12 to 13 m*M* $CaCl_2$ for heated yeast protein. The amount of the coagulated protein was higher in unheated protein concentrate than in the heated protein concentrate. Water-holding capacity of the fresh curds formed by calcium was higher than those from isoelectric precipitation. The heat treatment of the protein solution prior to precipitation did not influence the water-holding capacity of the curd. However, the water-holding capacity decreased when the curd was frozen and than thawed. The decrease was more prominent in the calcium precipitate than in the isoelectric precipitates and consequently, the water-holding capacity became identical for all samples after freeze-thawing.

The pattern of the force deformation curve of the hemispheres of protein curds as measured with the Instron Universal Testing Machine was similar to all samples, although the magnitude was different. The hardness of the calcium precipitated curd was only half of the isoelectric precipitated curd and heating does not affect its hardness. Upon freeze-thawing, the hardness of the calcium precipitated curd increased 3.5 times and that of the isoelectric precipitated curd twice.

The calcium curd had a much higher cohesiveness than the isoelectric curd, which is a similar phenomenon as it was observed in the case of soy protein curds. It is apparent that the cohesiveness of the protein curd is influenced by the protein interaction mechanism with higher interaction energy resulting in greater cohesiveness. One of the interesting results observed by authors[113] was that the heat treatment of the protein solution increased the cohesiveness of isoelectric precipitated curd but decreased that of calcium curd in both yeast and soybean proteins. Consequently, when the protein solution was heated there was no difference in the cohesiveness of the curds between the two precipitation methods used. This indicates that the cohesiveness of the curd is not entirely governed by the molecular structure (globular or unfolded) and the two interaction mechanisms (van der Waals forces or ionic bonding) studied. An inverse relationship exists between the adhesiveness and cohesiveness of yeast protein curds; the higher cohesiveness resulting in a lower adhesiveness.

The growing importance of extrusion technology in the food industry encouraged the research work associated with production of extruded products.[117-119] In most of the products starch and protein are the main components not only quantitatively but also as components determining the properties of the final product. Changes in protein content and/or type of protein greatly influence the properties of end product.[114,115] The effect of yeast protein concentrate on starch extrusion products was investigated by Lai et al.[116] The yeast protein concentrate used in experiments was prepared by an alkaline treatment and acid precipitation (see Figure 80). Varying amounts of yeast protein concentrate were added to wheat starch

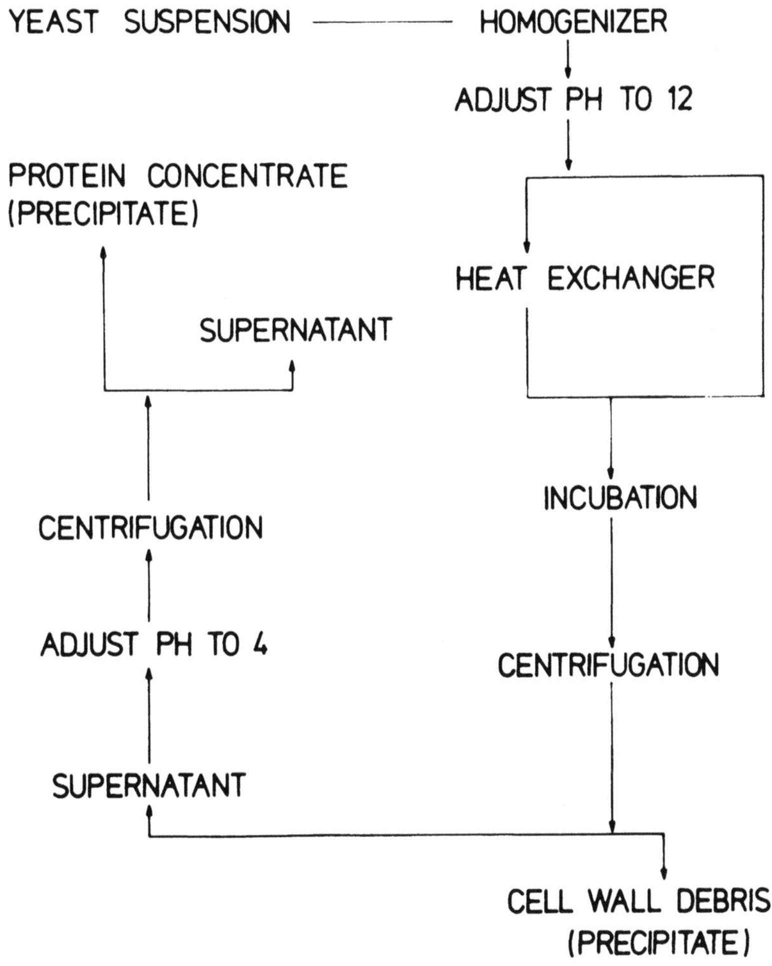

FIGURE 80. Production process for yeast protein concentrate (YCP). (From Faubion, J. M. and Hoseney, R. C., *Cereal Chem.*, 59, 533, 1982. With permission.)

and extruded. The effects of adding yeast protein concentrate were determined with changes in expansion, bulk density, breaking force and shear force of the extrudates. All these factors tended to decrease as the amount of yeast protein concentrate increased (Figure 81 and 82). Bulk density of extrudates containing yeast protein concentrate reached a minimum with a 2% addition and remained constant with higher levels of addition. The fact that a decrease in expansion was not accompanied by an increase in bulk density indicates that the production rate was altered. The addition of yeast protein concentrate in an amount of 2% decreased both breaking force and shear force of the extrudate. An addition level greater than 2% did not affect either parameter. Scanning electron micrographs showed a decrease in cell size which was roughly proportional to the yeast protein added. The addition of yeast protein concentrate also changed the distribution of cell size of extruded starch. Peripheral cells in the extruded starch and extrudate containing 1% of yeast protein preparation were relatively expanded and/or elongated, whereas those in extrudates containing high levels (6 and 10%) of yeast protein concentrate were small and round.

To determine the effects of yeast protein concentrate lipids, mixtures of starch and bound yeast lipids, defatted yeast protein concentrate and starch, and reconstituted yeast protein concentrate and starch were extruded. The addition of bound lipids of yeast protein preparation reduced the expansion of extruded starch without affecting bulk density, but increased

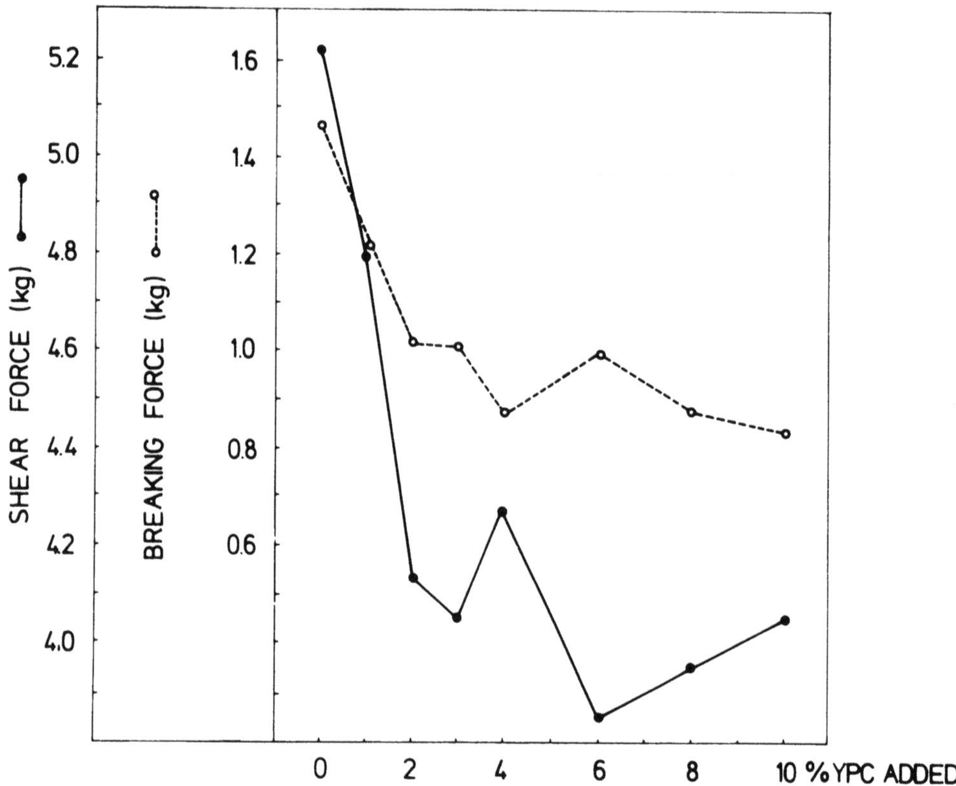

FIGURE 81. Shear force and breaking force of extrudates containing graded amounts of yeast protein concentrate (YPC). (From Faubion, J. M. and Hoseney, R. C., *Cereal Chem.*, 59, 533, 1982. With permission.)

the breaking force of the extrudate. Defatting of yeast protein concentrate did not significantly influence expansion, bulk density or breaking force of the extruded starch yeast protein concentrate mixture. The ultrastructure of extrudates containing intact and defatted yeast protein concentrate was very similar.

Reconstruction of the yeast protein concentrate by recombining defatted protein product with extracted lipids and starch did not restore the original effect of intact yeast protein concentrate. When compared with the mixture of starch and intact yeast protein preparation, the reconstituted yeast protein concentrate and starch mixture had higher bulk density. Examination by scanning electron microscopy of the ultrastructure of extrudates containing reconstituted yeast protein concentrate showed bands that consisted of either multiple layered structure or small thickwall cells. Areas of small and elongated cells were also often seen. These three features were not found in extrudates containing intact yeast protein concentrates. The process for preparing a condensed milk product was patented by Kruseman et al.[120]

After reducing RNA content, single-cell yeast proteins are suspended in water (pH = 7.2) and subjected to enzymatic hydrolysis (*B. cereus* protease) at 40°C. To inactivate enzyme suspension is boiled, centrifuged and supernatant dried. The protein material is mixed with lactose, butter oil, sucrose and water and the mixture pasteurized. The stability in hot coffee may be increased by adding basic salt, such as sodium or potassium carbonate or bicarbonate.

Protein-enriched beverages can be produced according to the patented procedure of Scharf and Schlingmann.[121] A beverage enriched with protein is obtained by extraction of microbial protein to reduce its content of nucleic acids and lipids. Enzymatic hydrolysis using one or more endoproteases, and isolation of the water-soluble fraction by membrane separation are then performed in protein processing.

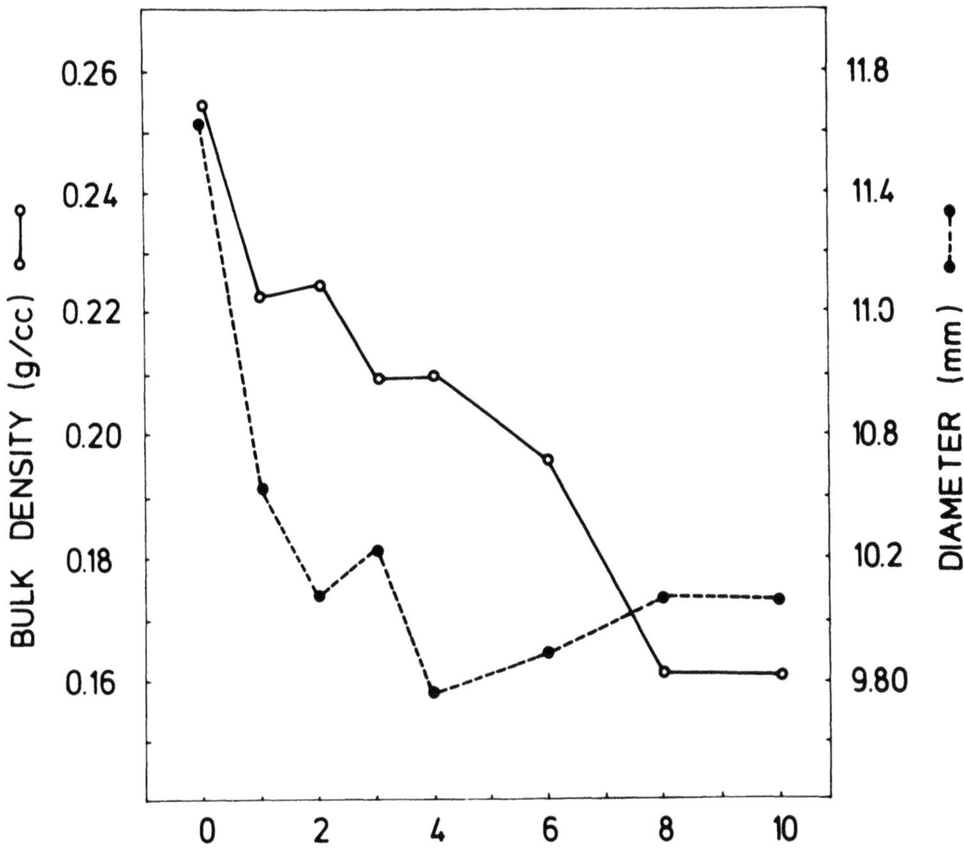

FIGURE 82. Expansion and bulk density of extrudates containing yeast protein concentrate (YCP). (From Faubion, J. M. and Hoseney, R. C., *Cereal, Chem.*, 59, 533, 1982. With permission.)

IV. IMPROVEMENT OF FUNCTIONALITY BY PROTEIN MODIFICATION

A. ENZYMATIC PROTEIN MODIFICATION

The fundamental aspects of enzymatic modifications of proteins are of interest since potential applications for nutritional and functional improvements of food proteins appear to be numerous and promising. Improvements in food protein quality by chemical or enzymatic modification, genetic methods, supplementation with free limiting essential amino acids should not be considered as competitive, but rather complementary methods.

Protein functionality can be altered or extended by enzymatic action to partially hydrolyze the polypeptide backbone, incorporate intermolecular or intramolecular cross-links, or attach specific groups to the protein.[123]

There are several advantages to utilizing enzymes to perform desired modifications of protein structure and function. Enzymes are generally specific in terms of the reactions they catalyze. Therefore, there is little potential for undesirable side reactions which, combined with the mild conditions necessary for enzyme catalysis, result in minimal tendency to form toxic byproducts. Furthermore, enzymes are effective catalysts at low concentration. Enzymatic catalyzed protein modifications can be divided into two groups, hydrolytic and nonhydrolytic reactions. *In vivo* enzymatic protein modifications occur under rather restricted conditions in contrast with *in vitro* modifications which are carried out under less specific conditions. Although only 20 amino acids are incorporated into proteins via translation, post-translational nonhydrolytic enzyme catalyzed reactions lead to 135 known modifications of the amino acid residues of proteins.[122] These modifications can be classified into six types of reactions: phosphorylation, glycosylation, hydroxylation, acylation, methylation and cross-linking. *In vivo* modifications are catalyzed by specific enzymes and are important for expressing biological activity. From these types of reactions only three have been achieved *in vitro* with enzymes — phosphorylation, cross-linking and tyrosine oxidation. Nonhydrolytic post-translational enzymatic modifications of proteins will not be treated further in this chapter. For reference see reviews of Whitaker et al.,[123] Whitaker[124] and Wold.[125] Hydrolytic modification *in vivo* and *in vitro* is the single most frequently occurring enzymatic protein modification.

Protease catalyzed reactions play an important role in biological systems like activation of enzymes by limited proteolysis, specific cleavage of polypeptide chain of presecretory proteins, protein-turnover, protein maturation, etc. *In vitro* hydrolytic modifications of protein include reactions that already have been applied to food systems or that represent potential uses for improvements of nutritional or functional properties of food proteins. Enzyme-catalyzed proteolytic reactions can be divided in two groups, the enzyme-catalyzed hydrolysis of peptide bonds and the enzyme-catalyzed peptide bond synthesis (plastein reaction).

1. Modification of Proteins Through Enzyme Catalyzed Peptide Hydrolysis

Proteases are the most widely used enzymes for modification of food protein. They hydrolyze selected peptide bonds to promote reduction of molecular weight, possible conformation changes, and enhanced hydrophobicity due to newly exposed amino and carboxyl groups. Effective hydrophobicity of certain globular proteins conceivable could be increased as well through exposure of apolar amino acid residues upon limited hydrolysis and subsequent unfolding of the polypeptide chain.

General effects of proteolysis on size, structure and polarity could result in dramatic changes of protein behavior. Specific properties of the reaction product are dependent upon the degree of hydrolysis which is influenced by the specific activity of the protease, physical and chemical properties of the protein substrate and reaction conditions.

Proteolysis increases solubility and decreases viscosity. In addition, proteolysis generally leads to reduced tendency of gelation, increases foam volume upon whipping, enhances thermal stability and decreases foam stability.[126] Emulsifying activity may be altered as well; however, the specific effect varies with the nature of the substrate.

A critical problem hindering the use of proteases for improving functional behavior is excessive hydrolysis. Although limited proteolysis often may yield a more functional protein, it is difficult to limit the degree of digestion to that which promotes desired performance. Enzymes may be inactivated thermally; however there is a risk of causing undesirable changes in the protein of interest such as insolubilization through denaturation and eventually full loss of functionality. A possible alternative for controlling the extent of proteolysis may be the use of immobilized enzymes. Reactors with immobilized enzymes should allow production of hydrolysates with the desired uniform degree of digestion, therefore more constant and reliable functional behavior. Another possibility was mentioned by Haard et al.[180] proposing the use of proteases isolated from marine fish. The poor thermal stability of fish proteases indicate that these enzymes can be inactivated intentionally by relatively mild heat treatment.

Proteolytic modification of proteins can be achieved not only by use of pure enzyme preparations but also by applying proteases of living microbial cells, particularly if enzymes are excreted by the cells. Although yeast cells generally have intracellular proteases, in given conditions due to partial permeabilization yeast proteases can be excreted in the medium. The proteolysis of food proteins *in situ* is mainly solved by living microbial cells.

The preparation of cheeses and soy derivatives, chillproofing of beer, gluten modification in doughs and production of protein hydrolysates represent the major food production technologies where yeast proteases are involved. Kang et al.[127] reported on utilization of yeast culture (*S. fragilis* and *D. hansenii*) for soft cheese production. Yeast was added to pasteurized milk inoculated with lactic acid starter culture before renning. The resulting yeast cheese had a soft texture and compared with a similar cheese made with lactic acid bacteria, only ripened more quickly. Detana et al.[128] added *Debaromyces hansenii* to ewes's milk at the same time as the cheese starter for manufacture of Pecosino romano cheese. Utilization of yeast resulted in a more rapid proteolysis and overall ripening.

Hagrass et al.[129] investigated the effect of yeast supplementation to the curd at concentrations of 1, 2, 3 or 4%. Addition of dry yeast up to 1% increased the water-holding capacity. At the same time soluble tyrosine and soluble tryptophan concentration increased. Authors concluded that the ripening of Ras cheese may be accelerated by the addition of 1% dry yeast. Application of yeast in the production of fermented soy products results in enhanced flavor. Higashi et al.[130] investigated salt tolerance of miso and soy sauce yeast. Fermentation of red pepper paste Kochujang with mixed cultures, of yeast strains (1) *Saccharomyces rouxii* and *Torulopsis versatilis*, (2) *Saccharomyces rouxii* and *Torulopsis etchelsii* and (3) *Torulopsis versatilis* and *Torulopsis etchellsii* were investigated. Best results were achieved with mixed culture 3 as it has the best liquefying and saccharogenic amylase activities and amino *N* also increased dramatically and reached the highest value (220 mg %) in comparison to 1, 2 and control Kochujang without starter.[132]

Use of proteases appears promising in the area of protein-supplemented beverages. Complete solubility achieved with partial enzymatic hydrolysis will permit protein fortification of acidic soft drinks and citrus juices. In addition, the thermal stability of the hydrolysates will minimize solubility loss upon beverage sterilization. Yeast protein hydrolyzed by proteinases (e.g., papain) can be used for the production of protein beverages.[133]

Formation of bitter peptides upon hydrolysis of certain proteins is another problem associated with enzyme catalyzed protein hydrolysis. It is generally accepted by researchers that the tendency for a protein to form bitter peptides is related to the content of hydrophobic amino acids. Casein and soy isolate are well known examples of proteins for developing bitter hydrolysates.[126]

The correlation between the composition (structure) of peptides obtained by hydrolysis and the bitter taste of some peptides was intensively investigated in different research laboratories. Otagiri et al.[181] investigated the taste of peptides including arginine, proline and phenylalanine residues. It was started by the authors that peptides containing arginine, proline and phenylalanine generally have a bitter taste. It has also been found that the hydrophobic amino acid is generally located at the C-terminal of bitter peptide and the basic one on the N-terminal. Very strong bitter taste was observed when the sequence of peptide arginine-proline (Arg-Pro) units were present, e.g., an octa-peptide having an Arg-Arg-Pro-Pro-Pro-Phe-Phe-Phe sequence was found to have the strongest bitter taste.

Bitter tasting tetrapeptides were studied by Nosho et al.[182] Sixteen synthetic model-peptides were investigated. The Arg-Pro-Phe-Phe tetrapeptide had a strong bitter taste, however the peptides Arg-Pro-Gly-Gly and Gly-Gly-Arg-Pro were tasteless.

Halász et al.[183] investigated the taste effect of peptide groups isolated from brewer's yeast hydrolysate using gel chromatography. Six fractions were separated using Biogel-P4 columns (see Figure 83). The fractions were tested organoleptically. Fractions 4 and 5 were found to have weak and strong bitter taste, respectively. The amino acid composition and C-terminal amino acids of fractions were also detemined. The fractions 2 and 3 having a meaty flavor contained high quantities of basic amino acids, glutamic acid and aspartic acid. The dominating C-terminal amino acids were alanine, leucine, isoleucine and valine. No acidic amino acid was found at the C-terminal position. The strongly bitter fraction 5 had high phenylalanine content both in total amino acid composition and in C-terminal groups.

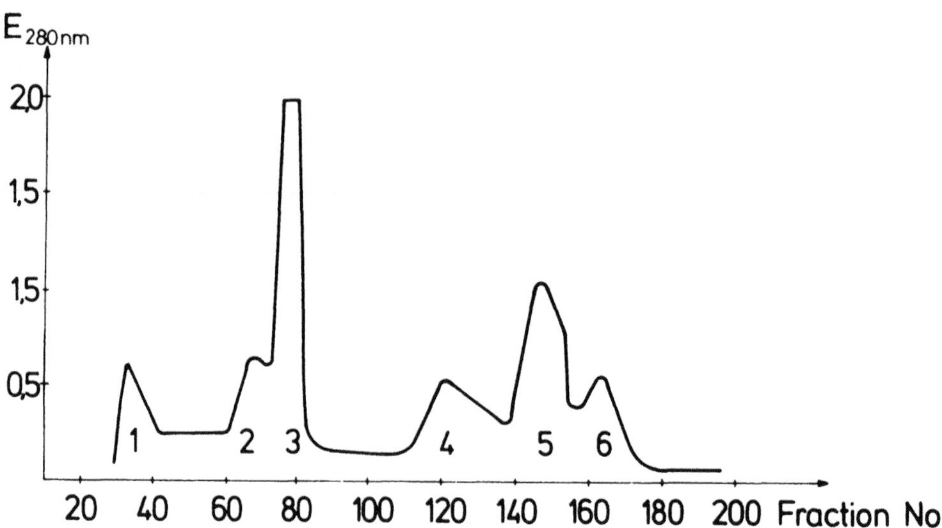

FIGURE 83. Fractionation of brewer's yeast autolysate on Biogel-4 column.

Bitterness is generally considered to be due to high hydrophobicity, especially if the C-terminal residue is hydrophobic. The results reported above confirm these views.

Enzymatic hydrolysis of yeast (by external enzymes or by its own intracellular enzymes during autolysis) must also be optimized for taste quality of autolysates got at different pH values. The biochemistry of yeast autolysis and use of yeast autolysates are treated in Chapter 5 of Part I and Chapter 9 of Part II.

To eliminate or at least to reduce bitterness to an acceptable level, different procedures may be used. Most commonly, activated carbon, carboxypeptidase-A, leucinaminopeptidase, ultrafiltration or plastein reaction is used for such purposes.

Due to the hydrophobic nature of the bitter peptide hydrophobic chromatography on the phenolic resins (e.g., Duolite S-761) hexyl-Sepharose B may be suitable to remove the bitter peptides and amino acids including phenylalanine. Solvent extraction of hydrolysates with azeotropic secondary butanol is very effective in reducing bitterness. Some examples of debittering hydrolysates of different proteins are shortly reviewed by Fox et al.[184] Limited hydrolysis of a protein product may improve certain functional properties such as whipping, emulsifying capacity, taste and solubility. These improvements are dependent on the enzyme and on the principle by which the reaction is controlled. The preferable way of controlling the modification is by the application of the pH-stat technique, by which the base consumption used for maintaining pH during the proteolytic reaction is directly converted to a degree of hydrolysis.[134] The degree of hydrolysis is the proportion between the number of peptide bonds cleaved and the total number of peptide bonds in the intact protein. Proteolytically modified proteins which have been thoroughly digested are low molecular protein hydrolysates. Such products often have less pronounced foaming or emulsifying properties than proteins, which have been only slightly hydrolyzed.[134] However, a need for this kind of protein product appears in the beverage industry for enrichment of soft drinks with protein and in the meat industry for pumping of whole meat cuts with low-cost proteins. Important properties of low molecular protein hydrolysates are a bland taste and a complete solubility over the wide pH-range used in foods.

In the production of low molecular protein hydrolysates hyperfiltration may be used for concentration and/or desalination. Instead of using the controlled batch hydrolysis and solids separation processes, the separation of peptides may be performed from an enzyme-substrate reaction mixture under continuous ultrafiltration in a so called membrane reactor.[134]

2. Enzymatic Peptide Synthesis

Enzymatic peptide synthesis is another example of proteinase catalyzed protein modification. In this reaction often partially hydrolyzed proteins are the substrate for further peptide modification. Potential uses of such types of reactions are removal of color, flavor, and pigments, removal of bitter peptides, removal of unwanted amino acids (e.g., phenylalanine), increased solubility, whippability, surface active properties, geling activity, increase of nutritive value by incorporating limiting essential amino acids, etc.

The protease catalyzed reaction has the following steps (see Figure 84): (1) formation of the enzyme-substrate complex by association and (2) a covalent acyl-enzyme intermediate with release of the other part of the polypeptide chain (P_1). The intermediate acyl-enzyme is hydrolyzed under normal conditions, releasing the other part of the original peptide P_2 (degradation, hydrolysis). Two additional reactions may occur depending on the presence and concentration of other nucleophiles in addition to water. The synthesis of new peptide bonds may occur by the amino group of a peptide competing with water for the acyl group of the acyl-enzyme intermediate (synthesis; a transpeptidation reaction), or the enzyme react with a P_2-like compound to form an acyl-enzyme intermediate (reversal of hydrolysis) followed by transpeptidation as shown for synthesis. Under physiological conditions, the equilibrium conditions in a protease-catalyzed reaction largely favors proteolysis, the cleavage of peptide bonds. However, according to the principle of microscopic reversibility,[135] the proteases indeed possess the ability to catalyze the synthesis of peptide bonds as well. Consequently, it is the equilibrium point of the reaction not the nature of the enzyme, that decides whether bonds will be synthesized or broken. If the equilibrium position can be shifted toward synthesis then formation of peptide bonds may proceed to a significant extent as it has been confirmed by a number of successful protease-catalyzed peptide synthesis.[136-140]

As mentioned before, it is well established that under physiological conditions, the equilibrium position in protease-catalyzed reactions is far over in the direction of proteolysis. The predominant contribution to the energetic barrier to proteosynthesis is accounted for by ionization-neutralization effects. The implication, from the point of view of peptide bond formation at physiological pH values, is that significant thermodynamic work is required for the transfer of protonated α-amino/group of one reactant to the deprotonated α-carboxyl group of the other reactant.[141]

In chemical terms, the formation or cleavage of a peptide can be considered as an electrophylic acylation of or an acyl group transfer to — an α-amino group of an amino acid or a water molecule, respectively.

An energetically favorable transition state can be obtained if, prior to reaction, either the nucleophilicity of the acyl group acceptor or the electrophilicity of the α-carboxyl carbon of the acyl group donor is increased. The second alternative is generally adapted in peptide synthase chemistry, the electron density of the α-carboxyl group is decreased by introducing an electronegative substituent which either replaces a hydrogen or a hydroxyl group of the carboxyl moiety. This so called carboxyl-activation provides an energy-rich intermediate of high acyl group transfer potential which readily undergoes aminolysis or hydrolysis, thus generating or destroying peptide bonds.[141]

A typical form of an activated acyl group donor is the so called active ester form RCCO-R′ (R′ = alkyl, acyl, aryl). In *in vitro* protein synthesis the peptidyl-t RNA represents this active form which exhibits a strong acylating power.[142]

In protease catalyzed peptide bond synthesis this carboxyl activation is also achieved, in this case the energy-rich intermediate exhibiting a high acyl group transfer potential is represented by the covalent acyl-enzyme complex. As the enzyme and its substrate is linked to each other through an ester bond, chymotryptic and other serine protease catalyzed reactions proceed via an active ester analogous to many chemical peptide syntheses.

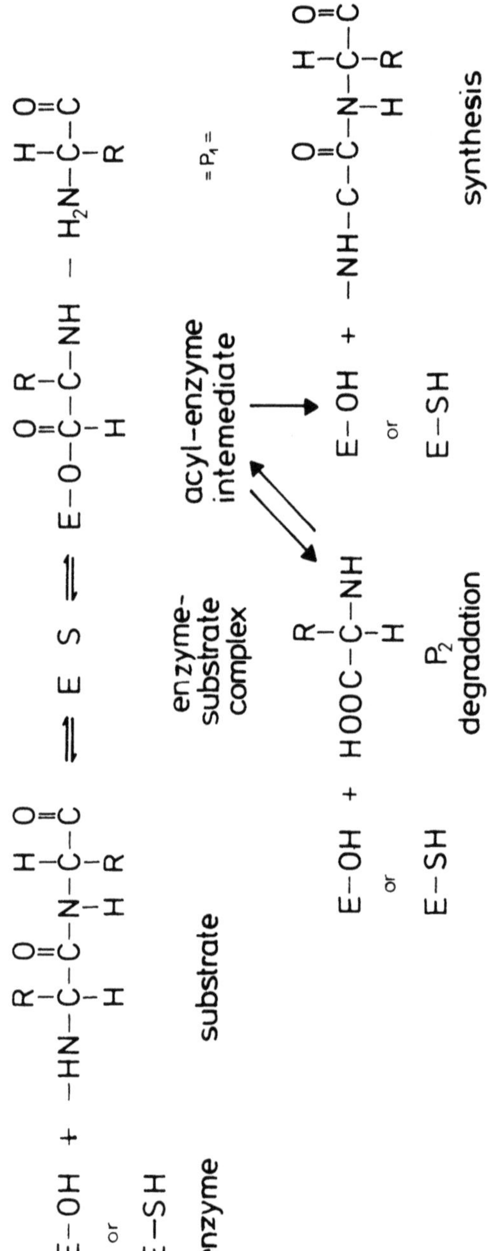

FIGURE 84. The protease catalyzed peptide reaction-mechanism.

The enzyme-mediated process is an intramolecular reaction, this intramolecular character contributes conically to its efficiency. The catalytic pathway followed by serine proteases such as α-chymotrypsin, trypsin, subtilisin, carboxy-peptidase Y appear similar to each other. The mode of activation is through the covalent bond of their substrates to a serine residue to form an *O*-acyl complex. Cysteine proteases papain, ficin and bromelain, covalently bind their substrates through an activated thio-ester.[143]

A highly active acyl-enzyme complex can also be generated via an anhydride linkage between the β- and γ-carboxyl groups, respectively, of the protease-inherent amino dicarboxylic acid residue of the protease and the α-carboxyl group of its substrate. This type of activation is analogous to the so called mixed anhydride method frequently employed during chemical peptide synthesis. Proteases which favor this mechanism are pepsin, carboxypeptidase A and thermolysin.[143]

The activated acyl-protease complexes of both ester and anhydride type are at a high energy level and can therefore readily undergo deacylation by nucleophilic attack. If the nucleophile is the α-amino group of an amino acid residue, the activated complex is aminolyzed, the acyl-group is transferred to the amino acid, and the synthesis of a peptide bond takes place. The partition of the acyl-protease complexes between destruction and synthesis does not only depend on the concentration of the nucleophilic agents but is also strongly influenced by the nucleophilic strength of the competing agents. In this respect, the amino acid — especially C-terminally protected ones — have an unequivocal advantage over water.[140,143] The main barrier to peptide bond synthesis comes from the energy required for the proton transfer depends crucially on the ionization equilibria of the reactants. The highly endergonic characters of the proton transfer is caused by the strong acidity and basicity, of the α-carboxyl and α-amino groups, which can be attributed to the zwitterionic nature of free amino acids and peptides. A favorable pH shift of the ionogenic groups may be achieved by reducing the zwitterionic character which can be done by introduction of α-amino- and α-carboxyl-protecting groups. Another possibility to shift ionic equilibria in favor of peptide synthesis is to add organic solvents to the reaction mixture. Peptide synthesis has been performed in solvents from co-solvents with up to 50% water to nearly anhydrous solvents.[145,146] By lowering the dielectric constants of the medium, the organic solvent diminishes the hydration of ionic groups, this predominately effects carboxyl functions.

To increase hydrophobicity and solubility in organic solvents, amino acid derivatives should be used. However, enzyme stability must also be considered in these systems. For example, in butanediol with water contents of 10% v/v chymotrypsin immobilized to Sepharose or chymotrypsin immobilized to chitin in 95% acetonitrile show poor stability. Much more stable enzyme preparations are possible in water-immiscible solvents with low water contents.[146] Organic solvent also has influence on thermal stability of the enzyme. Enzymes inactivate at high temperatures in aqueous media due to both the partial unfolding of the enzyme molecule and covalent alterations in the primary structure. For these inactivation processes water is required. Thus enzymatic catalysis in organic solvents obeys the conventional chemical property of enhanced reaction rates at elevated temperatures.

A general shift in substrate specificity takes place when enzymes are suspended in low-water environments. Organic solvents also affect the binding of substrates to the active site by altering the apparent K_m values. Hydrophylic and amide amino acids underwent peptide synthesis in benzene much more rapidly than did hydrophobic amino acids, contrary to the same reaction in water or biphasic systems.[146]

Chain length of the reactants influences both the acidity and the basicity of a peptide. The energy required for the proton transfer decreases with increasing chain length. Prospects of peptide synthesis will generally improve with growing chain length.

The formation of a peptide bond represents a condensation process during which two molecules are connected to each other via an amide bond to generate a new molecule. The

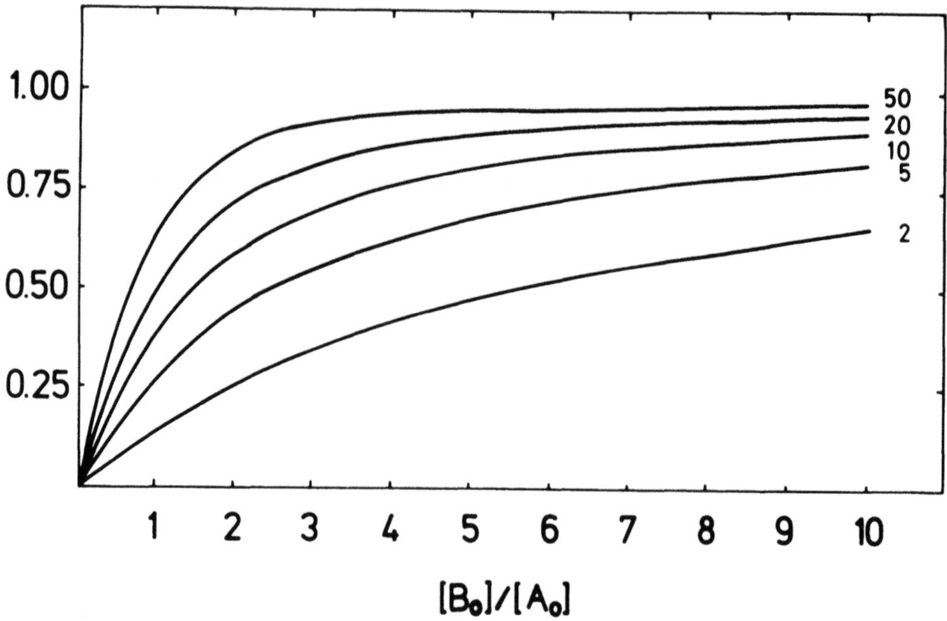

FIGURE 85. The effect of concentration ratio A to B and reaction time on product formation. (From Kullman, W., *Enzymatic Peptide Synthesis*, CRC Press, Boca Raton, FL, 1987. With permission.)

degree of synthesis α, the relative amount of synthesis of a dipeptid A-B from the constituent amino acids A and B can be deduced from the low of mass action according to Kullmann.[141]

$$\alpha = K_{syn} \frac{([A_0] + [B_0]) + 1}{2K_{syn}[A_0]} - \frac{\sqrt{(K_{syn})^2[A] - [B_0]^2 + 2K_{syn}[A_0] + [B_0] + 1}}{2K_{syn}[A_0]}$$

The effect of concentration ratio A to B on the degrees of synthesis can be of utmost importance.

The synthesis of a peptide bond is accompanied by the release of a water molecule. So the presence of organic solvent causes, in addition to the ionic equilibrium shift, a drastic reduction of water concentration and consequently, a shift of the chemical equilibrium in favor of peptide synthesis.

In the course of protease — catalyzed process, the target product RCO-NHR′ merely represents an intermediate form, which is object to subsequent secondary hydrolysis. The concentration of the product is a function of time and has a maximum (see Figure 85) where the yield of the intermediate may exceed the equilibrium level. It is of utmost importance to stop the reaction at the kinetic optimum.[141] This statement is confirmed by experimental results of Kuhl et al.[147] who determined that the yield of Ac-Leu-Phe-Leu-NH_2 synthesis is a function of reaction time and depending on the organic solvent utilized highest values were after 20 and 45 min reaction times resp. further increase of time causes decrease of yield.

The same research group[147] investigated the effect of salts on enzyme-catalyzed peptide synthesis and found that ammonium salts at a concentration of 0.1 mmol improved yield of the reaction.

3. The Plastein Reaction

In 1886 Danilewsky[148] discovered that an insoluble protein-like substance is formed if a small amount of pepsin or pancreatin is added to a concentrated solution of peptone or to a neutralized proteolytic digest of ovalbumin or other proteins. In 1898 Van't Hoff[149] suggested that trypsin might possess capacity to catalyze the synthesis of proteins from degradation products originally generated by its own proteolytic action. The term plastein was used by Savjalov[150] in 1901 to designate the precipitate resulting from the addition of rennin to a partial-hydrolysate of fibrin (peptone). Numerous studies on the subject of plasteins were published in subsequent years but it was not until the 1960s that the mechanism of plastein formation was elucidated by using well-characterized oligopeptides as substrates for the protease-catalyzed reaction.

Determann and co-workers,[151,152] Determann and Köhler[153] reported molecular weight distribution of enzymatic condensation products, the specificity of pepsin in the enzymatic modification of prolin-containing peptides. Experiments in which the pentapeptide L-tyrosyl-L-leucyl-glycyl-L-glutamyl-L-phenylalanin was the substrate and pepsin the catalyzer the dimer, tetramer and pentamer of the original peptide could be detected in the plastein product.[151]

Efficiency of the pepsin-catalyzed peptide bond formation is highly influenced by the character of C-terminal amino acid. Similar to the pepsin-catalyzed hydrolysis also in the plastein reaction C-terminal phenylalanine and tyrosine gives the highest synthesis rate while alanine at the C-terminal prevents the condensation in the plastein reaction.

Pepsin has a significant stereospecificity in peptide bond cleavage, the same was confirmed for plastein formation. Determann et al.[152] found a more pronounced specificity of pepsin in peptide condensation than in peptide-bond hydrolysis.

Chain length of the substrate is of importance for peptide bond synthesis, Determann and Köhler[153] stated that monomers have to consist of at least four amino acids to act as substrate in the plastein reaction. Horowitz and Haurowitz[154] studied plastein formation catalyzed by chymotrypsin. Besides specificity of the enzymic reaction they stated that free amino acids were not incorporated only their derivatives like ethyl esters, which were bound in peptide linkage. Since the α-amino nitrogen remained unchanged during plastein reaction it was concluded that transpeptidation reactions are primarily responsible for chymotrypsin catalyzed plastein synthesis.

Basic requirements for the plastein reaction are different from those optimal for hydrolysis:

1. Substrate has to be a low molecular weight peptide or peptide mixture[153-155]
2. Extremely high substrate concentration (20 to 50%) can serve as driving forces for this reverse reaction[156]
3. The reaction system should be kept between pH 4 to 6, irrespective of the optimal pH values of proteinases in their usual proteolysis reaction[157]

Highly hydrophilic protein hydrolysates are not effective substrates because the products are still soluble and eliminated with difficulty from the reaction system.[158] But highly hydrophobic protein hydrolysates are also ineffective because the products easily precipitate before growing sufficiently.[159] Concerning the exact mechanism of plastein reaction there is controversy. Because of the heterogenous peptide substrate (partial protein hydrolysate) there are difficulties in examining an insoluble material such as plastein. A few theories explaining the mechnaism have been proposed.

Yamashita et al.[161] proposed that plastein reaction is mainly a polycondensation reaction based on the fact that they demonstrated an actual peptide-enzyme intermediate. If plastein reaction is merely a formation of new peptide bonds a decrease in the total number of free amino groups should take place. As mentioned, Horowitz and Haurowitz[154] were unable to find any decrease of free amino groups during the plastein formation. A similar observation was made by v. Hofsten.[162] v. Hofsten and Lalasidis,[163] and Monti and Jost[164] could not detect significant molecular weight increase during plastein formation. Hajós and Halász,[165] investigating changes in molecular weight distribution of plastein products in comparison to their substrates, stated that during plastein reaction transpeptidation occurs which results in lower and higher molecular weight products than those of substrate. In case of a papain-catalyzed reaction a relatively greater amount of high molecular weight (30 kDa) product could be detected which hint at polycondensation.

Results that amino acid esters and not free amino acids are incorporated into plastein seems to confirm that plastein is formed essentially by transpeptidation, i.e., by the transfer of amino acyl residues from donor peptides to acceptor peptides or amino acids.

Indications that both reactions may occur are published by Determann and Köhler[153] who were able to detect condensation product and transpeptidation product from different pentapeptides and also a homolog of the monomer.[155] Some authors[166,167] believe that aggregates produced during plastein reaction are not a result of new covalent bonds but due to hydrophobic and ionic bonds. Results of v. Hofsten[162] on gel formation of whey-based plastein catalyzed by esterase showed that the amount of low molecular peptides increased indicating that the gel formation is independent of an increase in peptide chain length. Furthermore, gels produced under optimal conditions were soluble in 8 *M* urea or 6 *M* guanidine at neutral pH which indicate that the gels are not held together by covalent bonds. Aso et al.[160,168] also demonstrated that molecular chains assembled during plastein reaction and turned reaction mixture turbid could be more or less solved and turbidity clarified by treatment with urea, guanidine hydrochloride, sodium dodecylsulfate and some alcohols. Based on these results it was concluded that hydrophobic forces are a major factor in plastein chain assembly.

Apparently, different reactions are involved in the plastein reaction. If they are condensation reactions, a decrease of free amino groups would have to occur which is not stated. Transpeptidation is hard to accept as the sole factor responsible for the formation of insoluble or gel-forming products.[168] It seems most likely that the plastein reaction is an entropy-driven process, increase in entropy of water acting as the driving force[170] after an initial concentration of suitable peptides have been formed by either condensation, transpeptidation or both.[169]

Plastein formation is characterized according to Arai et al.[171] by plastein productivity (α) insolubility (β) and β/α values. Plastein productivity is given as

$$\alpha = \frac{10\%\text{TCA} - \text{insoluble nitrogen}}{\text{total substrate nitrogen}} \times 100$$

Insolubility

$$\alpha = \frac{20\%\ \text{acetone} - \text{insoluble nitrogen}}{\text{total substrate nitrogen}} \times 100$$

It was shown that hydrophilic substrates where β/α values are lower than 0.4 and hydrophic substrates where β/α is higher than 0.6 produced plastein in low yields.[171] Substrates which could be characterized by β/α approximates 0.5 produce plastein much more efficiently.

Substrates which are a 50:50 mixture of hydrophilic and hydrophobic groups have plastein productivity (α) higher than the arithmetic average of the α values of the individual substrates, with a β/α value approaching 0.5. The plastein productivity of yeast protein hydrolysate (*Candida* sp.) was found to be 86.5% and β/α = 0.47.[171] Best results were published for ovalbumin hydrolyzate where α was more than 90% with β/α value of 0.49. Therefore, with relative little knowledge of different hydrolyzates it is possible to find mixtures giving maximum plastein yield (90% or more).

In the presence of NaCl increase of plastein yield could be achieved.[172] At a lower concentration (0.1 *M*) it was considered to be a result of enzyme activation and at the higher concentration (0.8 *M*) a salting out effect of products occurs. The favorable effect of salt addition was also stated by v. Hofsten[162] and is in agreement with findings related to enzyme-catalyzed peptide synthesis.[147]

The term plastein has different meanings in various publications. It is not only referred to as the whole formed gel but also to this or that gel fraction separated on molecular weight or solubility. The plastein reaction takes place in solution of peptides with a molecular weight between 500 to 2000 g/mol, at a concentration exceeding 10% and pH between 4 and 7.

The structure of plasteins resembles that of denatured proteins. X-ray diffraction studies on plasteins revealed the presence of linear polypeptides with close packed chains. Structures resembling the secondary and tertiary structure of the proteins from the hydrolyzates of which they have been obtained have not been detected in the plastein investigated. The fact that the structure of the initial protein is not restored on plastein formation is confirmed by the absence of enzymatic activity in plasteins obtained from enzyme protein hydrolysates (pepsin and trypsin) and the same manner hormone activity disappeared in plasteins obtained from insulin.

In several aspects plasteins react like proteins — they form colored compounds with ninhydrin and give complex compounds with Cu^{2+} — can be precipitated with trichloroacetic acid and hydrolyzed by various proteolytic enzymes. Despite this protein-like behavior plasteins do not have an ordered structure. Belikov and Gololobov[185] suggest that plasteins consist of a conglomerate of intertwined polypeptide chains with different lengths and structures.

In the wide range of experimental conditions transpeptidation was shown to be an only chemical reaction causing the plastein structure formation. This standpoint is supported by the estimations of the number-average molecular weights of the reaction mixture and by the measurements of the quantity of TCA-insoluble fractions. Transpeptidation results in an increase of the reaction mixture viscosity. This process has two steps as it was published by Gololobov et al.[186] and these two steps have different characteristic times. During the first step, which lasts for several minutes, enzyme-catalyzed transpeptidation occurs. The second step takes a longer time and is characterized by the noncovalent interaction of the polypeptides formed during the first step. Investigations by transmission and scanning electron microscopy shows that in plastein reactions the aggregation of peptides takes place. The study of peptide bond thermodynamics of model peptides and kinetics of protease-catalyzed acyl-transfer shows that transpeptidation is the result of kinetic reasons.[186]

The nutritional quality of plasteins is of great importance. *In vitro* digestibility of plastein showed comparable digestibility for the original protein and the resulted plastein.[173] Rat studies showed an average *in vivo* digestibility of 90.3%, similar to that of soy protein isolates. Results of Noack and Hajós[174] and Hajós et al.[188] confirmed that the casein-based enzymatic modified product digested under physiological conditions showed no impairment in digestibility in comparison to casein. Enzymatic protein modification by plastein reaction was proposed by Yamashita et al.[187] to prepare a low-phenylalanine, high tyrosine product satisfying the special requirements for phenylketonuria patients. Fish protein concentrate and soy protein isolate were used as protein sources. Each substrate component was hydrolyzed subsequently by pepsin and pronase. Aromatic amino acids were removed by absorption to Sephadex G-15. The aromatic acid free hydrolyzates were substrates for the plastein reaction where tyrosine ethyl ester and tryptophan ethyl ester were incorporated. The resulting plasteins contained no free amino acids and were almost tasteless. Low phenylalanine (<0.25%) and high tyrosine (>7.6%) content of the plasteins make them acceptable as food materials for patients with phenylketonuria.

Plasteins from combined protein hydrolyzates with complementary essential amino acid profiles seem to be a more economic way of nutritional value improvement than incorporation of pure amino acid esters. Yamashita et al.[173] first studied the production of combined plastein with soy protein hydrolyzate and hydrolyzed wool keratin and with hydrolyzed soy protein and hydrolyzed ovalbumin. Wool keratin and ovalbumin are rich in sulfur amino acids which is the first limiting essential amino acid for soy protein. Plastein reaction can be catalyzed either by soluble or immobilized proteinases. Pallavicini et al.[175] compared the effect of soluble α-chymotrypsin with α-chymotrypsin immobilized on chitin on plastein formation from low molecular weight peptides of a peptic digest of soybean protein. The plasteins were finger-printed on silica gel by a combination of electrophoresis and TLC, eluted, hydrolyzed and amino acid composition determined. The results indicated that the plasteins from soluble α-chymotrypsin were richer in glycine, valine, leucine and serine than the plastein prepared by immobilized α-chymotrypsin. Plasteins prepared by immobilized α-chymotrypsin catalyzed were rich in glutamic acid, lysine, alanine and threonine, suggesting a much more hydrophilic plastein than prepared from the soluble form of the enzyme. Könnecke et al.[176] investigated immobilized trypsin for peptide synthesis and were able to reutilize the enzyme five times. Peptides prepared by this technique were invariably obtained in good yields. Immobilized papain proved to be less successful than soluble enzyme. Obviously the cleavage of the ester bond of the carboxyl component did not proceed faster than the secondary hydrolysis of the newly formed peptide bonds. Consequently, the product yields were rather unsatisfactory.[175] Best results were achieved with immobilized thermolysin, which gave a markedly improved product yield of 60%.

Enzyme catalyzed peptide bond synthesis may result in a oligomerization process of the incorporated amino acid. As investigations of Gaertner and Puigserver[177] stated the polymethionyl-casein type peptide has a good *in vitro* digestibility. The plastein reaction proved to be a convenient means for the elimination of the bitter taste from enzymatic protein hydrolyzates. It has been found that, when the bitter hydrolyzates are used as substrates of the plastein reaction, the resulting products are tasteless.

It is thought that the disappearance of the bitter taste in the plastein reaction is due to enzyme-catalyzed transpeptidation. This hypothesis could be confirmed by the reappearance of bitter taste after enzymatic hydrolysis of plastein product. Plasteins enriched on glutamic acid have no taste, like ordinary plasteins. Specificity of such products is that even after hydrolysis by endopeptidases the bitter taste does not reappear. Explanation for this unique behavior can be found in the tasteless peptide fraction of glutamic-acid enriched plastein hydrolysates. Principal component of this peptide fraction is diglutamic acid which is capable of masking the bitter taste not only of peptides but also of other bitter substances such as caffeine. Consequently, in this case the bitter taste disappears as a result of the masking effect of diglutamic acid and not as a result of the formation of new peptides. This debittering activity also acts in the mixture of diglutamic acid-containing peptide fraction and bitter peptide fraction.

The plastein reaction can be useful not only in preparation of protein-like substances with increased protein value or tasteless products from bitter hydrolyzates, but also to improve rheological properties. Plasteins provide the possibility of use for preparation of protein pastes and gels with high food value as fillers. Shimada et al.[189] reported improved surface properties or enzymatically modified proteins. Partially hydrolyzed gelatin and leucine hexyl ester and leucine dodecyl ester, respectively, were the substrates for plastein reaction catalyzed by papain. Plastein products had an average molecular weight of 7500 kDa and hexyl or dodecyl contents 1:1 mol/7500 g resp. Rheological properties were compared with control gelatin based plastein without luecine ester incorporation (GH), bovine serum albumin, casein, Tween-60 and Amplutol-20 BS.

It has been demonstrated that the rate of reduction in surface tension observed for proteins depends on their structure; flexible molecules as on casein can give rapid reduction in surface

TABLE 71
Physical Properties Investigated for Enzymatically Modified Proteins and Control Samples in Aqueous Systems at Various pH Values

Item investigated	pH	EMG-6	EMG-12	GH	BSA	Casein	Tween-60	Amphitol-20BS
Whippability	1	3.8	3.6	2.6	3.5	2.8	2.8	5.2
	3	3.9	3.6	2.7	3.5	2.6	2.7	5.4
	5	4.1	3.6	2.6	3.8	0.6	2.6	5.6
	7	4.1	3.1	2.5	3.8	4.0	2.8	5.2
	9	4.1	2.6	2.4	3.8	4.4	2.9	5.4
	11	3.9	2.6	2.0	4.0	4.4	2.9	6.0
Foam stability (%)	1	46	10	12	20	50	20	16
	3n	46	10	10	24	18	20	16
	5	46	10	10	20	30	20	14
	7	47	14	10	28	32	20	18
	9	45	16	14	32	32	20	18
	11	40	20	8	28	32	20	2
Emulsifying activity index (m^2/g)	1	20	22	a	12	7	24	15
	3	23	22	a	12	a	24	21
	5	23	26	a	15	a	23	13
	7	25	26	a	13	12	24	20
	9	24	23	a	13	14	23	19
	11	17	21	a	12	17	26	18
Emulsion stability (%)	1	78	100	0	60	74	77	72
	3	80	100	0	61	50	75	72
	5	79	100	0	57	51	77	53
	7	80	100	0	59	73	79	64
	9	79	100	0	60	70	77	66
	11	79	100	0	64	88	77	66

[a] Not determined because of insolubility or unstableness.

tension. The hydrophilic moieties of the modified gelatin products are considered to have flexible structures because the starting material, gelatin, is flexible. Such structures of the plastein will be responsible for the rapid reduction in the surface tension. Since flexible structures of the enzymatically modified gelatin and GH are considered to be similar in a hydrophilic moiety, amphiphilic structures by incorporation of leucine alkyl ester to gelatin will be responsible for the difference between the rate of reduction in surface tension observed for the different plastein products.

An amphiphilic structure of the protein molecule has been assessed by surface hydrophobicity, suggesting the larger the surface hydrophobicity is, the better the surface properties are. The localization of the hydrophobic region also contributes to surfactancy of the proteins. The gelatin-leucine ester plastein has excellent foam stability, and emulsion stability and high emulsifying activity. Whippability is similar to those of BSA (see Table 71). In spite of their proteinaceous nature, the plastein type surfactants behave as if they were nonionic and low-molecular weight, synthetic surfactants.

4. Application of Plastein Reaction to Modify Yeast Proteins

Arai et al.[171] investigated the plastein reactivity of a *n*-paraffin assimilating yeast (*Candida* sp.). Fujimaki et al.[178] applied plastein reaction for baker's yeast and hydrocarbon assimilating yeast. The main aim of this research work was to get impurity-free product from crude protein preparation. Impurities were not only possible, but contamination by some hydrocarbons, various types of flavor compounds, and lipids may cause secondarily, objectionable flavors, coloring and fluorescent compounds. Several of these components could not be removed by a simple extraction procedure as they are tightly adsorbed or bond to proteins. So an efficient removal is only possible after a partial hydrolysis. However, hydrolysis can cause bitter taste and destroys some important functional properties of the protein. To overcome this problem plastein synthesis can be a solution. In the first step the protein isolate is partially hydrolyzed. Hydrolysate is then extracted to remove fat content. Plastein synthesis was carried out of a concentrated hydrolysate catalyzed by α-chymotrypsin, papain, pepsin, trypsin and Bioprase. To get a bland protein-like fraction, the incubation mixture was treated with aqueous ethanol to remove taste substances, e.g., amino acids and oligopeptides. Bernarski and Leman[179] made enzymatic refining of yeast protein for *Candida utilis* and *Saccharomyces fragilis* yeasts. Protein hydrolysis occurred by using pepsin followed by protein resynthesis with α-chymotrypsin. The possibility of modifying the physicochemical properties and improving the biological value of proteins by combining field bean or yeast hydrolysate with whey hydrolysate or blood plasma protein hydrolysate before enzymic resynthesis was examined. The procedure resulted in a decrease of purine substances and improved solubility in water at pH range of 2 to 9.

Not only refining of yeast protein — removal of color, flavor, lipids pigment content, etc. — can be solved by plastein reaction but there is a great potential for improvement of protein biological value. It is well known from animal feeding experiments that supplementation of free essential amino acids in calculated amounts to fit a well-balanced amino acid composition failed to staisfy requirements due to the higher biological (nutritional) efficiency of protein-bound amino acids. Published works of Hajós[190] and Hajós and Halász[191] confirmed that in the proteinase catalyzed reaction amino acid esters are incorporated into peptides by covalent bonds.

Good results were achieved with α-chymotrypsin in case of methionine enrichment. The incorporated methionine is predominantly bound to the C-terminal of the peptide according to partial hydrolysis results by carboxyl-peptidase and aminopeptidase resp. Yeast protein is rich in essential amino acids but imbalanced for sulfur-containing amino acids. Methionine can substitute cysteine as essential amino acid, but cysteine cannot supply as a methyl donor so an enrichment with methionine is more valuable. Halász[192] and Halász et al.[193] succeeded in increasing covalently bound methionine enrichment of globulin-type yeast protein fraction and high molecular weight peptide fraction of yeast autolysate. Globulin-type yeast protein fraction was prepared after cell wall destruction by ultrasonication, insoluble parts were separated by centrifugation. Crude protein that got by freeze-drying was used for the plastein reaction at concentrations of 20 and 40%; α-chymotrypsin was added in a ratio of 1:100, methionine ethyl ester acted as amino acid component. After plastein reaction according to Hajós and Halász[165] and Hajós et al.[140] excess unincorporated methionine was eliminated by dialysis.

Methionine enrichment of the plastein product was investigated by amino acid analysis. The fact that the amino acid increase was due to covalently bound methionine was confirmed by exoproteinase terminal group splitting. Amino acid analysis of C- and N-terminal methionine content was carried out by ion exchange-TLC. As densitograms show in comparison to the substrate methionine content increase and amino acid incorporation was mainly at the C-terminal similarily to the findings in case of casein based substrate (Figures 86 and 87). As methionine is preferently bound to the C-terminal end of the peptide carboxypeptidase could be an efficient catalyzer for amino acid incorporation by plastein reaction. Halász[192] and Halász et al.[193] not only used commercial SIGMA carboxypeptidase enzyme but also crude carboxypeptidase-Y from brewer's yeast to catalyze methionine incorporation. Crude enzyme solution was prepared from cell free extract of brewer's yeast, endopeptidase activity of which had been inhibited by cell-own proteinase inhibitors. Amino acid analysis confirmed

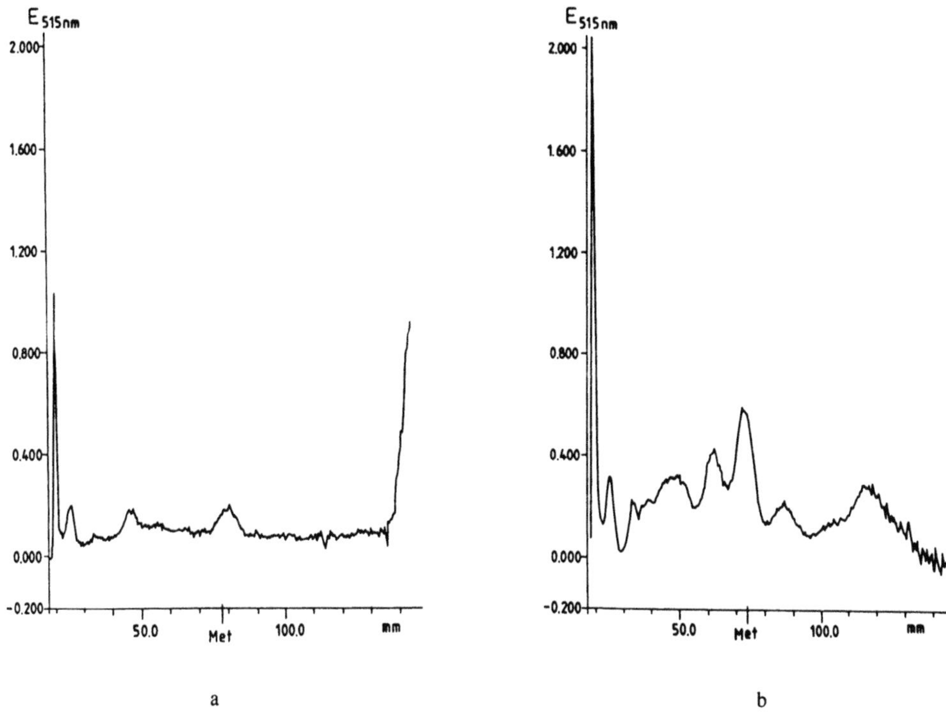

FIGURE 86(a). Densitogram of amino acids of the 60 min LAP hydrolysate of the reference sample, (b) densitogram of the amino acids of the 60 min LAP hydrolysate of the methionine-enriched yeast protein product (α-chymotrypsin catalyzed plastein reaction).

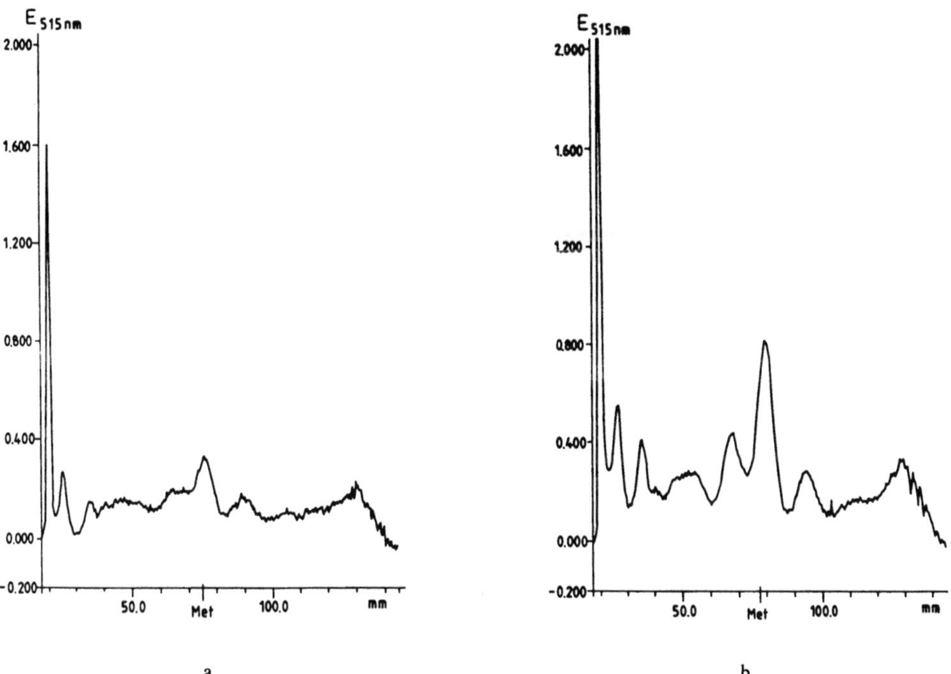

FIGURE 87(a). Densitogram of amino acids of the 60 min carboxypeptidase hydrolysate of the reference sample, (b) densitogram of the amino acids of 60 min carboxypeptidase hydrolysate of the methionine enriched yeast protein product (α-chymotrypsin catalyzed plastein reaction).

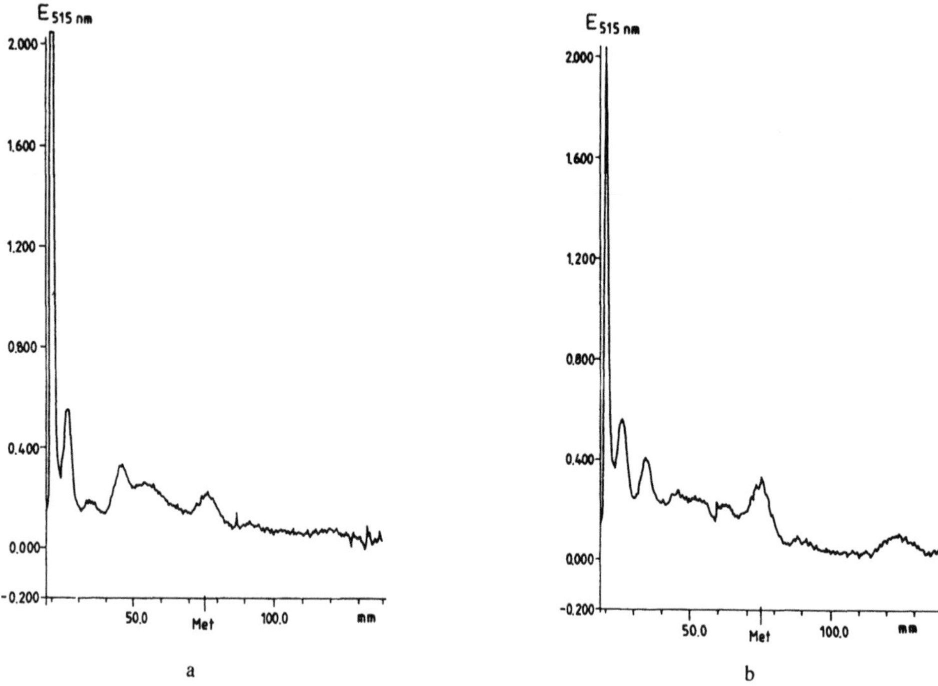

FIGURE 88(a). Densitogram of the amino acids of the 60 min LAP hydrolysate of the reference sample, (b) densitogram of the amino acids of the 60 min LAP hydrolysate of the methionine enriched yeast protein product (yeast C-4 catalyzed plastein reaction).

the efficiency of carboxypeptidase and stated that crude yeast carboxypeptidase could be used to catalyze plastein reaction. Results of splitting terminal amino acids by either aminopeptidase or carboxypeptidase showed, that in the case of plastein formed by carboxypeptidase methionine incorporation was very specific exclusively at the C-terminal of the peptide (see Figures 88 and 89).

B. CHEMICAL MODIFICATION OF PROTEINS

The specific modification of proteins by treatment with chemicals is widely practiced in fundamental studies on protein structure and function.[194,195,199,200] In food protein research, such studies are mainly connected with the identification of interacting groups in complex protein systems containing other nonprotein components such as carbohydrates, lipids, phytic acid, minerals, etc. In food systems the objective of chemical modification could also be the change of nutritive value of proteins or modification of functional properties. The high number of reactive groups in food proteins and their diversity opens many possibilities for chemical modification (Table 72 and 73). From a theoretical point of view the possible reactions may be divided into three groups: (1) chemical introduction of substituents; (2) modification of amino acid residues and (3) cleavage of peptide bonds (hydrolysis) and disulfide bonds of protein.

1. Chemical Introduction of Substituents

Many proteins have less-than-ideal amino acid profiles from a nutritional view point. Generally, the limiting amino acid in single cell proteins (including yeast proteins) is methioine. There are different ways to balance the amino acid profile of the diet. The simplest way is to mix the deficient protein with another rich in that amino acid. Supplementing of proteins with free amino acids is also possible but this has several limitations including loss of the free amino acids during food processing and cooking, flavor and color deterioration during storage, and possibly inefficient absorption. Modification of the amino acid profile

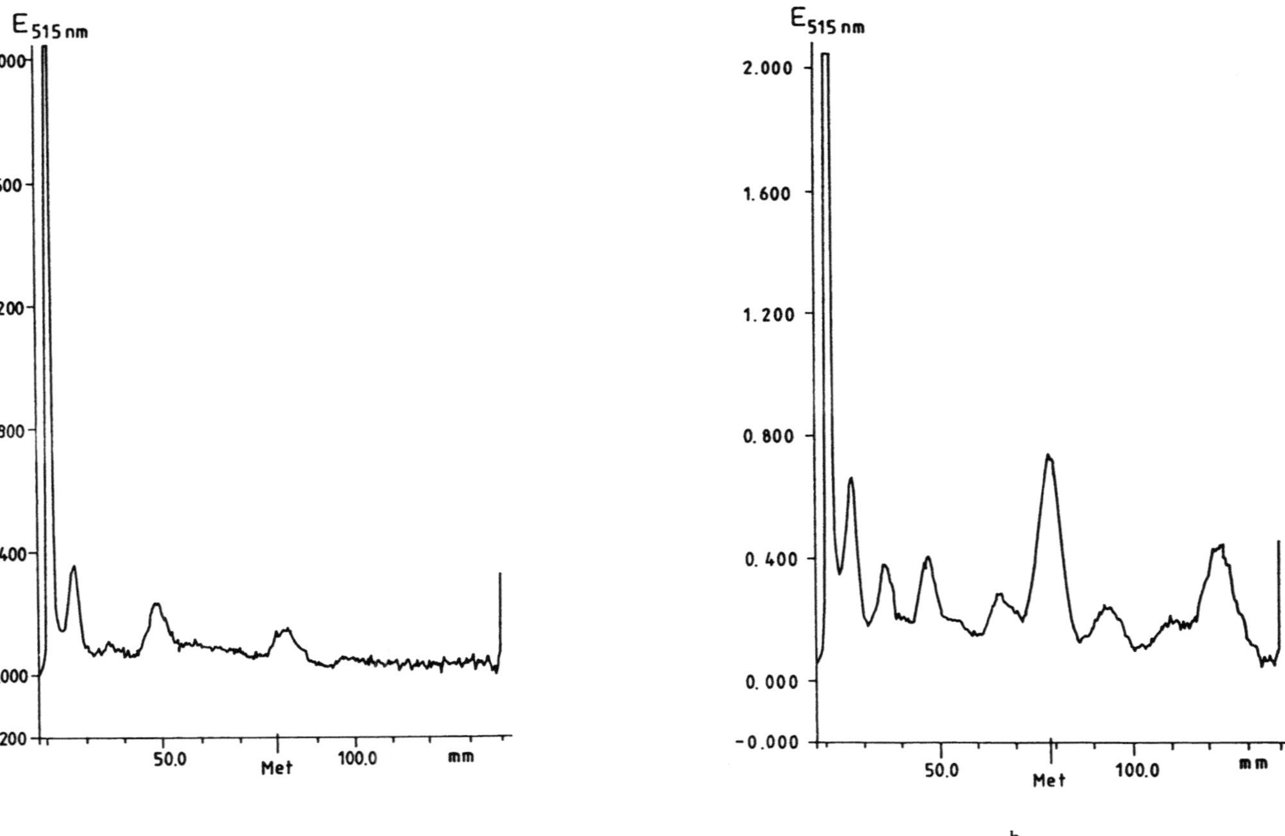

FIGURE 89(a). Densitogram of the amino acids of the 60 min carboxypeptidase hydrolysate of the reference sample, (b) densitogram of the amino acids of the 60 min carboxypeptidase hydrolysate of the methionine enriched yeast protein product (yeast C-4 catalyzed plastein reaction).

TABLE 72
Some Chemical Reactions of Amino Group of Amino Acids and Proteins

Reaction (reagent)	Derivative	Note
Acetylation (acetyc anhydride)	Acetylated protein CH_3-CO-NH-R	This reagent also reacts with thiol, imidazol, aliphatic and aromatic hydroxil groups
Reductive alkylation (formaldehyde/$NaBH_4$)	CH_3, H | N–R / CH_3 Alkylated protein	
Succinylation (succinic anhydride)	-OOC$(CH_2)_2$-CONHR Succinylated protein	Introduction of negative charge
Maleylation (maleic anhydride)	α- and ε-amino group derivatives	
Citraconylation (2-methyl maleic) anhydride)	*N*-acyl derivatives of lysine	
Reaction with aldehydes	R-N = CH – R^3	
Phosphorylation ($POCl_3$)	ε-*N*-phosphorylated protein	

TABLE 73
Some Chemical Reactions of the Carboxyl Group of Amino Acids and Proteins

Reaction	Derivative	Note
Esterification	R – $COOR^3$	
Amidation	R – CO NH_2	
Carbodiimide reaction	Acyl-derivative of carboxyl group	The reaction is suitable for covalent attach of amino acids to protein

via plastein reaction was discussed in a previous part of this chapter. Finally, there exists a possibility of covalent attachment of amino acids to protein chains.

The yeast protein is deficient in methionine so an enrichment in methionine can be interesting both from a practical and theoretical point of view. A method suitable for covalent attachment of methionine was realized on casein via active *N*-hydroxysuccinimide esters of amino acids through isopeptide bonds to the amino group of lysine.[196,197] The nutritional value of this protein was improved by covalent attachment of methionine. The isopeptide bond of ε-*N*-L-methionyl-L-lysine appeared to be readily hydrolyzed *in vivo* possibly by intestinal aminopeptidases. The carbodiimide condensation reaction was used for covalently binding methionine and tryptophan to soybean protein.[198] The optimal conditions of the reaction were also investigated. Of the factors investigated the pH, protein concentration, carbodiimide concentration, amino acid concentration and reaction temperature were found to significantly influence the tryptophan-binding efficiency to soy protein hydrolysate. In the case of methionine pH, protein concentration, carbodiimide concentration activation time and reaction time were the most important factors influencing covalent binding to the polypeptide chain.

Under the best conditions found, the methionine and tryptophan contents of methionine- and tryptophan-bound soy protein hydrolysate were increased 7.7- and 18.0-fold, respectively. An *in vitro* pepsin-pancreatin digestion test demonstrated that the bound amino acids were readily released. If nonhydrolyzed soy protein isolate was used a product with 95 to 99% protein recovery was obtained and its methionine and tryptophan content was increased 6.3- or 11.3-fold, respectively. High digestibility was still maintained for these products. Gel chromatography of products compared with original isolate demonstrated that carbodiimide reaction caused an increase in the molecular weights of soy protein fractions.

From the point of view of safety of modified protein isolates the potential health hazards must also be studied. In the experiments mentioned above 1-ethyl-3-(3-dimethyl-aminopropyl) carbodiimide hydrochloride (EDC) was used as reagent. The excess of reagent can easily be removed by washing with a dilute acid or water. The mammalian toxicity of carbodiimides is low (LD_{50} 2.6 g/kg in rats), nevertheless it was suggested that the safety and acceptability of a so modified product must be carefully evaluated.

The presence of carbohydrate moieties in proteins considerably modifies their functional properties, especially solubility, viscosity, hydration and gel-forming characteristics, undoubtedly owing to the hydrophilic groups of the carbohydrates. A covalent attachment of carbohydrates is possible via reductive alkylation of lysine in the presence of cyanoborohydride.[201,221] Monosaccharides were much more effectively coupled than the disaccharide. In an experiment with lactose, glucose and fructose it was observed that after 120 h treatment the degree of modification of lysine in casein was 17, 80 and 62%, respectively.

Reductive alkylation in the presence of sodium cyanoborohydride had been used successfully to couple glucose, fructose, maltose, cellobiose, melibiose and lactase to different proteins.[223,224] *In vitro* digestibility of modified protein decreased significantly. Fortification of diets containing modified protein with lysine only partially restored the nutritive value, suggesting that antinutritional factors, in addition to the blocking of lysine, were involved. All three sugar-casein derivatives caused some diarrhea, the lactose derivative having the more severe effect. Although formation of stable protein-sugar complexes is relatively simple, the disadvantages mentioned above make doubtful if such modification could be useful in the practice.

2. Modification of Amino Acid Side Chains

Intentional chemical alteration of food proteins by derivatization of functional groups on the side chains offers numerous opportunities for the development of new functional ingredients and improved processing procedures. The structure of the proteins is determined by their amino acid sequence as influenced by the immediate environment.

The secondary, tertiary and quaternary structure of the proteins is mostly stabilized by noncovalent interactions; hydrogen bonds, ionic bonds, hydrophobic interactions between side chains of continuous and opposed amino acid residues. Covalent disulfide bonds may be important in the maintenance of tertiary and quaternary structures.

Because the side chains of component amino acids and their involvement in noncovalent interactions markedly influence the structure and function of proteins, the chemical modification of specific functional groups should facilitate alteration of functional properties. This also may provide useful information concerning the role of specific amino acid groups in conformation and functionality of proteins.

With regard to side chain activity amino acid side chains may be divided into three groups: (1) electron rich; (2) electron deficient and (3) neutral.[202] Chemical modification predominantly involves nucleophilic or reductive reactions of electron-rich side chains. There are several types of electron-rich nucleophilic groups: (1) the nitrogen of the amino-, imidazole- and guanidino groups of lysine, histidine and arginine, respectively; (2) the oxygen nucleophiles of the hydroxil, phenolic- and carboxyl groups of serine, threonine, tyrosine, glutamic- and aspartic acid, respectively, and (3) the sulfur nucleophile of cysteine and thioether group of methionine. The reactivity of nucleophilic groups are greatly influenced by pH, accessibility of the side chain, the size of the modifying agent, and reaction conditions. Alfa amino acid thiol groups are reactive above pH 7.5, the ϵ-amino group of lysine have pH 8.5 where they are mostly not protonated, and the phenolic group of tyrosin above pH 10. The most widely used chemical modifications are acylation of amino, hydroxyl and phenolic residues; alkylation of amino, indole, phenolic, sulfhydryl and thioether groups and esterification of carboxyl group.

The α-amino group is readily acylated, however normally the ϵ-amino group of lysine is more active. The phenolic groups of tyrosine are readily acylated compared with the other amino acid residues. The tyrosine and phenolic groups, however, have a higher pH and are usually more protected from reaction (buried in the interior of the protein) than the free amino groups. Acylation of histidine and cysteine residues is seldom observed because the reaction products hydrolyze in aqueous solution. Serine and threonine hydroxyl groups are weak nucleophiles and are not easily acylated in aqueous solution.

The number of nucleophilic residues acylated is influenced by the level of acylating reagent used; the amino acid composition of the protein, accessibility of amino acid (protein size and conformation) and the reaction condition (temperature, pH, reaction time, concentration of reagent, etc.).

In the framework of modification by acylation a significant volume of work has been done on the protection of ϵ-amino groups of lysine.

Acylation may prevent the Maillard reaction between amino groups of lysine and carbohydrates. Transamidation, condensation reactions with dehydroalanine forming cross-links, and carbonyl amine interventions which all may involve free ϵ-amino groups of lysine can be hindered.

Different reagents may be used for acylation, among them the acetic anhydride, and succinic anhydride are most widely utilized. However, not only acetylation and succinylation is possible but also formylation, propionylation, etc. Some acid anhydrides cause reversible modifications of both the α- and ϵ-amino groups.[200] For such purposes maleic anhydride, 2-methylmaleic (citraconic) anhydride, 2,3-dimethylmaleic anhydride is used.

In the practice of acylation of protein preparations to be used for food purposes the acetylation has been investigated most thoroughly. Generally, acetylation enhances the thermal stability and reduces thermal coagulation and precipitation. Some other functional properties can also be modified. Investigating acetylated broad bean protein isolates Schneider et al.[203] observed an increase in emulsifying capacity (see Figure 90) and geling properties (see Table 74) of isolate. The changes were dependent on the degree of acetylation. Acetylation of yeast protein isolates increased the solubility of the protein preparation. Only slight changes were observed in the other functional properties.

Both α- and ϵ-amino groups of proteins can be modified by succinylation. Due to the fact that succinic acid has two carboxylic groups, after acylation a negatively charged new group will be introduced in the protein. Electrostatic repulsions resulting from the succinate anions alter the conformation of the protein and penetration of water molecules is physically easier because of the expanded loosened state of polypeptides. The structural instability of succinylated proteins is also associated with their negative charge and the replacement of short range attractive forces in the native molecule with short range repulsive forces with subsequent unfolding of the polypeptide chains. Dissociation into subunits increased viscosity and swelling was also observed.

The pH solubility profile of succinylated yeast proteins revealed the typical increased solubility between pH 4 to 6. The succinylated samples became increasingly insoluble below the isoelectric point as the degree of succinylation was increased.[204] Significantly succinylated yeast proteins were very stable to heat precipitation above pH 5, i.e., they remained soluble at temperatures up to 100°C. As the degree of succinylation was increased the rate of precipitation of the derivatized protein increased in the neighborhood of the isoelectric point and such larger protein flocs were obtained, facilitating their recovery. The emulsifying activity of succinylated yeast proteins was significantly improved in comparison to unmodified yeast protein and some other proteins (Table 74). It was reported that succynilation of yeast proteins increased emulsion viscosity but decreased emulsion stability.[205] The succinylation was successfully used for increasing the protein yield extracting yeast proteins for production yeast protein isolates and concentrates. The succinylation was a successful process for removing the nucleic acids and producing low RNA protein preparation.

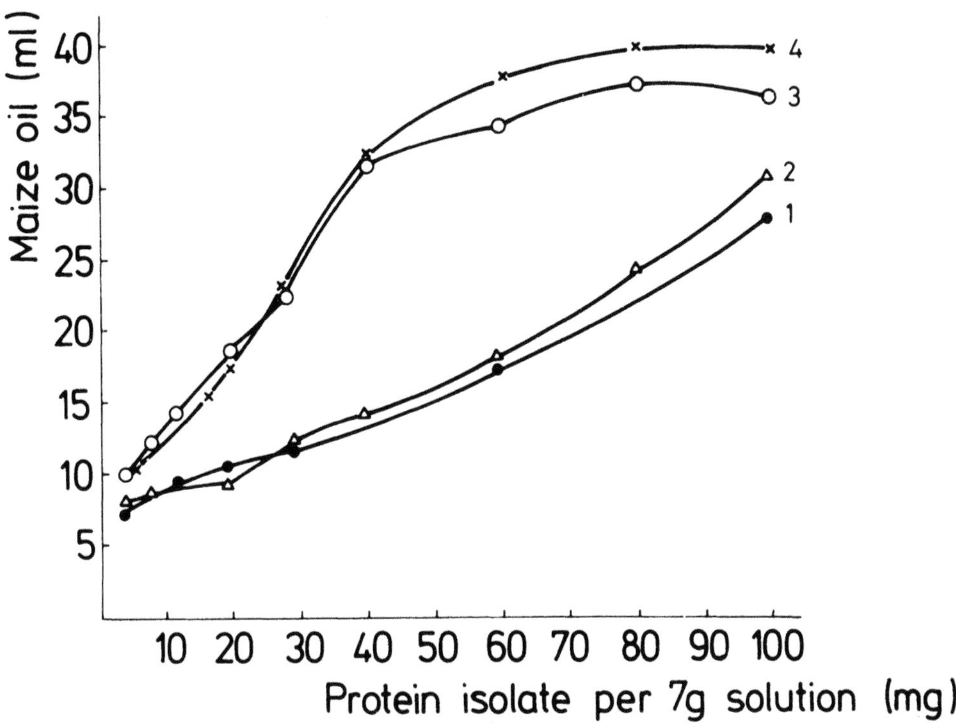

FIGURE 90. Emulsion characteristics of unmodified and acetated broad bean protein isolates; 1 — unmodified protein isolate, 2 — acetylation degree ~26%, 3 — acetylation degree ~80% and 4 — acetylation degree ~97%. (From Schneider, Ch., Muschiolik, G., Schultz, M., and Schmandke, H., *Die Nahrung*, 30(3/4), 429, 1986. With permission.)

TABLE 74
Emulsifying Activity Index Values for Some Protein Preparation

Protein preparation	Emulsifying activity index (m^2 g^{-1}) pH 6.5	pH 8.0
Yeast protein (Brewer's yeast)	15	82
Yeast protein (Baker's yeast)	18	85
Yeast protein *(Rhodotorula glutinis)*	27	104
Yeast protein (20% succynilated)	110	181
Yeast protein (60% succynilated)	232	301
Yeast protein (80% succynilated)	308	275
Soy protein isolate	112	187
Sodium caseinate	108	175

Succinylation during extraction significantly improved the foaming capacity of yeast protein isolate. The succinylated protein was almost equivalent to ovalbumin whereas the unmodified yeast was very inferior and failed to retain gas efficiently during expansion. Maximum foam strengths of succinylated yeast proteins and ovalbumin occurred slightly above the isoelectric points of these two proteins. The ovalbumin foam was significantly stronger than the succinylated yeast protein at similar concentrations of protein. A dissociation of high molecular weight protein components of yeast were observed after succinylation. The succinylated proteins were susceptible to hydrolysis by both α-chymotrypsin and trypsin.

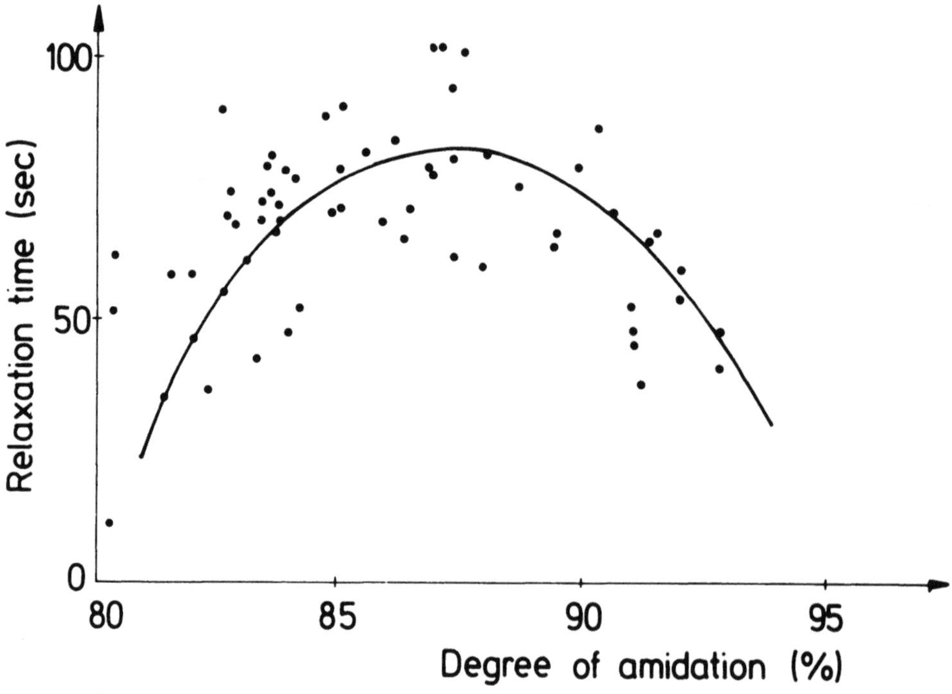

FIGURE 91. Correlation between the amidation degree of gluten and its rheological properties.

Nevertheless, the disadvantage of succinylation is that the final product of such process is succinylated protein because removal of succinyl groups cannot be made under mild conditions.

Recently a lot of reversible modifying agents were found and applied to different proteins including yeast proteins.[217-219] Citraconic anhydride and dimethylmaleic anhydride were used for such purposes. An increased yield and low nucleic acid content of protein isolates was achieved. The deacylation of the proteins was realized by incubation of the citraconylated (dimethylmaleinated) proteins in a slightly acidic solution (dispersion) at 30 to 50°C. The details of the procedure were treated deeper in Chapter 5 of Part I. Studies made with phosphoproteins, particularly with casein, revealed that many of the specific biological and functional properties of these proteins are associated with their phosphorylated side chains. Introduction of phosphate groups increases the net negative protein charge and would probably change the solubility or other functional properties. It is relatively easy to phosphorylate proteins using $POCl_3$.

Huang and Kinsella[210,211] and Damodaran and Kinsella[212] investigated the effect of phoshorylation on yeast proteins. They reported that phosphorylation of lysil groups of protein facilitated preparation of a phosphoprotein concentrate with low nucleic acid content. It was also stated that phosphorylation slightly increased the solubility of yeast proteins and significantly the water-holding capacity. An increase in viscosity of solutions was also observed. Phosphorylation of yeast proteins using sodium trimetaphosphate (STMP) was also reported.[213] The phosphorylation resulted in good removal of nucleic acids and production of low RNA protein concentrate. Similar changes as mentioned above were observed in functional properties of the phosphorylated proteins.

Other chemical modifications such as esterification of carboxylic groups or amidation of these groups are used mostly in investigations of protein structure. The effect of such modifications on the functional properties of food proteins was poorly investigated. Only in the case of gluten proteins was it shown that amidation or desamidation has a significant effect on functional properties of wheat gluten.[195,214,125] In Figure 91 a correlation between amidation degree and some rheological properties of gluten are shown.

TABLE 75
Metabolization of Acylated Lysine and Acylated Proteins and Peptides in Animals

Derivative	Degree of metabolization (%)	Test animal
ε-*N*-formyl lysine	50	Rat
ε-*N*-acetyl lysine	50	Rat, chicken
ε-*N*-propionyl lysine	70	Chicken
α-*N*-propionyl lysine	0	Rat
ε-*N*-succinylated lysine	30	Rat
ε-*N*-(γ-glutamyl)lysine	100	Rat
ε-*N*-(α-glutamyl)lysine	100	Rat
ε-*N*-glycil lysine	80	Rat
α-*N*-glycil lysine	100	Rat
ε-*N*-(acetyl-glycil)lysine	0	Rat
Acetylated casein	70	Rat
Succinylated casein	15	Rat
Acetylated lactalbumin	43—77	Rat

The effect of esterification of carboxylic groups using methanol or ethanol caused only small changes in the functional properties.[214] Acylation of proteins with acetyc anhydride or succinic anhydride was used for reduction of phytic acid content of protein isolates.[215]

As an interesting example of experiments with chemical modification it can be mentioned that Mosher[217] patented the preparation of a "roasted meat" flavoring agent by reacting *S*-acetyl-mercaptosuccinic acid, or its anhydride with a vegetable protein hydrolysate and/or autolyzed yeast.

The flavor characteristics of the reaction products of *S*-acetyl-mercaptosuccinic acid and vegetable protein hydrolysate can be modified by the addition of yeast autolysate to either the reaction mixture or the product. Addition of formaldehyde or glutaraldehyde as cross-linking reagents[218] was reported to be used to promote fibrous formation of the protein and investigated the potential for forming a continuous and uniform filament by spinning/extrusion.

Chemical modification of proteins by formation of derivatives is attractive as a means of producing proteins which have a broad spectrum of functional properties that are useful to food manufacturers. However, chemical modification of food proteins faces three major problems:[194] (1) health hazards, since many of the reagents used are toxic; (2) reduction of the nutritive value of proteins since modification frequently involves essential amino acids which may be rendered biologically unavailable and (3) large-scale availability of reagents at an economic price.

For chemically derivatized proteins to be successfully accepted as food ingredients they must be digestible, nontoxic, and ideally the modified amino acid residues should be available nutritionally. Theoretically succinylated and acetylated proteins are not toxic while both succinic- and acetic acid are normal metabilites in the tricarboxylic acid cycle. The same can be suggested concerning phosphorylated proteins although the balance of phosphorus and other minerals (e.g., calcium) and the availability of phosphorus in yeast protein preparation must be taken into consideration.

The nutritional effects of proteins with acylated lysine residues were tasted more widely.[194,218,219] The experiments with different acylated proteins and ε-N-acetyl-lysine, ε-*N*-formyl-lysine, ε-propionil-lysine, ε-*N*-glutamyl-lysine, *N*-glycil-lysines showed that most of the acylated lysine derivatives are metabolized by experimental animals. However, the degree of metabolization was generally much lower than that of unmodified proteins and amino acids resp. Generally, acetylated proteins have a higher nutritive value than succinylated-, formylated- and propionylated derivatives. Some data concerning the degree of metabolization are summarized in Table 75.

Summarizing the disadvantages of chemical modification from the point of view of nutrition it must be noted that it is unlikely that a modified protein would be consumed as a significant, much less a sole source of dietary protein for any particular population group.[218] Highly functional proteins may well facilitate the incorporation of nutritionally superior, but functionally inferior, proteins into a variety of food, — if their toxicity could be absolutely excluded — appealing to a broad range of consumers. The losses in nutritive value can also be reduced if the future research work will be successful in developing new reagents that react specifically with nonlimiting amino acids. Research to develop such functional derivatives should be encouraged.

Finally, it must also be noted in connection with the economic production of such protein derivatives that not only the costs of reagents and additional steps in technology must be taken in mind but also the high costs associated with proving to regulatory agencies that the products of these reactions are nontoxic.

3. Cleavage of Peptide Bonds (Hydrolysis) and Disulfide Bonds

Among the secondary bonds playing an important role in stabilizing the structure of the native proteins the role of disulfide bonds is undoubtful. Many examples show that the splitting of disulfide bonds by reduction (e.g., with β-mercapto-ethanol, dithiothreitol, sodium bisulfite, etc.) or by oxidation (e.g., performic acid) causes big changes in the properties of proteins. The functionality of several proteins is markedly influenced by the presence and reactivity of disulfide bonds. Through oxidation/reduction and thiol-disulfide interchange reactions, new molecular conformations and several intermolecular species of protein polymers may be formed, and in several instances these are essential for the functional performance of proteins in foods and in the fabrication of new foods. The formation of viscoelastic dough from wheat protein and its physical properties, i.e., strength, stability and extensibility, may involve interactions between thiol groups on the gliadin and glutenin molecules.[214] Production of fibers in some cases is also associated with formation of new disulfide bonds. Partial splitting of disulfide bonds may increase the flexibility of protein molecules which influences the foaming and emulsifying properties of proteins. The splitting of disulfide bonds could be associated with dissociation of quaternary structures, loosening of the conformation of the protein molecules influencing the solubility and water-absorbing and water-holding capacity of the proteins.

Certain physical properties of protein might be modified radically by covalent attachment of new sulfhydril groups. Chemical thiolation of proteins has been accomplished with several reagents, including polythioglycolides, *N*-acetyl-homocysteine thiolactone (*N*-AHTL), and *S*-acetyl mercaptosuccinic anhydride (*S*-AMSA). The reactions of the two latter reagents with protein are shown in Figures 92 and 93. As it was shown in experiments made with whey proteins[219,220] thiolation of amino groups reduced the isoelectric point. Using oxidizing agents (e.g., potassium iodate) in the thiolated β-lactoglobulin both intramolecular and intermolecular disulfide bonds were formed leading to the formation of polymeric products with molecular weights exceeding 600 kDa. The novel β-lactoglobulin polymers exhibited certain unique characteristics, such as enhanced heat stability, increased viscosity, foaming ability and gel-forming ability. In the presence of calcium ions, the polymers formed a strong, transparent, heat-stable gel. The chemical hydrolysis of proteins, particularly the partial hydrolysis, is a frequently used method to modify the functional properties of proteins. Although an alkaline hydrolysis of proteins is also possible the damage of proteins caused by severe alkaline conditions (for more details see Chapter 5, Part I) does not allow wider practical use of this process in food industry. Acid hydrolysis of different proteins is often used. Some partially hydrolyzed proteins are widely used in bakery items, soups and comminuted meats. Application of hydrolysates as flavor enhancers and sources of flavors in sauces, meats and gravies is also well known.

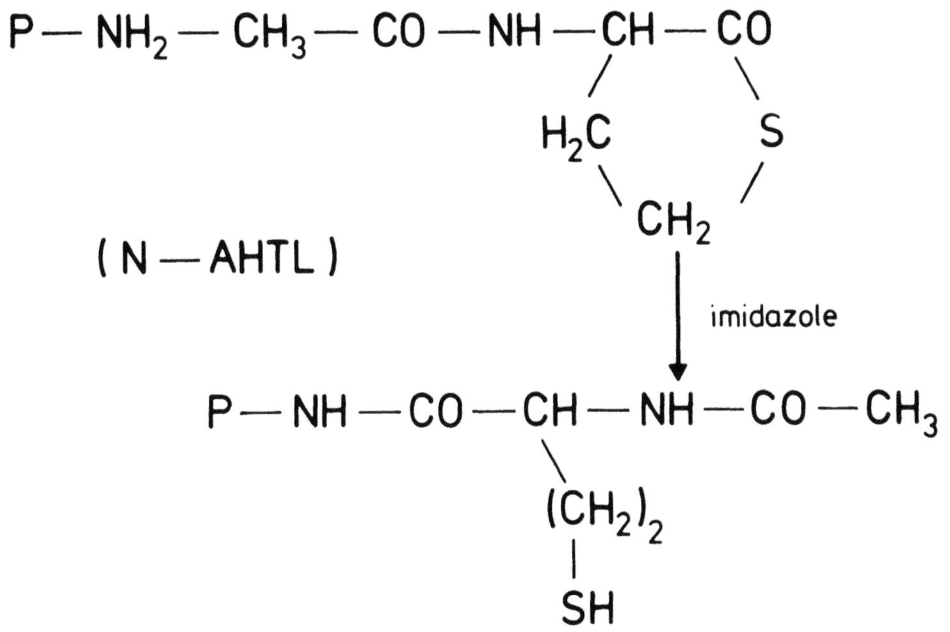

$$P-NH_2-CH_3-CO-NH-CH-CO$$
$$H_2C \quad S$$
$$CH_2$$

(N—AHTL)

imidazole

$$P-NH-CO-CH-NH-CO-CH_3$$
$$(CH_2)_2$$
$$SH$$

FIGURE 92. Thiolation of proteins using *N*-acetyl homocysteine thiolactone (*N*-AHTL).

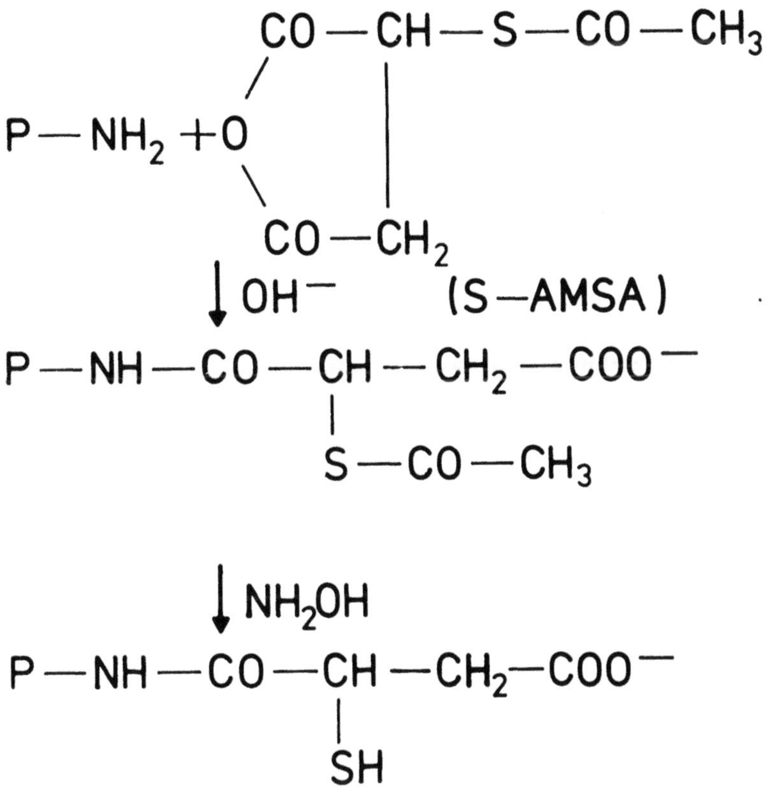

$$CO-CH-S-CO-CH_3$$
$$P-NH_2+O$$
$$CO-CH_2$$

OH^- (S—AMSA)

$$P-NH-CO-CH-CH_2-COO^-$$
$$S-CO-CH_3$$

NH_2OH

$$P-NH-CO-CH-CH_2-COO^-$$
$$SH$$

FIGURE 93. Thiolation of proteins using *S*-acetyl-mercaptosuccinic-anhydride (*S*-AMSA).

The partial hydrolysis of protein leads to a formation of polypeptides of various lengths.[225] The physicochemical properties of the mixtures of peptides — either alone or together with the macromolecular protein fraction — are evidently different from those of the starting material. It is possible to produce protein hydrolysates which are distinctive with respect to such useful functional properties as:[225] high solubility in acid systems; high solubility and low viscosity in beverages; good emulsifying properties, also at acid pH, where common food proteins usually perform poorly in that respect, and high foaming capacity.

As a quantitative measure of the hydrolytic degradation the degree of hydrolysis (DH) is used. DH is defined as the percentage of peptide bonds cleaved during the reaction. DH can be measured in a number of ways during the process, either on line using the pH-stat technique or off line by osmometry. There is however, one issue which should not be neglected in the field of protein hydrolysis; the problem of bitterness. During the hydrolytic reaction, small hydrophobic peptides are formed, and these peptides are the cause of bitter off-flavor, which for nearly two decades now has been perceived as the main obstacle to a wide spread use of hydrolysates in food.

In the case of hydrolysis reaction, DH increases from zero and usually follows a smooth curve with a downward curvature. Together with the rise of DH properties the system changes stepwise. Many food proteins are poorly soluble and the solubility increases with DH. Due to increased flexibility of hydrolyzed products their better solubility and surface activity the foaming and emulsifying properties will also be improved.

4. Physical Modification of Protein Isolates

Changes of protein functional performance through physical means can be achieved by thermal treatment, biopolymer complexing or texturization. Heat treatment of proteins over a given temperature causes denaturation. Partial denaturation, or combining partially denatured with native proteins, has been suggested as a technique for intentional modification of functionality.[226] More severe thermal treatment will cause full protein denaturation accompanied by a loss of aqueous solubilty and overall functional behavior.

Complexing between proteins and acidic polysaccharides often results in a more stable protein conformation, reflected by greater resistance to heat denaturation and thermotropic gelation.[227,228] Particularly production of meat analogs and texturization of proteins may be successfully used combinations of such polysaccharides and protein isolates. Coagulating agents such as carboxy-methyl-cellulose, carrageenan, polyphosphates, sulfonated lignins and chitosan may be used for protein recovery from solutions. The complex of the polymer and proteins might possess certain interesting and useful attributes.

Gurov and Nuss[228] investigated emulsifying properties of protein-dextran sulfate mixtures. The stability of emulsions formed with protein-carbohydrate complex was markedly higher than that produced with pure protein isolate. Although the mode of action of the polysaccharide is not clear yet, it appears reasonable to assume that the complex particles are similar to chain polymers (graft or block), forming at the liquid interfaces more elastic and firm layers (probably gel-like) than that of the free globular proteins.

Interesting results were reported by Braudo et al.[229] associated with investigation of multicomponent protein gels. Model experiments with gelatin-dextran and gelatin-agarose mixtures showed such functionally important properties of gels as critical concentration of gel formation (Figure 94), the elasticity of the gel (Figure 95) and the fusion temperature of the mixed gel (Figure 96). Production of mixtures of different proteins may also be a tool for production of protein preparations having different functional properties.[230]

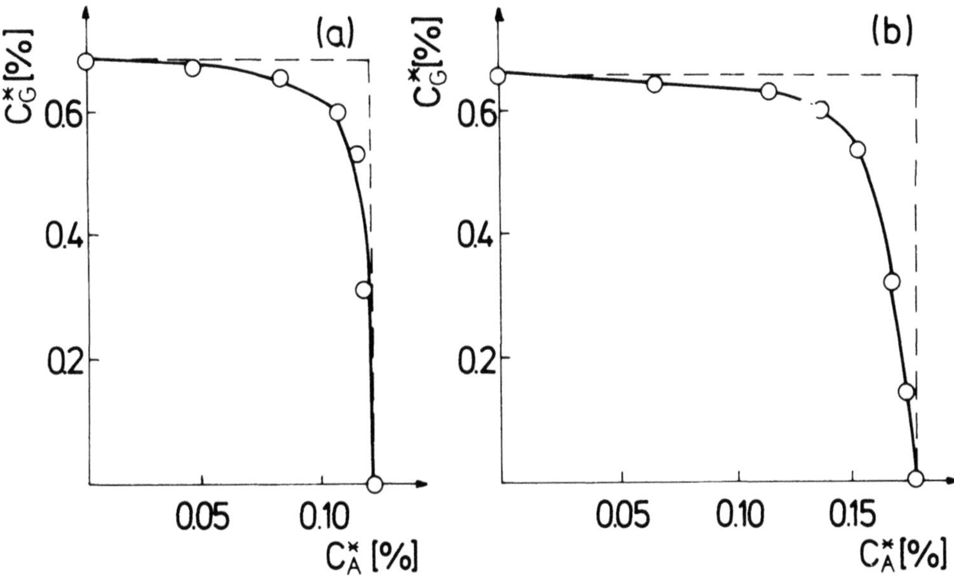

FIGURE 94. The dependence of the critical concentration of gel formation on the composition of a gelatin; (G) — agarose, (A) — water system, (a) in water, at pH 4.7 corresponding to the isoionic point of the gelatin and (b) in a 0.1 *M* phosphate buffer solution, at pH 7.0. (From Braudo, E. E., Gottlieb, A. M., Phashina, I. G., and Tolstoguzov, V. B., *Nahrung*, 30(3/4), 355, 1987. With permission.)

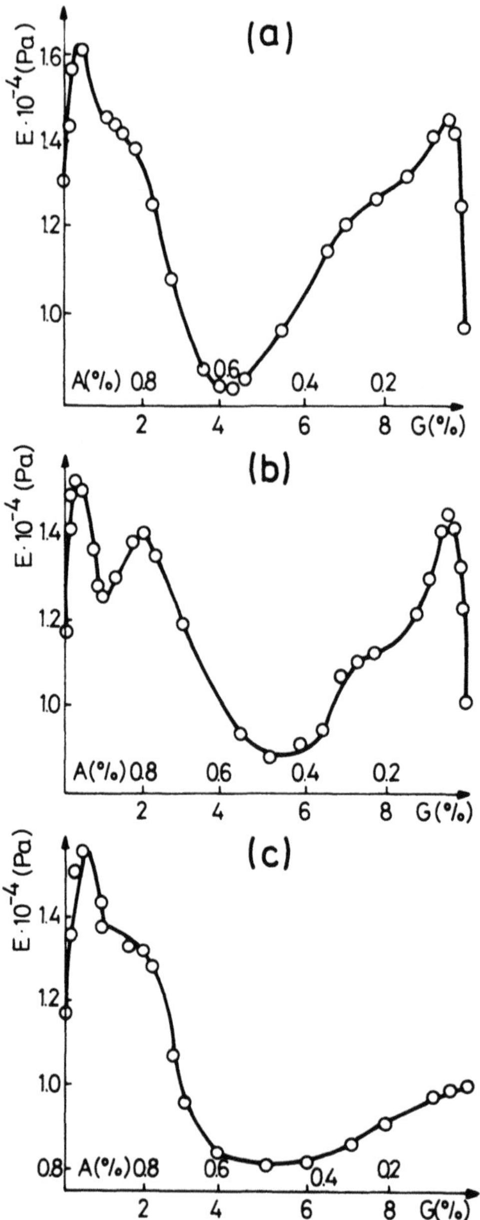

FIGURE 95. The dependence of the YOUNG modulus of mixed gelatin; (G) — agarose, (A) — water gels on the composition of the system: (a) in water, at pH 4.7 corresponding to the isoionic point of the gelatin; (b) in a 0.1 *M* phosphate buffer solution, at pH 7.0 and (c) in a 0.1 *M* solution of NaCl, at pH 4.7. (From Braudo, E. E., Gottlieb, A. M., Plashina, I. G., and Tolstoguzov, V. B., *Nahrung*, 30(3/4), 355, 1986. With permission.)

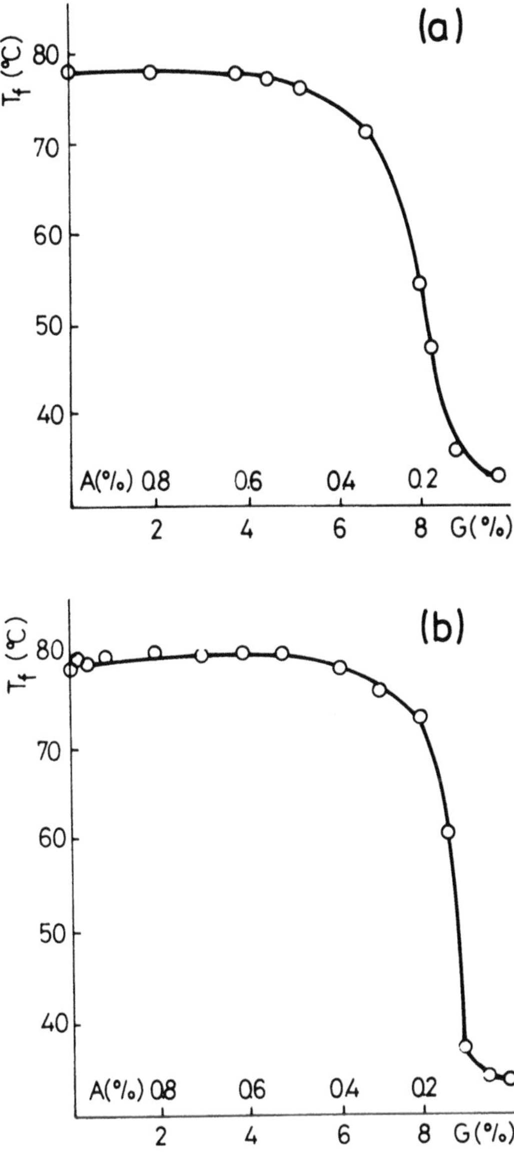

FIGURE 96. The dependence of the fusion temperature of mixed gelatin (G) — agarose — (A) gels on the composition of the system: (a) in water at pH 4.7, corresponding to the isoionic point of gelatin and (b) in a 0.1 *M* phosphate buffer solution, at pH 7.0. (From Braudo, E. E., Gottlieb, A. M., Plashina, I. G., and Tolstoguzov, V. B., *Nahrung*, 30(3/4), 355, 1986. With permission.)

REFERENCES

1. **Schlingmann, M., Faust, U., and Scharf, V.,** Bacterial proteins, in *Developments in Food Proteins-3*, Hudson, B. J. F., Ed., Elsevier Applied Science Publisher, London and New York, 1984, 139.
2. **Seeley, R. D.,** Unconventional protein resources. Functional aspects of SCP are a key to potential markets, *Food Prod. Dev.*, 9(9), 46, 1975.

3. **Tusé, D.,** Single-cell protein: current status and future prospects, *CRC Crit. Rev. Food Sci. Nutr.*, Vol. 19, Issue 4, 1984, 273.
4. **Kinsella, J. E.,** Functional properties of proteins in foods: a survey, *CRC Crit. Rev. Food Sci. Nutr.*, April 1976, 219.
5. **Guzman-Juarez, M.,** Yeast protein, in *Developments in Food Proteins-2*, Hudson, B. J. F., Ed., Elsevier Applied Science Publisher, New York and London, 1983, 262.
6. **El Pour, A.,** Protein functionality in foods, Cherry, J. P., Ed., ACS Symp. Series 147, Washington, D.C., 1981, 177.
7. **Rutkowski, A. and Gwiazda, A.,** Functional properties of plant proteins in meat systems, *Nahrung*, 30, 375, 1986.
8. **Nakai, S., LiChan, E., and Hayakowa, S.,** Contribution of protein hydrophobicity to its functionality, *Nahrung*, 30, 327, 1986.
9. **Bikbov, T. M., Grinberg, V. Ja., Grinberg, N. V., Varfolomeyeva, E. P., and Likhodzeivskaya, I. B.,** Thermotropic gelation of proteins, *Nahrung*, 30, 369, 1986.
10. **Polyakov, V. I., Popello, I. A., Grinberg, V. Ja., and Tolstoguzov, V. B.,** Thermodynamic compatibility of proteins in aqueous medium, *Nahrung*, 30, 365, 1986.
11. **Lin, C. F.,** Interaction of sulphated polysaccharides with proteins, in *Food Colloids*, Graham, H. D., Ed., AVI Publishing Westport, CT, 1977, 320.
12. **Barudo, E. E., Gotlieb, A. M., Plashina, I. G., and Tolstoguzov, V. B.,** Protein-containing multicomponent gels, *Nahrung*, 30, 355, 1986.
13. **Becker, P.,** *Encyclopedia of Emulsion Technology*, Marcel Dekker, New York, 1983.
14. **Friberg, S. E. and Venable, R. V.,** Microemulsion, in *Encyclopedia of Emulsion Technology*, Vol. 1., Becker, P., Ed., Marcel Dekker, New York, 1983, 287.
15. **Tadros, T. E. and Vincent, B.,** Emulsion stability, in *Encyclopedia of Emulsion Technology*, Vol. 1., Becker P., Ed., Marcel Dekker, New York, 1983, 129.
16. **Lásztity, R.,** *The Physical-Chemistry of Foods*, (in Hungarian), Tankönyvkiadó, Budapest, 1965.
17. **Graham, D. E. and Phillips, G.,** The conformation of proteins at interfaces and their role in stabilizing emulsions, in *Theory and Practice of Emulsion Technology, Chapter 5*, Smith, A. L., Ed., Academic Press, London, 1976.
18. **Haque, Z. and Kito, M.,** Lipophilization of α_{s1}-casein. 3. Purification and physicochemical properties of novel amphiphatic fatty-acyl peptides, *J. Agric. Food Chem.*, 32, 1392, 1984.
19. **Kinsella, J. E.,** Milk proteins physicochemical and functional properties, *CRC Crit. Rev. Food Sci Nutr.*, 21(3), 197, 1984.
20. **Mitchell, J. R.,** Foaming and emulsifying properties of proteins, in *Developments in Food Proteins-4*, Hudson, B. J. F., Ed., Elsevier Applied Science Publishers, London, 1986, 291.
21. **Kato, A. and Nakai, S.,** Hydrophobicity determined by a fluorescence probe method and its correlation with surface properties of proteins, *Biochim. Biophys. Acta*, 624, 13, 1980.
22. **Voutsinas, L. P., Cheung, E., and Nakai, S.,** Relationships of hydrophobicity to emulsifying properties of heat denaturated proteins, *J. Food Sci.*, 48, 26, 1983.
23. **Kinsella, J. E.,** Functional properties of proteins in foods a survey, *CRC Crit. Rev. Food Sci. Nutr.*, 7, 219, 1976.
24. **Kato, A., Tsutsui, N., Kobayashi, K., and Nakai, S.,** Effects of partial denaturation on surface properties of ovalbumine and lysozyme, *Agric. Biol. Chem.*, 45, 2755, 1981.
25. **Haque, Z. and Kinsella, J. E.,** Emulsifying properties of food protein: development of a standardized emulsification method, *J. Food Sci.*, 54, 39, 1989.
26. **Haque, Z. and Kinsella, J. E.,** Emulsifying properties of food proteins: bovine serum albumin, *J. Food Sci.*, 53, 416, 1988.
27. **Haque, Z. and Kinsella, J. E.,** Emulsifying properties of food proteins: effect of the association state of bovine serum albumin on emulsifying activity, *Agric. Biol. Chem.*, 52(5), 1141, 1988.
28. **Haque, Z. and Kinsella, J. E.,** Interaction between kappacasein and beta-lactoglobulin: predominance of hydrophobic interactions in the initial stages of complex formation, *J. Dairy Res.*, 55, 67, 1988.
29. **Kim, S. H. and Kinsella, J. E.,** Surface active properties of food proteins: effects of reduction of disulfidedbonds on film properties and foam stability of glycinin, *J. Food Sci.*, 52(1), 128, 1987.
30. **Franzen, K. L. and Kinsella, J. E.,** Parameters effecting the binding of volatile flavor compounds in model food systems. I. Proteins, *J. Agric. Food Chem.*, 22, 675, 1974.
31. **Damodaran, S. and Kinsella, J. E.,** Interaction of carbonyls with soy protein: Thermodynamic effects, *J. Agric. Food Chem.*, 29, 1949, 1981.
32. **Damodaran, S. and Kinsella, J. E.,** Interaction of carbonyls with soy protein: conformational effects, *J. Agric. Food Chem.*, 29, 1253, 1981.
33. **Voilley, A., Fares, K., Lorient, D., and Simatos, D.,** Interactions between proteins and aroma compounds, in *Functional Properties of Food Proteins*, Lásztity R. and Ember-Kárpáti M., Eds., MÉTE Publishing, Budapest, Hungary, 1988. 44.

34. **Lásztity, R.,** *The Chemistry of Cereal Proteins,* CRC Press, Boca Raton, FL, 1984.
35. **Hutton, C. W. and Campbell, A. M.,** Water and fat absorption, in *Protein Functionality in Foods,* Cherry, J. P., Ed., American Chemical Society, Washington, D.C., 1981, 177.
36. **Fennema, O. R.,** *Food Chemistry,* Marcel Dekker, New York, 1985.
37. **Colas, B., Courthaudon, J. C., Le Meste, M., and Simatos, D.,** Functional properties of caseinates. The role of the flexibility of the protein and its hydration level on surface properties, in *Functional Properties of Food Proteins,* Lásztity R. and Ember-Kárpáti, M., Eds., MÉTE Publishing, Budapest, 1988, 194.
38. **Walstra, P.,** Formation of emulsions, in *Encyclopedia of Emulsions Technology,* Vol. 1, Becker, P., Ed. Marcel Dekker, New York, 1983, 57.
39. **Tornberg, E. and Lundh, S.,** Functional characterization of protein stabilized emulsions: standard emulsifying procedure, *J. Food Sci.,* 5, 1553, 1978.
40. **Haque, Z. and Kinsella, J. E.,** Interaction between K-casein and lactoglobulin: effect of calcium, *Agric. Biol. Chem.,* 51, 1977, 1987.
41. **Nakamura, R. and Sato, Y.,** Studies on the foaming property of the chicken egg white. Part C. On the role of ovomucin in the egg white foaminess (the mechanism of foaminess), *Agric. Biol. Chem.,* 28, 530, 1964.
42. **Wanishen, R. D. and Kinsella, J. E.,** Foaming properties of proteins: evaluation of a column aeration apparatus using ovalbumin, *J. Food Sci.,* 44, 1398, 1979.
43. **Muschiolik, G. and Schmandke, M.,** Zur Herstellung und Charakterisierung von Lebensmittelschaumen aus Proteinrohstoffen, *Nahrung,* 28, 289, 1984.
44. **Gassmann, B., Kroll, J., and Cifuentes, S.,** Determination of foaming properties of proteins, *Nahrung,* 31(4), 321, 1987.
45. **Mansvelt, J. W.,** The use of foams in foods and food production, in *Foams,* Akers, J. R., Ed., Academic Press, London, 1976, 283.
46. **Rauch, W.,** Eier, in *Handbuch Lebensm.,* Vol. III/2., Schormüller, J., Ed., Springer Verlag, Berlin, 1968, 915.
47. **Doerfer, J. and Zimmermann, R.,** *Bäcker Konditor,* 27, 100, 1973.
48. **Behnke, U., Schultz, M., Ruttloff, H., and Schmandke, H.,** Veranderungen der funktionellan Eigenschaften von *Vicia faba* Proteinsisolaten durch partielle enzymatische Hydrolyse, *Nahrung,* 28, 313, 1982.
49. **Schwenke, K. D., Rauschal, E. J., and Robowski, D.,** Functional properties of plant proteins. Part IV. Foaming properties of modified proteins from faba beans, *Nahrung,* 27, 335, 1983.
50. **Schwenke, K. D., Rubowsky, D., and Augustat, D.,** Über Samenproteine: 8 Mitt. Einflus von Elektrolytgehalt und pH-Wert auf die Löslichkeit von Globulien aus den Samen von Sonnenblumen *(Helianthus annuus)* und aus Ackerbohnen *(Vicia Faba), Nahrung,* 425, 1978.
51. **de Vilbiss, E. D., Holsinger, V. H., Posati, L. P., and Pallausch, M. J.,** Whey protein concentrate foams, *Food Technol.,* 28(3), 40, 1974.
52. **Kitabatake, N. and Doi, E.,** Surface tension and foaming of protein solutions, *J. Food Sci.,* 47, 1218, 1982.
53. **Min, D. B. and Thomas, E. L.,** A study of physical properties of dairy whipped topping mixtures, *J. Food Sci.,* 42, 221, 1977.
54. **Grominger, H. and Miller, R.,** Preparation and aeration properties of an enzyme modified succinylated fish protein, *J. Food Sci.,* 40, 327, 1975.
55. **Richert, S. H.,** Physical-chemical properties of whey protein foams, *J. Agric. Food Chem.,* 27, 665, 1979.
56. **Quendt, H., Moldenhauer, A., and Linke, L.,** *Bäcker Konditor,* 24, 83, 1976.
57. **Lawhon, J. T. and Cater, C. M.,** Effect of processing method and pH of precipitation on the yields and functional properties of protein isolates from glandless cottonseed, *J. Food Sci.,* 36, 372, 1971.
58. **Shioya, T., Kako, M., Taneya, S., and Sone, T.,** Influence of time thickening on the whippability of creams, *J. Texture Stud.,* 12, 185, 1981.
59. **Bujnov, A. A., Ginsburg, A. S., and Syroedov, V. I.,** The influence of the depth of fermentation on the foaming properties of fish protein hydrolysates (in Russian), *Izv. Vyssh. Uchelon. Zaved. Pishch. Technol.,* 1, 132, 1978.
60. **Kato, A., Takabashi, A., Matsudomi, N., and Kobayashi, K.,** Determination of foaming properties of proteins by conductivity measurements, *J. Food Sci.,* 48, 62, 1983.
61. **Luther, H., Kraatz, K. H., Buettner, W., and Schuster, E.,** Process for recovery of protein for human nutrition, GDR Patent, DD 207 849, 1984.
62. **Tavaux, C. and Canard, P.,** Manufacture of a food product based on yeasts, French Patent Application, FR. 2 494 963, A 1, 1982.
63. **Turbatović, L., Trimić, Z., Polić, M., and Modić, P.,** Manufacture and use of brewer's yeast protein, *Technol. Mesa,* 24(2), 53, 1983.
64. **Tonnios, F. G.,** Process for preparation of low-nucleic acid protein concentrates from baker's yeast, *Int. Z. Lebensm. Verfahrenstechnik,* 34(1), 7, 10, 151, 1983.

65. **Bogacheva, T. Ja., Grinberg, V. J., and Tolstoguzov, V. B.,** Use of polysaccharides to remove lipids from the protein globulin fraction of baker's yeast, *Carbohydr. Polymers,* 2(3), 163, 1982.
66. **Cooney, C. C., Chokyun, Rha, and Tannenbaum, S. R.,** Single-cell protein: engineering, economies, and utilization in foods, *Adv. Food Res,* 26, 1, 52, 1980.
67. **Huang, F. and Rha, C. K.,** Rheological properties and spinnability of single cell protein/additive mixtures, *J. Food Sci.,* 43(3), 772, 1978.
68. **Huang, F. and Rha, C. K.,** Formation of single cell protein filament with hydrocolloids, *J. Food Sci.,* 43(3), 780, 1978.
69. **Morr, C. V.,** Food emulsifyers from waste products-derived proteins, in *Developments in Food Science 19, Food Emulsifiers Chemistry, Technology, Functional Properties and Applications,* Charalambous, G. and Doxastakis, G., Eds., Elsevier, Amsterdam, 1989, 205.
70. **Vananuvat, P. and Kinsella, J. E.,** Some functional properties of protein isolates from yeast, *Saccharomyces fragilis, J. Agric. Food Chem.,* 23, 613, 1975.
71. **Vananuvat, P. and Kinsella, J. E.,** Extraction of protein, low in nucleic acid, from *Saccharomyces fragilis,* grown continuously on crude lactose, *J. Agric. Food Chem.,* 23, 216, 1975.
72. **Hassan, I. M., Allam, M. H., and Abd El Rahman, N. R.,** Functional properties of yeast protein isolates from different yeasts, *Res. Bull. Fac. Agric. Ain Sham. University,* Cairo, No. 1647, 1981, 21.
73. **Sato, Y. and Hayakawa, M.,** Whipping property of the protein isolated from *Hansenula* yeast grown on methanol, *Lebensm. Wiss. Technol.,* 12, 41, 1979.
74. **Lewić, P. N., Lawford, H. P., Klingerman, A., and Lawford, G. R.,** *Biotechnol. Lett.,* 4(7), 441, 1986.
75. **Asano, M. and Shibasaki, K.,** Effect of preparation methods of yeast protein isolate on emulsifying capacity, *J. Jpn. Soc. Food Sci. Technol.,* 29(1), 31, 1982.
76. **Sung, H. Y., Chen, H. J., and Chuan, S. J.,** Preparation of yeast protein isolates with low nucleic acid content and improved functionality through chemical phosphorylation, Proc. 6th Int. Congr. Food Sci. Technol., 2, 7, 1983.
77. **Chen, S. H., Chen, H. J., and Sung, H. Y.,** Studies on the protein isolates and nucleic acids from brewer's yeast, *J. Chin. Agric. Chem. Soc.,* 23(3/4), 318, 1985.
78. **Duwe, H., Sonnenhalb, W., Rothe, K. P., and Rosahl, B.,** Extraktion von Lipiden aus mikrobiellen Eiweissisolaten, *Nahrung,* 30(7), 667, 1986.
79. **Tolstoguzov, V. B.,** Concentration and purification of proteins by means of two-phase systems: membraneless osmosis process, *Food Hydrocolloids,* 2(3), 195, 1988.
80. **Murray, E. D., Maurice, T. J., Barker, L. D., and Meyers, C. D.,** Method for the preparation of protein isolate, FRG Patent Application 3 022 532, 1981.
81. **Luther, H. and Petzold, G.,** Process for production of protein for human nutrition, GDR Patent 153 462, 1982.
82. **Shay, L. K.,** Functional protein products, U.S. Patent US 4, 564 329, 1986.
83. **Cheng, A. C. and Chang, W. H.,** Functional properties of protein products from brewer's yeast, *J. Chin. Agric. Chem. Soc.,* 22(3/4), 219, 1984.
84. **Cheng, A. C. and Chang, W. H.,** Extraction and recovery of proteins from brewer's yeast *Saccharomyces uvarum, J. Chin. Agric. Chem. Soc.,* 22(3/4), 207, 1986.
85. **Damodaran, S.,** Removal of nucleic acids from yeast nucleoprotein complexes by sulfitolysis, *J. Agric. Food Chem.,* 34(1), 26, 1986.
86. **Aoki, H. and Nagamori, N.,** Emulsifying properties of single cell proteins, *J. Jpn. Soc. Food Sci. Technol.,* 27(11), 550, 1980.
87. **Akin, C. and Darrington, F. D.,** Process for texturizing proteinaceous materials, U.S. Patent 4 161 546, 1979.
88. **Schmandke, H., Schmidt, C., and Muschiolik, G.,** Textures protein isolate from yeast grown on ethanol., *Nahrung,* 22(2), 207, 1978.
89. **Shen, J. L. and Morr, C. V.,** Physicochemical aspects of texturization: fiber formation from globular protein, *J. Am. Oil Chem. Soc.,* 56, 63A, 1979.
90. **Akin, C.,** Method for stretching a coagulable extrudate, U.S. Patent, US 4 154 856, 1979.
91. British Petroleum Co. Ltd., Textured microbial protein, British Patent, 1 587 150, 1981.
92. **Asano, M., Satoh, N., and Shibasaki, K.,** Extraction of proteins from yeast *(Saccharomyces cerevisiae)* using a vibrogen cell mill and low alkali concentration and some properties of the protein, *J. Jpn. Soc. Food Sci. Technol.,* 27(4), 172, 1980.
93. **Onuma Okezie, B. and Kosikowski, F. V.,** Extractability of protein from yeast cells grown on cassava hydrolysate, *Food Chem.,* 7(1), 7, 1981.
94. **Idemitsu Industries, K. K.,** Gelling agent, Japanese Examined Patent 5 608 581 1981.
95. **Idemitsu Industries, K. K.,** Gelling agent, Japanese Examined Patent 5 608 599, 1981.
96. Standard Oil company, Process for producing functional yeast protein, U.K. Patent 1 578 235, 1980.
97. **Achor, I. M., Richardson, T., and Draper, N. R.,** Effect of treating *Candida utilis* with acid or alkali, to remove nucleic acids, on the quality of the protein, *Agric. Food Chem.,* 29, 27, 1981.

98. **Schmandke, H. and Schmidt, G.**, Zur Untersuchung und Beeinflussung funktioneller Eigenschaften von Proteinrohstoffen und Proteinfasern: 5. Mitt. Zur Löslichheit von Äthanol-Hefeprotein und daraus hergestellten Proteinfasern in Abhängigheit von pH-Wert, *Nahrung*, 22(7), 643, 1978.
99. **Lawford, G. R., Kligerman, A., and Williams, T.**, Production of high quality edible protein from *Candida* yeast grown in continuous culture, *Biotechnol. Bioeng.*, 21, 1163, 1979.
100. **Huang, Y. T. and Kinsella, J. E.**, Functional properties of phosphorylated yeast protein: solubility, water holding capacity, and viscosity, *J. Agric. Food Chem.*, 34, 670, 1986.
101. **Roshkova, Zl., Dukiandjiev, S., and Pavlova, K.**, Biochemical characterization of yeast protein isolates, *Nahrung*, 30(3/4), 402, 1986.
102. **Kawasaki, S. and Nomura, D.**, Effect of additives on the spinnability of isolated baker's yeast protein, *J. Agric. Food Chem.*, 53(5), 143, 1979.
103. **Bednarski, W.**, Technology of enzymic modification of proteins from unconventional sources, *Zesz. Nauk. Akad. Roln. Tech. Olsztynie Technol. Zywn.*, No. 14, 3, 1979.
104. **Kawasaki, S. and Nomura, D.**, Viscoelastic properties of concentrated solutions of yeast protein, soybean protein and casein, *J. Jpn. Soc. Food Sci. Technol.*, 26(2), 52, 1979.
105. **Safronova, A. M., Shaternikov, V. A., Visotskij, V. G., Cholakova, A., and Nestorov, N.**, Biological criteria for the use of new protein sources in meat products (in Russian), *Vopr. Pitan.*, 38, 1985.
106. **Vanchikova, O. and Schimunek, Z.**, Use of non conventional protein sources in human nutrition (in Czech), *Prum. Potravin*, 33(8), 468, 1982.
107. **Gajger, O. and Ratkovic, D.**, Methods and economic grounds for making proteins based on brewer's yeast and blood plasma in Yugoslavia, *Technol. Mesa*, 21(1), 40, 1980.
108. **Turubatoviȁ, L., Benko, C., and Nadj, J.**, Possibility of using Belvit-S as a component of spice mixtures, *Technol. Mesa*, 21(2), 58, 1980.
109. **Turobatović, L., Perović, M., Jakovijević, M., and Jetić, L.**, Possibility of using protein preparations of different origin in isolated systems, *Technol. Mesa*, 21(2), 46, 1980.
110. **Kocovski, T., Turubatović, L., Lazić, M., and Sibalić, S.**, Stability and organoleptic properties of paste containing added Belvit-S and various ratios of water to fat, *Technol. Mesa*, 21(2), 53, 1980.
111. **Trumić, Z., Milosevski, V., Ćirić, M., and Polić, M.**, Use of Belvit-S and sodium caseinate for making frankfurter type sausages, *Technol. Mesa*, 21(3), 84, 1980.
112. **Turubstović, L., Kocovski, T., Śibalić, S., and Trtica, G.**, Optimal utilization of Belvit-S in meat paste manufacture, *Technol. Mesa*, 21(3), 88, 1980.
113. **Perović, M., Turubatović, L., Jakovijević, M., and Sekis, S.**, Possible utilization of the protein mixtures from brewer's yeast based protein preparations and blood based protein products in processing of meat products, *Technol. Mesa*, 21(2), 43, 1980.
114. **Tsintsadze, T. D., Cherl-Ho Lee, and Chokyun Rha**, Microstructure and mechanical properties of single cell protein curd, *J. Food Sci.*, 43, 625, 1979.
115. **Faubion, J. M. and Hoseney, R. C.**, High-temperature, short time extrusion cooking of wheat starch and flour. I. Effect of moisture and flour type on extrudate properties, *Cereal Chem.*, 59, 529, 1982.
116. **Faubion, J. M. and Hoseney, R. C.**, High-temperature shorttime extrusion cooking of wheat starch and flour. II. Effect of protein and lipid on extrudate properties, *Cereal Chem.*, 59, 533, 1982.
117. **Lai, C. S., Davis, B., and Hoseney, R. C.**, The effect of a yeast protein concentrate and some of its components on starch extrusion, *Cereal Chem.*, 62(4), 293, 1985.
118. **Fichtali, J. and van de Voort, F. R.**, Fundamental and practical aspects of twin screw extrusion, *Cereal Foods World*, 34(11), 921, 1989.
119. **Midden, T. M.**, Twin screw extrusion of corn flakes, *Cereal Foods World*, 34(11), 941, 1989.
120. **Dziezak, J. D.**, Single- and twin screw extruders in food processing, *Food Technol.*, 43(4), 164, 1989.
121. **Kruseman, J., Bretschy, P. Y., Hidalgo, J., and Rham, O.**, Process for preparing a condensed milk, British Patent 1525 52b, 1978.
122. **Scharf, V. and Schlingmann, M.**, Protein enriched beverage, GDR Patent Application DE 3 314 428 Al, 1981.
123. **Whitaker, J. R. and Puigserver, A. J.**, Fundamentals and applications of enzymatic modifications of proteins: an overview, in *Advances in Chemistry Series No. 198 Modification of Proteins, Food, Nutritional and Pharmacological Aspects*, Feeney, R. E. and Whitaker, J. R., Eds., American Chemical Society, 1982, 57.
124. **Whitaker, J. R.**, Food proteins: enzymatic Modification of proteins applicable in foods, in Improvement Through Chemical and Enzymatic Modification, *Adv. Chem. Ser.*, 1977, 160, 95.
125. **Wold, V. R.**, Chemical deteroration of proteins, ACC Symp. Ser., 123, 49, 1980.
126. **Kester, J. J. and Richardson, T.**, Modification of whey proteins to improve functionality, *J. Dairy Sci.*, 67, 2757, 1984.
127. **Kang, K. H., Lee, K. M., Choe, Y. J., and Lee, J. Y.**, Studies of making a yeast cheese with *Debaromyces hansensii* and *Saccharomyces fragilis*, *Korean J. Anim. Sci.*, 22(2), 101, 1980.

128. **Detana, P., Fatichenti, F., Farris, G. J., Mocquot, G., Lodi, R., Todesco, R., and Cecchi, L.,** Metabolization of lactic and acetic acids in Pecorino Romano cheese made with a combined starter of lactic acid bacteria and yeast, *Lait,* 64(640/641), 380, 1984.
129. **Hagrass, H. E. A., Sultan, N. E., and Hammad, Y. A.,** Chemical properties of Ras cheese during ripening as affected by the addition of native dry yeast and yeast autolysate, *Egypt. J. Dairy Sci.,* 12(1), 55, 1984.
130. **Higashi, K., Yamamoto, Y., and Yoshi, H.,** Study on salt tolerance of miso and soy-sauce yeast. I. Effect of sodium chloride on cell wall content and chemical composition of the cell wall, *J. Jpn. Soc. Food Sci. Technol.,* 31(4), 225, 1984.
131. **Oike, K., Yeneyama, T., Miyazaki, K., and Negeshi, M.,** Organic acids in miso, Res. Rep. Nogane State Lab. Food Technol., No. 10, 28, 1982.
132. **Lee, T. S., Yang, K. J., Park, Y. J., and Yu, J. H.,** Studies on the brewing of Kochujang (red pepper paste) with the addition of mixed cultures of yeast strains, *Korean J. Food Sci. Technol.,* 12(4), 313, 1980.
133. Dai-Nippon Seito, Co. Ltd, Protein beverage, Japanese Examined Patent 5 411 379, 1979.
134. **Sejr Olsen, J. and Adler-Nissen, J.,** Application of ultra- and hyperfiltration during production of enzymatically modified proteins, in Synthetic Membranes Vol. 11, No. 154, 1981, 133.
135. **Tolman, R. C.,** *Principles of Statistical Mechanisms,* Oxford University Press, 1938, 163.
136. **Jakubke, H. D. and Kuhl, P.,** Proteasen als Biokatalysatoren für die Peptidsynthese, *Pharmazie,* 37, 89, 1982.
137. **Fruton, J. S.,** Proteinase-catalysed synthesis of peptide bonds, *Adv. Enzymol. Relat. Areas Mol. Biol.,* 53, 239, 1982.
138. **Jakubke, H. P., Kuhl, P., and Könnecke, A.,** Grundprinzipien der proteaskatalysierten Knüpfung der Peptidbindung, *Angew. Chem.,* 97, 79, 1985.
139. **Kullman, W.,** Proteases as catalysts in peptide synthetic chemistry. Shifting the extent of peptide bond synthesis from a "quantite neglibeable" to a "quantite considerable", *J. Protein Chem.,* 4, 1, 1985.
140. **Hajós, Gy., Éliás, I., and Halász, A.,** Methionine enrichment of milk protein by enzymatic peptide modification, *J. Food Sci.,* 53(3), 739, 1988.
141. **Kullman, W.,** *Enzymatic Peptide Synthesis,* CRC Press, Boca Raton, FL, 1987.
142. **Weisbach, H. and Pestka, S.,** *Molecular Mechanism of Protein Biosynthesis,* Academic Press, New York, 1977.
143. **Lowe, G.,** The cysteine proteinases, *Tetrahedron,* 32, 291, 1976.
144. **Morahira, K. and Oka, T.,** α-Chymotrypsin as the catalyst for peptide synthesis, *Biochem. J.,* 163, 531, 1977.
145. **Nilsson, K. and Mosbach, K.,** Peptide synthesis in aqueous-organic solvent mixtures with α-chymotrypsin immobilized to tresyl-chloride activated agarose, *Biotechnol. Bioeng.,* 26, 1146, 1984.
146. **Dordick, J. S.,** Enzymatic catalysis in monophasic organic solvents, *Enzyme Microb. Technol.,* 11(4), 194, 1989.
147. **Kuhl, P., Walpuski, J., and Jakubke, H. D.,** Untersuchungen zum Einfluss der Reaktionsbedingungen auf die α-Chymotrypsin katalysierte Peptidsynthese in wässerig-organischen Zweiphasensystem, *Pharmazie,* 37(11), 766, 1982.
148. **Danilewsky, B.,** The organoplastic forces of the organism (in Russian) (Harkow, 1886), in *Comprehensive Biochemistry,* Vol. 32, Elsevier, Amsterdam, 1977, 314.
149. **Van't Hoff, J. H.,** Über die Zunehmende Bedeutung der anorganischen Chemie, *Z. Anorg. Chem.,* 18, 1, 1898.
150. **Savjalov, W. W.,** Zur Theorie der Eiweissverdauung, *Pflüegers Arch. Gesamte Physiol.,* 85, 171, 1901.
151. **Determann, H., Eggenschwiller, S., and Michel, W.,** Molekular-gewichtsverteilung des enzymatischen Kondensationsprodukts, *Liebigs Ann. Chem.,* 690, 182, 1965.
152. **Determann, H., Hever, J., and Jaworek, D.,** Spezifität des Pepsins bei der Kondensationsreaktion, *Liebigs Ann. Chem.,* 690, 189, 1965.
153. **Determann, H. and Köhler, R.,** Prolin-haltige Monomere und ihre enzymatische Umwandlungsprodukte, *Liebigs Ann. Chem.,* 690, 197, 1965.
154. **Horowitz, J. and Haurowitz, F.,** Mechanism of plastein formation, *Biochem. Biophys. Acta,* 33, 231, 1959.
155. **Virtanen, A. I., Laaksonen, T., and Kantola, M.,** Dependence of the peptic-hydrolysis of zein at different pH on the protein concentration, *Acta Chem. Scand.,* 5, 316, 1951.
156. **Yamashita, M., Arai, S., Matsuyama, J., Gonda, M., Kato, H., and Fujimaki, M.,** Enzymatic modification of proteins in food stuffs. III. Phenomenal survey on α-chymotryptic plastein synthesis from peptic hydrolysate of soy protein, *Agric. Biol. Chem. Jpn.,* 33, 1484, 1970.
157. **Tsai, S. J., Jamashita, M., Arai, S., and Fujimaki, M.,** Effect of substrate concentration on plastein productivity and some rheological properties of the products, *Agric. Biol. Chem. Jpn.,* 36, 1045, 1972.
158. **Yamashita, M., Tsai, S. J., Arai, S., Kato, H., and Fujimaki, M.,** Enzymatic modification of proteins in foodstuffs. 5. Plastein yields and their pH dependence, *Agric. Biol. Chem. Jpn.,* 35, 86, 1971.

159. **Fujimaki, M., Yamashita, M., Arai, S., and Kato, H.,** Enzymatic protein modification of protein in foodstuffs. 1. Enzymatic proteolysis and plastein synthesis. Application for preparing bland protein-like substances, *Agric. Biol. Chem. Jpn.*, 34, 1325, 1970.
160. **Aso, K., Yamashita, M., Arai, S., and Fujimaki, M.,** Tryptophan-threonine and lysine-enriched plastein from zein, *Agric. Biol. Chem. Jpn.*, 38, 681, 1974.
161. **Yamashita, M., Arai, S., Tanimoto, S., and Fujimaki, M.,** Condensation reaction occurring during plastein formation by α-chymotrypsin, *Agric. Biol. Chem.*, 37, 953, 1973.
162. **v. Hofsten, B.** Enzymatic hydrolys av näringsproteines. STV Report No. 733229, University of Uppsala, Sweden, 1974.
163. **v. Hofsten, B. and Lalasidis, G.,** Protease-catalyzed formation of plastein products and some of their properties, *J. Agric. Food Chem.*, 24(3), 460, 1976.
164. **Monti, J. C. and Jost, R.,** Papain-catalysed synthesis of methionine enriched soy plasteins. Average chain length of the plastein peptides, *J. Agric. Food Chem.*, 27(6), 1284, 1979.
165. **Hajós, Gy. and Halász, A.,** Incorporation of L-methionine into casein hydrolysate by enzymatic treatment, *Acta Aliment.*, 11, 189, 1982.
166. **Edwards, J. H. and Shite, W. F.,** Characterization of plastein reaction products formed by pepsin, α-chymotrypsin and papain treatment of egg albumin hydrolysates, *J. Food Sci.*, 41, 490, 1978.
167. **Sukan, G. and Andrews, A. T.,** Application of the plastein reaction to caseins and to skim-milk powder, *J. Dairy Res.*, 49(2), 265, 1982.
168. **Aso, K., Yamashita, M., Arai, S., and Fujimaki, M.,** Hydrophobic force as a main factor contributing to plastein chain assembly, *J. Biochem.*, 76, 341, 1976.
169. **Eriksen, S. and Fagerson, I. S.,** The plastein reaction and its application: a review, *J. Food Sci.*, 41, 490, 1976.
170. **Lauffer, M. A.,** *Entropy-Driven Processes in Biology,* Spinger Verlag, New York, 1975.
171. **Arai, S., Yamashita, M., Aso, K., and Fujimaki, M.,** A parameter related to the plastein reaction, *J. Food Sci.*, 40, 342, 1975.
172. **Tanimoto, S., Yamashita, M., Arai, S., and Fujimaki, M.,** Salt effects on plastein formation by α-chymotrypsin, *Agric. Biol. Chem.*, 39, 1207, 1975.
173. **Yamashita, M., Arai, S., Kato, H., and Fujimaki, M.,** Enzymatic modification of proteins in foodstuffs. II. Nutritive properties of soy plastein and its bio-utility evaluation in rats, *Agric. Biol. Chem.*, 34, 1333, 1970.
174. **Noack, A. and Hajós, Gy.,** The enzymic *in vitro* digestion of methionine-enriched plastein, *Acta Aliment.*, 13(3), 205, 1984.
175. **Pallavicini, C., Dal Belin Peruffo, A., and Finley, J. W.,** Comparative study of soybean plasteins synthesized with soluble and immobilized α-chymotrypsin, *J. Agric. Food Chem.*, 31, 846, 1983.
176. **Könnecke, A., Hänser, M., Schellenberger, V., and Jakubke, H. D.,** Peptydsyntheses mit immobilisierten Enzymen. II. Immobilisiertes Trypsin, Thermolysin, and Papain, *Monatsh. Chem.*, 114, 433, 1983.
177. **Gaertner, H. and Puigserver, A.,** Digestion *in vitro* de le polymethionyl-caseine par les enzymes du tractus digestif, *Lait,* (Paris), 62, 607, 1982.
178. **Fujimaki, M., Utaka, K., Yamashita, M., and Arai, S.,** Production of higher quality plastein from single-cell protein, *Agric. Biol. Chem. Jpn.*, 37, 2303, 1973.
179. **Bernarski, W. and Leman, J.,** Detoxification and parallel modification of some functional properties of unconventional protein by enzymatic treatment, *Acta Aliment. Polonica,* 5(2), 165, 1979.
180. **Haard, N. F., Felthan, L. A. W., Helbig, N., and Squires, J. E.,** Modification of proteins with proteolytic enzymes from the marine environment, in *Modification of Proteins: Food, Nutritional, and Pharmacological Aspects,* Feeney, R. E. and Whitaker, J. R., Eds., Adv. Chem. Ser., American Chemistry Society, Washington, D.C., Ser. 198, 1982, 223.
181. **Otagiri, K., Nosho, Y., Shinoda, I., Fukui, H., and Okai, H.,** Studies on a model of bitter peptides including arginine, proline and phenylalanine residues. I., *Agric. Biol. Chem.*, 49, 1019, 1985.
182. **Nosho, Y., Otagiri, K., Shinoda, I., and Okai, H.,** Studies on a model of bitter peptides including arginine, poline and phenylalanine residues. II, *Agric. Biol. Chem.*, 49, 1829, 1985.
183. **Halász, A., Hajós, Gy., and Teleky-Vámosy, Gy.,** Investigation of taste of hydrolysates of brewer's yeast. Research Report No. 12, Central Research Institute of Food Industry, Budapest, 1987.
184. **Fox, P. F., Morissey, P. A., and Mulvihill, D. M.,** Chemical and Enzymatic Modification of Food Proteins, in *Developments in Food Proteins-1,* Hudson, B. J. F., Ed., Applied Science Publishers, London, 1982, 1.
185. **Belikov, V. M. and Golobolov, M. Y.,** Plastein, their preparation, properties and use in nutrition, *Nahrung,* 30, 281, 1986.
186. **Golobolov, M. Y., Antonova, T. U., and Belikov, V. M.,** Transpeptidation as the main chemical reaction causing the formation of plasteins, *Nahrung,* 30, 289, 1986.
187. **Yamashita, M., Arai, S., and Fujimaki, M.,** A low phenylalanine, high tyrosine plastein as an acceptable dietetic food, *J. Food Sci.*, 41, 1029, 1976.

188. **Hajós, Gy., Halász, A., and Békési, F.,** Designed protein modification by enzymatic technique, *Acta Aliment.*, 18(3), 325, 1989.
189. **Shimada, A., Yamamoto, I., Sase, A., Yamazaki, Y., Watanabe, M., and Arai, S.,** Surface properties of enzymatically modified proteins in aqueous systems, *Agric. Biol. Chem.*, 48, 2681, 1984.
190. **Hajós, Gy.,** Enzymic Modification of Food Proteins (in Hungarian) Ph.D thesis, Hungarian Academy of Sciences, Budapest, 1987.
191. **Hajós, Gy. and Halász, A.,** Enzymic modification of food proteins. Part 3. Covalent linkage of amino acids into the protein chain (in Hungarian), *Élelmez. Ipar*, 43, 429, 1989.
192. **Halász, A.,** Biochemical Principles of the Use of Yeast Biomass in Food Industry, D.Sc. thesis, Hungarian Academy of Sciences, Budapest, 1988.
193. **Halász, A., Szakács-Dobozi, M., Hajós, Gy., Mátrai, B., and Szalma-Pfeiffer, I.,** Characterization of *Saccharomyces cerevisiae* proteinases (in Hungarian), *Élelmez. Ipar*, 43, 322, 1985.
194. **Fox, P. F., Morissey, P. A., and Mulvihill, D. M.,** Chemical and enzymatic protein modifications of food proteins, in *Developments in Food Proteins-1*, Hudson, B. J. F., Ed., Applied Science Publishers, London, 1982, 1.
195. **Lásztity, R.,** Correlation between chemical structure and rheological properties of gluten, *Ann. Technol. Agric.*, 29(2), 339, 1980.
196. **Puigserver, A. J., Sen, L. C., Gonzales-Flores, E., Feeney, R. E., and Whitaker, J. R.,** Covalent attachment of amino acids to casein. 1. Chemical modification and rates of *in vitro* enzymatic hydrolysis of derivatives, *J. Agric. Food Chem.*, 27, 1098, 1979.
197. **Puigserver, A. J., Sen, L. C., Clifford, A. J., Fenney, R. E., and Whitaker, J. R.,** Covalent attachment of amino acids to casein. 2. Bioavailability of methionine and *N*-acetyl-methionine covalently linked to casein, *J. Agric. Food Chem.*, 27, 1286, 1979.
198. **Voutsinas, L. P. and Nakai, S.,** Covalent binding of methionine and tryptophan to soy protein, *J. Food Sci.*, 44, 1205, 1979.
199. **Batey, I. L.,** Chemical modification as a probe of gluten structure, *Ann. Technol. Agric.*, 29(2), 363, 1980.
200. **Neurath, H. and Hill, R. L.,** *The Proteins*, Vol. 2, 3rd ed., Academic Press, New York, 1976.
201. **Lee, H. S., Sen, L. D., Clifford, A. J., Whitaker, J. R., and Feeney, R. E.,** Preparation and nutritional properties of caseins covalently modified with sugars. Reductive alkylation of lysines with glucose, fructose, or lactose, *J. Agric. Food Chem.*, 27, 1094, 1979.
202. **Knowles, J. R.,** Chemical modification and the reactivity of amino acids in proteins, in *MTP, Intl. Rev. of Science and Biochemistry*, Series One, Chemistry of Macromolecules, Gutenfreund, Eds., Vol. 1, Uni Park, Baltimore, MD, 1974, 149.
203. **Schneider, Ch., Muschiolik, G., Schultz, M., and Schmandke, H.,** The influence of process conditions and acetylation on functional properties of protein isolates from broad beans *(Vicia Faba L. minor), Die Nahrung*, 30(3/4), 429, 1986.
204. **Vananuvat, P. and Kinsella, J. E.,** Succinylation of yeast protein. I. Preparation and Composition, *Biotech. Bioeng.*, 20, 324, 1978.
205. **McElwain, M. D., Richardson, T., and Amudson, C. H.,** Some functional properties of succinylated single cell protein concentrate, *J. Milk Food Technol.*, 28, 521, 1975.
206. **Pearce, K. P. and Kinsella, J. E.,** Emulsifying properties of proteins: evaluation of a turbidimetric technique, *J. Agric. Food Chem.*, 26, 75, 1978.
207. **Brinegar, A. and Kinsella, J. E.,** Reversible modification of lysine in soybean proteins, using citraconic anhydride: characterization of physical and chemical changes in soy protein isolate, the 7S globulin and lypoxygenase, *J. Agric. Food Chem.*, 28, 818, 1980.
208. **Palacian, E., Lopez-Rivas, A., Pintor-Tora, J. A., and Hernandez, F.,** *Mol. Cell. Biochem.*, 36, 163, 1981.
209. **Shetty, J. K. and Kinsella, J. E.,** Isolation of yeast protein with reduced nucleic acid level using reversible acylating reagents: some properties of the isolated proteins, *J. Agric. Food Chem.*, 30, 1166, 1982.
210. **Huang, Y. T. and Kinsella, J. E.,** Functional properties of phosphorylated yeast protein: solubility, water holding capacity and viscosity, *J. Agric. Food Chem.*, 34, 670, 1986.
211. **Huang, Y. T. and Kinsella, J. E.,** Phosphorylation of yeast proteins: reduction of ribonucleic acid and isolation of yeast protein concentrate, *Biotechn. Bioeng.*, 28, 1690, 1986.
212. **Damodaran, S. and Kinsella, J. E.,** Dissociation of yeast nucleoprotein complexes by chemical phosphorylation, *J. Agric. Food Chem.*, 32, 1030, 1984.
213. **Chen, S. H., Chen, H. J., and Sung, H. Y.,** Studies on the protein isolates and nucleic acids from brewer's yeast, *J. Chin. Agric. Chem. Soc.*, 23, 318, 1986.
214. **Lásztity, R.,** *The Chemistry of Cereal Proteins*, CRC Press, Boca Raton, FL, 1984.
215. **Ma, E. Y., Oomah, B. D., and Holme, J.,** Effect of deamidation and succinylation on some physicochemical and baking properties of gluten, *J. Food Sci.*, 51, 99, 1986.

218. **Guzman-Juarez, M.,** Yeast protein, *in Developments in Food Proteins-2*. Hudson, B. J. F., Ed., Applied Science Publisher, London, 1983, 263.
219. **Kinsella, J. E.,** Functional properties of proteins in food: a survey, *CRC Crit. Rev. Food Sci. Nutr.*, April 1976, 219.
220. **Kim, S. C., Olson, N. F., and Richardson, T.,** The effect of thiolation on β-lactoglobulin, *J. Dairy Sci.*, 66, (Suppl. 1.), 98, 1983.
221. **Kester, J. J. and Richardson, T.,** Modification of whey proteins to improve functionality, *J. Dairy Sci.*, 29, 2757, 1983.
222. **Marshall, J. J.,** Manipulation of the properties of enzymes by covalent attachment of carbohydrate, *Trends Biochem. Sci.*, 3, 79, 1983.
223. **Gray, G. R.,** The direct coupling of oligosaccharides to proteins and derivatized gels, *Arch. Biochem. Biophys.*, 163, 426, 1974.
224. **Schwartz, B. A. and Gray, C. R.,** Proteins containing reductively aminated disaccharides. Synthesis and chemical characterization, *Arch. Biochem. Biophys.*, 181, 542, 1977.
225. **Adler-Nissen, J.,** Enzymic modification of proteins effect on physico-chemical and functional properties, in *Functional Properties of Food Proteins,* Lásztity, R. and Ember-Kárpáti, M., Eds., MÉTE Publishing, Budapest, 1988, 31.
226. **Ryan, D. S.,** Determinants of the functional properties of proteins and protein derivatives in foods, in *Food Proteins: Improvement Through Chemical and Enzymatic Modification,* Feeney, E. R. and Whitaker, J. R., Eds., Adv. Chem. Ser. 160, American Chemistry Society, Washington, D.C., 1977, 67.
227. **Tolstoguzov, V. B., Braudo, E. E., and Gurov, A. N.,** On the protein functional properties and the methods of their control. Part 2. On the methods of control of functional properties of proteins and gel-forming systems, *Nahrung,* 25, 817, 1981.
228. **Gurov, A. N. and Nuss, P. V.,** Protein-polysaccharide complexes as surfactants, *Nahrung,* 30(3/4), 349, 1986.
229. **Braudo, E. E., Gottlieb, A. M., Plashina, I. G., and Tolstoguzov, V. B.,** Protein-containing multi-component gels, *Nahrung,* 30(3/4), 355, 1986.
230. **Polyakov, V. I., Popello, I. A., Grinberg, V. Ja., and Tolstoguzov, V. B.,** Thermodynamic compatibility of proteins in aqueous medium, *Nahrung,* 30(3/4), 365, 1986.

Chapter 9

USE OF YEAST AUTOLYSATES AND YEAST HYDROLYSATES IN FOOD PRODUCTION

The history of the use of hydrolyzed proteinaceous materials for food flavoring is hundreds of years old. The original process is said to have arisen in China. Later it spread to other countries in the Far East and by the Chinese and Japanese influence in Europe and North America. The original flavorings were produced using enzymes already present in their natural state. Development of the hydrolysis of proteins using acids began only in the previous century and now mainly hydrolyzed vegetable protein is produced in this way.

The importance of meat extract as a source of flavor has steadily declined not only because of improved methods of storing and transporting beef and hence, the decreasing availability of meat extract, but also because of the gradual development of superior meat flavors. While meat extract usage has declined the role of yeast autolysates has grown in importance. The products have become increasingly more sophisticated; partly through the development of strains of yeast with improved flavor properties and also due to the better understanding of which components are important for flavor and the development of techniques for growing and autolyzing of yeast with maximum retention of these components.[1-4] Yeast autolysate is used not only as flavoring material but in some cases also as an ingredient improving physical properties and nutritional value of food products.[5,6] They can also be used as a flavor carrier in seasoning formulations for processed meat products.[3] The use of yeast extracts as a source of nutrients in media for cultivation of fastidious microorganisms has been known for more than 60 years.[7]

I. PRODUCTION OF YEAST AUTOLYSATES

The principles and main steps of the production of yeast autolysates were treated in detail in Chapter 5 of Part I. Here some industrial applications and specific procedures will be treated.

The Bovril Food Ingredients Company has a process for the production of different yeast autolysates under the trade name Yeatex as reported by Tuley.[2] Brewer's yeast is used as raw material. The yeast is mixed with water and following the removal of carbon dioxide passes into one of either 4000 gallon autolyzers. Intracellular enzymes of yeast are activated and catalyze a digestive process of the yeast cells. After autolysis the temperature is raised to destroy the enzyme. As a result of autolysis a suspension is formed containing a lot of nutritionally valuable soluble components which are separated from hop, debris and cell wall material via centrifugal separation and counter-current washing techniques. The resulting liquor, low in total solids, has to be concentrated to an intermediate stage. This is achieved on a series of continuous multistage evaporators. Three types of evaporators are used: batch pan, plate and falling film. Finally, the extract is filtered through rotary vacuum filters to remove all possible debris before further processing. Column adsorption technique is used to remove color and bitterness from the extract. Concentrates with different solid content and dried powders can be prepared according to the customer's needs.

Although great progress was achieved in improving the technology of yeast autolysis in the seventies, the research work associated with further development still continues. Chrenova et al.,[9] to determine optimal activation conditions for yeast endoenzymes, (nucleases, proteases) investigated the effects of different metallic cations (Mg, Ca and Zn using 0.1 to 0.6% salt concentrations) on autolysis of baker's yeast over a temperature range of 45 to 60°C. Criteria for the evaluation of the process included an initial rate of autolysis, yield of protein degradation products and nucleic acid constituents. The results showed that the added ions have little effect on cell structure, but have a stabilizing effect on endoenzymes. Yields of protein and nucleic acid constituents were increased; Scheffeld et al.[10] studied a temperature-sensitive lysis mutant of *Saccharomyces cerevisiae JD 109*. This mutant normally stops growing at 27°C and partially lyses at a nonpermissive temperature (37°C). The rate of lysis was faster at pH 8 than at pH 5.6. The protein in the cells decreased from 45 to 35%.

A method for preparation of a yeast hydrolysate was developed by Basoppa et al.[11] In several procedures examined, wet compressed baker's yeast (*Saccharomyces cerevisiae* var. *ellipsoidues*) was suspended in water and treated using agents such as HCl, NaCl, H_2SO_4, NaOH and enzymes (xylanase, glucanase, papain) to hydrolyze or induce autolysis of the yeast cells. Hydrolysates and autolysates were analyzed for crude protein, moisture, NaCl, free amino *N*, trichloroacetic acid soluble protein, sensory properties (color, flavor, texture) and nutritional quality (using *Bacillus megatherium*). Results showed that hydrolysis with 0.1 *N* HCl, neutralization with sodium hydroxide, cation-exchange, vacuum filtration and vacuum drying gave an autolysate satisfactory nutrition but its organoleptic characteristics could be improved.

An enzymatic method for yeast autolysis was used in experiments of Knorr et al.[12] Treatment of intact cells with lytic enzymes, i.e., zymolase and lysozyme, markedly increased release of nitrogen and proteins during incubation. Addition of pancreatin or pronase during incubation of cells with lytic enzymes caused concurrent hydrolysis of yeast proteins. The precipitable yeast protein at pH 4.5 decreased from 73 to 21% within 60 min. This procedure is recommended by authors for commercial production of yeast autolysate.

Thermolysis parameters of *Rhodotorula glutinis 214*, a carotenoid producing yeast, were studied by Stabnikova et al.[13] Conditions governing the thermolysis of *Rhodotorula glutinis 214* were investigated. The effect of thermal regimes and holding periods on the extent of cell inactivation, biochemical changes in the biomass and the persistence and qualitative composition of carotenoid-type pigments was established. Optimal storage conditions for the biomass after thermolysis were also studied.

Thermolysis should be at 80°C for 1 h, this completely immobilizes the cells without causing any serious changes in biochemical indices of the biomass (e.g., contents of beta-carotene, torulin, torularhodin, proteins, amino acids, riboflavin and nucleic acids). The product should be stored at 0°C in an inert gas atmosphere to avoid loss of the carotenoid pigments.

Different chemical and physical treatments were used in order to increase the rate of autolysis.[14] Dilute alkali treatment with 0.01 *N* NaOH caused a considerable increase in soluble nitrogen extraction in the 6 h initial autolysis. Addition of 10% fresh yeast autolysate to yeast slurry could increase the autolysis rate from 65 to 83%. Microwave treatment of yeast slurry for 40 s raised the autolysis rate by about 15%. Extrusion of yeast cells under high pressures (16,000, 20,000 lb/bm^2) brought about a significantly higher autolysis after digestion for 9 h, but thereafter showed a gradual drop in soluble nitrogen extraction.

Yeast autolysate manufacture, its characteristics and use in prepared meals is discussed by Sommer.[23] Aspects considered include: yeast species used for autolysate preparation; autolysis methods; separation of cell walls from cell contents; concentration; composition of yeast autolysates (with special reference to nucleotides and other components important for flavor), color characteristics; flavor enhancing effect of yeast autolysates in comparison to that of other flavor enhancers and applications in foods including convenience foods, stock cubes, dried soups, seasonings, etc.

A process for producing functional protein hydrolysates was patented by Scharf et al.[24] The product based on microbial protein isolates contain (by weight) more than 90% protein, less than 2% nucleic acids and less than 1% lipids. The hydrolysates have 80 to 100% suspension ability, a foaming number of 4 to 7 (foam ability) a foam stability of a half-life value of 10 to 300 min, emulsification ability 300 to 500 ml/g protein a molecular weight of 125 to 150 Da. Hydrolysate is recommended for use in mayonnaise.

Yeast extracts with improved flavor are obtained according to a Japanese patent[25] by combining a water miscible organic solvent, preferably ethanol, *n*-propanol or *n*-butanol, with an edible yeast extract solution and decanting to remove soluble components. The final glutinous yeast extract is dehydrated in a vacuum dryer and water is added to give a paste or by further spray drying a powder.

Another Japanese patent[26] is dealing with production of yeast extract containing flavoring. The yeast extract contains flavoring 5′-nucleoside and has improved thickness and body in taste. Yeast cells are suspended in the process in the presence of autolysis stimulators at a pH of 6.0 to 6.6. The autolyzed suspension is heated to 90 to 110°C for 1 to 3 h to extract intracellular RNA and the following steps are performed: (1) hydrolyzing of extracted RNA with 5′-phosphodiesterase and (2) separating the resulting extract from the insoluble residue. The final product can be used as a component in seasoning compositions.

II. PRACTICAL EXAMPLES OF UTILIZATION OF YEAST AUTOLYSATES AND HYDROLYSATES

Experiments were conducted by Parks et al.[22] to determine effects of autolyzed yeast (AY) on frankfurter firmness, flavor and yield. Smokehouse yield of laboratory prepared frankfurters was not significantly affected by addition of AY at 1%. Commercially produced frankfurters containing 0 to 1.5% AY were subjected to sensory and yield evaluations. Products with 0.75 and 1.0% AY were more firm than controls. Vacuum packaged frankfurters containing AY, and held for 2 to 6 weeks at 2 to 5°C, showed less purge loss (moisture accumulation in the package) than controls. AY appears to enhance frankfurter flavor and firmness and to reduce purge loss on vacuum packaging.

Use of yeast extract in sausage manufacture is reported by Slyusarenko et al.[6] It was stated, based on experiments, that yeast extract may replace part of the meat in sausage-type smoked meat products. The nutritional properties of autolysates are acceptable both from the point of view of biological value and organoleptic properties. The yeast extract can be added to almost any type of smoked product. Optimum level of addition depends on the product, but is generally 1 to 2%. The extract may increase the protein content of some products.

It was observed in pilot plant experiments[15] that addition of yeast autolysate to meat before freezing prevents the undesired changes of the extractability of meat proteins occurring during storage in a frozen state. Not only the solubility and extractability of proteins remained at the initial level, but also some increase in water-holding capacity was observed. The flavor of low sodium foods could be enhanced by yeast autolysates.[16] Autolyzed yeast products (Zyest-70, Zyest-45, etc; Pure Culture, Inc.) prepared from *Candida utilis* yeast grown on grain alcohol were recommended for such purposes. The yeast is autolyzed by the action of enzymes. Since no salts are added Zyest products are therefore suitable for use in low sodium foods. Zyest-70 which is recommended for improving flavor of foods containing chicken and pork and in extended meats, masks the bitterness of KCl in low sodium foods as well as enhancing flavor profiles. Zyest-45 is recommended for use in low sodium poultry food systems and other high salt meat systems. Zyest-FM imparts an earthy flavor suitable for use in mushroom and cheese products.

A lot of patents deal with the production of yeast autolysate-based flavoring preparations. For example, a flavor enhancer for cheese-containing foods was described by Andres.[17] The flavor contribution is approximately the same as 6 to 7 times the weight of cheddar cheese. The preparation can be pasteurized without flavor loss. Konrad and Lieshe[18] patented a process for preparation of flavoring material with chicken flavor. Yeast cells were autolyzed by repeated variations of pH over the range 3.0 to 7.0 at 60°C and the autolysate was pasteurized. Cell residues were separated, and the hydrolysate was concentrated and dried. The product has a chicken-like flavor and is free from bitterness or "hydrolysate" flavor. It is recommended for use in soups, sauces, seasonings, bakery products and snacks. Toshinore et al.[19] reported a process for producing seasoning agent and food containing sauce. Suspension of brewer's yeast was extracted with water at 40 to 100°C for 5 to 120 min so that the extracts obtained are substantially free of decomposed cells. Such extracts were centrifuged to remove solids, the pH is lowered to 2 to 4 with HCl or food grade acids and then were sterilized at 60 to 90°C for 1 to 30 min. Finally, the extract was neutralized with NaOH to pH 4.5 to 7.0 and concentrated. Flavor enhancing ingredients can be used in combination with the yeast extracts including monosodium glutamate, nucleotides and hydrolyzed proteins. Use of various Zyest autolyzed yeast flavorings to reduce bitterness of potassium chloride, used as a NaCl replacement, and enhance food flavor is discussed by Murray.[28] Several examples of foods in which these products are effective are given by the author (e.g., sausages, margarine, snack foods, hams, soup mixes, etc.).

It was reported[29] that yeast autolysate plus hydrolyzed protein is available commercially as "Fretas A" which can be used at 0.15 or 0.3% in cultured milks or 0.5% in a cheese product to control lypolysis and enhance flavor in products without affecting the texture.

A short review is given by Sommer[30] about manufacture of yeast autolysate and application in frozen and instant dried meals and use in meat products. Addition of yeast extract (0.2% w/w) to buffalo's milk during cheese making resulted in greater acidity development, higher soluble/total protein ratios, greater amino acid N' released during pickling and enhanced flavor development after 20 days pickling.[31]

The method of increasing the organoleptic acceptability of shank meat is described by Bender et al.[32] The concurrent use of autolyzed yeast extract, sodium tripolyphosphate hydrated with a solution containing citrus juice solids greatly improves the organoleptic acceptability of comminuted meats.

To prevent the oxidation of fats the surface of melted fats (beef tallow, lard, bone fats) were treated with a thin layer (1 to 2 mm) of yeast autolysate or moistened with an extract of yeast autolysate.[20] The yeast autolysate had antioxidant properties due to its high content (230 to 250 mg%) of sulfhydryl compounds which formed a synergistic complex of water soluble, relatively easily oxidable, low-molecular weight, highly reactive compounds. The fats treated with yeast autolysate showed, even after 150 days storage, no changes or development of off-flavor.

Enzyme hydrolysis of single cell protein was made by Chen et al.[21] to produce a flavoring preparation. *Hansenula polymorpha ATC 26012* yeast (grown on methanol) was partially hydrolyzed by adding proteases and 5'-phosphodiesterases. The autolyzed yeast containing small peptides, and 5'-nucleotides can be used as seasoning ingredients in the food industry. Yeast autolysis was carried out at 55°C and pH 5.5 to 6.0 for about 4 to 20 h, then heated to 65°C for about 60 to 70 min. Threshold concentration of taste was 1.2 to 2.5; 2.5 to 2.8; 1.0 to 1.1; 0.5 to 0.7% of autolysate for spray-dried cells, freeze-dried cells, spray-dried cells with emulsifyer and fresh cell resp. The digestibility ranged from 78 to 92%.

Autolyzed yeast was used to reduce the bitter flavor imparted by potassium salts in potassium-containing salt substitutes or in foods which contain potassium salt.[27] Less acceptable grades of meat from the standpoint of flavor, such as shank meat, are thus upgraded in taste to that of chuck meat.

Preparation of a yeast extract from any of the following types of yeast is described by Rooij and Hakkaart:[33] *Kluyveromyces, Saccharomyces, Candida* and *Torula*. Enzymatic degradation of the yeast tissue is accompanied by fermentation using lactic acid bacteria, (*Lactobacillus acidophylus, L. plantarum, L. fermentum,* etc.) lactic streptococci and yeasts (*Saccharomyces rouxii* or *Saccharomyces cerevisiae*). After fermentation and enzymatic degradation, insoluble matter is removed by centrifugation or filtration leaving a yeast extract which can be mixed with a maltodextrin carrier and concentrated; spray-, freeze- or drum dried. The extract has no bitter taste and less pronounced or no yeasty flavor. A patented process for yeast extract food flavor was published by the authors mentioned above.[34]

Yeast is inactivated at 70 to 150°C for 5 to 120 min. It is than subjected to enzymatic degradation together with fermentation with *Lactobacillus acidophilus, Streptococcus lactis, Pediococcus pentosaceus, Saccharomyces rouxii* and *Saccharomyces cerevisiae* alone or in combination. The yeast extract produced has no bitter or yeasty flavor, contains 20 to 45% by weight protein, peptides and free amino acids, 8 to 20% lactic acid and can be concentrated or dried (by spray-, freeze- or drum drying). Potential applications include soups, gravies, meat products, cheese and confectionery. Levels of use vary widely from 0.15 to 5% (dry weight of yeast in ready-to-eat food).

REFERENCES

1. **Blake, T.,** Significant development of meat analogue flavouring agents, *Food Ind. S. Africa,* February 1988, 37.
2. **Tuley, L.,** Boost for Bovril, *Food, Flavouring, Ingredients, Packaging and Processing,* 8(8), 20, 1986.
3. **Schmidt, G.,** Bubbling over with yeast, *Food, Flavouring, Ingredients, Packaging and Processing,* 9(5), 25, 1987.
4. **Woollen, A.,** Flavouring the "natural" way, *Food Eng. Int.,* 11, (October), 33, 1986.
5. **Tisovski, S., Durisic, B., and Stojanovic, M.,** The use of yeast autolysate for augmentation of nutritive value of foods in modern conditions of nutrition, *Hrana Ishrana,* (Beograd), 26(5/8), 187, 1985.
6. **Slyusarenko, T. P., Tsimbalova, N. M., Ziyasintseva, S. P., and Lyapun, Yu. S.,** Use of yeast extract in sausage manufacture, *Izv. Vysshikh. Uchebn. Zaved., Pishch. Tekhnol.,* No. 3, 67, 1981.
7. **Ayers, S. H. and Rupp, P.,** Extracts of pure dry yeast for culture media, *J. Bacteriol.,* 5, 89, 1920.
8. **Orberg, P. K., Sandine, W. E., and Ayres, J. W.,** Autolysate of whey-grown *Kluyveromyces fragilis* as a substitute for yeast extract in starter culture media, *J. Dairy Sci.,* 67, 37, 1984.
9. **Chrenova, N. M., Besrukov, M. G., Kogan, A. S., and Sergejev, V. A.,** Autolysis of the yeast *Saccharomyces cerevisiae* induced by metal ions. I. Metals of group II, *Nahrung,* 25(9) 837, 1981.
10. **Schaffeld, G., Sinskey, A. J., and Rha, C. K.,** Release of single cell protein by induced cell lysis, *J. Food Sci.,* 47(6), 2072, 1982.
11. **Basoppa, S. C., Jaleel, S. A., Shankar Murthy, A., and Shreenivasa Murthy, V.,** Preparation of yeast hydrolysate — a flavoured food adjunct, *Indian Food Ind.,* 5(1), 23, 1986.
12. **Knorr, D., Shetty, K. J., Hood, L. F., and Kinsella, J. E.,** An enzymic method for yeast autolysis, *J. Food Sci.,* 44(5), 1362, 1979.
13. **Stabnikova, E. V., Slyusarenko, T. P., Khalyutina, T. Yu., and Koval, L. T.,** Determination of the thermolysis parameters of *Rhodotorula glutinis 214, Izv. Vysskyh. Uchebn. Zaved., Pishch. Technol.,* No. 4, 86, 1981.
14. **Choi, I. S. and Shim, K. M.,** Effect of some treatments on the autolysis of baker's yeast (S. cerevisiae), *J. Korean Soc. Food Nutr.,* 13(3), 313, 1984.
15. **Lásztity, R.,** Cryobiological aspects of long term storage of meats and changes of their quality, in Cryobiol. Freeze-Drying, Proc. Third Natl. School Cryobiol. Freeze-Drying, Tsvetkov, T., Ed., Central Problem Laboratory of Cryobiology and Freeze-Drying, Sofia, 1987, 199.
16. **Anon.,** Autolysed yeast enhance flavours in low sodium foods, *Food Dev.,* 15(9), 52, 1981.
17. **Andres, C.,** Flavour enhancer for cheese containing foods, Food Process. (U.S.A.), 43(4), 78, 1982.
18. **Konrad, G. and Lieske, B.,** Process for preparation of flavouring material with chicken flavour, German Democratic Republic Patent, DD 252 537, 1987.
19. **Toshinose, S., Tadaski, M., and Susumm, T.,** Process for producing seasoning agent, and foods containing sauce, European Patent Application EP 0247 500 A 2, 1987.
20. **Tulchevskij, M. G., Petrov, K. B., and Bal, L. V.,** Preventing oxidation of melted food fats of animal origin, *Pishch. Prom., Respub. Mezhved. Nauchno-Tekh. Sb.,* No. 30, 56, 1984.
21. **Chen, H. F., Yang, M. T., and Fong, H. Y.,** Enzymic hydrolysis of single cell protein, *J. Chin. Agric. Chem. Soc.,* 22(1/2), 119, 1984.
22. **Parks, L. L., Carpenter, J. A., and Reagan, J. O.,** Effects of an autolysed yeast on physical and sensory properties of frankfurters, *J. Food Qual.,* 9(4), 225, 1986.

23. **Sommer, R.,** Yeast autolysate — manufacture, characteristics and use in prepared meals, *Lebensmitteltechnik,* 16(1/2), 30, 1984.
24. **Scharf, U., Schlingmann, M., and Lipinsky, R. W.,** Functional protein hydrolysates, process for producing them, and food containing these hydrolysates, German Federal Republic Patent Application DE 3 143 947 A1, 1983.
25. **Kojin, K. K.,** Yeast extracts, Japanese Examined Patent 5 645 578, 1981.
26. **Tanekawa, T., Takashima, H., and Hachiya, T.,** Production of yeast extract containing flavouring, U.S. Patent 4 306 680, 1981.
27. **Shackelford, J. R.,** Salt substitutes having reduced bitterness, U.S. Patent 4 297 375, 1981.
28. **Murray, D. G.,** Flavourings for foods with reduced salt levels, *Dairy Food Sanitation,* 3(9), 331, 1983.
29. **Akatsuka, S.,** Elimination of rancid odor in fermented milk products, U.S. Patent 4 435 431, 1984.
30. **Sommer, R. A.,** Yeast autolysates as flavour enhancers in prepared meals and meat products, *Lebensmitteltechnik,* 15(5), 229, 1983.
31. **Said, M. R. and Mohran, M. A.,** Acceleration of white soft cheese ripening by using some bacterial growth factors, *Egypt. J. Dairy Sci.,* 12(2), 229, 1984.
32. **Bender, F. G., Everson, C. W., and Swartz, W. E.,** Method of increasing the organoleptic acceptability of shank meat, U.S. Patent 4 500 559, 1985.
33. **Rooij, J. F. M. and de Haakart, M. J. J.,** Food flavours, European Patent Application EP 0191513, A1, 1986.
34. **Rooij, J. F. and de Haakart, M. J. J.,** Yeast extract food flavour, U.K. Patent Application GB 2171585 A, 1986.

Appendix

APPENDIX

OTHER USES OF YEAST BIOMASS

I. NEW TRENDS IN UTILIZATION OF YEAST BIOMASS

From an industrial point of view the yeast biomass is a fascinating raw material, containing different compounds suitable for production of a wide range of different products which could be utilized in production of foods, feeds, biochemicals, etc. Although the technology of yeast growing on different media had a rapid progress and achieved a high level, the post fermentation treatment of the yeast biomass, separation of the cell components and different compounds occurring in yeast cell in order to produce a variety of different products is, in many cases, only in experimental stages at a laboratory or pilot plant level. So it is understandable that the current research work is associated particularily with the fractionation of yeast biomass in order to extract subfractions with useful functional characteristics and with further processing for obtaining derivatives useful for different purposes.[1] In such a way a waste-free technology can be developed utilizing all the parts and/or components of the yeast biomass. Similar tendencies are observed in some other field, e.g., total utilization of cereal crops in agriculture and industry.[2,3] Using waste-free technology three groups of products can be produced from the yeast biomass:

1. Whole yeast products — yeast slurry or yeast paste; dried inactive yeast (flake, powder); yeast autolysate (liquid, paste, granules, powder) and yeast extracts.
2. Separated parts (components) of yeast cell — protein isolates and concentrates; cell wall polysaccharides; nucleic acids; enzymes; lipids (fat); pigments (carotenoids); peptides and amino acids; antibiotics; vitamins; sterols; aliphatic acids and mineral components.
3. Derivatives of yeast cell components — chemically modified proteins (acylated, phosphorylated, etc.); enzymatically modified proteins (partial hydrolysates, proteins with covalently attached essential amino acids, etc.); physically modified proteins (partly denaturated-, texturized proteins); nucleotides, nucleosides; flavoring products and flavor substances; salt replacers; immobilized enzymes and substances encapsulated inside the yeast (flavors, pharmaceuticals, etc.).

All the above mentioned possibilities underline the useful nature of the yeast organism (yeast biomass) in today's food industry, although some of the products outlined above have yet to be exploited successfully commercially. Probably further development in yeast biomass processing will help in the solving of problems associated with the economy of the yeast biomass processing.

The main fields of the practical application of yeast biomass in food production such as use of inactive dried yeast, of yeast protein concentrates and isolates and of autolysates and extracts were treated in Part I of this book and in the previous chapters of this book (Part II). In the following, a brief survey of some other potential uses will be given and in addition a short review about the problems of acceptance of new products and ingredients resp.

II. PRODUCTION AND APPLICATION OF YEAST ENZYMES

The importance of microorganisms as a source of commercially useful chemicals, antibiotics and enzymes has been recognized for a very long time. Among the microorganisms suitable for production of enzymes on a commercial scale yeast is one of the most important. Although a number of enzymes is excreted by the cell (extracellular enzymes) a much larger proportion of the potentially useful microbial products is retained within the cells. A vast majority of the enzymes known, for example, are intracellular. The isolation of intracellular material requires that the cell either be genetically engineered so that what would normally be an intracellular product is excreted into the environment, or it must be desintegrated by physical, chemical or enzymatic means to release its contents in the surrounding medium. The most commonly used methods of disruption were recently reviewed by Chisti and Moo-Young.[4] General aspects of enzyme extraction and purification of enzymes were treated in Chapter 5 of Part I.

Among the intracellular enzymes produced commercially invertase (used in confectionery) extracted from *Saccharomyces cerevisiae,* β-galactosidase (used for hydrolysis of lactose in dairy products) extracted from *Kluyveromyces fragilis* and *Saccharomyces lactis* and glucose-6-phosphate dehydrogenase (used in clinical analysis) can be mentioned. Due to the high costs of the commercially available enzyme preparations efforts were made to find mutants having hyperproduction of a given enzyme. Such yeast mutants can be used on a commercial scale in whole cell (whole biomass) form if their cell wall is permeabilized for the enzyme produced in high quantity. One example of such a use of whole yeast cells for enzymatic hydrolysis of lactose will be shortly described in the forthcoming part of the Appendix.

Among the enzymes of yeast the carbohydrate fermenting (hydrolyzing) enzymes were investigated most thoroughly. Studies on fermentation of sugars (sucrose, glucose, maltose, raffinose) by intracellular and cell wall enzyme preparations from five *Saccharomyces carlsbergensis* varieties were described by Cihosz et al.[5] The enzyme preparations were isolated from yeast cultured on wort- or molasses-based media; effects of two cell disintegration methods (pressure and acetone disintegration) were also considered. The results showed that enzyme preparations from the different yeast varieties differed considerably in activity; cell wall extracts showed higher activity than intracellular extracts; preparations from yeasts cultured on wort media were more active than those from yeasts cultured on molasses, and pressure disintegration tended to give higher enzymatic activity than acetone disintegration.

Fermentation activity of immobilized yeast cells was used for evaluation of protein hydrolysates by Tchorbanov and Lazarova.[6] *Kluyveromyces fragilis 20-C* cells were immobilized and their activity in different commercial protein hydrolysates of meat industry origin (peptones) and milk industry origin (tryptone, lactalbumin and casein hydrolysates) was determined. The test proved sensitive to hydrolysate concentration, origin and degree of hydrolysis. Many redoxi enzymes were isolated from yeasts such as alcohol dehydrogenase,[7] formaldehyde dehydrogenase,[8] aldehyde dehydrogenase,[9] NADPH: Cytochrome P-450 reductase.[10,11] Although yeast cells contain considerable amounts of proteases and nucleases, no yeast proteases and nucleases are produced on a commercial scale.

In the following, some examples of production and food use of yeast enzymes will be reviewed. Some of the enzymes produced from yeast may be used in food analysis for solving special problems associated with determination of microcomponents, e.g., yeast α-glucosidase and β-fructosidase may be applied for determination of different oligosaccharides containing α- or β-glycosidic bonds. It was also reported that the α- and β-glucosidases can potentially be used to characterize cariogenic potential of sugar substitutes.[12,13]

Preparation and properties of β-galactosidase confined in cells of *Kluyveromyces* sp. were reported by Champluvier et al.[14] β-D-galactosidase (β-galactosidase, lactase, EC 3.2.1.23) preparations have applications in the hydrolysis of lactose in milk, whey and whey permeate and are valuable since lactose hydrolysis products (glucose and galactose) are sweeter and more soluble than the native disaccharide, and are more readily fermented. In the report mentioned above β-galactosidase was immobilized in *Kluyveromyces lactis CBS 683* to produce a biocatalyst. A chloroform-ethanol mixture was used for cell permeabilization. β-galactosidase was retained inside the ghosts, by treatment with glutaraldehyde (0.4%). The biocatalyst produced could be stored in suspension at 4°C for 14 months with 20% deactivation. The biocatalyst was used and recovered by centrifugation 7 times without loss of enzymatic activity (hydrolysis of 5% lactose solution in buffer at 30°C; conversion approximately 80%).

Maintenance of hyperproduction of an exo-inulase (β-D-fructofuranoside fructohydrolase, E.C. 3.2.1.26) by a *Kluyveromyces fragilis* mutant (ATC 52466) was studied by Tsang and Grootwassink[15] in continuous culture using lactose as the fermentation substrate. Lactose was chosen as an inexpensive carbon source which was readily available as a waste product (in whey) from cheese manufacture. It is suggested that the exo-inulase may be used as an alternative to invertase in sucrose hydrolysis and in the manufacture of high fructose syrups and alcohol from fructans found in roots or tubers in chicory, Jerusalem artichoke and dahlia. After initial phenotypic adaptation cultures lacking detectable residual lactose showed rapid loss of enzyme yield. Displacement of the original hyperproducing strain by a wild-type cell population with two-fold the affinity for lactose (resulting from increased lactase, β-D-galactoside galactohydrolase = E.C. 3.2.1.23 activity) could be prevented by shifting from carbon to nitrogen limitation using large excess of lactose.

The possibility of total hydrolysis of inuline in plant extracts using yeast inulinase was also studied.[16,17] Chicory and Jerusalem artichokes were used as raw materials. It was stated that enzymatic hydrolysis does not produce dark colored fractions. The process seems to be suitable for the preparation of high fructose syrups. Immobilized inulinase was also used by Kin et al.[17] for hydrolysis of inulin containing extracts. The commercial utility of whey is enhanced by lactose hydrolysis, thereby overcoming its low solubility, low sweetness and narrow fermentation range. It renders whey more suitable for incorporation in dairy, bakery and confectionary products. The one time use of soluble lactase (β-galactosidase, E.C. 3.2.1.23) by direct addition to dairy products is too expensive due to the high price of commercial enzyme preparations. Alternatively, immobilized enzyme bioreactor systems have been developed.[18,19] Only clarified acidic product streams such as cottage cheese ultrafiltration permeate can be treated on an industrial scale. Whole milk or whey causes operational problems related to hygiene and clogging.

Intact yeast cells do not hydrolyze external lactose at an appreciable rate due to the presence of the cell membrane creating a permeability barrier. For the practical exploitation of the hydrolytic capacity of the intracellular lactase *in situ,* the cell membrane requires permeabilization. A low-cost industrial grade lactase was developed for one-time use in dairy products by Brodsky and Grootwassink.[20] The preparation contained the entire biomass of a selected hyperproducing strain of the yeast *Kluyveromyces fragilis.* Intracellular lactase was made freely accessible to its substrate by permeabilization of the cell membrane with food compatible reagents.

In 0.1 *M* phosphate plus 0.4% methyl paraben (methyl-*p*-hydroxybenzoic acid), permeabilization was complete in 0.5 to 1 h at 50°C, with 90% activity recovery from 180 g/l of cells. Whole cell lactase contained no viable cells and was free of proteolytic activity. Its pH optimum of 5.5 to 6.0 proved suitable for lactose hydrolysis concurrent with cottage cheese fermentation. Hydrolyzed whey was used in ice cream and baker's yeast production. Extracellular polygalacturonases are produced first by fermentation of fungi. However, some yeast strains such as *Kluyveromyces lactis* (NRRL-1137) are also good producers of these enzymes.[21] Saccharomyces cerevisiae sp. (NRRL-Y-1253) is capable of producing α-galactosidase on glucose, mellibiose or galactose medium.[22] The enzyme preparation may be used for raffinose hydrolysis in the low greens of sucrose crystallization, stachyose and raffinose hydrolysis in soybean meal and soybean milk.

Protease from brewer's yeast were isolated and characterized because of their importance both in reduction of protein yield from brewer's yeast, an industrial byproduct with potential human food application, and as a possible useful enzyme in the food processing industry by Woods and Kinsella.[23] Lyophilized yeast was rehydrated and autolyzed, proteases partially purified by $(NH_4)_2SO_4$ precipitation and hydroxyapatite chromatography. pH optima were observed at 3 and 6 to 8 the two activities were due to distinct proteases. The neutral protease was further studied. It was stable at 37°C but not at 45°C and was inactivated in 5 min at 70°C. The protease was most stable at pH 6.0. The enzyme was not affected by EDTA, was slightly inhibited by 10 m*MN*-ethylmaleimide and was completely inactivated by 1 m*M* PMSF. Effects of ionic strength, metal cation, and urea on enzyme activity were also reported.

A simple method was reported[24] for isolation of yeast ribonuclease from brewer's yeast. A lipoxygenase from baker's yeast was isolated and characterized by Schechter and Grossmann.[25] The mitochondrial fraction of yeast cell was used for the extraction and the raw enzyme was purified by affinity chromatography.

Two commercial preparations of chymosin derived from *Kluyveromyces lactis* by genetic engineering were prepared from residual yeasts.[26] Both enzymes were identical in biochemical tests to bovine chymosin. One of the two preparations tended to give a shorter renneting time than calf rennet and both tended to give a firmer curd. There were, however, no significant differences in general composition or ripeness of Eidam and Tilsit cheeses made with calf rennet and those made with the genetically derived chymosin.

Transformed yeasts, containing DNA at least are a copy of a fragment coding for 1,4-β-*N*-acetylmuranidase, are described by Oberto and Davison.[27] Lysozymes can be produced by growing the transformed yeasts, and these have applications in cheese manufacture, meat products, wine, sake and milks for pediatric use.

III. NUCLEIC ACIDS, NUCLEOTIDES AND NUCLEOSIDES

Yeast cells contain a significant amount of nucleic acids, mainly ribonucleic acids (see Chapter 3, Part I). As a byproduct of the production of low-RNA protein preparations nucleic acids and nucleotides are obtained in a ratio depending on the process used for RNA reduction (a survey of methods of RNA reduction is given in Chapter 5, Part I). Nucleotides and nucleosides together with amino acids and other compounds play a very important role in the formation of the flavor of roasted meat and meat products.

For practical flavoring of meat products, soups, gravy mixes, stock cubes, etc. yeast autolysates (yeast extracts) — containing nucleotides, nucleosides and amino acids — are used widely. Particularly in the eighties the status of yeast autolysates as a rich source of natural, savory flavor increased rapidly. The inherent savory nature of inactive yeast products makes them an ideal base on which more complete flavors can be built by the addition of other ingredients. Thus a flavor with an identifiable origin (e.g., meat, cheese) can be produced. It is also possible to modify flavor at the time of manufacture by incorporating other flavor-imparting ingredients so that defined flavor develops during processing. The process of autolysis, production and use of autolysates is reviewed in Chapter 5, Part I and Chapter 9, Part II.

Some nucleotides, nucleosides and amino acids are well known as flavor enhancers. Several Oriental groups have traditionally cooked foods with the marine alga *Laminaria japonica* having flavor enhancing properties. Later it was determined that sodium glutamate is the flavor enhancing agent. Subsequently monosodium glutamate (MSG) has been employed as a flavor enhancer in amounts considerably greater than obtainable by cooking with seaweed.[28] The overall world production of MSG raises over 20,000 t per year. In 1968, first Kwok[29] described a constellation of effects beginning 15 to 20 min after the first dish was eaten in a Chinese restaurant. He suggested that the high sodium content of Chinese

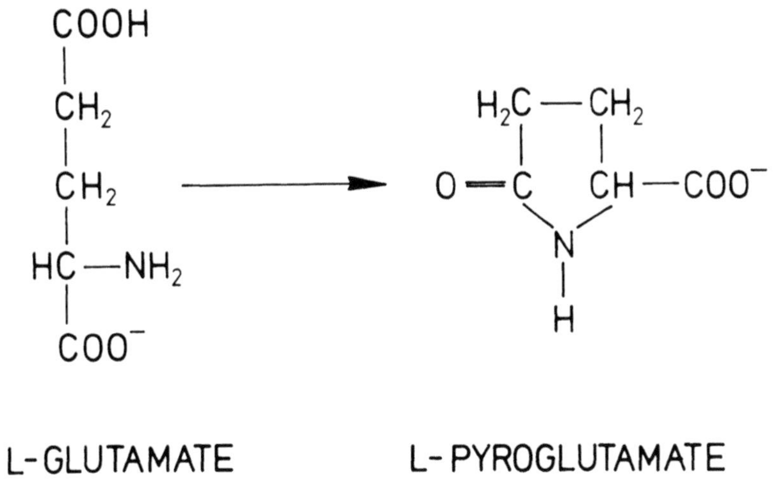

$$\begin{array}{c} COOH \\ | \\ CH_2 \\ | \\ CH_2 \\ | \\ HC-NH_2 \\ | \\ COO^- \end{array} \longrightarrow \text{(cyclic: } H_2C-CH_2,\ O{=}C,\ CH-COO^-,\ N-H)$$

L-GLUTAMATE L-PYROGLUTAMATE

FIGURE 97. The cyclization of L-glutamate to L-pyroglutamate.

food might be responsible. Later these effects (first of all changes in facial cutaneous blood flow, flushing) were named ''Chinese restaurant syndrome''.[28,30,31] Although early accounts indicated a pharmacologic, or toxic, type of reaction, subsequent evidence introduced reasonable doubt, while in some studies employing substantial amounts of MSG, a toxic reaction to MSG was not elicited.[32-34] Several workers have speculated that the reaction is due to MSG in the presence of another agent or is due solely to a second agent.[35] The speculations included a cyclization product of glutamic acid, pyroglutamate (See Figure 97) too while under conditions of cooking formation of pyroglutamate is possible.[36] More recently Wilkin and Richmond[36] reported that no flushing was provoked among the 24 people tested, 18 of whom gave a positive history of Chinese restaurant syndrome flushing. These results indicate that MSG-provoked flushing, if it exists at all, must be rare. MSG and its cyclization product, pyroglutamate, may provoke edema and associated symptoms. The nonvolatile flavor compounds and flavor potentiators of meat extracts were intensively investigated. Inosine-5-monophosphate (IMP) (Figure 98) is a major component of meat extracts. It is considered a key flavor component of meat extracts.[37] Other nonvolatile components with taste properties include amino acid peptides and other nucleotides. The high content of IMP in meat extracts is due to the presence of adenosine-5′-triphosphate (ATP) as a major mononucleotide in the muscle cells of living animals. After slaughter there is a rapid transformation of this nucleotide to adenosine-5′-monophosphate (AMP), which is then deaminated to IMP. Another nucleotide being a natural flavor enhancer of meat extracts is guanosine-5′-monophosphate (GMP). The commercial scale production and use of these compounds (IMP, GMP) begun in the early sixties in Japan and the U.S.[38] A good review about these and other flavor potentiators is given by Maga.[38] AMP and GMP may be used in much lower quantities than MSG to achieve a flavor enhancing effect.

A disadvantage of nucleotides is their lower heat stability.[39,40] Recent developments in yeast technology allow production of the RNA metabolites (including IMP and GMP) in economical quantities. They are excellent flavor enhancers replacing MSG and also give a possibility of reduction of sodium in foods.[41]

The importance of flavor precursors and flavor enhancers in the formation of meat flavor may be easily recognized by the fact that a large number of potential meat flavoring compositions, which are based on heat treatment of flavor precursors and addition of flavor enhancers, have been awarded in the last decade. Some of the patents are based on Maillard reactions between sugars and amino acids. Others make use of the strong flavor potentiation of nucleotide derivatives, along with sugar-amine reactions. IMP and GMP are the preferred ribonucleotides.

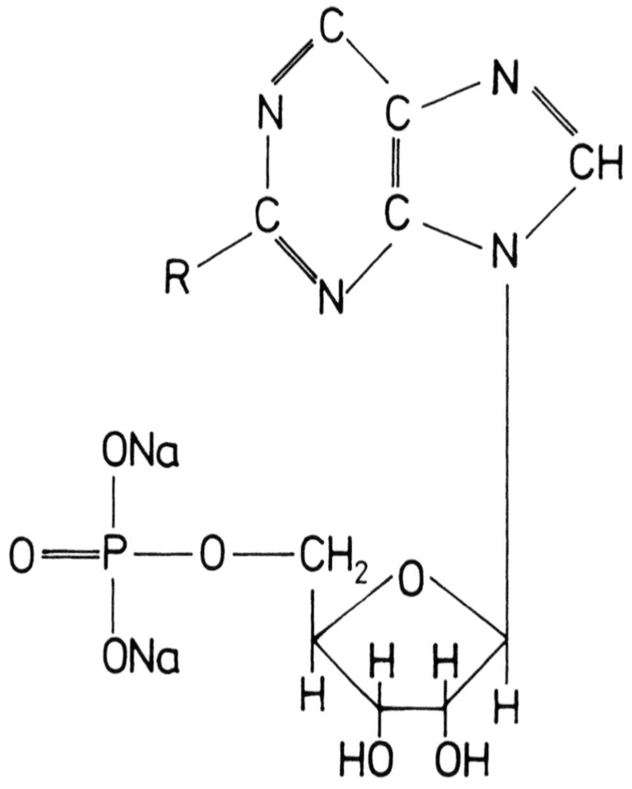

FIGURE 98. Chemical formulae of IMP and GMP.

TABLE 76
Lipid Content of Some Oleaginous Yeasts

Yeast species	Lipid content (% cell dry weight)	Ref.
Rhodotorula glutinis	48—72	42, 43, 44, 45
Candida sp. 107	44	46
Endomyces vernalis	28—65	47, 48
Rhodotorula gracilis	20—60	49
Lypomyces lipoferus 199	28	49
Cryptococcus tericolus	14—65	49, 50, 51

IV. LIPIDS AND CARBOHYDRATES

The investigation of yeast lipids has a growing tendency in the last years in part due to the numerous unanswered questions surrounding lipid metabolism, but also due to the possible commercial application of yeast lipids. As it was surveyed in Chapter 3 of Part I, the lipid content of the majority of yeasts is relatively low and does not exceed 18 to 20% of the cell dry weight. However, there exists yeast species that have lipid contents ranging from 20 to 80% of the dry weight. Recently an excellent review about lipids of yeasts and their metabolism was given by Ratledge and Evans.[42] Some lipid-producing species are listed in the Table 76. Due to the fact that the total lipid content and the ratio of neutral and polar lipids can be influenced by selection of yeast species and growth conditions, such olecoginous yeast may be used for the production of fat and phospholipids.[42,45,52,54]

Among the carbohydrates of yeast the polysaccharides of the yeast cell wall are the most important from the point of view of possible practical use. Some of these polysaccharides were reviewed in Chapter 4, Part I and a recent detailed review was published by Farkas.[53] The cell wall material separated during production of protein isolates or yeast autolysates has good emulsifying properties. The mannoprotein, which is a major complement of cell wall, has been found to be a more effective bioemulsifier.[55] The mannoprotein can be extracted in a high yield from the whole cells of baker's yeast by two methods: (1) autoclaving in neutral citrate buffer and (2) digestion with zymolase, a β-1,3-glucanase. Heat extracted emulsifier is purified by ultrafiltration and contains 44% carbohydrate and 17% protein. Treatment of preparation with proteases eliminates the emulsification. The preparation was successfully used for stabilization of different oil in water emulsions. Spent yeast from the manufacture of beer and wine is also a possible source for large-scale production. Experiments in our laboratory associated with complex utilization of yeast biomass confirmed the good emulsifying properties of mannoproteins and other cell wall preparations both from *Saccharomyces cerevisiae* and *Saccharomyces carlsbergensis.*

A large number of microbial species produce extracellular polysaccharides that can be recovered from the media of growth. Most of the polysaccharides that have found industrial applications are produced by bacteria, but a few of these polymers from yeasts or yeast-like organisms also have shown potential utility in foods or are used in industrial processes.[56]

Several species of the genus *Hansenula* produce extracellular phosphomannans.[57] In spite of attractive viscosity and gel properties for possible use as food additives, such application of these water-soluble gums has been hampered because of their sensitivity to salts, shear, and heat, as well as their instability in products at low pH.[58]

V. VITAMINS

Characterized by a relatively high content of protein, vitamins and minerals, yeast can be incorporated into many foods as an ingredient for nutritional fortification. Although yeast are capable of synthesizing most vitamins *de novo,* they are also able to incorporate vitamins actively from their environment, and the presence of specific transport systems for thiamin, riboflavin, vitamin B_6 and biotin is therefore indicated. Above all, accumulation of thiamin in yeast is extraordinarily high. *Kloeckera apiculatas* has been reported to accumulate thiamin to the level of one tenth the dry cell weight.[59]

In certain strains of *Saccharomyces cerevisiae* and *Saccharomyces uvarum (carlsbergensis),* high doses of ultraviolet radiation induced stable mutants excrete thiamine from their living cells during the ethanol production.[60] Both laboratory and pilot plant fermentations in 10 and 4% Plato hopped worts showed the suitability of selected mutants for the production of thiamine-rich beers which fulfilled all quality requirements and contained 0.67 to 0.80 and 0.22 to 0.33 mg thiamine hydrochloride per liter, respectively.

Various products sold in health food shops in Sweden were analyzed for vitamin B_{12} content. Great variability in vitamin content was revealed, especially in foods containing yeast.[61]

Both theoretically and practically an interesting fact is that some strains of *Saccharomyces carlsbergensis* contain not only vitamin D_3 but also the metabolite of vitamin D_3 the 25-hydroxy-vitamin D_3. An analysis of a calcium-containing brewer's yeast preparation confirmed the presence of 25-hydroxy vitamin D_2,[62] showing that yeast cells are able to synthesize these active metabolites of vitamin D_2 and D_3.

Some yeasts, principally the colored ones, produce carotenoid pigments. In the red yeasts, species of *Rhodotorula* and *Rhodosporium,* the most abundant carotenoids are torulene and torularhodin (see Figure 99). Other carotenoids may also occur (e.g. α- and β-carotenes, phytoene, etc.).[42] *Phaffia rhodozyma,* a pigmented, fermentative species has

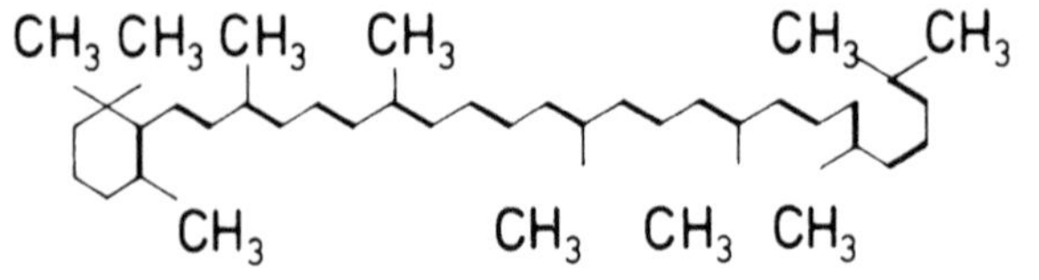

TORULENE

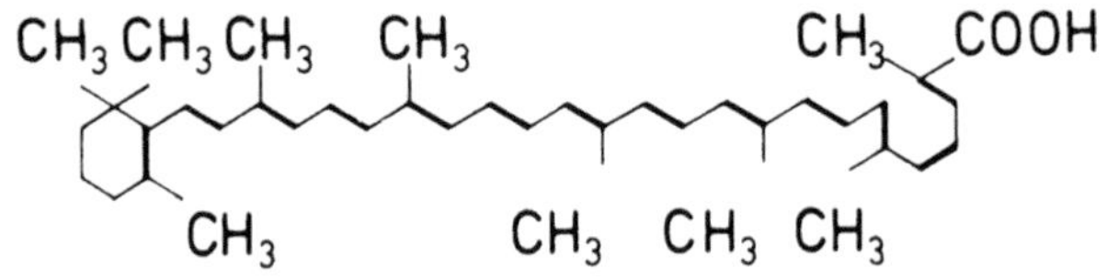

TORULARHODIN

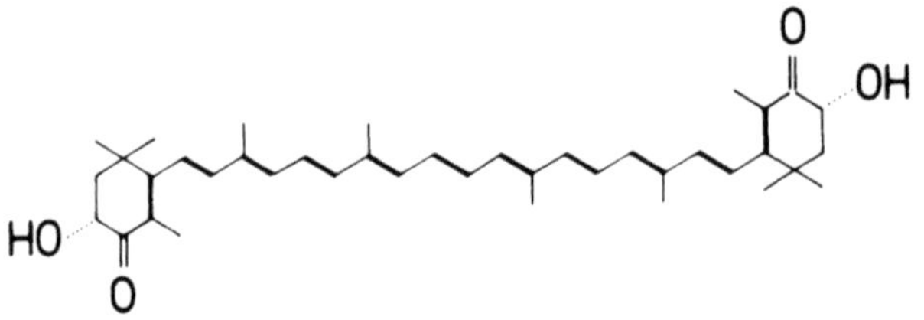

ASTAXANTHIN

FIGURE 99. Torulin, thorularodine and astaxanthin.

potential use in biotechnology because its carotenoids are mainly astaxanthin (see Figure 99), a pigment which is also responsible for the orange to pink color of salmonid flesh and the reddish color of boiled crustacean shells. Feeding of pen-reared salmonids with a diet containing this yeast induces pigmentation of the white muscle.[63,64] Laying hens deposit astaxanthin in egg yolks when fed broken *Phaffia rhodozyma* cells in their diet.[65] High β-carotene producing strains of *Rhodoturula glutinis* were investigated and optimal conditions of fermentation were determined by Biacs and Kövágó.[66]

VI. OTHER COMPOUNDS

Yeasts grown on media containing elevated amounts of metallic cations can accumulate these metals in the cells. This phenomenon may be used for preparation of food ingredients suitable for fortification of foods being poor in miocroelements. Korhola et al.[67] have studied the incorporation and distribution of selenium in baker's yeast with radioactive selenium (^{75}Se). Analysis of the protein fraction of selenium yeast has shown that selenium is present in all the major soluble proteins. Selenomethionine was identified as the major selenium-containing compound in the protein fraction as well as in whole cell (Figure 100). The selenium yeast is recommended for fortification of selenium poor foods as in Finland.[68]

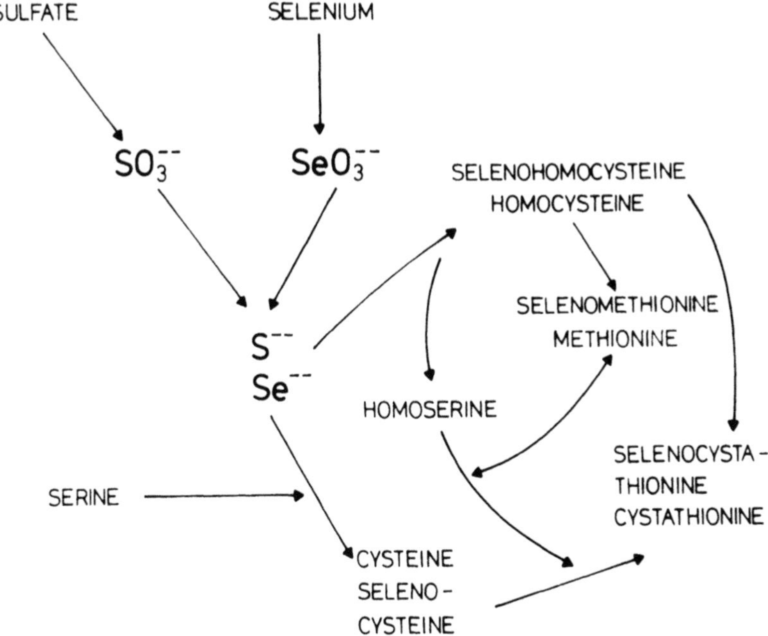

FIGURE 100. Biosynthesis of sulfur- or selenium-containing amino acids.

Germanium-containing yeast and its production was patented by Kehlbeck.[69] The yeast, preferably a *Saccharomyces* yeast is cultured in a medium containing germanium salt, e.g., citrate or lactate. Germanium is taken up by yeast which can be brewer's-, wine- baking- or food yeast.

Yeast autolysates contain a relatively high amount of free amino acids, so they are a potential source of production of some amino acids or amino acid concentrates.[70,71]

The novel process known as biological microencapsulation opens new fields for the use of yeast. The substance to be deposited is absorbed by live yeast cells and encapsulated. At the end of the process the cells are dead. Applications in the food industry may be numerous. For example, mustards or garlics in which a subtle profile of yeast cells could be an advantage is one possibility. Essential oils encapsulate particularly well.[72]

VII. CONSUMER ACCEPTABILITY OF NOVEL PROTEIN PRODUCTS

Extensive research and technological developments have made possible the production of a wide range of edible materials that could provide an abundant and relatively inexpensive source of good quality proteins. The novel and/or nonconventional protein can be utilized in human food in two distinct forms: (1) as functional ingredients which embody desirable characteristics influencing the quality and acceptability of a food product and (2) as protein sources in their own right, as substitutes or extenders for another food such as meat, meat products, milk, cheese, etc.

Although there is a growing market for nontextured new protein preparations as functional ingredients which are used at a low level in manufacture of many food products, the growth of the retail market for textured new protein products has been slow. These facts show that aside from the nutritive value, safety and price factors, the fundamental problem of consumer acceptability of novel protein sources in the markets of both developed and developing countries is of major practical concern. New food items may face formidable barriers to acceptance by the food manufacturers and the retail trade as well as the consumer.

Acceptance of food is a complex dynamic process and is well illustrated both by reference to the development of the marketing of margarine products,[73] and by current consumer reactions to the use of novel proteins.

From the second half of the nineteenth century, the major factors responsible for the growth of the consumer acceptance of margarine was the consistently lower price of margarine. Moreover the early margarines were inferior to butter in terms of taste and texture and upper and middle class consumers maintained an unfavorable attitude toward the lower priced substitutes. Due to the subsequent improvements in the products the quality of newer margarines not only achieved that of butter but also from the point of view of nutrition and health hazards became better than butter (beneficial effect of unsaturated fatty acids). As a result, the butter consumption declined in developed countries and the margarine market has increased.

It seems that the acceptance of novel protein foods depends first of all on their functional properties[74] and organoleptic quality. The nutritional advantages of many novel protein foods are known in wide circles of population and most consumers are favorably disposed toward the increased use of plant and microbial proteins. The poor reputation of the products is associated with the poor quality of many of products. The main task for the food manufacturing industry is improving the organoleptic quality and overcoming the negative attitudes through appropriate educational and marketing policies.

REFERENCES

1. **Seeley, R. D.,** Functional aspects of SCP are a key to potential markets, *Food Prod. Dev.*, 8(9), 46, 1975.
2. **Rexen, F. and Munck, L.,** Cereal Crops for Industrial Use, The Commission of the European Communities, Copenhagen, 1984.
3. **Pomeranz, Y. and Munck, L., Eds.,** *Cereals a Renewable Resource, Theory and Protein,* AACC Publishers, St. Paul, MN, 1981.
4. **Chisti, Y. and Moo-Young, M.,** Disruption of microbial cells for intracellular products, *Enzyme Microb. Technol.*, 8, 194, 1986.
5. **Cihosz, G., Rymaszewski, J., Poznanski, S., and Grabska, J.,** Fermentation activity of extracts from cultures of selected species of the yeast *Saccharomyces carlsbergensis, Monatschrift Brauerei,* 33(6), 221, 1980.
6. **Tchorbanov, B. and Lazarova, G.,** Evaluation of protein hydrolysates using the fermentation activity of immobilized yeast cells, *Biotechnol. Appl. Biochem.*, 10(3), 309, 1988.
7. **Scopes, R. K., Griffith-Smith, K., and Millar, D. G.,** Rapid purification of yeast alcohol dehydrogenase, *Anal. Biochem.*, 118, 284, 1981.
8. **Schütte, H., Flossdorf, J., Sahm, H., and Kula, M. R.,** Purification and properties of formaldehyde dehydrogenase and formate dehydrogenase from *Candida boidinii, Eur. J. Biochem.*, 62, 151, 1976.
9. **Bostian, K. A. and Betti, G. F.,** Rapid purification and properties of potassium activated aldehyde dehydrogenase from *Saccharomyces cerevisiae, Biochem. J.*, 173, 773, 1978.
10. **Azari, M. R. and Wiseman, A.,** Purification and characterization of the cytochrome P-448 component of a benzo*(a)*pyrene hydroxylase from *Saccharomyces cerevisiae, Anal. Biochem.*, 122, 129, 1982.
11. **King, D. J., Azari, M. R., and Wiseman, A.,** Studies on the properties of highly purified cytochrome P-448 and its dependent activity benzo (a) pyrene hydroxylase from *Saccharomyces cerevisiae, Xenobiotica,* 14, 187, 1984.
12. **Siebert, G. and Ziesenitz, S. C.,** Preliminary test using yeast enzymes. I. Yeast α-glucosidase, *Z. Ernährungswiss.*, 25(4), 242, 1986.
13. **Ziesenitz, S. C.,** Test methods for sugar substitutes. Preliminary test using yeast enzymes. II. Yeast β-fructosidase, *Z. Ernährungswiss.*, 25(4), 248, 1986.
14. **Champluvier, B., Kamp, B., and Rouxhet, P. G.,** Preparation and properties of α-galactosidase confined in cells of *Kluyveromyces* sp., *Enzyme Microbiol. Technol.*, 10(10), 611, 1988.
15. **Tsong, E. W. T. and Grootwassink, J. W. D.,** Stability of exoinulase production on lactose in batch and continuous culture of a *Kluyveromyces fragilis* hyperproducing mutant, *Enzyme Microbiol. Technol.*, 10(5), 297, 1988.

16. **Guirand, J. P. and Gabay, P.,** Enzymatic hydrolysis of plant extracts containing inulin, *Enzyme Microbiol. Technol.*, 3(4), 305, 1981.
17. **Kim, W. Y., Byun, S. M., and Uhm, T. B.,** Hydrolysis on inulin from Jerusalem artichoke by inulinase immobilized on aminoethylcellulose, *Enzyme Microbiol. Technol.*, 4(4), 223, 1982.
18. **Coughlin, R. W. and Charles, M.,** Applications of lactase and immobilized lactase, in *Immobilized Enzyme for Food Processing,* Pitcher, W., Jr., Ed., CRC Press, Boca Raton, FL, 1980, chap. 6., 153.
19. **Sproessber, B. and Plainer, H.,** Immobilized lactase for processing of whey, *Food Technol.*, 37(10), 93, 1983.
20. **Prodsky, J. A. and Grootwassink, J. W. D.,** Development and evaluation of whole cell yeast lactase for use in dairy processing, *J. Food Sci.*, 51(4), 897, 1986.
21. **Elias, A. N., Foda, M. S., and Attia, L. A.,** Formation of extracellular polygalacturonase and pectin-methylesterase activities by fungi and yeasts, *Egypt. J. Microbiol.*, 18(1-2), 225, 1984.
22. **Olivieri, R., Pausolli, P., Fascetti, E., and Cioffoletti, P.,** Process for the production of α-galactosidase and uses of the enzyme obtained, European Patent EP 008126281, 1986.
23. **Woods, F. C. and Kinsella, J. E.,** Isolation and properties of protease from *Saccharomyces carlsbergensis, J. Food Biochem.*, 4(2), 79, 1980.
24. **Shetty, J. K., Weaver, R. C., and Kinsella, J. E.,** Ribonuclease isolated from yeast *(Saccharomyces carlsbergensis):* characterization and properties, *Biotechnol. Bioeng.*, 23(5), 953, 1981.
25. **Schechter, G. and Grossmann, S.,** Lipoxygenase from baker's yeast: purification and properties, *Int. J. Biochem.*, 15(11), 1295, 1983.
26. **Prokopek, D., Meisel, H., Frister, H., Krussh, V., Renther, H., Schlimme, E., and Teuber, M.,** Making Eidam and Tilsit cheese using genetically engineered bovine chymosin from *Kluyveromyces lactis, Kiel. Milchwissensch. Forschungsber.*, 40(1), 43, 1988.
27. **Oberto, J. and Davison, J. R. N.,** Transformed yeasts producing lysozyme, plasmids used for this transformation and uses thereof, European Patent Application EP 0184575 A2, 1986.
28. **Reif-Lahser, L.,** Possible significance of adverse reactions to glutamate in humans, *Fed. Proc.*, 35, 2205, 1976.
29. **Kwok, R. H. M.,** Chinese restaurant syndrome, *N. Engl. J. Med.*, 278, 796, 1968.
30. **Kerr, G. R., Wu Lee, M., and El-Lozy, M.,** Prevalence of the "Chinese restaurant syndrome", *J. Am. Diet. Assoc.*, 75, 29, 1979.
31. **Ghadimi, H., Kumar, S., and Abaci, F.,** Studies on monosodium glutamate ingestion, *Biochem. Med.*, 5, 417, 19.
32. **Bazzano, G., D'Elia, J. A., and Olson, R. E.,** Monosodium glutamate: feeding of large amounts in man and gerbils, *Science,* 169, 1208, 1970.
33. **Morselli, P. L. and Gerattini, S.,** Monosodium glutamate and the Chinese restaurant syndrome, *Nature,* 227, 611, 1970.
34. **Rosenblum, I., Bradley, J. D., and Coulston, F.,** Single and double blind studies with oral monosodium glutamate in man, *Toxicol. Appl. Pharmacol.*, 18, 367, 1971.
35. **Gore, M. E. and Salmon, P. R.,** Chinese restaurant syndrome: fact or fiction?, *Lancet,* 1, 251, 1980.
36. **Wilkin, J. K. and Richmon, M. D.,** Does monosodium glutamate cause flushing (or merely "glutamania")?, *J. Am. Acad. Dermatol.*, 15, 225, 1986.
37. **Dwivedi, B. K.,** Meat Flavor, in *Fenaroli's Handbook of Flavor Ingredients,* 2nd ed., CRC Press, Boca Raton, FL, 1975, 812.
38. **Maga, J. A.,** Flavor potentiators, *CRC Crit. Rev. Food Sci. Nutr.*, 18, 231, 1983.
39. **Nguyen, T. T. and Sporns, P.,** Decomposition of the flavour enhancers monosodium glutamate, inosine-5′-monophosphate and guanosine-5′-monophosphate during canning, *J. Food Sci.*, 50(3), 812, 1985.
40. **Shaoul, O. and Sporns, P.,** Hydrolytic stability at intermediate pHs of the common purine nucleotides in food, inosine-5′-monophosphate, guanosine-5′-monophosphate and adenosine-5′-monophosphate, *J. Food Sci.*, 52(3), 810, 1987.
41. **Anon.,** Salt alternatives brew in marketplace, *Prepared Foods,* 157(5), 242, 1988.
42. **Ratledge, C. and Evans, C. T.,** Lipids and their metabolism, in *The Yeasts, Vol. 3. Metabolism and Physiology of Yeast,* Rose, A. H. and Harrison, J. S., Eds., Academic Press, London, 1989, 367.
43. **Ratledge, C.,** *Prog. Ind. Microbiol.*, 16, 119, 1982.
44. **Yoon, S. H. and Rhee, J. S.,** *Process Biochem.*, 18(5), 2, 1983.
45. **Yoon, S. H. and Rhee, J. S.,** Lipid from yeast fermentation: effects of cultural conditions on lipid production and its characteristics of *Rhodotorula gblutinis, J. Am. Oil Chem. Soc.*, 60(7), 1281, 1983.
46. **Moreton, R. S.,** Modification of fatty acid composition of lipid accumulating yeasts with cyclopropene fatty acid desaturase inhibitors, *Appl. Microbiol. Biotechnol.*, 22, 41, 1985.
47. **Krylova, N. I., Portnova, I. A., and Dedynkina, E. G.,** Lipid synthesis depending on the growth rate of yeasts-producers of lipids, *Mikrobiologije (USSR),* 53, 275, 1984.
48. **Witter, B., Debuch, H., and Steiner, M.,** Die Lipide von Endomycopsis vernalis bei verschiedener Stickstoff-Ernáhrung, *Arch. Microbiol.*, 101, 321, 1974.

49. **Oleshko, V. S.,** Growth of lipide-producing yeasts on wood hydrolysate (in Russian), *Gidroliz. Lesokhim. Prom.*, No. 4, 11, 1984.
50. **Hansson, L. and Dostalek, M.,** Influence of cultivation conditions on lipid production by *Cryptococcus albidus, Appl. Microbiol. Biotechnol.*, 24, 12, 1986.
51. **Hansson, L. and Dostalek, M.,** Lipid formation by *Cryptococcus albidus* in nitrogen-limited and in carbon-limited chemostat cultures., *Appl. Microbiol. Biotechnol.*, 24, 187, 1986.
52. **Hamid, S., Shakir, M., and Batthy, M. K.,** Synthetic fat from industrial byproducts. I. Microbial fats from molasses, *Fette, Seifen, Anstrichm.*, 83(1), 30, 1981.
53. **Farkas, V.,** Polysaccharide metabolism, in *The Yeasts, Volume 3, Metabolism and Physiology of Yeasts*, 2nd ed., Rose, A. H. and Harrison, J. S., Eds., Academic Press, London, 1989, 317.
54. **Misra, S., Ghooh, A., and Dutto, J.,** Production and composition of microbial fat from *Rhodotorula glutinis, J. Sci. Food Agric.*, 35(1), 59, 1984.
55. **Cameron, D. R., Cooper, D. G., and Neufeld, R. J.,** The mannoprotein of *Saccharomyces cerevisiae* is an effective bioemulsifier, *Appl. Environ. Microbiol.*, 54(6), 1420, 1988.
56. **Phaff, H. J.,** Ecology of yeasts with actual and potential value in biotechnology, *Microb. Ecol.*, 12, 31, 1986.
57. **Slodki, M. E. and Cadmus, M. C.,** Production of microbial polysaccharides, *Adv. Appl. Microbiol.*, 23, 19, 1978.
58. **Kang, K. S. and Cottrell, I. W.,** Polysaccharides, in *Microbial Technology*, 2nd ed., Vol. I, Peppler, H. J., and Perlman, D., Eds., Academic Press, New York, 1979, 417.
59. **Umezova, C. and Kishi, T.,** Vitamin metabolism, in *The Yeasts, Vol. 3, Metabolism and Physiology of Yeasts*, 2nd ed., Rose, A. H. and Harrison, J. S., Eds., Academic Press, London, 1989, 457.
60. **Silhankova, L.,** Yeast mutants excreting vitamin B_1 and their use in the production of thiamine-rich beers, *J. Inst. Brew*, 91 (3-4), 78, 1985.
61. **Janne, K.,** Vitamin B_{12} in health food products, *Vor Föda*, 34(5), 217, 1982.
62. **Garabedian, M., Grimberg, R., and Lenoir, G.,** Une source d'apport exogine de vitamine D: la levure de biére. *(Saccharomyces cerevisiae), Arch. Fr. Rediatr.*, 41, 495, 1984.
63. **Johnson, E. A., Conclin, D. E., and Lewis, M. J.,** The yeast *Phaffia rhodozyma* as a dietary pigment source for salmonoids and crustaceans, *J. Fish. Res. Board. Can.*, 34, 2417, 1977.
64. **Johnson, E. A., Ville, T. G., and Lewis, M. J.,** *Phaffia rhodozyma* as an astaxanthin source in animal diets, *Aquaculture*, 20, 123, 1980.
65. **Johnson, E. A., Lewis, M. J., and Gran, C. R.,** Pigmentation of egg yolks with astaxanthin from the yeast *Phaffia rhodozyma, Poultry Sci.*, 59, 1777, 1980.
66. **Biacs, P. and Kövágó, Á.,** Investigation of carotene production of *Rhodotorula* yeasts (in Hungarian), *Szeszipar*, 21, 151, 1972.
67. **Korhola, M., Vainio, A., and Eddman, K.,** Selenium yeast, *Ann. Clin. Res.*, 18, 65, 1986.
68. **Koivistonen, P. and Hattunen, J. K.,** Selenium in food and nutrition in Finland. An overview on research and action, *Ann. Clin. Res.*, 18, 13, 1986.
69. **Kehlbeck, H.,** Yeast containing a metal and its production, German Federal Republic Patent Application, DE 3345 211 A1, 1985.
70. **Popova, V. A., Ostrowskij, D. I., Klyukvin, A. N., and Petrenkova, M. M.,** Yeast autolysate — a source of amino acids (in Russian), *Gidroliz. Lesokhim. Prom.*, No. 4, 9, 1984.
71. **Higashimura Shayu, K. K.,** Shoyn production, Japanese Examined Patent JP 5755 388 B2, 1982.
72. **Schmidt, G.,** Bubbling over the yeast, *Food Ingredients, Flavouring, Packag. Process.*, (5), 25, 1987.
73. **Richardson, D. P.,** Consumer acceptability of novel protein foods, in *Developments in Food Proteins-1*, Hudson, B. J. F., Ed., Applied Science Publisher, London, 1982, 217.
74. **Seeley, R. D.,** Functional aspects of SCP are a key to potential market, *Food Prod. Dev.*, 9(9), 46, 1975.

INDEX

A

B

C

D

E

F

G

H

O

P

Q

R

S

Milton Keynes UK
Ingram Content Group UK Ltd.
UKHW051947071024
449327UK00026B/2212

9 780367 450724